Introduction to numerical linear algebra and optimisation

Introduction to numerical linear algebra and optimisation

PHILIPPE G. CIARLET
Université Pierre et Marie Curie, Paris

with the assistance of
Bernadette Miara and Jean-Marie Thomas
for the exercises

Translated by A. Buttigieg, S.J.
Campion Hall, Oxford

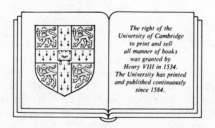

The right of the
University of Cambridge
to print and sell
all manner of books
was granted by
Henry VIII in 1534.
The University has printed
and published continuously
since 1584.

CAMBRIDGE UNIVERSITY PRESS

Cambridge

New York   Port Chester

Melbourne   Sydney

Published by the Press Syndicate of the University of Cambridge
The Pitt Building, Trumpington Street, Cambridge CB2 1RP
40 West 20th Street, New York, NY 10011–4211, USA
10 Stamford Road, Oakleigh, Melbourne 3166, Australia

Originally published in French as *Introduction à l'analyse numérique matricielle et à l'optimisation* and *Exercises d'analyse numérique matricielle et d'optimisation* in the series *Mathématiques appliquées pour la maîtrise*, editors P.G. Ciarlet and J.L. Lions, Masson, Paris, 1982 and © Masson, Editeur, Paris, 1982, 1986

First published in English by Cambridge University Press 1989 as
Introduction to numerical linear algebra and optimisation
Reprinted 1991

Printed in Great Britain at the University Press, Cambridge

British Library cataloguing in publication data
Ciarlet, Philippe, G.
Introduction to numerical linear algebra and optimisation. –
(Cambridge texts in applied mathematics; v. 2).
1. Equations – Numerical solutions
2. Mathematical optimisation
I. Title II. Miara, Bernadette
III. Thomas, Jean-Marie IV. Introduction à l'analyse
numérique matricielle et à l'optimisation. *English*
512.9′42 QA218

Library of Congress cataloguing in publication data
Ciarlet, Philippe, G.
[Introduction à l'analyse numérique matricielle et à l'optimisation. English]
Introduction to numerical linear algebra and optimisation/Philippe G. Ciarlet, with the assistance of Bernadette Miara and Jean-Marie Thomas for the exercises; translated by A. Buttigieg, p. cm.
Translation of: Introduction à l'analyse numérique matricielle et à l'optimisation and Exercices d'analyse numérique matricielle et d'optimisation.
Bibliography: p.
Includes index.
ISBN 0 521 32788 1. ISBN 0 521 33984 7 (pbk.)
1. Algebra, Linear. 2. Numerical calculations. 3. Mathematical optimization.
I. Miara, Bernadette. II. Thomas, Jean-Marie. III. Title. IV. Title:
Numerical linear algebra and optimisation.
QA184.C525 1988
512′.5–dc 19 87–35931 CIP

ISBN 0 521 32788 1 hardback
ISBN 0 521 33984 7 paperback

TM

I dedicate this English edition to Richard S. Varga

Contents

1 Preface to the English edition

My main purpose in writing this textbook was to give, within reasonable limits, a thorough description, and a rigorous mathematical analysis, of some of the most commonly used methods in Numerical Linear Algebra and Optimisation.

Its contents should illustrate not only the remarkable efficiency of these methods, but also the interest *per se* of their mathematical analysis. If the first aspect should especially appeal to the more practically oriented readers and the second to the more mathematically oriented readers, it may be also hoped that both kinds of readers could develop a common interest in these two complementary aspects of Numerical Analysis.

This textbook should be of interest to advanced undergraduate and beginning graduate students in Pure or Applied Mathematics, Mechanics, and Engineering. It should also be useful to practising engineers, physicists, biologists, economists, etc., wishing to acquire a basic knowledge of, or to implement, the basic numerical methods that are constantly used today.

In all cases, it should prove easy for the instructor to adapt the contents to his or her needs and to the level of the audience. For instance, a three hours per week, one-semester, course can be based on Chapters 1 to 6, or on Chapters 7 to 10, or on Chapters 4 to 8.

The mathematical prerequisites are relatively modest, especially in the first part. More specifically, I assumed that the readers are already reasonably familiar with the basic properties of matrices (including matrix computations) and of finite-dimensional vector spaces (continuity and differentiability of functions of several variables, compactness, linear mappings). In the second part, where various results are presented in the more general settings of Banach or Hilbert spaces, and where differential calculus in general normed vector spaces is often used, all relevant definitions and results are precisely stated wherever they are needed. Besides, the text is written in such a way that, in each case, the reader not familiar with these more abstract situations can, without any difficulty, 'stay in finite-dimensional spaces' and thus ignore these generalisations (in this spirit, weak convergence is used for proving only one 'infinite-dimensional' result, whose elementary 'finite-dimensional' proof is also given).

This textbook has some features which, in my opinion, are worth mentioning.

The *combination in a single volume of Numerical Linear Algebra and Optimisation*, with a progressive transition, and many cross-references, between these two themes;

A *mathematical level slowly increasing with the chapter number*;

A considerable space devoted to *reviews of pertinent background material*;

A *description of various practical problems*, originating in *Physics, Mechanics,* or *Economics,* whose numerical solution requires methods from Numerical Linear Algebra or Optimisation;

Complete proofs are given of each theorem;

Many *exercises* or *problems* conclude each section.

The *first part* (Chapters 1 to 6) is essentially devoted to *Numerical Linear Algebra*. It contains:

A review of all those results about *matrices* and *vector or matrix norms* that will be subsequently used (Chapter 1);

Basic notions about the *conditioning* of linear systems and eigenvalue problems (Chapter 2);

A review of various *approximate methods* (finite-difference methods, finite element methods, polynomial and spline interpolations, least square approximations, approximation of 'small' vibrations) that eventually lead to the solution of a linear system or of a matrix eigenvalue problem (Chapter 3);

A description and a mathematical analysis of some of the fundamental *direct methods* (Gauss, Cholesky, Householder; cf. Chapter 4) and *iterative methods* (Jacobi, Gauss–Seidel, relaxation; cf. Chapter 5) for solving *linear systems*;

A description and a mathematical analysis of some of the fundamental methods (Jacobi, Givens–Householder, QR, inverse method) for computing the *eigenvalues and eigenvectors of matrices* (Chapter 6).

The *second part* (Chapters 7 to 10) is essentially devoted to *Optimisation*. It contains:

A thorough review of all relevant prerequisites about *differential calculus in normed vector spaces* (Chapter 7) and about *Hilbert spaces* (Chapter 8);

A progressive introduction to Optimisation, through analyses of *Lagrange multipliers*, of *extrema* and *convexity* of real functions, and of *Newton's method* (Chapter 7);

A description of various linear and nonlinear problems whose approximate solution leads to *minimisation problems in $\mathbb{R}^n$, with or without constraints* (Chapters 8 and 10);

A description and mathematical analysis of some of the *fundamental algorithms of Optimisation theory* – relaxation methods, gradient methods (with optimal, fixed, or variable, parameter), conjugate gradient methods, penalty methods (Chapter 8), Uzawa's method (Chapter 9), simplex method (Chapter 10);

An introduction to *duality theory* – Farkas lemma, Kuhn and Tucker relations, Lagrangians and saddle-points, duality in linear programming (Chapters 9 and 10).

More complete descriptions of the topics treated are found in the *introductions* to each chapter.

Important results are stated as *theorems, which thus constitute the core of the text* (there are no lemmas, propositions, or corollaries).

Although the many *remarks* may be in principle skipped during a first reading, they should nevertheless prove to be helpful, by mentioning various special cases of interest, possible generalisations, counter-examples, etc.

The numerous *exercises* and *problems* that conclude each section provide often important, and sometimes challenging to prove, additions to the text.

In addition to '*local*' *references* (about a specific result, a particular extension, etc.) found at some places, *references of a more general nature are listed by subject and commented upon in a special section, titled* '*Bibliography and comments*', at the end of the book. The reader interested by more in-depth treatments of the various topics considered here, or by the *practical implementation* of the methods, should definitely refer to this section.

While I wrote this text, many colleagues and students were kind enough to make various comments, remarks, suggestions, etc., that substantially contributed to its improvement. In this respect, particular thanks are due to Alain Bamberger, Claude Basdevant, Michel Bernadou, Michel Crouzeix, David Feingold, Srinivasan Kesavan, Colette Lebaud, Jean Meinguet, Annie Raoult, Pierre-Arnaud Raviart, François Robert, Ulrich

Tulowitzki, Lars Wahlbin. Above all, my sincere thanks are due to Bernadette Miara and Jean-Marie Thomas, who not only carefully read the entire manuscript, but also significantly contributed to devising many exercises and problems.

It is also my pleasure to thank David Tranah of Cambridge University Press, and the translator, Alfred Buttigieg, S.J., whose friendly and efficient co-operation made this edition possible.

In 1964, at Case Institute of Technology (now Case Western Reserve University), I had the honour of having an outstanding teacher, who communicated to me his enthusiasm for Numerical Analysis. It is indeed a great privilege to dedicate this English edition to this teacher: Richard S. Varga.

Philippe G. Ciarlet
July 1988

A summary of results on matrices

Introduction

The purpose of this chapter is to recall, and to prove, a number of results relating to matrices and finite-dimensional vector spaces, of which frequent use will be made in the sequel.

It is assumed that the reader is familiar with the elementary properties of finite-dimensional vector spaces (and, in particular, with the theory of matrices). In section 1.1, we give the central definitions and notation relevant to these properties, as also the notion of *the partitioning of a matrix*, which is of outstanding importance in the area of the Numerical Analysis of Matrices.

In order to make this volume as 'self-contained' as possible, all results which are required subsequently are proved: in particular, *the reduction of a general matrix to triangular form, the diagonalisation of normal matrices* (Theorem 1.2-1), and *the equivalence of a matrix to the diagonal matrix of its singular values* (Theorem 1.2-2). (In this respect, it is relevant to point out that we will have no call to make use of Jordan's theorem.) We then examine (Theorem 1.3-1) the *characterisations of the eigenvalues of symmetric or Hermitian matrices* through the use of *Rayleigh's quotient*, and notably the characterisations in terms of 'min-max' and 'max-min'.

We next review the *vector norms* which are the most frequently utilised in the Numerical Analysis of Matrices. These are particular cases of the '*l_p-norms*' (Theorem 1.4-1). We then determine the corresponding *subordinate matrix norms* (Theorem 1.4-2), an example of a matrix norm which is not subordinate to a vector norm being given in Theorem 1.4-4. A reminder is given in Theorem 1.4-5 of the conditions for the invertibility of matrices of the form I + B, and it is shown (Theorem 1.4-3) that *the spectral radius of a matrix is the lower bound of the values of its norms*. This last result is in turn used to prove *two results about the sequence of successive powers of a matrix* (Theorems 1.5-1 and 1.5-2). These play a fundamental role in the study of iterative methods for the solution of linear systems, which are studied in Chapter 5.

1.1 Key definitions and notation

Let V be a vector space of finite dimension n, over the field $\mathbb{R}$ of real numbers, or the field $\mathbb{C}$ of complex numbers; if there is no need to distinguish between the two, we will speak of the field $\mathbb{K}$ of *scalars*.

A *basis* of V is a set $\{e_1, e_2, \ldots, e_n\}$ of n linearly independent vectors of V, denoted by $(e_i)_{i=1}^n$, or quite simply by (e_i) if there is no risk of confusion. Every vector $v \in V$ then has the unique representation

$$v = \sum_{i=1}^n v_i e_i,$$

the scalars v_i, which we will sometimes denote by $(v)_i$, being the *components* of the vector v relative to the basis (e_i). *As long as a basis is fixed unambiguously, it is thus always possible to identify V with $\mathbb{K}^n$*; that is why it will turn out to be just as likely for us to write $v = (v_i)_{i=1}^n$, or simply (v_i), for a vector v whose components are v_i.

In matrix notation, the vector $v = \sum_{i=1}^n v_i e_i$ will *always* be represented by the *column vector*

$$v = \begin{pmatrix} v_1 \\ v_2 \\ \vdots \\ v_n \end{pmatrix},$$

while v^{T} and v^* will denote the following *row vectors*:

$$v^{\mathrm{T}} = (v_1 v_2 \cdots v_n), \quad v^* = (\bar{v}_1 \bar{v}_2 \cdots \bar{v}_n),$$

where, in general, $\bar{\alpha}$ is the complex conjugate of α. The row vector v^{T} is the *transpose* of the column vector v, and the row vector v^* is the *conjugate transpose* of the column vector v.

The function $(\cdot, \cdot): V \times V \to \mathbb{K}$ defined by

$$(u, v) = v^{\mathrm{T}} u = u^{\mathrm{T}} v = \sum_{i=1}^n u_i v_i \quad \text{if} \quad \mathbb{K} = \mathbb{R},$$

$$(u, v) = v^* u = \overline{u^* v} = \sum_{i=1}^n u_i \bar{v}_i \quad \text{if} \quad \mathbb{K} = \mathbb{C},$$

will be called the *Euclidean scalar product* if $\mathbb{K} = \mathbb{R}$, the *Hermitian scalar product* if $\mathbb{K} = \mathbb{C}$ and the *canonical scalar product* if the underlying field is left unspecified. When it is desired to keep in mind the dimension of the vector space, we shall write

$$(u, v) = (u, v)_n.$$

Let V be a vector space which is provided with a canonical scalar product. Two vectors u and v of V are *orthogonal* if $(u, v) = 0$. By extension, the vector v is said to be *orthogonal to the subset U of V* (in symbols, $v \perp U$), if the vector v is orthogonal to all the vectors in U. Lastly, a set $\{v_1, \ldots, v_k\}$ of vectors belonging to the space V is said to be *orthonormal* if

$$(v_i, v_j) = \delta_{ij}, \quad 1 \leqslant i, j \leqslant k,$$

where δ_{ij} is the *Kronecker delta*: $\delta_{ij} = 1$ if $i = j$, $\delta_{ij} = 0$ if $i \neq j$.

Let V and W be two vector spaces over the same field, equipped with bases $(e_j)_{j=1}^n$ and $(f_i)_{i=1}^m$ respectively. *Relative to these bases,* a linear transformation

$$\mathscr{A} : V \to W$$

is represented by the *matrix* having *m rows* and *n columns*:

$$A = \begin{pmatrix} a_{11} & a_{12} & \cdots & a_{1n} \\ a_{21} & a_{22} & \cdots & a_{2n} \\ \vdots & \vdots & & \vdots \\ a_{m1} & a_{m2} & \cdots & a_{mn} \end{pmatrix},$$

the *elements a_{ij}* of the matrix A being defined uniquely by the relations

$$\mathscr{A}e_j = \sum_{i=1}^m a_{ij} f_i, \quad 1 \leqslant j \leqslant n.$$

Equivalently, the jth *column vector*

$$\begin{pmatrix} a_{1j} \\ a_{2j} \\ \vdots \\ a_{mj} \end{pmatrix}$$

of the matrix A represents the vector $\mathscr{A}e_j$ relative to the basis $(f_i)_{i=1}^m$. We call

$$(a_{i1} a_{i2} \cdots a_{in})$$

the ith *row vector of the matrix* A.

A matrix with m rows and n columns is called a *matrix of type (m, n)*, and the vector space over the field $\mathbb{K}$ consisting of matrices of type (m, n) with elements in $\mathbb{K}$ is denoted by $\mathscr{A}_{m,n}(\mathbb{K})$ or simply $\mathscr{A}_{m,n}$. A column vector is then a matrix of type $(m, 1)$ and a row vector a matrix of type $(1, n)$. A matrix is called *real* or *complex* according as its elements are in the field $\mathbb{R}$ or the field $\mathbb{C}$.

A matrix A with elements a_{ij} is written as

$$A = (a_{ij}),$$

the first index i *always* designating the row and the second, j, the column. Given a matrix A, $(A)_{ij}$ denotes the element in the ith row and jth column.

The *null matrix* and the *null vector* are represented by the same symbol 0.

Given a matrix $A \in \mathscr{A}_{m,n}(\mathbb{C})$, $A^* \in \mathscr{A}_{n,m}(\mathbb{C})$ denotes the *adjoint* of the matrix A and is defined uniquely by the relations

$$(Au, v)_m = (u, A^*v)_n \quad \text{for every } u \in \mathbb{C}^n, \quad v \in \mathbb{C}^m,$$

which imply that $(A^*)_{ij} = \bar{a}_{ji}$. In the same way, given a matrix $A = \mathscr{A}_{m,n}(\mathbb{R})$, $A^T \in \mathscr{A}_{n,m}(\mathbb{R})$ denotes the *transpose* of the matrix A and is defined uniquely by the relations

$$(Au, v)_m = (u, A^Tv)_n \quad \text{for every } u \in \mathbb{R}^n, \quad v \in \mathbb{R}^m,$$

which imply that $(A^T)_{ij} = a_{ji}$.

Remarks

(1) One could also define the transpose of a complex matrix. However, that would provide a concept of limited interest, since the function $u, v \to \sum_{i=1}^n u_i v_i$ is not a scalar product in $\mathbb{C}^n$.

(2) The notation A^T has been given preference over the notation ${}^tA$, this latter being more suitably linked to the notion of a dual basis. The notation A^T keeps in mind the dependence of the notion of transpose upon a particular scalar product, the canonical scalar product.

To the composition of linear transformations there corresponds the multiplication of matrices. If $A = (a_{ik})$ is a matrix of type (m, l) and $B = (b_{kj})$ of type (l, n), their product AB is the matrix of type (m, n) defined by

$$(AB)_{ij} = \sum_{k=1}^l a_{ik}b_{kj}.$$

Recall that $(AB)^T = B^TA^T$, $(AB)^* = B^*A^*$.

Let $A = (a_{ij})$ be a matrix of type (m, n). We shall use the term *submatrix* of A for every matrix of the form

$$\begin{pmatrix} a_{i_1j_1} & a_{i_1j_2} & \cdots & a_{i_1j_q} \\ a_{i_2j_1} & a_{i_2j_2} & \cdots & a_{i_2j_q} \\ \vdots & \vdots & & \vdots \\ a_{i_pj_1} & a_{i_pj_2} & \cdots & a_{i_pj_q} \end{pmatrix},$$

provided the integers i_k and j_l satisfy

$$1 \leqslant i_1 < i_2 < \cdots < i_p \leqslant m, \quad 1 \leqslant j_1 < j_2 < \cdots < j_q \leqslant n.$$

Let $A = (a_{ij})$ be the matrix representing a linear transformation from V into W and let

$$V = V_1 \oplus V_2 \oplus \cdots \oplus V_N, \quad W = W_1 \oplus W_2 \oplus \cdots \oplus W_M$$

be decompositions of the spaces V and W into the direct sum of subspaces V_J and W_I, of dimensions n_J and m_I respectively, each spanned by a set of basis vectors. With this decomposition of the spaces V and W is associated the *partitioning of the matrix* A:

$$A = \begin{pmatrix} A_{11} & A_{12} & \cdots & A_{1N} \\ A_{21} & A_{22} & \cdots & A_{2N} \\ \vdots & \vdots & & \vdots \\ A_{M1} & A_{M2} & \cdots & A_{MN} \end{pmatrix} = (A_{IJ})$$

each submatrix A_{IJ}, of type (m_I, n_J), representing a linear transformation from the space V_J into the space W_I. What is of interest in these partitionings is the fact that some of the operations defined on matrices remain *formally* the same, 'the coefficients a_{ij} being replaced by the submatrices A_{IJ}'. However, care is required over the *order* of the factors!

Thus, let $A = (A_{IK})$ and $B = (B_{KJ})$ be two matrices, of type (m, l) and (l, n) respectively, partitioned into blocks, the *partitioning corresponding to the index K being the same* for each matrix. *The matrix* AB *then admits the following partitioning*

$$AB = (C_{IJ}), \quad \text{with} \quad C_{IJ} = \sum_K A_{IK} B_{KJ},$$

and in this way one is said to have carried out the *block multiplication* of the two matrices.

In the same way, let v be a vector in the space V and let $v = \sum_{J=1}^{N} v_J$, $v_J \in V_J$, be the (unique) representation associated with the decomposition of the space V into a direct sum. The vector $Av \in W$ then has the representation

$$Av = \sum_{I=1}^{M} w_I, \quad \text{with} \quad w_I = \sum_{J=1}^{N} A_{IJ} v_J,$$

as the unique representation associated with the decomposition of the space W into a direct sum. This is equivalent to considering *the vectors v and Av as partitioned into blocks*

$$v = \begin{pmatrix} v_1 \\ v_2 \\ \vdots \\ v_N \end{pmatrix}, \quad Av = \begin{pmatrix} w_1 \\ w_2 \\ \vdots \\ w_M \end{pmatrix}, \quad w_I = \sum_{J=1}^{N} A_{IJ} v_J,$$

the last equation embodying *the block multiplication of the matrix A by the vector v*.

A matrix of type (n, n) is said to be *square*, or *a matrix of order n* if it is desired to make explicit the integer n; it is convenient to speak of a matrix as *rectangular* if it is not necessarily square. One denotes by

$$\mathscr{A}_n = \mathscr{A}_{n,n} \quad \text{or} \quad \mathscr{A}_n(\mathbb{K}) = \mathscr{A}_{n,n}(\mathbb{K})$$

the ring of square matrices of order n, with elements in the field $\mathbb{K}$.

Unless anything is said to the contrary, *the matrices to be considered up to the end of this section will be square.*

If $A = (a_{ij})$ is a square matrix, the elements a_{ii} are called *diagonal elements*, and the elements $a_{ij}, i \neq j$, are called *off-diagonal elements*. The *identity matrix* is the matrix

$$I = (\delta_{ij}).$$

A matrix A is *invertible* if there exists a matrix (which is unique, if it does exist), written as A^{-1} and called *the inverse* of the matrix A, which satisfies $AA^{-1} = A^{-1}A = I$. Otherwise, the matrix is said to be *singular*. Recall that if A and B are invertible matrices

$$(AB)^{-1} = B^{-1}A^{-1}, \quad (A^T)^{-1} = (A^{-1})^T, \quad (A^*)^{-1} = (A^{-1})^*.$$

A matrix A is

symmetric if A is real and $A = A^T$,
Hermitian if $A = A^*$,
orthogonal if A is real and $AA^T = A^TA = I$,
unitary if $AA^* = A^*A = I$,
normal if $AA^* = A^*A$.

A matrix $A = (a_{ij})$ is *diagonal* if $a_{ij} = 0$ for $i \neq j$ and is written as

$$A = \text{diag}(a_{ii}) = \text{diag}(a_{11}, a_{22}, \ldots, a_{nn}).$$

The *trace* of a matrix $A = (a_{ij})$ is defined by

$$\text{tr}(A) = \sum_{i=1}^{n} a_{ii}.$$

Let $\mathfrak{S}_n$ be the group of permutations of the set $\{1, 2, \ldots, n\}$. To every element $\sigma \in \mathfrak{S}_n$ there corresponds the *permutation matrix*

$$P_\sigma = (\delta_{i\sigma(j)}).$$

Observe that *every permutation matrix is orthogonal.*

The *determinant* of a matrix A is defined by

$$\det(A) = \sum_{\sigma \in \mathfrak{S}_n} \varepsilon_\sigma a_{\sigma(1)1} a_{\sigma(2)2} \cdots a_{\sigma(n)n},$$

where $\varepsilon_\sigma = 1$, resp. -1, if the permutation σ is even, resp. odd.

The *eigenvalues* $\lambda_i = \lambda_i(A)$, $1 \leq i \leq n$, of a matrix A of order n are the n roots, real or complex, simple or multiple, of the *characteristic polynomial*

$$p_A : \lambda \in \mathbb{C} \to p_A(\lambda) = \det(A - \lambda I)$$

of the matrix A. The *spectrum* of the matrix A is the subset

$$\text{sp}(A) = \bigcup_{i=1}^{n} \{\lambda_i(A)\}$$

of the complex plane. We recall the relations

$$\text{tr}(A) = \sum_{i=1}^{n} \lambda_i(A), \quad \det(A) = \prod_{i=1}^{n} \lambda_i(A),$$
$$\text{tr}(AB) = \text{tr}(BA), \quad \text{tr}(A+B) = \text{tr}(A) + \text{tr}(B),$$
$$\det(AB) = \det(BA) = \det(A)\det(B).$$

The *spectral radius* of the matrix A is the non-negative number defined by

$$\varrho(A) = \max\{|\lambda_i(A)|: 1 \leqslant i \leqslant n\}.$$

To every eigenvalue λ of a matrix A there corresponds (at least) one vector p satisfying

$$p \neq 0 \quad \text{and} \quad Ap = \lambda p,$$

called an *eigenvector* of the matrix A, *corresponding to the eigenvalue* λ. If $\lambda \in \text{sp}(A)$, the vector subspace

$$\{v \in V: Av = \lambda v\}$$

(of dimension at least 1) is called the *eigenspace corresponding to the eigenvalue* λ.

By convention, in every partitioning $A = (A_{IJ})$ of a *square* matrix, the *diagonal submatrices* A_{II} will always be *square*.

Given two vector spaces V and W of finite (but not necessarily equal) dimension, the *rank of the linear transformation* $\mathscr{A}: V \to W$ is equal to the dimension of the vector subspace

$$\text{Im}(\mathscr{A}) = \{\mathscr{A}v \in W: v \in V\}.$$

If the spaces V and W are equipped with bases, relative to which the transformation $\mathscr{A}$ is represented by a matrix A, the rank of $\mathscr{A}$ is also equal to the largest order of the (square) invertible submatrices of A. That is why the rank of $\mathscr{A}$ is also called the *rank of the matrix* A. It is denoted by $r(A)$.

Finally we make a general remark, which is to hold in all that follows: *whenever it is 'reasonably' clear, no mention will be made of the sets of indices.* And so, if $A = (a_{ij})$ is a matrix of type (m, n) we shall write

$$\max_{i} \left\{ \min_{j} a_{ij} \right\}, \quad \text{in place of} \quad \max_{1 \leqslant i \leqslant m} \left\{ \min_{1 \leqslant j \leqslant n} a_{ij} \right\},$$

or

$$p_i^* p_j = \delta_{ij}, \quad \text{in place of } p_i^* p_j = \delta_{ij}, \quad 1 \leqslant i, j \leqslant n,$$

if it is clear that the indices i and j belong to the same set $\{1, 2, \ldots, n\}$, etc.

Exercises

1.1-1. Let A be an invertible matrix whose elements, as well as those of A^{-1}, are all non-negative. Show that there exists a permutation matrix P and a matrix $D = \text{diag}(d_i)$, with d_i positive, such that $A = PD$ (the converse is obvious).

1.1-2. Let $A = (a_{ij})$ be a matrix of type (m, n). Show that

$$\max_i \left\{ \min_j a_{ij} \right\} \leqslant \min_j \left\{ \max_i a_{ij} \right\}.$$

1.1-3. Let A and B be two square matrices of the same order. Show that the matrices AB and BA have the same characteristic polynomial.

1.1-4. Let a, b and c be given scalars. Find the eigenvalues and eigenvectors of the matrix

$$A = \begin{pmatrix} a & b & & & & \\ c & a & b & & & \\ & c & a & b & & \\ & & \ddots & \ddots & \ddots & \\ & & & c & a & b \\ & & & & c & a \end{pmatrix}.$$

1.1-5. Let $A = (a_{ij})$ be a complex matrix of order n.

(1) Show that (*Gerschgorin–Hadamard theorem*)

$$\text{sp}(A) \subset \bigcup_{i=1}^{n} \left\{ z \in \mathbb{C} : |z - a_{ii}| \leqslant \sum_{j \neq i} |a_{ij}| \right\}.$$

(2) Show that, if the matrix A is *strictly diagonally dominant*, that is to say, if

$$|a_{ii}| > \sum_{j \neq i} |a_{ij}|, \quad 1 \leqslant i \leqslant n,$$

then it is invertible.

(3) Show that, if the matrix A is strictly diagonally dominant,

$$|\det(A)| \geqslant \prod_{i=1}^{n} \left(|a_{ii}| - \sum_{j \neq i} |a_{ij}| \right).$$

1.1-6. A matrix $A \in \mathscr{A}_n(\mathbb{C})$ is said to be *reducible* if there exists a permutation matrix $P \in \mathscr{A}_n(\mathbb{R})$ such that the matrix $P^T A P$ is block upper triangular:

$$P^T A P = \begin{pmatrix} A_{11} & A_{12} \\ 0 & A_{22} \end{pmatrix}.$$

Otherwise, the matrix A is said to be *irreducible*.

(1) Show that a necessary and sufficient condition for a matrix $A = (a_{ij}) \in \mathscr{A}_n(\mathbb{C})$ to be irreducible is that, for *every* ordered pair (i, j), $1 \leqslant i, j \leqslant n$, it is possible to find

a sequence $(a_{p_k, p_{k+1}})_{k=0}^s$ of non-zero elements of A (the integer s depending on the particular pair involved), such that $p_0 = i$, $p_{s+1} = j$.

(2) Let $A = (a_{ij}) \in \mathscr{A}_n(\mathbb{C})$ be a matrix. By Gerschgorin's theorem (Exercise 1.1-5), every eigenvalue of A belongs to the set

$$\sigma(A) \overset{\text{def}}{=} \bigcup_{i=1}^n \left\{ z \in \mathbb{C} : |z - a_{ii}| \leqslant \sum_{j \neq i} |a_{ij}| \right\}.$$

Show that, if the matrix A is irreducible and if an eigenvalue λ of A belongs to the boundary of the set $\sigma(A)$, then *all* the circles $\{z \in \mathbb{C} : |z - a_{ii}| = \sum_{j \neq i} |a_{ij}|\}$, $1 \leqslant i \leqslant n$, pass through the point λ.

(3) Let $A = (a_{ij}) \in \mathscr{A}_n(\mathbb{C})$ be an irreducible matrix such that

$$\begin{cases} |a_{ii}| \geqslant \sum_{j \neq i} |a_{ij}|, & 1 \leqslant i \leqslant n, \\ |a_{i_0 i_0}| > \sum_{j \neq i_0} |a_{i_0 j}| & \text{for at least one index } i_0 \in \{1, 2, \ldots, n\}. \end{cases}$$

Show that the matrix A is invertible (this, then, is an extension of the result of Exercise 1.1-5(2)). Show that if the further assumption is made that $a_{ii} > 0, 1 \leqslant i \leqslant n$, then all the eigenvalues of the matrix A have positive real part.

Remark. Important developments of this exercise are given in Exercises 3.1-6 and 3.1-7.

1.1-7. (1) Let A and B be two square matrices of the same order. Is it possible to have $AB - BA = \lambda I$, with λ some non-zero scalar?

(2) Let A and B be two square matrices of the same order. Is it possible to have $AB + BA = 0$?

1.1-8. Let $A = \begin{pmatrix} A_{11} & A_{12} \\ A_{21} & A_{22} \end{pmatrix}$ be a square matrix partitioned into blocks. Assuming the submatrix A_{11} to be invertible, show that

$$\det A = \det A_{11} \det(A_{22} - A_{21} A_{11}^{-1} A_{12}).$$

1.1-9. Let $a_i, 1 \leqslant i \leqslant n$, be any scalars. Show that the eigenvalues of the matrix

$$A = \begin{pmatrix} a_1 & a_2 & \cdots & a_{n-1} & a_n \\ a_n & a_1 & a_2 & \cdots & a_{n-1} \\ & a_n & a_1 & & \\ \vdots & \cdots & \cdots & \cdots & \vdots \\ a_3 & \cdots & a_n & a_1 & a_2 \\ a_2 & a_3 & \cdots & a_n & a_1 \end{pmatrix},$$

called a *circulant matrix*, are of the form

$$\lambda_{l+1} = a_1 + a_2 \xi_l + a_3 \xi_l^2 + \cdots + a_n \xi_l^{n-1}, \quad 0 \leqslant l \leqslant n-1,$$

where $\xi_l = e^{2i\pi l/n}$.

1.1-10. Let $A_k, 1 \leqslant k \leqslant m$, be matrices of order n satisfying

$$\sum_{k=1}^m A_k = I.$$

Show that the following conditions are equivalent.

(i) $A_k = (A_k)^2$, $1 \leqslant k \leqslant m$,

(ii) $A_k A_l = 0$ for $k \neq l$, $1 \leqslant k, l \leqslant m$,

(iii) $\sum_{k=1}^{m} r(A_k) = n$,

denoting, in general, by $r(B)$ the rank of the matrix B.

1.2 Reduction of matrices

Let V be a vector space of finite dimension n and let $\mathscr{A}: V \to V$ be a linear transformation, represented by a (square) matrix $A = (a_{ij})$ relative to a basis (e_i). Relative to another basis (f_i), the same transformation is represented by the matrix

$$B = P^{-1}AP,$$

where P is the invertible matrix whose jth column vector consists of the components of the vector f_j in the basis (e_i).

Since the same linear transformation $\mathscr{A}$ can in this way be represented by different matrices, depending on the basis that is chosen, the problem arises of finding a basis relative to which the matrix representing the transformation is 'as simple as possible'. Equivalently, given a matrix A, there arises the task of finding, among all *matrices similar to the matrix* A, that is to say, those which are of the form $P^{-1}AP$, with P invertible, those which have a form that is 'as simple as possible'. And that is the problem of the *reduction of a matrix*.

The most 'convenient' case occurs when there exists an invertible matrix P such that the matrix $P^{-1}AP$ is diagonal, in which case the matrix A is said to be *diagonalisable*. It is to be observed that, *in that case, the diagonal elements of the matrix* $P^{-1}AP$ *are the eigenvalues* $\lambda_1, \lambda_2, \ldots, \lambda_n$ *of the matrix* A, and that *the jth column vector of the matrix* P *consists of the components* (relative to the same basis as that used for the matrix A) *of an eigenvector corresponding to* λ_j; this follows from the equivalence

$$P^{-1}AP = \operatorname{diag}(\lambda_i) \Leftrightarrow Ap_j = \lambda_j p_j, \quad 1 \leqslant j \leqslant n.$$

In other words, *a matrix is diagonalisable if and only if there exists a basis of eigenvectors*.

There are matrices which are not diagonalisable (Exercise 1.2-1). For such matrices *Jordan's theorem* gives the simplest form among all similar matrices; the reader interested in this result is referred to the Bibliography and comments given at the end of the book. For our purposes, the result which follows, whose proof is much more simple, is adequate for all future needs.

We recall first the following definitions. A matrix $A = (a_{ij})$ of order n is *upper triangular* if $a_{ij} = 0$ for $i > j$ and *lower triangular* if $a_{ij} = 0$ for $i < j$. If there is no need to distinguish between the two, the matrix is simply called *triangular*.

Theorem 1.2-1

(1) *Given a square matrix* A, *there exists a unitary matrix* U *such that the matrix* $U^{-1}AU$ *is triangular.*

(2) *Given a normal matrix* A, *there exists a unitary matrix* U *such that the matrix* $U^{-1}AU$ *is diagonal.*

(3) *Given a symmetric matrix* A, *there exists an orthogonal matrix* O *such that the matrix* $O^{-1}AO$ *is diagonal.*

Proof

(i) We first prove the property (1) for matrices U which *are not necessarily unitary*. The property is true for $n = 1$; suppose it is true for all matrices of order $n - 1$. Let $\mathscr{A}: V \to V$ be the linear transformation associated with the matrix A. This transformation possesses at least one eigenvector f_1 corresponding to the eigenvalue λ. Let $\varepsilon_2, \ldots, \varepsilon_n$ be vectors such that $(f_1, \varepsilon_2, \ldots, \varepsilon_n)$ is a basis of V. Then,

$$\mathscr{A} f_1 = \lambda f_1, \quad \mathscr{A} \varepsilon_j = \alpha_j f_1 + \mathscr{B} \varepsilon_j, \quad 2 \leqslant j \leqslant n,$$

where $\mathscr{B}$ is a linear transformation defined on the subspace W spanned by the vectors $\varepsilon_2, \ldots, \varepsilon_n$.

By the induction hypothesis, there exists a basis $(f_i)_{i=2}^n$ of W with $f_i = \sum_{j=2}^n \gamma_{ij} \varepsilon_j$, relative to which the transformation $\mathscr{B}$ is represented by an upper triangular matrix. From the equalities

$$\mathscr{A} f_1 = \lambda f_1, \quad \mathscr{A} f_i = \left(\sum_{j=2}^n \alpha_j \gamma_{ij} \right) f_1 + \mathscr{B} f_i, \quad 2 \leqslant i \leqslant n,$$

it follows that the transformation $\mathscr{A}$ is represented by an upper triangular matrix relative to the basis (f_i).

(ii) Using the *Gram–Schmidt orthogonalisation process*, an *orthonormal* basis $(u_i)_{i=1}^n$ is constructed such that

$$u_j = \sum_{k=1}^j \gamma_{kj} f_k, \quad 1 \leqslant j \leqslant n,$$

$$f_i = \sum_{l=1}^i \beta_{li} u_l, \quad 1 \leqslant i \leqslant n.$$

Since by (i)

$$\mathscr{A} f_i = \sum_{i=1}^{j} b_{ij} f_i, \quad 1 \leqslant j \leqslant n,$$

it follows that $\mathscr{A} u_j$ is a linear combination of the vectors $u_1, \ldots, u_j$, so that the transformation $\mathscr{A}$ is still represented by an upper triangular matrix relative to the basis (u_i) as well. As the basis (u_i) is orthonormal, the corresponding change matrix is unitary.

(iii) Set

$$T = (t_{ij}) = U^{-1}AU = U*AU.$$

If the matrix A is normal ($A*A = AA*$), so is the matrix T, since

$$T*T = U*A*UU*AU = U*A*AU.$$

Since the matrix T is upper triangular,

$$\sum_{k=1}^{n} |t_{1k}|^2 = (TT*)_{11} = (T*T)_{11} = |t_{11}|^2, \quad \text{so that} \quad t_{1k} = 0, \quad 2 \leqslant k \leqslant n,$$

$$\sum_{k=2}^{n} |t_{2k}|^2 = (TT*)_{22} = (T*T)_{22} = |t_{22}|^2, \quad \text{so that} \quad t_{2k} = 0, \quad 3 \leqslant k \leqslant n,$$

etc., which proves that the matrix T is diagonal.

(iv) If the matrix A is symmetric, the eigenvector f_1 and the eigenvalue λ of (i) are real and the previous arguments remain valid upon replacing everywhere 'unitary' by 'orthogonal' and the 'adjoint' by the 'transpose'.

$\square$

Remarks

(1) The matrices U satisfying the conditions of the statement are not unique (consider, for example, $A = I$).

(2) The diagonal elements of the triangular matrix $U^{-1}AU$ of (1), or of the diagonal matrix $U^{-1}AU$ of (2), or of the diagonal matrix of (3), are the *eigenvalues* of the matrix A. Consequently, they are real numbers if the matrix A is Hermitian or symmetric and complex numbers of modulus 1 if the matrix is unitary or orthogonal.

(3) It follows from (2) that *every Hermitian or unitary matrix is diagonalisable by a unitary matrix*.

(4) The preceding argument shows that, if O is an orthogonal matrix, there exists a unitary matrix U such that $D = U*OU$ is diagonal (the diagonal elements of D having modulus equal to 1), but the matrix U is not, in general, real, that is to say, orthogonal. Some hints are to be found on this subject in Exercise 1.2-2.

The *singular values* of a square matrix are the positive square roots of the eigenvalues of the Hermitian matrix A*A (or $A^T A$, if the matrix A is real). These latter are always non-negative, since from the relation $A*Ap = \lambda p$, $p \neq 0$, it follows that $(Ap)*Ap = \lambda p*p$. It is to be observed also that *the singular values are all strictly positive if and only if the matrix A is invertible.* In fact,

$$Ap = 0 \Rightarrow A*Ap = 0 \Rightarrow p*A*Ap = (Ap)*Ap = 0 \Rightarrow Ap = 0.$$

Two matrices A and B of type (m, n) are said to be *equivalent* if there exists an invertible matrix Q of order m and an invertible matrix P of order n such that

$$B = QAP.$$

Of course, this is a *more general* notion than that of the similarity of matrices. In fact, it can be shown that every square matrix is equivalent to a diagonal matrix.

Theorem 1.2-2

If A is a real, square matrix, there exist two orthogonal matrices U and V such that

$$U^T AV = \text{diag}(\mu_i),$$

and, if A is a complex, square matrix, there exist two unitary matrices U and V such that

$$U*AV = \text{diag}(\mu_i).$$

In either case, the numbers $\mu_i \geq 0$ are the singular values of the matrix A.

Proof

In order to fix ideas, let us suppose that the matrix A is complex. By Theorem 1.2-1, there exists a unitary matrix V such that

$$V*A*AV = \text{diag}(\mu_i^2),$$

the numbers $\mu_i \geq 0$ being the singular values of the matrix A. Denoting by f_j the jth column vector of the matrix AV, this matrix equality can also be written as

$$f_i^* f_j = \mu_i^2 \delta_{ij}, \quad 1 \leq i, j \leq n.$$

Let $\{\mu_1, \mu_2, \ldots, \mu_r\}$ be the set (perhaps empty) of the zero singular values; then

$$f_j = 0, \quad 1 \leq j \leq r.$$

If we set

$$u_j = \mu_j^{-1} f_j, \quad r + 1 \leq j \leq n,$$

then it follows at once that

$$u_i^* u_j = \delta_{ij}, \quad r+1 \leqslant i,j \leqslant n.$$

This set of vectors may be 'extended' (to a basis) by vectors u_i, $1 \leqslant i \leqslant r$, such that

$$u_i^* u_j = \delta_{ij}, \quad 1 \leqslant i,j \leqslant n.$$

The matrix U, whose jth column vector is u_j, is the required matrix: on the one hand, the relation given immediately above shows that it is a unitary matrix and, on the other,

$$(U^*AV)_{ij} = u_i^* f_j = \begin{cases} 0 = \mu_j \delta_{ij}, & 1 \leqslant j \leqslant r, \\ \mu_j u_i^* u_j = \mu_j \delta_{ij}, & r+1 \leqslant j \leqslant n. \end{cases}$$

The proof is similar if the matrix A is real.   □

Exercises

1.2-1. Show that the matrix

$$\begin{pmatrix} 0 & 1 & & & \\ & 0 & 1 & & \\ & & \ddots & \ddots & \\ & & & 0 & 1 \\ & & & & 0 \end{pmatrix}$$

is not diagonalisable.

1.2-2. Let O be an orthogonal matrix. Show that there exists an orthogonal matrix Q such that

$$Q^{-1}OQ = \begin{pmatrix} 1 & & & & & & & & & \\ & \ddots & & & & & & & & \\ & & 1 & & & & & & & \\ & & & -1 & & & & & & \\ & & & & \ddots & & & & & \\ & & & & & -1 & & & & \\ & & & & & & \cos\theta_1 & \sin\theta_1 & & \\ & & & & & & -\sin\theta_1 & \cos\theta_1 & & \\ & & & & & & & & \ddots & \\ & & & & & & & & & \cos\theta_r & \sin\theta_r \\ & & & & & & & & & -\sin\theta_r & \cos\theta_r \end{pmatrix}.$$

1.2-3. This exercise constitutes an extension of Theorem 1.2-2 to rectangular matrices. Let A be a complex matrix (in order to fix ideas) of type (m, n). Show that there exists a unitary matrix U of order m and a unitary matrix V of order n

such that

$$U^*AV = \begin{pmatrix} \mu_1 & & & & \\ & \mu_2 & & & \\ & & \ddots & & 0 \\ & & & \mu_r & \\ & 0 & & & 0 \end{pmatrix},$$

where the numbers $\mu_i > 0$, $1 \leqslant i \leqslant r$, are, by definition, the *singular values* of the matrix A. A notable result is to show the link which exists between the singular values of the matrix A and the eigenvalues of the square matrix A^*A; show, too, that the integer r is equal to the rank of the matrix A.

1.2-4. Let A be a real matrix of order n. Show that a necessary and sufficient condition for the existence of a unitary matrix U of the same order and of a real matrix B (of the same order) such that $U = A + iB$ (in other words, such that the matrix A is the 'real part' of the matrix U) is that all the singular values of the matrix A should be not greater than 1.

1.3 Special properties of symmetric and Hermitian matrices

In order to fix ideas, we shall consider in what follows Hermitian matrices, with the understanding that all that is said in this section applies equally well to symmetric matrices, if throughout 'Hermitian', 'unitary', 'complex' and 'adjoint' are replaced by 'symmetric', 'orthogonal', 'real' and 'transpose' respectively.

Recall that *all the eigenvalues of a Hermitian matrix are real*, and that *every Hermitian matrix is diagonalisable by a unitary matrix* (Theorem 1.2-1). There exist, moreover, various remarkable *characterisations* of the eigenvalues of a Hermitian matrix: these are the subject-matter of Theorem 1.3-1 below. In order to state them, we need a preliminary definition.

Let A be a square matrix representing a linear transformation of a space V over the field $\mathbb{C}$, equipped with its Hermitian scalar product. The *Rayleigh quotient* of the matrix A is the function

$$R_A : V - \{0\} \rightarrow \mathbb{C}$$

defined by

$$R_A(v) = \frac{(Av, v)}{(v, v)} = \frac{v^*Av}{v^*v}, \quad v \neq 0.$$

It is to be observed that, *if the matrix A is Hermitian, the Rayleigh quotient R_A takes on real values*. Furthermore,

$$R_A(\alpha v) = R_A(v) \quad \text{for every} \quad \alpha \in \mathbb{C} - \{0\}.$$

Consequently, every property involving the set of values taken by the Rayleigh quotient as the vector v runs through a subspace $U \subset V$ can be studied equally well over the unit sphere $\{v \in U : v^*v = 1\}$ of the same subspace; this applies particularly to the properties established in the result which follows. In order to avoid cumbersome expressions, from now on it will not be stated explicitly, when writing $R_A(v)$, that the argument v must not be zero.

Theorem 1.3-1

Let A *be a Hermitian matrix of order n, with eigenvalues*

$$\lambda_1 \leqslant \lambda_2 \leqslant \cdots \leqslant \lambda_n,$$

the associated eigenvectors $p_1, p_2, \ldots, p_n$ satisfying

$$p_i^* p_j = \delta_{ij}.$$

For $k = 1, \ldots, n$, let V_k denote the subspace of V spanned by the vectors p_i, $1 \leqslant i \leqslant k$, and let $\mathscr{V}_k$ denote the set of subspaces of V of dimension k. Furthermore, set

$$V_0 = \{0\}, \mathscr{V}_0 = \{V_0\}.$$

The eigenvalues then have the following characterisations, for $k = 1, 2, \ldots, n$.

(1) $$\lambda_k = R_A(p_k),$$

(2) $$\lambda_k = \max_{v \in V_k} R_A(v),$$

(3) $$\lambda_k = \min_{v \perp V_{k-1}} R_A(v),$$

(4) $$\lambda_k = \min_{W \in \mathscr{V}_k} \max_{v \in W} R_A(v),$$

(5) $$\lambda_k = \max_{W \in \mathscr{V}_{k-1}} \min_{v \perp W} R_A(v).$$

Furthermore,

(6) $$\{R_A(v) : v \in V\} = [\lambda_1, \lambda_n] \subset \mathbb{R}.$$

Proof

Let U be the unitary matrix whose column vectors consist of the eigenvectors $p_1, p_2, \ldots, p_n$, so that

$$U^*AU = \operatorname{diag}(\lambda_l) \overset{\text{def}}{=} D,$$

and let v be a non-zero vector of V. Setting $v = Uw$, it follows that

$$R_A(v) = \frac{v^*Av}{v^*v} = \frac{w^*U^*AUw}{w^*U^*Uw} = \frac{w^*Dw}{w^*w} = R_D(w).$$

Since every vector $v \in V_k$ is of the form $v = \sum_{i=1}^{k} \alpha_i p_i$, the corresponding vector w is accordingly of the form

$$w = \begin{pmatrix} \alpha_1 \\ \vdots \\ \alpha_k \\ 0 \\ \vdots \\ 0 \end{pmatrix},$$

as can be seen at once from the equality $v = Uw$. Hence

$$R_A \left\{ \sum_{i=1}^{k} \alpha_i p_i \right\} = \frac{\sum_{i=1}^{k} \lambda_i |\alpha_i|^2}{\sum_{i=1}^{k} |\alpha_i|^2},$$

which proves (1) and (2). In the same way, since every vector v orthogonal to V_{k-1} is of the form $v = \sum_{i=k}^{n} \alpha_i p_i$, equality (3) also follows. By (2),

$$\lambda_k = \max_{v \in V_k} R_A(v) \geqslant \inf_{W \in \mathcal{V}_k} \max_{v \in W} R_A(v),$$

and it is enough, therefore, to prove the converse inequality in order to establish (4), that is to say, we have to show that

$$\lambda_k \leqslant \max_{v \in W} R_A(v) \quad \text{for every} \quad W \in \mathcal{V}_k.$$

Defining the vector space

$$V_{k-1}^{\perp} = \{ v \in V : v \perp V_{k-1} \},$$

which has dimension $n - k + 1$, it will be enough to show that if W is any subspace of V of dimension k, then the subspace $W \cap V_{k-1}^{\perp}$ contains vectors other than the zero vector, that is

$$\dim(W \cap V_{k-1}^{\perp}) \geqslant 1.$$

It will then be possible to conclude from (3) that

$$v \neq 0 \quad \text{and} \quad v \in W \cap V_{k-1}^{\perp} \Rightarrow \lambda_k \leqslant R_A(v) \leqslant \max_{v \in W} R_A(v).$$

Now since

$$\dim(W \cap V_{k-1}^{\perp}) = \dim(W) + \dim(V_{k-1}^{\perp}) - \dim(W + V_{k-1}^{\perp}),$$

where

$$W + V_{k-1}^{\perp} = \{ z \in V : z = w + v, w \in W, v \in V_{k-1}^{\perp} \},$$

the relations

$$\dim(W) = k, \quad \dim(V_{k-1}^{\perp}) = n - k + 1, \quad \dim(W + V_{k-1}^{\perp}) \leqslant \dim(V) = n,$$

show that $\dim(W \cap V_{k-1}^{\perp}) \geqslant 1$.

Equation (5) is proved analogously. Lastly, equation (6) follows from the inequalities

$$\lambda_1 \leqslant R_A(v) \leqslant \lambda_n \quad \text{for every } v \in V - \{0\}$$

(which, in turn, follow from (2) and (3)), from the continuity of the restriction of the function R_A to the unit sphere $\{v \in V : v^*v = 1\}$ of V and, lastly, from the connectedness of the same unit sphere.   □

Remarks

(1) As particular cases of the characterisations (3) and (2), we have

$$\lambda_1 = \min\{R_A(v) : v \in V\},$$
$$\lambda_n = \max\{R_A(v) : v \in V\}.$$

(2) As a study of the Rayleigh quotient of a non-Hermitian matrix, see Exercise 1.3-5.

(3) Properties (4) and (5) are due to E. Fischer, and R. Courant subsequently extended them to elliptic operators. That is why they often go by the name *the Courant–Fischer theorem*.

(4) As a remarkable application of this theorem see Theorem 2.3-2. □

To conclude, we recall some definitions: a *Hermitian* matrix A is *positive definite* if

$$v^*Av > 0 \quad \text{for every } v \in V - \{0\},$$

and *non-negative definite* if

$$v^*Av \geqslant 0 \quad \text{for every } v \in V.$$

By arguments similar to those given towards the beginning of the proof of the last theorem, it can be readily shown that *a Hermitian matrix is positive definite, or non-negative definite, if and only if all its eigenvalues are positive, or non-negative, respectively.*

Remark

The term 'non-negative matrix' denotes a matrix whose elements are non-negative (in principle, there ought to be no fear of confusion...), and we shall use it in section 3.1. The two ideas are related: see Exercise 1.3-2.

                    □

Exercises

1.3-1. Let (a_{ij}) and (b_{ij}) be two non-negative definite Hermitian matrices of the same order.

(1) Show that $\sum_{i,j} a_{ij} b_{ij} \geqslant 0$.

(2) More generally, show that the Hermitian matrix $(a_{ij} b_{ij})$ is also non-negative definite.

1.3-2. Let $A = (a_{ij})$ be a square matrix whose elements satisfy $a_{ii} \geqslant 0$, $a_{ij} \leqslant 0$ if $i \neq j$. Show that the following two properties are equivalent.

(1) The matrix A is invertible and the elements of A^{-1} are all non-negative.

(2) There exists a diagonal matrix D, with positive diagonal elements, and a number α such that

$$\alpha > 0 \quad \text{and} \quad (v, DAv) \geqslant \alpha(v, v) \quad \text{for every } v \in \mathbb{R}^n.$$

In other words, the matrix DA (in general, not symmetric) is also 'positive definite'.

1.3-3. Let A and B be two Hermitian matrices. Show that they possess a common basis of eigenvectors if and only if $AB = BA$.

1.3-4. (1) Let D be a positive definite Hermitian matrix. Show that there exists a unique positive definite Hermitian matrix B (symmetric, if D is symmetric) such that $D = B^2$.

(2) Show that every invertible matrix A may be written uniquely in the form $A = UB$, U a unitary matrix (orthogonal, if A is real) and B a positive definite Hermitian matrix (symmetric, if A is real).

(3) Show likewise that every invertible matrix A may be written uniquely in the form $A = CU'$, U' a unitary matrix (orthogonal, if A is real) and C a positive definite Hermitian matrix (symmetric, if A is real), and that, moreover, $U = U'$.

(4) Examine the continuity of the functions $A \to B$, $A \to C$ and $A \to U$ (which, by question (3), are well-defined when A belongs to the set of invertible matrices).

Remarks. The matrix B appearing in question (1) is called the *square root* of the matrix D. The factorisations $A = UB$ and $A = CU$ of questions (2) and (3) are called the *polar factorisations* of the matrix A: they are, in effect, a generalisation of the well-known factorisation $z = \rho e^{i\theta}$ of complex numbers. It should be observed, moreover, that the eigenvalues of the matrix B, which coincide with those of the matrix C, are just the *singular values* of the matrix A (in fact, we have $A^*A = B^2 = U^*C^2U$).

1.3-5. The aim in this problem is to study the properties of the Rayleigh quotient of a general matrix (not necessarily Hermitian); it is, therefore, complementary to Theorem 1.3-1. Let A be a complex matrix of order n. Set

$$F(A) = \{v^*Av \in \mathbb{C} : v \in \mathbb{C}^n, v^*v = 1\}.$$

Prove the following properties (recall that sp(A) denotes the set of eigenvalues of A).

(1) $\text{sp}(A) \subseteq F(A)$.

(2) If U is a unitary matrix, $F(A) = F(U^*AU)$.

(3) The set $F(A)$ is compact and convex.

(4) The boundary $\partial F(A)$ of $F(A)$ is a union of algebraic curves. The only points at which $\partial F(A)$ does not have a tangent are the eigenvalues of A.

(5) If the matrix A is normal, the set $F(A)$ is the closed convex hull of sp(A) (hence a polygon).

(6) The set $F(A)$ is a segment of the real axis if and only if the matrix A is Hermitian.

1.3-6. Let A be an invertible symmetric matrix. Prove the identity

$$v^{\mathrm{T}}A^{-1}v = \frac{\det(A + vv^{\mathrm{T}})}{\det A} - 1 \quad \text{for every vector } v.$$

1.3-7. Let $b = (b_i)$ be a non-zero vector of $\mathbb{R}^n$. Determine the eigenvalues and the eigenspaces of the symmetric matrix $(b_i b_j)$.

1.3-8. Let f_i, $1 \leqslant i \leqslant n$, be real functions which are continuous over the interval $[0,1]$. Show that these functions are linearly independent if and only if the symmetric matrix $A = (a_{ij})$, where $a_{ij} = \int_0^1 f_i(x) f_j(x) \, dx$, is positive definite.

1.3-9. Let A be a symmetric matrix with eigenvalues λ_i, and whose corresponding eigenvectors p_i, $1 \leqslant i \leqslant n$, form an orthonormal set. Show that

$$A = \sum_{i=1}^{n} \lambda_i p_i p_i^{\mathrm{T}}.$$

1.3-10. Denoting by $\mathscr{S}_n(\mathbb{R})$ and $\mathscr{O}_n(\mathbb{R})$ the sets of symmetric and orthogonal matrices respectively, of order n, show that a function

$$\mathscr{H} : \mathscr{S}_n(\mathbb{R}) \to \mathscr{S}_n(\mathbb{R})$$

satisfies

$$\mathscr{H}(QBQ^{\mathrm{T}}) = Q\mathscr{H}(B)Q^{\mathrm{T}}$$

for every $Q \in \mathscr{O}_n(\mathbb{R})$ and for every $B \in \mathscr{S}_n(\mathbb{R})$, if and only if it is of the form

$$\mathscr{H} : B \in \mathscr{S}_n(\mathbb{R}) \to \mathscr{H}(B) = \sum_{k=0}^{n-1} \beta_k(B) B^k$$

where the functions $\beta_k : B \in \mathscr{S}_n(\mathbb{R}) \to \mathbb{R}$ are symmetric functions of the *principal invariants* of B, that is to say, of the coefficients of the characteristic polynomial of the matrix B (*Rivlin–Ericksen theorem*).

1.3-11. Let A and B be two positive definite Hermitian matrices. Show that the Hermitian matrix

$$C = A^{-1} + B^{-1} - 4(A + B)^{-1}$$

is also positive definite.

1.4 Vector and matrix norms

Let V be a vector space over the field $\mathbb{K}$ of scalars. A *norm* on V is a function $\|\cdot\| : V \to \mathbb{R}$ which satisfies the following properties:

$$\|v\| = 0 \Leftrightarrow v = 0, \quad \text{and} \quad \|v\| \geqslant 0 \quad \text{for every } v \in V,$$

$$\|\alpha v\| = |\alpha| \|v\| \quad \text{for every } \alpha \in \mathbb{K} \quad \text{and } v \in V,$$

$$\|u + v\| \leqslant \|u\| + \|v\| \quad \text{for every } u, v \in V,$$

the last property being known as the *triangle inequality*. A norm on
V will also be called a *vector norm*. When various spaces are being
considered, the notation $\|\cdot\|_V$ will occasionally be used to recall the space
V under consideration. Finally, we call a vector space which is provided
with a norm a *normed vector space*.

Let V be a finite-dimensional space. The following three norms are the
ones most commonly used *in practice*:

$$\|v\|_1 = \sum_i |v_i|,$$

$$\|v\|_2 = \left(\sum_i |v_i|^2\right)^{1/2} = (v, v)^{1/2},$$

$$\|v\|_\infty = \max_i |v_i|,$$

the norm $\|\cdot\|_2$ being called the *Euclidean norm*. It is easy to verify directly
that the two functions $\|\cdot\|_1$ and $\|\cdot\|_\infty$ are indeed norms (the justification
for the notation $\|\cdot\|_\infty$ is given in Exercise 1.4-1). As for the function $\|\cdot\|_2$,
it is a particular case of the following more general result.

Theorem 1.4-1

*Let V be a finite-dimensional vector space. For every real number $p \geqslant 1$, the
function $\|\cdot\|_p$ defined by*

$$\|v\|_p = \left(\sum_i |v_i|^p\right)^{1/p}$$

is a norm.

Proof

The proof of the case $p = 1$ being immediate, we confine ourselves to the
case $p > 1$; we can then write q (also > 1) for the real number which satisfies

$$\frac{1}{p} + \frac{1}{q} = 1.$$

The proof depends on the following preliminary result: if α and β are
non-negative, then

$$\alpha\beta \leqslant \frac{\alpha^p}{p} + \frac{\beta^q}{q}.$$

In fact, let x and y be any two real numbers and θ a number satisfying
$0 < \theta < 1$. The convexity of the exponential function implies that

$$e^{(\theta x + (1-\theta)y)} \leqslant \theta e^x + (1-\theta)e^y.$$

The case $\alpha\beta = 0$ being trivial, let us suppose that $\alpha > 0$ and $\beta > 0$. To obtain the required result, replace θ by $1/p$, x by $p \log \alpha$ and y by $q \log \beta$.

Let u and v be two vectors of V. By the inequality given above,

$$\frac{|u_i v_i|}{\|u\|_p \|v\|_q} \leqslant \frac{1}{p} \frac{|u_i|^p}{\|u\|_p^p} + \frac{1}{q} \frac{|v_i|^q}{\|v\|_q^q} \quad \text{for every } i,$$

Hence, summing,

$$\sum_i |u_i v_i| \leqslant \|u\|_p \|v\|_q.$$

In order to establish that the function $\|\cdot\|_p$ is a norm, it is enough to prove the triangle inequality, since the other properties are evident. Now, for every index i, we can write

$$(|u_i| + |v_i|)^p = |u_i|(|u_i| + |v_i|)^{p-1} + |v_i|(|u_i| + |v_i|)^{p-1},$$

so that, summing and using the inequality above,

$$\sum_i (|u_i| + |v_i|)^p \leqslant (\|u\|_p + \|v\|_p) \left(\sum_i (|u_i| + |v_i|)^{(p-1)q} \right)^{1/q}.$$

The desired triangle inequality then follows from the equation $(p-1)q = p$.

$\square$

For $p > 1$ and $1/p + 1/q = 1$, the inequality

$$\sum_i |u_i v_i| \leqslant \left(\sum_i |u_i|^p \right)^{1/p} \left(\sum_i |u_i|^q \right)^{1/q}$$

is called *Hölder's inequality*. Hölder's inequality for $p = 2$,

$$\sum_i |u_i v_i| \leqslant \left(\sum_i |u_i|^2 \right)^{1/2} \left(\sum_i |v_i|^2 \right)^{1/2},$$

is called *the Cauchy–Schwarz inequality*, or *the inequality of Bunyakovskiĭ* (especially in the Soviet Union). The triangle inequality for the norm $\|\cdot\|_p$,

$$\left(\sum_i |u_i + v_i|^p \right)^{1/p} \leqslant \left(\sum_i |u_i|^p \right)^{1/p} + \left(\sum_i |v_i|^p \right)^{1/p},$$

is called *Minkowski's inequality*.

The norms defined above are *equivalent*, this property being a particular case of the equivalence of norms in a finite-dimensional space. Recall that two norms $\|\cdot\|$ and $\|\cdot\|'$, defined over the same vector space V, are *equivalent* if there exist two constant C and C' such that

$$\|v\|' \leqslant C \|v\| \quad \text{and} \quad \|v\| \leqslant C' \|v\|' \quad \text{for every } v \in V.$$

Let $\mathscr{A}_n$ be the ring of matrices of order n, with elements in the field $\mathbb{K}$.

A *matrix norm* is a function $\|\cdot\|:\mathscr{A}_n \to \mathbb{R}$ which satisfies the following properties.

$$\|A\| = 0 \Leftrightarrow A = 0 \quad \text{and} \quad \|A\| \geqslant 0 \quad \text{for every } A \in \mathscr{A}_n,$$

$$\|\alpha A\| = |\alpha| \, \|A\| \quad \text{for every } \alpha \in \mathbb{K}, \quad A \in \mathscr{A}_n,$$

$$\|A + B\| \leqslant \|A\| + \|B\| \quad \text{for every } A, B \in \mathscr{A}_n,$$

$$\|AB\| \leqslant \|A\| \, \|B\| \quad \text{for every } A, B \in \mathscr{A}_n.$$

The ring $\mathscr{A}_n$ being itself a vector space of dimension n^2, the first three properties above are nothing other than those of a vector norm, considering a matrix as a vector with n^2 components. The last property is evidently special to square matrices!

The result which follows gives a particularly simple means of constructing matrix norms. *Given a vector norm $\|\cdot\|$ on $\mathbb{C}^n$, the function $\|\cdot\|:\mathscr{A}_n(\mathbb{C}) \to \mathbb{R}$ defined by*

$$\|A\| = \sup_{\substack{v \in \mathbb{C}^n \\ v \neq 0}} \frac{\|Av\|}{\|v\|} = \sup_{\substack{v \in \mathbb{C}^n \\ \|v\| \leq 1}} \|Av\| = \sup_{\substack{v \in \mathbb{C}^n \\ \|v\| = 1}} \|Av\|,$$

is a matrix norm, called the *subordinate matrix norm* (subordinate to the given vector norm). Evidently this is just one particular case of the usual definition of the norm of a linear transformation. But it is necessary to be aware that *there exist matrix norms which are not subordinate to any vector norm.* An example will be given in Theorem 1.4-4.

To establish that the function defined above has the required properties, observe that the number $\|A\|$ is well-defined, since $\sup_{\|v\|=1} \|Av\| < +\infty$ (from the continuity of the function $v \to \|Av\|$ over the unit sphere, which is compact, since the space is finite-dimensional). The other properties of a matrix norm are quite readily verified.

It follows from the definition of a *subordinate* norm that

$$\|Av\| \leqslant \|A\| \, \|v\| \quad \text{for every } v \in \mathbb{C}^n,$$

and that *the norm $\|A\|$ can also be defined by*

$$\|A\| = \inf\{\alpha \in \mathbb{R} : \|Av\| \leqslant \alpha \|v\| \quad \text{for every } v \in \mathbb{C}\}.$$

Furthermore, since the unit sphere is compact, there exists (at least) one vector u such that

$$u \neq 0 \quad \text{and} \quad \|Au\| = \|A\| \, \|u\|.$$

Finally, we observe that *a subordinate norm always satisfies*

$$\|I\| = 1.$$

Remark

We do not introduce the norm

$$|A| \overset{def}{=} \sup_{v \in \mathbb{R}^n - \{0\}} \frac{\|Av\|}{\|v\|} \leqslant \|A\|.$$

in order to avoid having to take certain precautions. It is possible, in actual fact, to construct vector norms and real matrices for which $|A| < \|A\|$ (this is not the case for the matrix norms subordinate to the vector norms $\|\cdot\|_1$, $\|\cdot\|_2$, $\|\cdot\|_\infty$; we will establish in the proof of Theorem 1.4-2 that, if the matrix A is real, the upper bound of the ratio $\|Av\|/\|v\|$ is attained for real vectors). This approach allows us to avoid one particular complication in the proof of Theorem 1.4-3 (where the vector p could be complex).

Let us now calculate each of the subordinate norms of the vector norms $\|\cdot\|_1$, $\|\cdot\|_2$, $\|\cdot\|_\infty$. To simplify the notation, we will omit, from now on, any indication that the upper bounds are to be evaluated over the set of non-zero vectors of $\mathbb{C}^n$.

Theorem 1.4-2

Let $A = (a_{ij})$ *be a square matrix. Then*

$$\|A\|_1 \overset{def}{=} \sup \frac{\|Av\|_1}{\|v\|_1} = \max_j \sum_i |a_{ij}|,$$

$$\|A\|_2 \overset{def}{=} \sup \frac{\|Av\|_2}{\|v\|_2} = \sqrt{\varrho(A^*A)} = \sqrt{\varrho(AA^*)} = \|A^*\|_2,$$

$$\|A\|_\infty \overset{def}{=} \sup \frac{\|Av\|_\infty}{\|v\|_\infty} = \max_i \sum_j |a_{ij}|.$$

The norm $\|\cdot\|_2$ *is invariant under unitary transformations:*

$$UU^* = I \Rightarrow \|A\|_2 = \|AU\|_2 = \|UA\|_2 = \|U^*AU\|_2.$$

Furthermore, if the matrix A is normal,

$$AA^* = AA^* \Rightarrow \|A\|_2 = \varrho(A).$$

Proof

For every vector v,

$$\|Av\|_1 = \sum_i \left| \sum_j a_{ij} v_j \right| \leqslant \sum_j |v_j| \sum_i |a_{ij}| \leqslant \left\{ \max_j \sum_i |a_{ij}| \right\} \|v\|_1.$$

To show that the number $\max_j \sum_i |a_{ij}|$ is in fact the smallest number α

for which the inequality $\|Av\|_1 \leqslant \alpha \|v\|_1$ holds for every vector v, let us construct a vector u (which, of course, depends on the matrix A) for which the following equality holds:

$$\|Au\|_1 = \left\{ \max_j \sum_i |a_{ij}| \right\} \|u\|_1.$$

It is enough to consider the vector u with components

$$u_i = 0 \quad \text{for} \quad i \neq j_0, \quad u_{j_0} = 1,$$

where j_0 is an index satisfying

$$\max_j \sum_i |a_{ij}| = \sum_i |a_{ij_0}|.$$

In the same way,

$$\|Av\|_\infty = \max_i \left| \sum_j a_{ij} v_j \right| \leqslant \left(\max_i \sum_j |a_{ij}| \right) \|v\|_\infty.$$

Let i_0 be an index satisfying

$$\max_i \sum_j |a_{ij}| = \sum_j |a_{i_0 j}|.$$

The vector u with components

$$u_j = \frac{\overline{a_{i_0 j}}}{|a_{i_0 j}|} \quad \text{if} \quad a_{i_0 j} \neq 0, \quad u_j = 1 \quad \text{if} \quad a_{i_0 j} = 0,$$

satisfies

$$\|Au\|_\infty = \left\{ \max_i \sum_j |a_{ij}| \right\} \|u\|_\infty,$$

which establishes the result for the norm $\|\cdot\|_\infty$.

Since

$$\|A\|_2^2 = \sup \frac{v^* A^* A v}{v^* v} = \sup R_{A^* A}(v),$$

Theorem 1.3-1 allows us to affirm that the upper bound of the Rayleigh quotient of the Hermitian matrix A*A is the largest eigenvalue of this matrix, which also happens to be its spectral radius since it is non-negative definite.

We next show that $\varrho(A^*A) = \varrho(AA^*)$. If $\varrho(A^*A) > 0$, there exists a vector p such that

$$p \neq 0 \quad \text{and} \quad A^* A p = \varrho(A^* A) p,$$

so that we certainly have $Ap \neq 0$ ($\varrho(A^*A) > 0$). Since, therefore,

$$Ap \neq 0 \quad \text{and} \quad AA^*(Ap) = \varrho(A^*A) Ap,$$

it follows that

$$0 < \varrho(A^*A) \leqslant \varrho(AA^*),$$

and hence that $\varrho(AA^*) = \varrho(A^*A)$ since $(A^*)^* = A$. If $\varrho(A^*A) = 0$, then also $\varrho(AA^*) = 0$; otherwise, the previous argument would show that $\varrho(A^*A) > 0$. Hence, in either case, it follows that

$$\|A\|_2^2 = \varrho(A^*A) = \varrho(AA^*) = \|A^*\|_2^2.$$

The invariance of the norm $\|\cdot\|_2$ under unitary transformations is nothing more than the interpretation of the equalities

$$\varrho(A^*A) = \varrho(U^*A^*AU) = \varrho(A^*U^*UA) = \varrho(U^*A^*UU^*AU).$$

Lastly, if the matrix A is normal, there exists a unitary matrix U (Theorem 1.2-1) such that

$$U^*AU = \mathrm{diag}\,(\lambda_i(A)) \overset{\mathrm{def}}{=} D.$$

Accordingly,

$$A^*A = (UDU^*)^*UDU^* = UD^*DU^*,$$

which proves that

$$\varrho(A^*A) = \varrho(D^*D) = \max_i |\lambda_i(A)|^2 = (\varrho(A))^2. \qquad \square$$

Remarks

(1) *The norm* $\|A\|_2$ *is nothing other than the largest singular value of the matrix* A (see section 1.2).

(2) If a matrix A is Hermitian, or symmetric (and hence normal), we have $\|A\|_2 = \varrho(A)$.

(3) If a matrix A is unitary, or orthogonal (and hence normal), we have $\|A\|_2 = \sqrt{\varrho(A^*A)} = \sqrt{\varrho(I)} = 1$.

(4) From the *practical* point of view, it is worth observing that, while the norms $\|A\|_1$ and $\|A\|_\infty$ are calculated very easily merely from a knowledge of the elements of the matrix A, the same is not true for the norm $\|A\|_2$.

(5) It is easy to convince oneself, simply by an examination of the proof given above, that the expressions given for $\|A\|_1$, $\|A\|_2$, $\|A\|_\infty$ remain valid *even if the matrix* A *is rectangular*. But, of course, the functions so obtained are no longer matrix norms in the sense accepted here, since the multiplication of these matrices does not make sense in general. They are only norms in the vector space of rectangular matrices of a given type.

$$\square$$

As a result of Theorem 1.4-2, we know that there are matrices A and norms $\|\cdot\|$ (subordinate, as it turns out) which satisfy the equality

$\|A\| = \varrho(A)$, namely, the norms $\|\cdot\|_2$ and normal matrices. But there exist matrices for which it is certainly not possible to find matrix norms (subordinate or not) satisfying this equality. It is enough to consider, for example, the matrix

$$A = \begin{pmatrix} 0 & 1 \\ 0 & 0 \end{pmatrix},$$

for which it is always the case that $\varrho(A) = 0 < \|A\|$, since $A \neq 0$. While it is true that it is possible to find cases where equality is never attained, it will be proved, nevertheless, that, *for a given matrix*, it is always possible to approximate its spectral radius *as closely as we please from above*, with the help of a matrix norm chosen appropriately. This result plays a fundamental role in the study of the convergence of sequences of matrices (see section 1.5).

Theorem 1.4-3

(1) *Let* A *be any square matrix and* $\|\cdot\|$ *any matrix norm, subordinate or otherwise. Then*

$$\varrho(A) \leqslant \|A\|.$$

(2) *Given a matrix* A *and any number* $\varepsilon > 0$, *there exists at least one subordinate matrix norm such that*

$$\|A\| \leqslant \varrho(A) + \varepsilon.$$

Proof

Let p be a vector satisfying

$$p \neq 0, \quad Ap = \lambda p, \quad |\lambda| = \varrho(A),$$

and let q be a vector such that the matrix pq^T is not null. Since

$$\varrho(A)\|pq^T\| = \|\lambda pq^T\| = \|Apq^T\| \leqslant \|A\|\,\|pq^T\|,$$

by the last of the properties of matrix norms, the inequality $\varrho(A) \leqslant \|A\|$ is proved.

Now let A be a given matrix. There exists an invertible matrix U (see Theorem 1.2-1; that the matrix U is unitary is of no consequence here) such that the matrix $U^{-1}AU$ is upper triangular; for example,

$$U^{-1}AU = \begin{pmatrix} \lambda_1 & t_{12} & t_{13} & \cdots & & t_{1n} \\ & \lambda_2 & t_{23} & \cdots & & t_{2n} \\ & & \ddots & & & \vdots \\ & & & & \lambda_{n-1} & t_{n-1,n} \\ & & & & & \lambda_n \end{pmatrix},$$

the scalars λ_i being the eigenvalues of the matrix A. With every scalar $\delta \neq 0$, we associate the matrix

$$D_\delta = \operatorname{diag}(1, \delta, \delta^2, \ldots, \delta^{n-1}),$$

so that

$$(UD_\delta)^{-1}A(UD_\delta) = \begin{pmatrix} \lambda_1 & \delta t_{12} & \delta^2 t_{13} & \cdots & \delta^{n-1} t_{1n} \\ & \lambda_2 & \delta t_{23} & \cdots & \delta^{n-2} t_{2n} \\ & & \ddots & & \vdots \\ & & & \lambda_{n-1} & \delta t_{n-1,n} \\ & & & & \lambda_n \end{pmatrix}.$$

Given the number $\varepsilon > 0$, let us fix the number δ so that

$$\sum_{j=i+1}^{n} |\delta^{j-i} t_{ij}| \leqslant \varepsilon, \quad 1 \leqslant i \leqslant n-1.$$

Then the function

$$\|\cdot\|: B \in \mathscr{A}_n \to \|B\| = \|(UD_\delta)^{-1}B(UD_\delta)\|_\infty,$$

which, of course, *depends on the matrix* A *and the number* ε, has the required property. For we have, on the one hand,

$$\|A\| \leqslant \varrho(A) + \varepsilon,$$

by the choice of δ and the definition of the matrix norm $\|\cdot\|_\infty$ ($\|(c_{ij})\|_\infty = \max_i \sum_j |c_{ij}|$), and, on the other, it is indeed a matrix norm; for it can be verified that it is the matrix norm subordinate to the vector norm

$$v \in \mathbb{K}^n \to \|(UD_\delta)^{-1}v\|_\infty. \qquad \square$$

An important example of a matrix norm which is *not subordinate* is given in the following theorem.

Theorem 1.4-4

The function $\|\cdot\|_E: \mathscr{A}_n \to \mathbb{R}$ *defined by*

$$\|A\|_E = \left\{ \sum_{i,j} |a_{ij}|^2 \right\}^{1/2} = \{\operatorname{tr}(A^*A)\}^{1/2}$$

for every matrix $A = (a_{ij})$ *of order n is a matrix norm which is not subordinate* (*for* $n \geqslant 2$), *is invariant under unitary transformations,*

$$UU^* = I \Rightarrow \|A\|_E = \|AU\|_E = \|UA\|_E = \|U^*AU\|_E,$$

and satisfies

$$\|A\|_2 \leqslant \|A\|_E \leqslant \sqrt{n}\,\|A\|_2 \quad \text{for every } A \in \mathscr{A}_n.$$

Proof

The function $\|\cdot\|_E$ is nothing other than the Euclidean norm (hence the notation) over the vector space $\mathscr{A}_n$, of dimension n^2. Thus it is enough to prove the fourth property of matrix norms, which is a simple consequence of the Cauchy–Schwarz inequality:

$$\|AB\|_E^2 = \sum_{i,j} \left| \sum_k a_{ik}b_{kj} \right|^2 \leqslant \sum_{i,j} \left\{ \sum_k |a_{ik}|^2 \right\} \left\{ \sum_l |b_{lj}|^2 \right\}$$

$$= \left\{ \sum_{i,k} |a_{ik}|^2 \right\} \left\{ \sum_{j,l} |b_{lj}|^2 \right\} = \|A\|_E^2 \|B\|_E^2.$$

This norm is certainly not a subordinate norm, since $\|I\|_E = \sqrt{n}$. If U is a unitary matrix,

$$\|A\|_E^2 = \text{tr}(A*A) = \text{tr}(U*A*AU) = \|AU\|_E^2 = \text{tr}(A*U*UA) = \|UA\|_E.$$

Finally, the inequalities in the statement of the theorem follow from the inequalities

$$\varrho(A*A) \leqslant \text{tr}(A*A) \leqslant n\varrho(A*A). \qquad \square$$

Remark

Unlike the subordinate matrix norm $\|\cdot\|_2$, the norm $\|\cdot\|_E$ lends itself readily to a *quick* calculation. In that lies one of its principal sources of interest, since it provides, in particular, a bound for the norm $\|\cdot\|_2$ (see, in this respect, the numerical example of section 2.2). $\qquad \square$

We end with a theorem which gathers together some useful properties of matrices of the form $I + B$.

Theorem 1.4-5

(1) *Let* $\|\cdot\|$ *be a subordinate matrix norm and* B *a matrix satisfying*

$$\|B\| < 1.$$

Then the matrix $I + B$ *is invertible and*

$$\|(I + B)^{-1}\| \leqslant \frac{1}{1 - \|B\|}.$$

(2) *If a matrix of the form* $I + B$ *is singular, then necessarily*

$$\|B\| \geqslant 1$$

for every matrix norm, subordinate or not.

Proof

(1) Since

$$(I + B)u = 0 \Rightarrow \|u\| = \|Bu\|,$$

$$\|B\| < 1 \quad \text{and} \quad u \neq 0 \Rightarrow \|Bu\| < \|u\|,$$

for the corresponding vector norm, it follows that

$$(I + B)u = 0 \Rightarrow u = 0.$$

The matrix $I + B$ is therefore invertible, so that one can write

$$(I + B)^{-1} = I - B(I + B)^{-1},$$

with the result that

$$\|(I + B)^{-1}\| \leqslant 1 + \|B\| \|(I + B)^{-1}\|,$$

which gives the required inequality.

(2) To state that the matrix $I + B$ is singular amounts to stating that -1 is an eigenvalue of B; so that, applying Theorem 1.4-3, it follows that $\|B\| \geqslant \varrho(B) \geqslant 1$. $\qquad\qquad\square$

Exercises

1.4-1. Let V be a finite-dimensional vector space. Show that

$$\lim_{p \to \infty} \|v\|_p = \|v\|_\infty \quad \text{for every } v \in V.$$

1.4-2. Given a diagonalisable matrix A, does a matrix norm $\|\cdot\|$ exist for which $\rho(A) = \|A\|$?

1.4-3. (1) Let A be an invertible matrix. Evaluate

$$\inf\{\|A - B\|_2 : B \text{ singular}\}.$$

(2) Let A be a singular matrix. Evaluate

$$\inf\{\|A - B\|_2 : B \text{ invertible}\}.$$

Hint. It will be of great help to introduce the singular values of the matrix A.

(3) What topological property may be deduced about the set of invertible matrices of a given order, regarded as a subset of the set of all matrices of the same order?

1.4-4. Let $A = (a_{ij})$ be a *strictly diagonally dominant* matrix, in the sense that

$$0 < \min_i \left\{ |a_{ii}| - \sum_{j \neq i} |a_{ij}| \right\} \overset{\text{def}}{=} \delta.$$

Prove that (such a matrix being necessarily invertible; cf. Exercise 1.1-5)

$$\|A^{-1}\|_\infty \leqslant \delta^{-1}.$$

1.4-5. The aim in this exercise is to characterise *subordinate* matrix norms. All matrices considered belong to one and the same ring $\mathscr{A}_n$.

(1) If $\|\cdot\|$ and $\|\cdot\|'$ are two matrix norms, we define the following relation

$$\|\cdot\| \leqslant \|\cdot\|' \Leftrightarrow \|A\| \leqslant \|A\|' \quad \text{for every } A \in \mathscr{A}_n.$$

Show that this provides a partial ordering of the set $\mathscr{N}$ of matrix norms defined over the ring $\mathscr{A}_n$.

(2) Let $\|\cdot\|$ and $\|\cdot\|'$ be two matrix norms subordinate to the vector norms $|\cdot|$ and $|\cdot|'$, respectively. If $\|A\| \leqslant \|A\|'$ for all matrices $A \in \mathscr{A}_n$ of *rank* 1, show that there exists a constant c such that

$$|v| = c|v|' \quad \text{for every vector } v.$$

(3) Deduce, from (2), that:

two subordinate matrix norms which are comparable (that is, which satisfy either $\|\cdot\| \leqslant \|\cdot\|'$ or $\|\cdot\|' \leqslant \|\cdot\|$) are equal;

if a matrix norm is subordinate to two vector norms $|\cdot|$ and $|\cdot|'$, then $|v| = c|v|'$ for every vector v;

if two subordinate matrix norms are distinct ($\|A\| \neq \|A\|'$ for at least one matrix $A \in \mathscr{A}_n$), then they are not comparable.

(4) Let $\|\cdot\|'$ be any matrix norm. Show that there exists (at least) one subordinate matrix norm $\|\cdot\|$ satisfying $\|\cdot\| \leqslant \|\cdot\|'$.

(5) Prove that a matrix norm $\|\cdot\|$ is subordinate if and only if it is a minimal element of the set $\mathscr{N}$, that is to say, if and only if

$$\|\cdot\|' \in \mathscr{N} \quad \text{and} \quad \|\cdot\|' \leqslant \|\cdot\| \Rightarrow \|\cdot\| = \|\cdot\|'.$$

(6) Show that there exist matrix norms $\|\cdot\|$ satisfying $\|I\| = 1$, which are yet not subordinate.

1.4-6. Let A be a Hermitian matrix. Find a necessary and sufficient condition for the function $v \to (v^*Av)^{1/2}$ to be a norm.

1.4-7. Prove that the function

$$v \in \mathbb{K}^n \to \|v\|_p = \left(\sum_{i=1}^n |v_i|^p \right)^{1/p}$$

is not a norm when $0 < p < 1$ (unless $n = 1$).

1.4-8. Prove *Jensen's inequality*: if $0 < p < q$, then

$$\left(\sum_{i=1}^n |v_i|^q \right)^{1/q} \leqslant \left(\sum_{i=1}^n |v_i|^p \right)^{1/p}.$$

1.4-9. Find the *smallest* constants C for which

$$\|v\| \leqslant C \|v\|' \quad \text{for every } v \in \mathbb{K}^n,$$

when the distinct norms $\|\cdot\|$ and $\|\cdot\|'$ are chosen from the set $\{ \|\cdot\|_1, \|\cdot\|_2, \|\cdot\|_\infty \}$ (there are thus 6 cases to consider).

1.4-10. Do there exist any values of the number $\alpha > 0$ for which the function

$$A = (a_{ij}) \in \mathscr{A}_n \to \alpha \max_{i,j} |a_{ij}|$$

is a matrix norm?

1.4-11. Let $\|\cdot\|$ be any matrix norm. Does there exist a vector norm $|\cdot|$ satisfying

$$|Av| \leqslant \|A\| |v|$$

for every matrix A and every vector v?

1.4-12. Let A be a real matrix and $\|\cdot\|$ a vector norm over $\mathbb{C}^n$. Define

$$\|A\| = \sup_{v \in \mathbb{C}^n - \{0\}} \frac{\|Av\|}{\|v\|},$$

$$|A| = \sup_{v \in \mathbb{R}^n - \{0\}} \frac{\|Av\|}{\|v\|}.$$

(1) Show that $\|A\| = |A|$ if the vector norm is any of the norms $\|\cdot\|_p$, $p = 1, 2, \infty$ (refer to the proof of Theorem 1.4-2). Is the result also true for the norms $\|\cdot\|_p$, for the remaining values of $p \geqslant 1$?

(2) Give an example of a matrix A and a vector norm for which $|A| \leqslant \|A\|$.

(3) Prove that $\rho(A) < |A|$ (a fact which is not a consequence of the proof of Theorem 1.4-3).

1.5 Sequences of vectors and matrices

A *sequence* (infinite) of elements $x_0, x_1, \ldots,$ of a set X will be denoted by $(x_k)_{k \geqslant 0}$, or even simply by (x_k) if there is no risk of confusion. In a vector space V, equipped with a norm $\|\cdot\|$, a sequence (v_k) of elements of V is said to *converge to an element* $v \in V$, which is the *limit of the sequence* (v_k), if

$$\lim_{k \to \infty} \|v_k - v\| = 0,$$

and one writes

$$v = \lim_{k \to \infty} v_k.$$

If the space is finite-dimensional, the equivalence of norms shows that *the convergence of a sequence is independent of the norm chosen*. The particular choice of the norm $\|\cdot\|_\infty$ shows that *the convergence of a sequence of vectors is equivalent to the convergence of n sequences* (n being equal to the dimension of the space) *of scalars consisting of the components of the vectors*.

By considering the set $\mathscr{A}_{m,n}(\mathbb{K})$ of matrices of type (m, n) as a vector space of dimension mn, one sees in the same way that *the convergence of a sequence of matrices of type (m, n) is independent of the norm chosen*, and that *it is equivalent to the convergence of mn sequences of scalars consisting of the elements of these matrices*.

The result which follows gives necessary and sufficient conditions for the convergence of the particular sequence consisting of the successive powers of a given (square) matrix to the null matrix (further results are to be found in Exercise 1.5-4). From these conditions can be derived the fundamental criterion for *the convergence of iterative methods* for the solution of linear systems of equations (Theorem 5.1-1).

Theorem 1.5-1
Let B *be a square matrix. The following conditions are equivalent:*

(1) $$\lim_{k \to \infty} B^k = 0,$$

(2) $$\lim_{k \to \infty} B^k v = 0 \text{ for every vector } v,$$

(3) $$\varrho(B) < 1,$$

(4) $\quad \| B \| < 1$ *for at least one subordinate matrix norm* $\| \cdot \|$.

Proof
$(1) \Rightarrow (2)$. Let $\| \cdot \|$ be a vector norm and $\| \cdot \|$ the corresponding subordinate matrix norm. Given a vector v, the inequality

$$\| B^k v \| \leqslant \| B^k \| \, \| v \|$$

shows that $\lim_{k \to \infty} B^k v = 0$.
$(2) \Rightarrow (3)$. If $\varrho(B) \geqslant 1$, one can find a vector p such that

$$p \neq 0, \quad Bp = \lambda p, \quad |\lambda| \geqslant 1.$$

Hence the sequence of vectors $(B^k p)_{k \geqslant 1}$ cannot converge to 0 (since $B^k p = \lambda^k p$).
$(3) \Rightarrow (4)$. This is an immediate consequence of Theorem 1.4-3.
$(4) \Rightarrow (1)$. It is enough to apply the inequality

$$\| B^k \| \leqslant \| B \|^k,$$

using the matrix norm given in (4). $\qquad \square$

The result which follows is also useful for the study of iterative methods, as regards the *rate* of convergence. And it is no more than a particular case (the finite-dimensional case) of a result in Functional Analysis which holds in Banach spaces; see, for example, Taylor (1958), Theorem 5.2-E.

Theorem 1.5-2
Let B *be a square matrix and let* $\| \cdot \|$ *be any matrix norm. Then*

$$\lim_{k \to \infty} \| B^k \|^{1/k} = \varrho(B).$$

Proof

Since $\varrho(B) \leqslant \| B \|$ (Theorem 1.4-3) and since $\varrho(B) = \{\varrho(B^k)\}^{1/k}$, it follows at once that

$$\varrho(B) \leqslant \| B^k \|^{1/k} \quad \text{for every } k.$$

We shall next establish that, for every $\varepsilon > 0$, there exists an integer $l = l(\varepsilon)$ such that

$$k \geqslant l \Rightarrow \| B^k \|^{1/k} \leqslant \varrho(B) + \varepsilon,$$

which will give the required result. Suppose, then, that $\varepsilon > 0$ is given; the matrix

$$B_\varepsilon = \frac{B}{\varrho(B) + \varepsilon}$$

satisfies $\varrho(B_\varepsilon) < 1$, so that, from Theorem 1.5-1, it follows that $\lim_{k \to \infty} B_\varepsilon^k = 0$. Consequently, there exists an integer $l = l(\varepsilon)$ such that

$$k \geqslant l \Rightarrow \| B_\varepsilon^k \| = \frac{\| B^k \|}{\{\varrho(B) + \varepsilon\}^k} \leqslant 1,$$

which is the required result. $\qquad\qquad\qquad\qquad\qquad\qquad\qquad\qquad\square$

Exercises

1.5-1. Let (U_k) be a sequence of unitary matrices. Show that there exists a subsequence which converges to a unitary matrix.

1.5-2. Let A be a square matrix such that the sequence $(A^k)_{k \geqslant 1}$ converges to an invertible matrix. Find A.

1.5-3. Let B be a square matrix satisfying $\| B \| < 1$. Prove that the sequence $(C_k)_{k \geqslant 1}$, where

$$C_k = I + B + B^2 + \cdots + B^k,$$

converges and that

$$\lim_{k \to \infty} C_k = (I - B)^{-1}.$$

1.5-4. Prove that the equivalent conditions (1)–(4) of Theorem 1.5-1 are also equivalent to the following:

(5) The matrix $(I - B)$ is invertible and

$$(I - B)^{-1} = \lim_{k \to \infty} (I + B + \cdots + B^k).$$

(6) The matrix $(I - B)$ is invertible and all the eigenvalues of the matrix $(I + 2(B - I)^{-1})$ have negative real part.

(7) There exists a positive definite Hermitian matrix H such that the (Hermitian) matrix $(H - B^*HB)$ is positive definite (*Stein's theorem*).

(8) Given any matrix norm $\| \cdot \|$, there exists an integer l such that $\| B^l \| < 1$.

(9) There exists a constant c such that

$$c < 1, \quad |tr(B^k)| \leqslant nc^k \quad \text{for every} \quad k \geqslant 1,$$

where n is the dimension of the space.

1.5-5. (1) Let A be a square matrix. Set

$$B_k = I + \frac{A}{1!} + \frac{A^2}{2!} + \cdots + \frac{A^k}{k!}, k \geqslant 1.$$

Show that the sequence (B_k) converges. Its limit is denoted by e^A.

(2) Prove that

$$\det(e^A) = e^{tr(A)},$$

which shows that the matrix e^A is always invertible.

(3) Prove that

$$AB = BA \Rightarrow e^{A+B} = e^A e^B.$$

1.5-6. Let (A_k) be a sequence of matrices of type (m, n). Prove that the following conditions are equivalent:

(i) the sequence (A_k) converges;
(ii) for every vector $v \in \mathbb{R}^n$, the sequence of vectors $(A_k v)$ converges in $\mathbb{R}^m$.

1.5-7. The aim in this exercise is to prove directly (that is, without making use of the result of Theorem 1.4-3) the equivalence (proved in Theorem 1.5-1)

$$\lim_{k \to \infty} B^k = 0 \Leftrightarrow \rho(B) < 1.$$

Recall that, if B is a matrix of order n, there exists an invertible matrix P which reduces the matrix B to a block diagonal matrix, called the *Jordan canonical form* (for a proof, see, for example, Godement (1966), §35, or Strang (1980), Appendix B):

$$P^{-1}BP = \begin{pmatrix} \boxed{J_1} & & & \\ & \boxed{J_2} & & \\ & & \ddots & \\ & & & \boxed{J_r} \end{pmatrix}.$$

where each submatrix J_l, of order n_l, is of the form

$$J_l = \lambda_l I \quad \text{if } n_l = 1,$$

$$J_l = \begin{pmatrix} \lambda_l & 1 & & & \\ & \lambda_l & 1 & & \\ & & \ddots & \ddots & \\ & & & \lambda_l & 1 \\ & & & & \lambda_l \end{pmatrix} \quad \text{if } n_l \geqslant 2,$$

λ_l being, as a result, one of the eigenvalues of the matrix B; two distinct submatrices J_l may correspond to the same eigenvalue and/or have the same order. This result will be assumed for the rest of the problem.

(1) Prove that the matrix B is diagonalisable if and only if $n_l = 1$, $1 \leqslant l \leqslant r$.

(2) Assuming that $n_l \geqslant 2$, calculate explicitly the elements of the matrix J_l^k.

(3) Deduce the required equivalence.

(4) Show that the sequence of successive powers $(A^k)_{k \geqslant 0}$ of a matrix A converges if and only if either $\rho(A) < 1$ or $\rho(A) \leqslant 1$, and in the latter case the only eigenvalue of modulus 1 is precisely 1 *and* the dimension of the corresponding eigenspace is equal to its multiplicity (in other words, the submatrices J_l corresponding to the eigenvalue 1 are of order 1).

1.5-8. (1) Let P be a positive definite Hermitian matrix and B a given matrix. Show that *Lyapunov's equation*

$$B^*X + XB = -P$$

where X is a positive definite Hermitian matrix, has a solution if and only if

$$\text{sp}(B) \subset \{z \in \mathbb{C} : \text{Re}\, z < 0\}.$$

Hint. Use the result of Exercise 1.5-4(6).

(2) When the condition given above is satisfied, prove that

$$X = \int_0^\infty e^{B^*t} P e^{Bt} dt$$

(refer to Exercise 1.5-5 for the definition of the exponential of a matrix).

1.5-9. Let A be a square matrix. Prove that

$$\rho(A) = \limsup_{k \to \infty} |\text{tr}(A^k)|^{1/k}.$$

2

General results in the numerical analysis of matrices

Introduction

The *two fundamental problems of the Numerical Analysis of Matrices* are the solution of linear systems of equations and the calculation of the eigenvalues and eigenvectors of a matrix. In the following chapter we shall indicate the great variety of numerical problems which, in the final analysis, belong to this category.

In section 2.1, two types of error are described which occur in the solution of these problems: *rounding errors*, due to the limitations of computers, and *truncation errors*, which are a feature of *iterative methods* (as opposed to *direct methods*). After that, certain properties are examined which play a deciding role in the *choice of method* of solution, notably *the distribution of the zeros* of a matrix. Finally, some indication is given of the methods *currently available* for the solution of either problem, the choice being determined by the properties which may be possessed by the matrix in question, properties such as symmetry and positive definiteness, or having a band structure, etc.

A third source of error is inherent in what is called the *condition* of the problem: a problem is *ill-conditioned* when 'small' changes in the data (the elements of a matrix, or the components of a vector, etc.) produce 'large' changes in the result (the solution of a linear system, or the eigenvalues of a matrix, etc.), even if the actual calculations are carried out *exactly*, that is, without rounding or truncation errors. For this reason, in sections 2.2 and 2.3, we examine *the condition of a linear system* and *the condition of the eigenvalue problem*, illustrating our considerations with a number of particularly 'spectacular' numerical examples.

2.1 The two fundamental problems; general observations on the methods in use

The *two fundamental problems of the Numerical Analysis of Matrices* are the following.

(1) *The solution of a linear system.* Given an invertible matrix A and a vector b, find the vector u which is the solution of the linear system

$$Au = b.$$

(2) *The calculation of the eigenvalues and eigenvectors of a matrix.* Given a square matrix A, find its eigenvalues, or only a *subset* of them, and, *subsequently*, the corresponding eigenvectors; in other words, look for scalars λ and vectors p such that

$$p \neq 0, \quad Ap = \lambda p.$$

Remark

The elements of the matrix and the components of the vector in (1) are for the most part real numbers, so that all subsequent calculations involve only real numbers. In contrast, it is to be observed that complex numbers may arise in the solution of (2), even if the matrix involved is real. □

In solving a problem of type (1) or (2), the first step, quite obviously, is *the choice of a particular method* (which will depend on criteria that are given explicitly in the following chapters), capable of producing a *numerical result*. Now, *whatever the method*, this result is, in general, not exact. This is because certain errors are inevitable during the course of calculation. These are the *rounding errors*.

In a computer, calculations are most often carried out in *floating point*: with each real number is associated an ordered pair (a, b) (the numbers a and b being represented in binary form), preceded by the sign $+$ or $-$. This pair represents the number $(+a) \times (2^b)$ or $(-a) \times (2^b)$. The *mantissa* a satisfies $1/2 \leqslant a < 1$; in other words, the binary representation of the number a is always of the form $a = 0.1 \cdots$. The *exponent* b can be positive, zero or negative.

The *limitations* imposed by the computer are an upper bound for $|b|$ and the number t of binary digits (0 or 1) in the representation of the mantissa a. As a consequence of the limitation on $|b|$ it is not possible to deal with numbers too large in absolute value; in which case, the computer's 'word-length' is said to have been *exceeded*. Another consequence is the *replacement by zero* of numbers too small in absolute value. We illustrate by example the effect of the limitation imposed on the mantissa; suppose we wish to represent the number $x = 405.59$ in a computer for which $t = 13$. Since the number x can also be written as

$$405.59 = (2^{-1} + 2^{-2} + 2^{-5} + 2^{-7} + 2^{-9} + 2^{-10} + 2^{-13} + 0.44 \times 2^{-13})2^9,$$

it will be represented by the number

$$(+0.110\,010\,101\,100\,1) \times 2^{1001}$$

in the computer (supposing that 9 is smaller than the upper bound, in absolute value, of permissible exponents). The error in the mantissa is then 0.44×2^{-13}, and it is easy to see that, in the general case, *the error in the mantissa is bounded in absolute value by* 2^{-t}.

Accordingly, at each stage of a sequence of calculations, every datum and every result of an operation are necessarily *rounded* in the way indicated above: and the errors thus incurred in a sequence of calculations appropriate to a given method are called *rounding errors*.

Remarks
(1) If it turns out to be necessary to do so in some calculations (in order to achieve 'high' precision, say), one can augment the number t (for example, by doubling it; in that case the calculations are said to be carried out in *double precision*) by making use of basic characteristics of the computer. However, though this course of action has the effect of diminishing rounding errors, it does not, of course, eliminate them altogether.

(2) Similarly, in some computers, it is possible to increase the upper bound of permissible exponents.

(3) Lastly, the mantissa is occasionally calculated in the hexadecimal base. □

This gives rise to the idea that the effect of the rounding errors on the final result increases with the number of *elementary operations* executed by the given method; the elementary operations are those which a computer can carry out, namely, addition, subtraction, multiplication and division. The *count of elementary operations used*, or at least an estimate of the number, is therefore indispensable for the analysis of this type of error.

Remark
The count is equally important for the evaluation of the *time* required for a given set of calculations, or, in the final analysis, its *cost*; for, it has to be recognised, it is often that which is the deciding factor, sometimes requiring the rejection of a sequence of calculations which is 'too long', at others leading to the conclusion that the cost of the use of double precision (which invariably takes longer) is prohibitive, etc.

In this regard, it will be of help to keep in mind the following orders of magnitude (which are valid for the more recent computers): the time required for one multiplication is approximately four times that required for an addition (or subtraction), while the time required for a division is approximately ten times that required for an addition. □

While rounding errors are *always* present in any calculation, the second kind of error is to be found only in *certain* methods. As we shall see, methods for the solution of linear systems fall into two classes, direct methods and iterative methods. A *direct method* would produce an *exact* solution to a problem in a *finite* number of elementary operations *if no rounding errors were present*; the use of *Cramer's Rule* (which calculates determinants) is an example of a direct method (which, however, is not utilised in practice; we shall see why in section 4.2).

In an *iterative method*, on the other hand, the solution u to the problem is the limit of a sequence (u_k) of 'approximate' solutions u_k each of which has to be calculated. As it is obligatory to end the sequence for some integer k_0, a *truncation error* is thereby incurred, which is measured by the quantity $\|u - u_{k_0}\|$ (for a given vector norm $\|\cdot\|$).

Remark

Since, in principle, the exact solution u is unknown, one could only hope to have *estimates* for the quantity $\|u - u_{k_0}\|$. Similarly, the choice of the integer k_0 for the termination of calculation has more to do with intuition, experience or practical considerations (such as length of time...), than with a strict argument permitting the firm assertion that the truncation error $\|u - u_{k_0}\|$ is smaller than some pre-assigned value.... □

In the circumstances, it is clear that methods for the calculation of eigenvalues could only be iterative! For a straightforward verification shows that the arbitrary polynomial of degree n,

$$p: \lambda \in \mathbb{C} \to \lambda^n + a_1 \lambda^{n-1} + \cdots + a_{n-1}\lambda + a_n,$$

is nothing other, up to a factor of $(-1)^n$, than the characteristic polynomial of the matrix

$$\begin{pmatrix} -a_1 & -a_2 & -a_3 & \cdots & -a_{n-1} & -a_n \\ 1 & 0 & & & & \\ & 1 & 0 & & & \\ & & 1 & \ddots & & \\ & & & \ddots & 0 & \\ & & & & 1 & 0 \end{pmatrix}$$

called the *companion matrix of the polynomial.* Accordingly, the existence of a direct method for the calculation of eigenvalues would be equivalent to the affirmation that it is possible to calculate the roots of an arbitrary polynomial in a finite number of elementary operations. Now, even if the taking of a pth root is allowed to count as an elementary operation, this possibility would stand in contradiction to the famous *theorem of Abel* (a proof of which may be found, for example, in Herstein (1964), p. 214) regarding the impossibility of 'solving by radicals' polynomials of degree $\geqslant 5$.

Naturally, the *choice* of a method of solving a problem of type (1) or (2) depends on the *properties of the matrix* in question; for this reason we shall now review various properties to be taken into consideration in this regard.

One of the first characteristics to consider is the *number*, and, then, the *distribution*, of the *null elements* of a matrix, which, by misuse of language, are called the '*zeros*' of the matrix. It is by this feature that *sparse matrices* (with 'many' zeros) are distinguished from *full matrices* (with 'few' zeros).

If a matrix of order n is full, it is necessary to set aside n^2 *memory locations* to hold the elements of the matrix, and that imposes an essential constraint on the possible values of the integer n. On the other hand, there exist some 'very' sparse matrices whose zeros are arranged in a rather special fashion – *broadly speaking* parallel to the diagonal (various examples originating from concrete problems will be given in the following chapter); these are the *band matrices*, that is, matrices $A = (a_{ij})$ for which $a_{ij} = 0$ for a 'large' number of values of the integer $|i - j|$. Such matrices require a much smaller number of registers, the more so because the non-zero elements a_{ij} are quite often calculated in a simple fashion from the values of some known function (see the examples).

These considerations indicate why it is possible to solve linear systems of a high order when their matrices are banded. Generally, such systems lend themselves well to treatment by iterative methods, while direct methods are better kept for linear systems with full matrices (if for no other reason than the fact that the general, full matrix does not satisfy the hypotheses which guarantee the convergence of iterative methods...); and this despite the fact that currently an opposite tendency in usage may be discerned.

Within the class of band matrices, the following, in particular, can be distinguished: *diagonal matrices, tridiagonal matrices, block tridiagonal matrices, triangular matrices* (already defined), *block diagonal matrices, block triangular matrices, Hessenberg matrices.* These are represented in figure 2.1-1, with the notational conventions made evident. We recall that, for *block* diagonal, tridiagonal or triangular matrices, the *diagonal submatrices* A_{II} *are square*, by definition of the partitioning of a square matrix (section 1.1).

It would be well to keep in mind also the following two principles.

(1) *The problems involving the solution of linear systems which are the easiest to treat numerically are those whose matrices are symmetric and positive definite.* This point will be illustrated in the comparison of the methods of Gauss and Cholesky (sections 4.2 and 4.4), as regards direct methods, and in the hypotheses which guarantee convergence, as regards iterative methods (section 5.3).

(2) *The methods of calculating eigenvalues and eigenvectors which are the easiest to treat numerically are those whose matrices are symmetric* (the property of being positive definite does not here come into consideration, because the eigenvalues of the matrices A and $A + xI$ satisfy the relationship $\lambda(A) = \lambda(A + xI) - x$ for every scalar x). In this respect, comparison should be made between the methods of Jacobi and Givens–Householder, on the one hand (sections 6.1 and 6.2), and the QR algorithm for general matrices (section 6.3), on the other.

Figure 2.1-1. Examples of band matrices

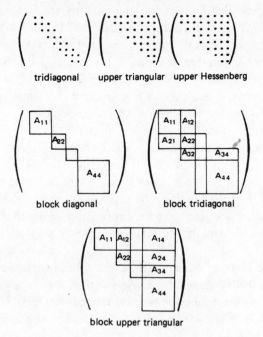

block upper triangular

We give now some indications of what it is *currently possible* to achieve in the numerical solution of the problems (1) and (2); however, we are concerned only with the *orders of magnitude*, and as regards calculations carried out by methods *currently* employed.

Quite an arsenal of methods is available for the solution of linear systems; and it is possible to solve without too great difficulty linear systems with full matrices of order $\leqslant 500$. If the matrices are very sparse, the use of methods which are suitable in this situation makes possible the treatment of matrices of the order of 100 000.

Efficient methods are likewise available for the calculation of eigenvalues and eigenvectors of full symmetric matrices of order $\leqslant 500$. It is quite a different matter for non-symmetric matrices: while it is possible, with greater or lesser success, to calculate their eigenvalues, the calculation of their eigenvectors is a problem that is not yet fully solved.

We end with the observation that the modern methods found in the Numerical Analysis of Matrices have been actually proved to be efficient, and *there* lies their *principal justification*; this is especially true of *all* the methods described in the rest of this work. It is vain to hope to find a new method without acquiring first a good knowledge, theoretical but, above all, *practical*, of the methods which are already in existence. Methods which are theoretically quite acceptable (for example, the use of Cramer's rule), or which are strikingly 'elegant', may turn out to be quite disappointing when it comes to solving a concrete problem!

Exercises

2.1-1. (1) Show that the solution of a linear system involving complex matrix and vector amounts to the solution of a linear system with real matrix and vector.

(2) Is it possible, in like fashion, to reduce the calculation of the eigenvalues of a complex matrix to the calculation of those of a real matrix?

2.1-2 (1) Show that it is possible to carry out the multiplication of two complex numbers in a way which involves the multiplication of only three real numbers.

(2) Show that it is possible to achieve the multiplication of two real matrices of order 2 in a way which involves only seven multiplications.

(3) Show that it is possible to achieve the multiplication of two matrices of order 3, whose elements belong to a *noncommutative field*, in a way which involves only 23 multiplications.

2.1-3. Let A and B be two matrices (A invertible) such that

$$AB = I + E.$$

Assuming that the norm $\| E \|$ is sufficiently small, obtain an *a priori* bound for $\| A^{-1} - B \|$ in terms of $\| B \|$ and $\| E \|$.

2.1-4. The identity (with A an invertible matrix of order n, $\beta \in \mathbb{R}$ and $u, v \in \mathbb{R}^n$)

$$(A + \beta uv^\mathsf{T})^{-1} = A^{-1} - \frac{\beta}{1 + \beta v^\mathsf{T} A^{-1} u} A^{-1} uv^\mathsf{T} A^{-1}$$

is sometimes used to calculate the inverse of a matrix which turns up in just the form of a 'perturbation' $A + \beta uv^\mathsf{T}$ of a matrix A whose inverse is known. In fact, this is a particular case of the more general identity (B, U and V being rectangular matrices)

$$(A + UBV)^{-1} = A^{-1} - A^{-1}U(I + BVA^{-1}U)^{-1}BVA^{-1}.$$

Prove this result, stating its domain of validity, as well as the types of the matrices B, U and V.

2.1-5. Given a matrix $C \in \mathscr{A}_n(\mathbb{C})$, set

$$C = A + iB, \quad \text{with } A, B \in \mathscr{A}_n(\mathbb{R}),$$

denoting by i the well-known complex number. Under hypotheses which should be stated, establish the relation

$$C^{-1} = (A + BA^{-1}B)^{-1} - iA^{-1}B(A + BA^{-1}B)^{-1},$$

whose interest lies in the fact that it makes possible the calculation of the inverse of a complex matrix through the use of methods for the calculation of the inverse of real matrices.

2.1-6. The object of this exercise is the study of an algorithm for the calculation of the smallest eigenvalue of a Hermitian matrix, based on the calculation of the smallest eigenvalues of a sequence of Hermitian matrices of lower order. Under certain conditions (notably if the matrices $D - \lambda I$ introduced below are 'easily invertible'), this algorithm can be of interest in dealing with matrices of 'too' high an order.

In what follows, A denotes a Hermitian matrix of order N, partitioned into blocks of the form $(n < N)$

$$A = \left(\begin{array}{|c|c|} \hline B & C \\ \hline C^* & D \\ \hline \end{array} \right), \quad \text{with } B \in \mathscr{A}_n(\mathbb{C}).$$

(1) Let p be a non-zero vector of $\mathbb{C}^N$, partitioned into blocks of the form

$$P = \left(\begin{array}{|c|} \hline q \\ \hline r \\ \hline \end{array} \right), \quad \text{with } q \in \mathbb{C}^n.$$

Assuming that $\lambda \notin \mathrm{sp}(D)$, establish the equivalence

$$Ap = \lambda p \Leftrightarrow \begin{cases} (B - C(D - \lambda I)^{-1} C^*)q = \lambda q, \\ r = -(D - \lambda I)^{-1} C^* q. \end{cases}$$

(2) Denoting by λ, *from here on*, the smallest eigenvalue of the matrix A, show

that λ is also the smallest real number which satisfies

$$\begin{cases} B(\lambda)q = \lambda q, \quad \text{where } B(\lambda) \overset{def}{=} B - C(D - \lambda I)^{-1}C^*, \\ q \in \mathbb{C}^n - \{0\}. \end{cases}$$

(3) Prove that $\lambda \leqslant \min\{sp(B)\} \overset{def}{=} \alpha$.

(4) Suppose, *from here on*, that the Hermitian matrix $D - \alpha I$ is positive definite. Define the sequence of real numbers $(\lambda_k)_{k \geqslant 0}$ by

$$\lambda_0 = \alpha \quad \text{and} \quad \lambda_{k+1} = \min\{sp(B(\lambda_k))\}, \, k \geqslant 0.$$

Prove the inequalities

$$\lambda_{2p+1} \leqslant \lambda \leqslant \lambda_{2p} \leqslant \alpha, \, p \geqslant 0.$$

To this end, one may use the identity (having first proved it)

$$B(\mu) = B(\lambda) - (\mu - \lambda)C(D - \mu I)^{-1}(D - \lambda I)^{-1}C^*,$$

showing that the Hermitian matrix $M_{2p}, \, p \geqslant 1$, is positive definite, where

$$M_k \overset{def}{=} C(D - \lambda_k I)^{-1}(D - \lambda_{k-1}I)^{-1}C^*, \, k \geqslant 1.$$

(5) Establish the bounds

$$(\lambda_{2p} - \lambda_{2p-1}) \leqslant (\lambda_{2p-2} - \lambda_{2p-1})\|M_{2p-1}\|_2, \, p \geqslant 1,$$
$$\|M_{2p-1}\|_2 \leqslant \|C((D - \alpha I)^{-1})^2 C^*\|_2, \, p \geqslant 1,$$

and deduce from them a sufficient condition for the convergence of the sequence (λ_k) to λ.

2.2 Condition of a linear system

Consider the linear system (this example is due to R.S. Wilson)

$$\begin{pmatrix} 10 & 7 & 8 & 7 \\ 7 & 5 & 6 & 5 \\ 8 & 6 & 10 & 9 \\ 7 & 5 & 9 & 10 \end{pmatrix} \begin{pmatrix} u_1 \\ u_2 \\ u_3 \\ u_4 \end{pmatrix} = \begin{pmatrix} 32 \\ 23 \\ 33 \\ 31 \end{pmatrix}, \quad \text{with solution} \quad \begin{pmatrix} 1 \\ 1 \\ 1 \\ 1 \end{pmatrix},$$

and consider the perturbed system, where the right-hand side has been 'very slightly' modified, the matrix staying unchanged,

$$\begin{pmatrix} 10 & 7 & 8 & 7 \\ 7 & 5 & 6 & 5 \\ 8 & 6 & 10 & 9 \\ 7 & 5 & 9 & 10 \end{pmatrix} \begin{pmatrix} u_1 + \delta u_1 \\ u_2 + \delta u_2 \\ u_3 + \delta u_3 \\ u_4 + \delta u_4 \end{pmatrix} = \begin{pmatrix} 32.1 \\ 22.9 \\ 33.1 \\ 30.9 \end{pmatrix}, \quad \text{with solution} \quad \begin{pmatrix} 9.2 \\ -12.6 \\ 4.5 \\ -1.1 \end{pmatrix}.$$

In other words, a relative error of the order of $1/200$ in the data (here, the components of the right-hand side) produces a relative error of the order of 10/1 in the result (the solution of the linear system), which represents an amplification of the relative errors of the order of 2000!

Now consider the following perturbed system, where it is the elements of the matrix which are 'very slightly' modified:

$$\begin{pmatrix} 10 & 7 & 8.1 & 7.2 \\ 7.08 & 5.04 & 6 & 5 \\ 8 & 5.98 & 9.89 & 9 \\ 6.99 & 4.99 & 9 & 9.98 \end{pmatrix} \begin{pmatrix} u_1 + \Delta u_1 \\ u_2 + \Delta u_2 \\ u_3 + \Delta u_3 \\ u_4 + \Delta u_4 \end{pmatrix} = \begin{pmatrix} 32 \\ 23 \\ 33 \\ 31 \end{pmatrix},$$

$$\text{with solution} \quad \begin{pmatrix} -81 \\ 137 \\ -34 \\ 22 \end{pmatrix}.$$

Here again, very small changes in the data (the elements of the matrix) completely alter the result (the solution of the linear system). And yet, the matrix A of the system has a 'good look' about it; it is symmetric, its determinant has the value 1 and the matrix

$$A^{-1} = \begin{pmatrix} 25 & -41 & 10 & -6 \\ -41 & 68 & -17 & 10 \\ 10 & -17 & 5 & -3 \\ -6 & 10 & -3 & 2 \end{pmatrix},$$

also looks harmless!

Remark

This example is all the more disquieting in that the errors in the data are of an order that is considered quite satisfactory in the experimental sciences. In the circumstances, it is clear that a user of the results of the Numerical Analysis of Matrices is not likely to have too good an impression of its qualities, if he is told that when the data of a linear system (which is only of order 4 and whose elements, moreover, are integers...) are known only to an accuracy of about 1/200, the solution, *even if calculated with exact arithmetic*, could be affected by a relative error which is 2000 times as great.... □

Let us now analyse this kind of phenomenon. In the first case, one is given an *invertible* matrix A, and a comparison needs to be made between the two *exact* solutions u and $u + \delta u$ of the systems

$$Au = b,$$
$$A(u + \delta u) = b + \delta b.$$

Let the same symbol $\|\cdot\|$ denote any vector norm and its subordinate matrix norm. From the equalities $\delta u = A^{-1}\delta b$ and $b = Au$, we conclude

$$\|\delta u\| \leqslant \|A^{-1}\|\|\delta b\|, \quad \|b\| \leqslant \|A\|\|u\|,$$

so that the relative error in the result, measured by the ratio $\|\delta u\|/\|u\|$, is bounded in terms of the relative error $\|\delta b\|/\|b\|$ in the datum b, as follows

$$\frac{\|\delta u\|}{\|u\|} \leqslant \{\|A\|\|A^{-1}\|\}\frac{\|\delta b\|}{\|b\|}.$$

In the second case, it is the matrix which changes, and the interest here lies in comparing the *exact* solutions u and $u + \Delta u$ of the systems

$$Au = b,$$

$$(A + \Delta A)(u + \Delta u) = b.$$

From the equality $\Delta u = -A^{-1}\Delta A(u + \Delta u)$, there results

$$\|\Delta u\| \leqslant \|A^{-1}\|\|\Delta A\|\|u + \Delta u\|,$$

which can also be written as

$$\frac{\|\Delta u\|}{\|u + \Delta u\|} \leqslant \{\|A\|\|A^{-1}\|\}\frac{\|\Delta A\|}{\|A\|}.$$

Consequently, the relative error in the result, measured this time by the ratio $\|\Delta u\|/\|u + \Delta u\|$, is again bounded in terms of the relative error $\|\Delta A\|/\|A\|$ in the datum A.

Remarks
(1) The reasoning given above is valid even if the matrix $A + \Delta A$ is singular, as long as there exists a vector $u + \Delta u$ which is a solution of the second system.

(2) If $\|\Delta A\|$ is 'sufficiently small', then there are grounds for expecting the ratio $\|\Delta u\|/\|u + \Delta u\|$ to be a good approximation to the more 'natural' relative error $\|\Delta u\|/\|u\|$. This point is made in more detail in Theorem 2.2-2.

In each of the two cases, it may be verified that the relative error in the result is bounded by the relative error in the data, *multiplied by the number* $\|A\|\|A^{-1}\|$. In other words, for a given relative error in the data, the relative error in the corresponding result *may* be larger by a factor proportional to this number; in actual fact, it will be shown (Theorems 2.2-1 and 2.2-2) that this number is optimal, the inequalities above being the 'best' possible. These considerations lead to the following definition.

Let $\|\cdot\|$ be a *subordinate* matrix norm and let A be an *invertible* matrix. The number

$$\text{cond}(A) = \|A\| \|A^{-1}\|$$

is called *the condition number of the matrix* A, relative to the given matrix norm.

Remark

In the Anglo-Saxon literature, this number is often denoted by $\kappa(A)$. ☐

The inequalities established earlier show that *the number* cond(A) *measures the sensitivity of the solution u of the linear system* Au = b *to variations in the data* A *and* b; a feature which is referred to as *the condition of the linear system in question*. The preceding, therefore, gives sense to a statement such as '*a linear system is well-conditioned or ill-conditioned*', according as the condition number of its matrix is 'small' or 'large'.

Remark

The condition of a linear system is merely a particular case of a notion that has wide-ranging applicability within Numerical Analysis. One may likewise examine the condition of the roots of a polynomial with regard to changes in its coefficients, or the condition of the eigenvalues or eigenvectors of a matrix with regard to changes in its elements. On this question, care needs to be taken to avoid erring over the following point. A linear system Au = b may be ill-conditioned, while the problem of finding the eigenvalues of the *same* matrix A may be well-conditioned (this point will be taken up again in the next section). ☐

The two results which follow resume, and complete, the inequalities previously obtained. They are stated for an arbitrary vector norm and for the corresponding subordinate matrix norm.

Theorem 2.2-1

Let A *be an invertible matrix and let* u *and* u + δu *be the solutions of the linear systems*

$$Au = b,$$
$$A(u + \delta u) = b + \delta b.$$

Suppose that $b \neq 0$. *Then the inequality*

$$\frac{\|\delta u\|}{\|u\|} \leqslant \text{cond}(A) \frac{\|\delta b\|}{\|b\|}$$

holds and is the best possible: that is, for a given matrix A, *it is possible to find vectors* $b \neq 0$ *and* $\delta b \neq 0$ *for which equality holds.*

Proof
The inequality has already been established. To show that equality may be attained, it is enough to observe that there exist vectors $u \neq 0$ and $\delta b \neq 0$ for which (and this is a property of subordinate matrix norms)

$$\| A^{-1} \delta b \| = \| A^{-1} \| \, \| \delta b \|, \quad \| Au \| = \| A \| \, \| u \|. \qquad \square$$

Theorem 2.2-2
Let A *be an invertible matrix and let* u *and* $u + \Delta u$ *be the solutions of the linear systems*

$$Au = b,$$
$$(A + \Delta A)(u + \Delta u) = b.$$

Suppose that $b \neq 0$. *Then the inequality*

$$\frac{\| \Delta u \|}{\| u + \Delta u \|} \leqslant \operatorname{cond}(A) \frac{\| \Delta A \|}{\| A \|}$$

holds and is the best possible: that is, for a given matrix A, *it is possible to find a vector* $b \neq 0$ *and a matrix* $\Delta A \neq 0$ *for which equality holds.*
Furthermore, we have the inequality

$$\frac{\| \Delta u \|}{\| u \|} \leqslant \operatorname{cond}(A) \frac{\| \Delta A \|}{\| A \|} \{ 1 + O(\| \Delta A \|) \}.$$

Proof
The first inequality has already been established. We recall that, for its proof, it is not necessary to suppose that the matrix $A + \Delta A$ is invertible; it is enough to suppose that the second linear system has (at least) one solution, which is written as $u + \Delta u$.

To show that equality may be attained, let w be any vector such that

$$w \neq 0, \| A^{-1} w \| = \| A^{-1} \| \, \| w \|,$$

and let β be any non-zero scalar. Now the vectors

$$\Delta u = - \beta A^{-1} w, u + \Delta u = w, b = (A + \beta I) w,$$

and the matrix

$$\Delta A = \beta I$$

satisfy exactly

$$Au = b, (A + \Delta A)(u + \Delta u) = b,$$
$$\| \Delta u \| = | \beta | \, \| A^{-1} w \| = \| \Delta A \| \, \| A^{-1} \| \, \| u + \Delta u \|.$$

If no eigenvalue of the matrix A is equal to the number $-\beta$, the matrix $A + \Delta A = A + \beta I$ is invertible and the vector b is non-zero.

To prove the last inequality, we shall restrict ourselves to the case where $\|\Delta A\| < \|A^{-1}\|^{-1}$, which is quite permissible, since we seek to establish a property for $\|\Delta A\|$ close to 0 (the matrix A is fixed; it is the matrices ΔA which vary). The matrix $I + A^{-1}\Delta A$ is then invertible, since

$$\|A^{-1}\Delta A\| \leqslant \|A^{-1}\| \|\Delta A\| < 1,$$

and (Theorem 1.4-5)

$$\|(I + A^{-1}\Delta A)^{-1}\| \leqslant \frac{1}{1 - \|A^{-1}\Delta A\|} \leqslant \frac{1}{1 - \|A^{-1}\| \|\Delta A\|}.$$

From the equalities $Au = b$ and $(A + \Delta A)(u + \Delta u) = b$, we obtain at once the equalities

$$\Delta u = -A^{-1}\Delta A(u + \Delta u), \quad u + \Delta u = (I + A^{-1}\Delta A)^{-1}u,$$

and, consequently,

$$\|\Delta u\| \leqslant \frac{\|A^{-1}\| \|\Delta A\|}{1 - \|A^{-1}\| \|\Delta A\|} \|u\|,$$

or

$$\frac{\|\Delta u\|}{\|u\|} \leqslant \operatorname{cond}(A) \frac{\|\Delta A\|}{\|A\|} \left\{ \frac{1}{1 - \|A^{-1}\| \|\Delta A\|} \right\},$$

which is just the kind of inequality being sought.   ☐

The condition numbers used in practice correspond to the three subordinate matrix norms introduced in section 1.4. They are denoted by

$$\operatorname{cond}_p(A) = \|A\|_p \|A^{-1}\|_p, \quad \text{for } p = 1, 2, \infty.$$

The result which follows gathers together properties of the condition number which are more or less evident, but which it is useful to record (obviously, the condition number is defined only for invertible matrices; this will not be repeated in the statement).

Theorem 2.2-3

(1) *For every matrix* A,

$$\operatorname{cond}(A) \geqslant 1,$$

$$\operatorname{cond}(A) = \operatorname{cond}(A^{-1}),$$

$$\operatorname{cond}(\alpha A) = \operatorname{cond}(A) \quad \text{for every scalar } \alpha \neq 0.$$

(2) *For every matrix* A,

$$\operatorname{cond}_2(A) = \frac{\mu_n(A)}{\mu_1(A)},$$

where $\mu_1(A) > 0$ and $\mu_n(A) > 0$ denote respectively the smallest and the largest of the singular values of the matrix A.

(3) *If the matrix A is normal*

$$\mathrm{cond}_2(A) = \frac{\max_i |\lambda_i(A)|}{\min_i |\lambda_i(A)|},$$

where the numbers $\lambda_i(A)$ are the eigenvalues of the matrix A.

(4) *The condition number* $\mathrm{cond}_2(A)$ *of a unitary or orthogonal matrix A is equal to* 1.

(5) *The condition number* $\mathrm{cond}_2(A)$ *is invariant under unitary transformations:*

$$UU^* = I \Rightarrow \mathrm{cond}_2(A) = \mathrm{cond}_2(AU) = \mathrm{cond}_2(UA) = \mathrm{cond}_2(U^*AU).$$

Proof

The properties in (1) follow from the properties of matrix norms; in particular,

$$AA^{-1} = I \Rightarrow 1 = \|I\| \leqslant \|A\| \|A^{-1}\|,$$

the matrix norm being subordinate by hypothesis.

From the definition of singular values and Theorem 1.4-2,

$$\|A\|_2^2 = \varrho(A^*A) = \max_i \lambda_i(A^*A) = \mu_n^2(A),$$

$$\|A^{-1}\|_2^2 = \varrho((A^{-1})^*A^{-1}) = \varrho(A^{-1}(A^{-1})^*)$$

$$= \max_i \lambda_i((A^*A)^{-1}) = \frac{1}{\min_i \lambda_i(A^*A)} = \frac{1}{(\mu_1(A))^2},$$

which proves (2).

Property (3) follows from the equality $\|A\|_2 = \varrho(A)$ satisfied by normal matrices (Theorem 1.4-2).

If A is a unitary or orthogonal matrix, the equality $\|A\|_2 = \sqrt{\varrho(A^*A)} = \sqrt{\varrho(I)} = 1$ implies (4). Finally, (5) derives from the invariance of the matrix norm $\|\cdot\|_2$ under unitary transformations (Theorem 1.4-2). $\quad\square$

We examine certain practical consequences of the preceding theorem. The inequality cond $(A) \geqslant 1$, together with the inequalities of Theorems 2.2-1 and 2.2-2, demonstrates that *a linear system* Au = b *is the better conditioned, the closer the number* cond (A) *is to unity* (it is understood, of course, that this is so relative to a particular subordinate matrix norm).

Again, on this question, property (4) shows that *linear systems* Au = b *with orthogonal (or unitary) matrices are very well conditioned*, since cond$_2$ (A) = 1. In the same way, property (5) shows that *orthogonal, or unitary, transformations preserve the condition number* cond$_2$ (A). *These considerations provide the justification for the use of orthogonal matrices* as 'auxiliary' matrices in various methods, for example, Householder matrices (cf. section 4.5).

While property (3) shows that a normal matrix has a 'large' condition number (for the norm $\|\cdot\|_2$) if and only if the *ratio* of the two extreme moduli of its eigenvalues is 'large', a general matrix may have a 'large' condition number even when all its eigenvalues are equal! See in this respect Exercise 2.2-2.

Remark

It is possible to give a *simple geometrical interpretation* of the ill-conditioning of a linear system A$u = b$ whose matrix is *normal*. To show this, let λ_1 and λ_n be two eigenvalues satisfying

$$|\lambda_1| = \min_i |\lambda_i(A)| \quad \text{and} \quad |\lambda_n| = \max_i |\lambda_i(A)|$$

and let p_1 and p_n be two corresponding eigenvectors. Then the choice

$$b = p_n, \quad \delta b = \lambda_1 p_1$$

leads to the equality

$$\frac{\|\delta u\|_2}{\|u\|_2} = \text{cond}_2(A) \frac{\|\delta b\|_2}{\|b\|_2}.$$

It is seen from this that if the condition number cond$_2$ (A) = $|\lambda_n|/|\lambda_1|$ is 'large', then a 'small' perturbation δb of a vector b, which is parallel to the vector p_n, in the direction of the vector p_1, produces a 'large' change in the solution. It is this that figure 2.2-1 is meant to suggest, where, in order to fix ideas, the vectors p_1 and p_n have been taken to have the same norm. □

Figure 2.2-1

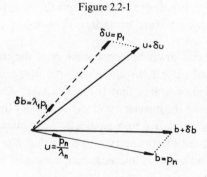

The property cond $(\alpha A) = $ cond (A) shows that it is futile to expect to be able to improve the condition of a linear system by multiplying all the equations by a scalar. On the contrary, it is indeed possible to diminish cond$_2$ (A) by multiplying *every* row and *every* column by a suitable number; this is the task of the *equilibration of a matrix* which may be stated as follows. Given a matrix A, find two invertible diagonal matrices D_1 and D_2 such that ($\mathcal{D}$ denoting the set of all diagonal matrices)

$$\text{cond}\,(D_1 A D_2) = \inf_{\Delta_1, \Delta_2 \in \mathcal{D}} \text{cond}\,(\Delta_1 A \Delta_2),$$

relative to a given matrix norm. With this step carried out, the solution of the system $Au = b$ is effected by solving the system

$$(D_1 A D_2)v = D_1 b,$$

and then calculating

$$u = D_2 v.$$

The problem of equilibration is very important in practice, but we shall not deal with it here; we refer the interested reader to the Bibliography and comments.

We take up again the numerical example given at the beginning of this section and work out the values of the norms $\|\cdot\|_2$. The eigenvalues of the matrix

$$A = \begin{pmatrix} 10 & 7 & 8 & 7 \\ 7 & 5 & 6 & 5 \\ 8 & 6 & 10 & 9 \\ 7 & 5 & 9 & 10 \end{pmatrix}$$

have approximately the numerical values

$$\lambda_1 \approx 0.01015 < \lambda_2 \approx 0.8431 < \lambda_3 \approx 3.858 < \lambda_4 \approx 30.2887,$$

so that

$$\text{cond}_2\,(A) = \frac{\lambda_4}{\lambda_1} \approx 2984.$$

Furthermore,

$$u = \begin{pmatrix} 1 \\ 1 \\ 1 \\ 1 \end{pmatrix}, \quad \delta u = \begin{pmatrix} 8.2 \\ -13.6 \\ 3.5 \\ -2.1 \end{pmatrix}, \quad b = \begin{pmatrix} 32 \\ 23 \\ 33 \\ 31 \end{pmatrix}, \quad \delta b = \begin{pmatrix} 0.1 \\ -0.1 \\ 0.1 \\ -0.1 \end{pmatrix},$$

so that

$$\frac{\|\delta u\|_2}{\|u\|_2} \approx \frac{16.397}{2} \approx 8.1985,$$

$$\mathrm{cond}_2(A)\frac{\|\delta b\|_2}{\|b\|_2} \approx 2984 \times \frac{0.2}{60.025} \approx 9.9424,$$

from which it may be seen that we are not far from equality!

It will have been observed that the preceding calculation of the number $\mathrm{cond}_2(A)$ depends on a prior knowledge of the extreme eigenvalues of the matrix A. In the absence of this information, the matrix norm $\|\cdot\|_E$ (*which is not subordinate*) can turn out to be useful, since it is easy to calculate, on the one hand, and

$$\|A\|_2 \leqslant \|A\|_E \leqslant \sqrt{n}\|A\|_2,$$

on the other (cf. Theorem 1.4-4), *provided, of course, that the matrix* A^{-1} *is known*....

In the example that concerns us, one would obtain in this way the bounds

$$\|A\|_2 = \lambda_4 \approx 30.2887 \leqslant 30.5450 \approx \|A\|_E,$$

$$\|A^{-1}\|_2 = \frac{1}{\lambda_1} \approx 98.5222 \leqslant 98.5292 \approx \|A^{-1}\|_E.$$

$$\mathrm{cond}_2(A) \approx 2984 < 3009 \approx \|A\|_E\|A^{-1}\|_E.$$

Remark

Because of the eigenvalues of the matrix A, it was foreseeable that the bounds would turn out so satisfactory. Why?

Exercises

2.2-1. Carry out the numerical analysis of the linear systems $Au = b$ and $A(u + \delta u) = b + \delta b$ given towards the beginning of this section, using the particular norm $\|\cdot\|_\infty$. Compare the results with the inequalities of Theorem 2.2-1.

2.2-2. Let A be the matrix of order 100,

$$A = \begin{pmatrix} 1 & 2 & & & \\ & 1 & 2 & & \\ & & \ddots & \ddots & \\ & & & 1 & 2 \\ & & & & 1 \end{pmatrix}.$$

Show that $\mathrm{cond}_2(A) \geqslant 2^{99}$.

2.2-3. Here is another example of a linear system which is ill-conditioned (the matrix is due to H. Rutishauser).

$$\begin{pmatrix} 10 & 1 & 4 & 0 \\ 1 & 10 & 5 & -1 \\ 4 & 5 & 10 & 7 \\ 0 & -1 & 7 & 9 \end{pmatrix} \begin{pmatrix} u_1 \\ u_2 \\ u_3 \\ u_4 \end{pmatrix} = \begin{pmatrix} 15 \\ 15 \\ 26 \\ 15 \end{pmatrix},$$

with solution $\begin{pmatrix} 1 \\ 1 \\ 1 \\ 1 \end{pmatrix}$.

In order to see this, consider the perturbed linear system, where the right-hand side has been 'slightly' modified,

$$\begin{pmatrix} 10 & 1 & 4 & 0 \\ 1 & 10 & 5 & -1 \\ 4 & 5 & 10 & 7 \\ 0 & -1 & 7 & 9 \end{pmatrix} \begin{pmatrix} u_1 + \delta u_1 \\ u_2 + \delta u_2 \\ u_3 + \delta u_3 \\ u_4 + \delta u_4 \end{pmatrix} = \begin{pmatrix} 16 \\ 16 \\ 25 \\ 16 \end{pmatrix},$$

with solution $\begin{pmatrix} 832 \\ 1324 \\ -2407 \\ 2021 \end{pmatrix}$!

The matrix A of the linear system is invertible, det $A = 1$ and

$$A^{-1} = \begin{pmatrix} 105 & 167 & -304 & 255 \\ 167 & 266 & -484 & 406 \\ -304 & -484 & 881 & -739 \\ 255 & 406 & -739 & 620 \end{pmatrix}.$$

Carry out the numerical analysis of this example, knowing that the smallest and the largest of the eigenvalues of the matrix have, respectively, the values

$$\lambda_1 \approx 0.0005343, \quad \lambda_4 \approx 19.1225.$$

2.2-4. Let $\mathscr{E}$ be the set of matrices of order 2, whose elements a_{ij} are integers satisfying $0 \leqslant a_{ij} \leqslant 100$. Show that

$$\mathrm{cond}_2(A) = \inf_{E \in \mathscr{E}} \mathrm{cond}_2(E), \quad \text{where} \quad A = \begin{pmatrix} 100 & 99 \\ 99 & 98 \end{pmatrix}.$$

In order to do this, the first step is to establish that, for a general matrix $A = (a_{ij})$ of order 2,

$$\mathrm{cond}(A) = \sigma + \{\sigma^2 - 1\}^{1/2}, \quad \text{with} \quad \sigma = \frac{\displaystyle\sum_{i,j=1}^{2} |a_{ij}|^2}{2|\det(A)|}.$$

As an application, solve the systems

$$Au = b \quad \text{and} \quad A(u + \delta u) = b + \delta b,$$

with

$$b = \begin{pmatrix} 199 \\ 197 \end{pmatrix} \quad \text{and} \quad \delta b = \begin{pmatrix} -0.0097 \\ 0.0106 \end{pmatrix}.$$

2.2-5. Prove the following relations (due to P.J. Erdelsky).

(1) $$\text{cond}(A) = \lim_{B \to A^{-1}} \sup \frac{\|AB - I\|}{\|BA - I\|},$$

(1') $$\text{cond}(A) = \lim_{B \to A^{-1}} \sup \frac{\|BA - I\|}{\|AB - I\|}.$$

In other words, if A is a matrix with large condition number, one and the same matrix B may turn out to be a 'good' approximation to a right inverse ($\|AB - I\|$ 'small') and at the same time a 'bad' approximation to a left inverse ($\|BA - I\|$ 'large'), or vice versa.

(2) For every $\varepsilon > 0$,

$$\frac{2\varepsilon}{\|A\|} \text{cond}(A) = \sup \{ \|L - R\| : \|LA - I\| \leqslant \varepsilon, \|AR - I\| \leqslant \varepsilon \}.$$

In other words, 'good' approximations to right or left inverses may differ considerably!

(3) $$\frac{\text{cond}(A)}{\|A\|} = \lim_{B \to A^{-1}} \sup \frac{\|B - A^{-1}\|}{\|BA - I\|},$$

(3') $$\frac{\text{cond}(A)}{\|A\|} = \lim_{B \to A^{-1}} \sup \frac{\|B - A^{-1}\|}{\|AB - I\|}.$$

In other words, 'good' approximations to right or left inverses may differ considerably from the 'true' matrix A^{-1}.

2.2-6. Calculate the solutions of the linear systems

$$\begin{pmatrix} 240 & -319.5 \\ -179.5 & 240 \end{pmatrix} \begin{pmatrix} u_1 \\ u_2 \end{pmatrix} = \begin{pmatrix} 3 \\ 4 \end{pmatrix},$$

$$\begin{pmatrix} 240 & -319 \\ -179 & 240 \end{pmatrix} \begin{pmatrix} u_1 + \Delta u_1 \\ u_2 + \Delta u_2 \end{pmatrix} = \begin{pmatrix} 3 \\ 4 \end{pmatrix},$$

and carry out the corresponding numerical analysis.

2.2-7. (*The condition of the problem of the inversion of a matrix*) Let A be a given invertible matrix.

(1) If $A + \delta A$ is an invertible matrix, prove the inequality

$$\frac{\|(A + \delta A)^{-1} - A^{-1}\|}{\|(A + \delta A)^{-1}\|} \leqslant \text{cond}(A) \frac{\|\delta A\|}{\|A\|}.$$

Is this the 'best' inequality possible?

(2) Prove the inequality

$$\frac{\|(A + \delta A)^{-1} - A^{-1}\|}{\|A^{-1}\|} \leqslant \text{cond}(A)\frac{\|\delta A\|}{\|A\|}(1 + O(\|\delta A\|)).$$

2.2-8. The symmetric matrix of order n

$$H = (h_{ij}), \quad \text{where } h_{ij} = \frac{1}{i+j-1}, \quad 1 \leqslant i, j \leqslant n,$$

is called the *Hilbert matrix*. An example of a problem in which it appears is given in Exercise 3.7-3.

(1) Show that this matrix is positive definite (hence invertible).

(2) The condition number $\text{cond}_2(H)$ of the Hilbert matrix is a function which *grows very rapidly* with n. One soon becomes convinced upon calculating it for the values $n = 2, 3, 4, 5$ (for $n = 5$, we have already reached $\text{cond}_2(H) = 4.77 \times 10^5$; for $n = 10$, $\text{cond}_2(H) = 1.60 \times 10^{13}\ldots$).

2.2-9. Over the set $\mathscr{I}$ of invertible matrices, we define the function

$$d: (A, B) \in \mathscr{I} \times \mathscr{I} \rightarrow d(A, B) = 1 - \frac{1}{\text{cond}(A^{-1}B)},$$

the condition number being relative to a general *subordinate* matrix norm. Show that this function is 'nearly' a metric, in the sense that it satisfies

$$0 \leqslant d(A, B) < 1,$$
$$d(A, B) = d(B, A),$$
$$d(A, C) \leqslant d(A, B) + d(B, C),$$

for every $A, B, C \in \mathscr{I}$.

2.2-10. (1) Let $\mathscr{S}_n$ be the set of symmetric matrices and $\mathscr{S}_n^+$ the subset of non-negative definite symmetric matrices. A matrix norm $\|\cdot\|$ is said to be *monotone* if

$$A \in \mathscr{S}_n^+ \ \& \ B - A \in \mathscr{S}_n^+ \Rightarrow \|A\| \leqslant \|B\|$$

Show that the norms $\|\cdot\|_2$ and $\|\cdot\|_E$ are monotone.

(2) More generally, show that if a matrix norm $\|\cdot\|$ is invariant under unitary transformations, that is, if $\|A\| = \|AU\| = \|UA\|$ for every unitary matrix U, then it is monotone.

(3) Let $\|\cdot\|$ be a monotone norm and $\text{cond}(\cdot)$ the condition number function associated with it. Prove that

$$A, B \in \mathscr{S}_n^* \Rightarrow \text{cond}(A + B) \leqslant \max\{\text{cond}(A), \text{cond}(B)\}$$

where $\mathscr{S}_n^*$ denotes the subset of positive definite symmetric matrices. In other words, the condition number of a positive definite symmetric matrix does not increase if another positive definite symmetric matrix is added to it.

2.2-11. (1) Let p be a non-negative integer. Define the square matrix A(p), of arbitrary order n, by the equations

$$(A(p))_{ij} = \binom{p+j-1}{i-1}, \; 1 \leqslant i,j \leqslant n.$$

Show that the inverse matrix $B(p) = (A(p))^{-1}$ has as its elements

$$(B(p))_{ij} = (-1)^{i+1} \sum_{l=0}^{n-j} \binom{p+l-1}{l} \binom{l+j-1}{i-1},$$

with the convention that

$$\binom{-1}{0} = 1 \quad \text{and} \quad \binom{r}{s} = 0 \quad \text{for } r < s \text{ or } s < 0.$$

It should be noted in passing that we are dealing with matrices whose elements are integers and whose inverses also have integer entries.

(2) Give an order of magnitude for the condition number

$$\text{cond}_1 (A(p)) = \| A(p) \|_1 \; \| B(p) \|_1$$

for $n = 10$, $p = 0,5,10,15,20$. It is useful to compare the approximate value $\text{cond}_1 (A(20)) \approx 10^{18}$ with the approximate value $\text{cond}_1 (H) \approx 10^{14}$ for the Hilbert matrix H of order 10 (cf. Exercise 2.2-8), whose condition number is in any case very large!

2.3 Condition of the eigenvalue problem

Consider the matrix of order n

$$A(\varepsilon) = \begin{pmatrix} 0 & & & & \varepsilon \\ 1 & 0 & & & \\ & 1 & 0 & & \\ & & \ddots & \ddots & \\ & & & 1 & 0 \end{pmatrix}.$$

All the eigenvalues of $A(\varepsilon)$ are equal to zero for $\varepsilon = 0$. But for $n = 40$ and $\varepsilon = 10^{-40}$, the eigenvalues all have the same modulus of 10^{-1} (they are the nth roots of the number ε). In other words, the change in the eigenvalues (measured by the distance function over the complex plane) is equal to the change in the parameter ε *multiplied by* 10^{39} !

There is another disquieting aspect to this phenomenon. The number $\varepsilon = 10^{-40}$ being automatically replaced by zero in the computer (because it is too small for the computer to hold; cf. section 2.1), the calculation of the matrix $A(10^{-40})$ for $n = 40$ is then *necessarily* affected by an error of the order 10^{-1} !

Again, one comes face to face with a problem where a 'small' change in the data (here, the elements of the matrix) produces a very 'large' change in the result (here, the eigenvalues of the matrix). This gives rise to the need of providing a measure for the *condition of the eigenvalue problem*; which is the object in Theorem 2.3-1, where consideration is limited to diagonalisable matrices and to particular matrix norms, and in Theorem 2.3-2, which 'improves upon' Theorem 2.3-1, should the matrices be symmetric or Hermitian.

Theorem 2.3-1

Let A *be a diagonalisable matrix,* P *a matrix such that*

$$P^{-1}AP = \operatorname{diag}(\lambda_i) \stackrel{\text{def}}{=} D,$$

and $\|\cdot\|$ *a matrix norm satisfying*

$$\| \operatorname{diag}(d_i) \| = \max_i |d_i|$$

for every diagonal matrix. Then, for every matrix δA,

$$\operatorname{sp}(A + \delta A) \subset \bigcup_{i=1}^{N} D_i,$$

where

$$D_i = \{ z \in \mathbb{C} : |z - \lambda_i| \leqslant \operatorname{cond}(P) \|\delta A\| \}.$$

Proof

Let λ be an eigenvalue of the matrix $A + \delta A$. If $\lambda = \lambda_j$ for some index j, then the result is evident.

If $\lambda \neq \lambda_j$, $1 \leqslant j \leqslant n$, the matrix $D - \lambda I$ is invertible and it is possible to write

$$P^{-1}(A + \delta A - \lambda I)P = D - \lambda I + P^{-1}(\delta A)P$$
$$= (D - \lambda I)\{ I + (D - \lambda I)^{-1}P^{-1}(\delta A)P \}.$$

The matrix $A + \delta A - \lambda I$ being singular, so too is

$$I + (D - \lambda I)^{-1}P^{-1}(\delta A)P,$$

so that (by Theorem 1.4-5)

$$1 \leqslant \|(D - \lambda I)^{-1}P^{-1}(\delta A)P\| \leqslant \|(D - \lambda I)^{-1}\| \, \|P^{-1}\| \, \|\delta A\| \, \|P\|.$$

Since

$$\|(D - \lambda I)^{-1}\| = \frac{1}{\min_i |\lambda_i - \lambda|},$$

because of the hypothesis regarding the matrix norms under consideration, there exists at least one index i such that

$$|\lambda - \lambda_i| \leqslant \|P^{-1}\| \, \|\delta A\| \, \|P\| = \operatorname{cond}(P)\|\delta A\|. \qquad \square$$

Remarks

(1) The matrix norms $\|\cdot\|_1$, $\|\cdot\|_2$, $\|\cdot\|_\infty$ all satisfy the hypothesis in the statement of the theorem.

(2) This last theorem is known as the *Bauer–Fike theorem*.

(3) An enhancement of this theorem is to be found in Exercise 2.3-1.

$\square$

Evidently, the condition number of a matrix (viz. P) enters into the question of the condition of an eigenvalue problem. However, *this matrix is not* the matrix of the problem itself (as is the case for linear systems), that is to say, it is not the matrix whose eigenvalues are being sought. Rather, *it is the condition of the diagonalising matrix, that is critically involved.* More precisely, it follows from Theorem 2.3-1 that, *if* A *is a diagonalisable matrix,* then

$$\mathrm{sp}(A + \delta A) \subset \bigcup_{i=1}^{n} \{z \in \mathbb{C} : |z - \lambda_i| \leqslant \Gamma(A) \|\delta A\|\},$$

where

$$\Gamma(A) = \inf\{\mathrm{cond}(P) : P^{-1}AP = \mathrm{diag}(\lambda_i)\}.$$

That is why the number $\Gamma(A)$ defined above is called *the condition number for the matrix* A, *relative to the eigenvalue problem.* It follows from Theorem 2.3-1 that *normal matrices are very well conditioned for the eigenvalue problem.* These being diagonalisable by a unitary matrix (Theorem 1.2-1), we have

$$AA^* = A^*A \Rightarrow \Gamma_2(A) \overset{\mathrm{def}}{=} \inf\{\mathrm{cond}_2(P) : P^{-1}AP = \mathrm{diag}(\lambda_i)\} = 1,$$

so that

$$AA^* = A^*A \Rightarrow \mathrm{sp}(A + \delta A) \subset \bigcup_{i=1}^{n} \{z \in \mathbb{C} : |z - \lambda_i| \leqslant \|\delta A\|_2\}.$$

Consider now the case of a *symmetric* matrix A with eigenvalues $\alpha_1 \leqslant \alpha_2 \leqslant \cdots \leqslant \alpha_n$, and let δA be a 'perturbation' matrix, also *symmetric*. Let $\beta_1 \leqslant \beta_2 \leqslant \cdots \leqslant \beta_n$ be the eigenvalues of the 'perturbed' matrix $B = A + \delta A$. Given the kth eigenvalue β_k of the matrix B, it follows from the preceding (a symmetric matrix is a particular kind of normal matrix) that there exists at least one eigenvalue α_i of A satisfying

$$|\beta_k - \alpha_i| \leqslant \|\delta A\|_2$$

(this result makes no assumption about the symmetry of the matrix δA; in contrast, this hypothesis will play an essential role in what follows).

The theorem given below leads to a 'stronger' result, specifically, that it is *precisely the kth* eigenvalue α_k which satisfies the inequality given above.

Theorem 2.3-2

Let A *and* B = A + δA *be two symmetric or Hermitian matrices with eigenvalues*

$$\alpha_1 \leqslant \alpha_2 \leqslant \cdots \leqslant \alpha_n, \quad \text{and} \quad \beta_1 \leqslant \beta_2 \leqslant \cdots \leqslant \beta_n,$$

respectively. Then

$$|\alpha_k - \beta_k| \leqslant \|\delta A\|_2, \quad 1 \leqslant k \leqslant n.$$

Proof

Let V_k denote the subspace spanned by the eigenvectors of the matrix A corresponding to the eigenvalues α_i, $1 \leqslant i \leqslant k$, and $\mathcal{V}_k$ the set of all k-dimensional subspaces of $\mathbb{K}^n$. By repeated application of Theorem 1.3-1,

$$\beta_k = \min_{U \in \mathcal{V}_k} \max_{v \in U} R_B(v) \leqslant \max_{v \in V_k} R_B(v) = \max_{v \in V_k} (R_A(v) + R_{\delta A}(v))$$

$$\leqslant \max_{v \in V_k} R_A(v) + \max_{v \in V_k} R_{\delta A}(v) = \alpha_k + \max_{v \in V_k} R_{\delta A}(v) \leqslant \alpha_k + \max_{v \in \mathbb{K}^n} R_{\delta A}(v)$$

$$\leqslant \alpha_k + \|\delta A\|_2,$$

since

$$\max_{v \in \mathbb{K}^n} R_{\delta A}(v) = \max_i \lambda_i(\delta A) \leqslant \varrho(\delta A) \leqslant \|\delta A\|_2.$$

Reversing the roles of A and B, one would similarly obtain

$$\alpha_k \leqslant \beta_k + \max_{v \in \mathbb{K}^n} R_{(-\delta A)}(v) \leqslant \beta_k + \|\delta A\|_2,$$

proving the theorem. □

Some observations on the condition of the eigenvector problem are to be found in Exercises 2.3-4 and 2.3-5.

Exercises

2.3-1. (Complement to Theorem 2.3-1) If there exists an integer m satisfying $1 \leqslant m \leqslant n$ such that the union $\bigcup_{i=1}^{m} D_i$ of the m disks D_i is disjoint from the union $\bigcup_{i=m+1}^{n} D_i$ of the remaining $n - m$ disks (it is always possible to assume that it is the first m and last $n - m$ disks which have this property), then the union $\bigcup_{i=1}^{m} D_i$ contains exactly m eigenvalues of the matrix A + δA.

2.3-2. The matrix

$$A = (a_{ij}) = \begin{pmatrix} -149 & -50 & -154 \\ 537 & 180 & 546 \\ -27 & -9 & -25 \end{pmatrix},$$

has the eigenvalues

$$\lambda_1 = 1, \lambda_2 = 2, \lambda_3 = 3;$$

hence, it is diagonalisable. The perturbed matrix $B = (b_{ij})$, where only the element a_{22} is changed, being replaced by the value $b_{22} = 180.01$, has the eigenvalues

$$\mu_1 \approx 0.207\,265\,65, \quad \mu_2 \approx 2.300\,834\,90, \quad \mu_3 \approx 3.501\,899\,44.$$

Carry out the numerical analysis of this example (due to G.J. Davis and C. Moler), comparing the quantities $|\mu_i - \lambda_i|$ with the bounds predicted by Theorem 2.3-1.

2.3-3. (1) Let A be a diagonalisable matrix, $\lambda_i \in \mathbb{R}$ a *simple* eigenvalue of A, and B any matrix. Show that, for $\varepsilon > 0$ sufficiently small, there exists a unique eigenvalue $\lambda_i(\varepsilon)$ of the 'perturbed' matrix $A + \varepsilon B$ satisfying

$$\lambda_i(\varepsilon) = \lambda_i + \varepsilon \frac{q_i^* B p_i}{q_i^* p_i} + O(\varepsilon^2),$$

where p_i and q_i are vectors which satisfy

$$\|p_i\|_2 = 1, \ A p_i = \lambda_i p_i,$$
$$\|q_i\|_2 = 1, \ A^* q_i = \lambda_i q_i.$$

Do we thus obtain once again a result of the text stated for Hermitian matrices?

(2) The inequality

$$\left| \frac{q_i^* B p_i}{q_i^* p_i} \right| \leqslant \frac{\|B\|_2}{\sigma_i}, \quad \text{where } \sigma_i = |q_i^* p_i|,$$

shows that the sensitivity of the ith eigenvalue to perturbations is 'measured' by the number σ_i^{-1} (in fact, it is possible to limit consideration to perturbations which are of the form $A + \varepsilon B$ with $\|B\|_2 = 1$). Prove that the number σ_i^{-1} is invariant under unitary transformations (of the matrix A) and that it satisfies the inequality

$$1 \leqslant \sigma_i^{-1} \leqslant \Gamma_2(A) \stackrel{\text{def}}{=} \inf\{\text{cond}_2(P) : P^{-1}AP = \text{diag}(\lambda_k)\}.$$

(3) Prove that the result of question (1) is still true even if the matrix A is not diagonalisable, while the assumption continues to be made that the eigenvalue λ_i is real and simple. To this end, good use can be made of the *Jordan canonical form* of the matrix A (cf. Exercise 1.5-7).

2.3-4. (*The condition of an eigenvector problem*) Calculate the eigenvalues and an orthonormal basis of eigenvectors of the matrix

$$A(\varepsilon) = \begin{pmatrix} 1 + \varepsilon \cos\dfrac{2}{\varepsilon} & -\varepsilon \sin\dfrac{2}{\varepsilon} \\ -\varepsilon \sin\dfrac{2}{\varepsilon} & 1 - \varepsilon \cos\dfrac{2}{\varepsilon} \end{pmatrix}.$$

Show that the eigenvectors cannot have a non-zero limit as $\varepsilon \to 0$.

2.3-5. (*The condition of an eigenvector problem*) Let A and B = A + δA be two symmetric matrices with eigenvalues

$$\alpha_1 \leqslant \alpha_2 \leqslant \cdots \leqslant \alpha_n \quad \text{and} \quad \beta_1 \leqslant \beta_2 \leqslant \cdots \leqslant \beta_n.$$

Let α_k be a simple eigenvalue of the matrix A and let a_k be an eigenvector (with $\|a_k\|_2 = 1$) corresponding to the eigenvalue α_k. Show that if

$$\|\delta A\|_2 < \Delta \overset{\text{def}}{=} \min_{i \neq k} |\alpha_i - \alpha_k|,$$

there exists an eigenvector b_k (with $\|b_k\|_2 = 1$) corresponding to the eigenvalue β_k, which satisfies

$$\|a_k - b_k\|_2 \leqslant \gamma(1 + \gamma^2)^{1/2}, \quad \text{with} \quad \gamma = \frac{\|\delta A\|_2}{\Delta - \|\delta A\|_2}.$$

Sources of problems in the numerical analysis of matrices

Introduction

While this chapter makes no attempt to be exhaustive, nevertheless an appreciable effort has been made to give an idea of the great variety of problems which eventually lead to the solution of one of the two fundamental problems of the Numerical Analysis of Matrices, and in particular those problems whose source lies in *approximation of linear partial differential equations arising in Physics*.

In this regard, it did not seem fruitless to make some brief observations on the physical aspects of the problems, on their mathematical analysis and, lastly, on their numerical analysis, especially whenever this last depends upon matrix analysis. Thus, in the case of *one-dimensional boundary-value problems* (introduced in section 3.1), we establish the *convergence* of the methods of approximation which are considered (cf. Theorems 3.1-2 and 3.4-3).

In sections 3.1, 3.2 and 3.3 some simple examples of *boundary-value problems* are considered (Poisson's equation, the heat equation and the wave equation). This gives rise to a description of *finite-difference methods*. Their use, in turn, involves the *solution of linear systems* with very 'sparse' matrices, whose zeros are arranged in a quite remarkable fashion, the matrices involved being examples of *tridiagonal* or *block tridiagonal* type. Related matrices which are, moreover, always symmetric are encountered in the *variational approximation* of these same problems (sections 3.4 and 3.5).

Problems involving the *calculation of eigenvalues and eigenvectors* of matrices are encountered in the approximation of the 'stationary solutions', or 'waves' of various physical systems. It is shown, in this respect, in section 3.6, that these problems very often make their appearance in the form of 'generalised' eigenvalue problems.

Other sources of linear systems are the problems of the *interpolation of data* by means of 'piecewise polynomial' functions, notably *spline functions*, as well as *problems of least-squares approximation*: these problems are encountered in section 3.7.

Finally, we mention some other examples, which get further treatment later in this work:

the solution of *non-linear systems of equations*, where the use of Newton's method leads to the solution of a sequence of 'tangential' systems of linear equations (section 7.5);

problems of the *minimisation of a quadratic functional with linear constraints* (section 7.2);

the implementation of *Uzawa's method* (section 9.4);

the implementation of the *simplex method* (section 10.3).

Some of the results of this chapter make use of notions (differentiability, pre-Hilbert spaces) which are defined precisely only later on; yet others (partial differential equations) lie outside the scope of this work. Readers who may be experiencing difficulty as a result may usefully turn to the Index or the Bibliography and comments.

3.1 The finite-difference method for a one-dimensional boundary-value problem

In what follows, $\mathscr{C}^m(I)$ will denote the vector space of real functions which are m times continuously differentiable over an interval $I \subset \mathbb{R}$, m being a non-negative integer. Moreover, f', f'' and $f^{(n)}$ for $n \geqslant 3$ will denote the successive derivatives of a function of a single real variable.

Consider the following problem. Given the functions c and $f \in \mathscr{C}^0([0, 1])$ and two constants α and β, find a function $u \in \mathscr{C}^2([0, 1])$ satisfying

$$\begin{cases} -u''(x) + c(x)u(x) = f(x), & 0 < x < 1, \\ u(0) = \alpha, \quad u(1) = \beta. \end{cases}$$

Such a problem is called a *boundary-value problem*, because the unknown function has to satisfy the *boundary conditions* $u(0) = \alpha$, $u(1) = \beta$ at the ends of the open interval over which the differential equation has to be satisfied.

Figure 3.1-1

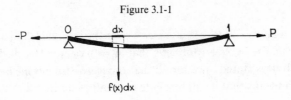

An example, of physical origin, where it is encountered is that of the bending of a beam of unit length, stretched along its axis by a force P, and subjected to a transverse load $f(x)\,dx$ per element dx, and simply supported at its ends 0 and 1 (figure 3.1-1). The bending moment $u(x)$ at the abscissa x then turns out to be the solution of a boundary-value problem of the type given above with $c(x) = P/(EI(x))$, where E is the Young's modulus of the material of which the beam is made and $I(x)$ is the principal moment of inertia of the cross-section of the beam at the abscissa x, with $\alpha = \beta = 0$.

If it may be supposed that the function c is non-negative over the interval $[0, 1]$ (and we shall indeed be making this hypothesis for the remainder of this section), then it can be shown (Exercise 3.1-1) that the problem has a unique solution, which will be denoted by φ.

Remark

The condition $c \geqslant 0$ is in no way a necessary condition for the existence and uniqueness of a solution. These are guaranteed (for example) by the weaker hypothesis

$$-\pi^2 < \inf\{c(x): 0 \leqslant x \leqslant 1\},$$

or even by

$$-(k+1)^2\pi^2 < \inf\{c(x): 0 \leqslant x \leqslant 1\} \leqslant \sup\{c(x): 0 \leqslant x \leqslant 1\} < -k^2\pi^2,$$

for some integer $k \geqslant 1$. However, it should be kept in mind that, in contrast to differential equations of the second order with *initial* conditions, that is, conditions of the type $u(0) = \alpha$, $u'(0) = \alpha'$, the existence and uniqueness of solutions cannot be guaranteed in general. It is easy to persuade oneself of this fact in the case of differential equations with constant coefficients (Exercise 3.1-2).   □

Except for the rarest of cases, there is no known 'formula' for calculating directly the value of the solution at a general point of the interval $[0, 1]$. As a result, there arises the problem of finding a way of *approximating the values of the solution as closely as may be desired*. One method of achieving this is *the finite-difference method* which will now be described.

Given an integer $N \geqslant 1$, set

$$h = \frac{1}{N+1},$$

and define a *uniform mesh* of *mesh-size* h over the interval $[0, 1]$ as the set of points $x_i = ih$, $0 \leqslant i \leqslant N+1$ (observe that $x_0 = 0, x_{N+1} = 1$), called the *nodes* of the mesh. It is stated once for all that the *mesh-size* h *is meant to tend to zero*, which means that it can be made as small as desired (in theory!). *The finite-difference method is a way of obtaining an approximation of the solution* φ *at the nodes of the mesh.* In other words, we look for a *vector*

$$u_h = \begin{pmatrix} u_1 \\ u_2 \\ \vdots \\ u_N \end{pmatrix} \in \mathbb{R}^N,$$

such that u_i *is 'close' to* $\varphi(x_i)$ *for* $i = 1, 2, \ldots, N$ (the values $\varphi(x_0) = \alpha$ and $\varphi(x_{N+1}) = \beta$ being known), *the accuracy of the approximation improving as h diminishes* (hopefully!).

Remark
In contrast, the method of variational approximation, which will be examined in section 3.4, provides an approximation at *each* of the points of the interval [0, 1]. $\square$

Suppose that the solution φ is four times continuously differentiable in the interval [0, 1]. By Taylor's formula, for $i = 1, 2, \ldots, N$, we can write

$$\varphi(x_{i+1}) = \varphi(x_i) + h\varphi'(x_i) + \frac{h^2}{2}\varphi''(x_i) + \frac{h^3}{6}\varphi^{(3)}(x_i) + \frac{h^4}{24}\varphi^{(4)}(x_i + \theta_i^+ h),$$

$$\varphi(x_{i-1}) = \varphi(x_i) - h\varphi'(x_i) + \frac{h^2}{2}\varphi''(x_i) - \frac{h^3}{6}\varphi^{(3)}(x_i) + \frac{h^4}{24}\varphi^{(4)}(x_i + \theta_i^- h),$$

$$\text{with } -1 < \theta_i^- < 0 < \theta_i^+ < 1,$$

so that

$$-\varphi(x_{i+1}) + 2\varphi(x_i) - \varphi(x_{i-1}) = -h^2\varphi''(x_i)$$
$$- \frac{h^4}{24}\{\varphi^{(4)}(x_i + \theta_i^+ h) + \varphi^{(4)}(x_i + \theta_i^- h)\}.$$

By the intermediate value theorem,

$$\varphi^{(4)}(x_i + \theta_i^+ h) + \varphi^{(4)}(x_i + \theta_i^- h) = 2\varphi^{(4)}(x_i + \theta_i h),$$
$$\text{with } |\theta_i| \leqslant \max\{\theta_i^+, -\theta_i^-\} < 1,$$

so that, finally,

$$-\varphi''(x_i) = \frac{-\varphi(x_{i-1}) + 2\varphi(x_i) - \varphi(x_{i+1})}{h^2} + \frac{h^2}{12}\varphi^{(4)}(x_i + \theta_i h),$$

with

$$|\theta_i| < 1, 1 \leqslant i \leqslant N.$$

To make the notation less cumbersome, we write

$$\varphi_i = \varphi(x_i), \quad c_i = c(x_i), \quad f_i = f(x_i), \quad 1 \leqslant i \leqslant N,$$

and express the fact that the function φ is the solution of the problem at the points x_i, $1 \leqslant i \leqslant N$, as follows, replacing the values $-\varphi''(x_i)$ by the expressions given above and taking account of the boundary conditions:

$$-\frac{\alpha}{h^2} + \frac{2\varphi_1 - \varphi_2}{h^2} + c_1\varphi_1 = f_1 - \frac{h^2}{12}\varphi^{(4)}(x_1 + \theta_1 h)$$

$$\frac{-\varphi_{i-1} + 2\varphi_i - \varphi_{i+1}}{h^2} + c_i\varphi_i = f_i - \frac{h^2}{12}\varphi^{(4)}(x_i + \theta_i h), \quad 2 \leqslant i \leqslant N-1,$$

$$\frac{-\varphi_{N-1} + 2\varphi_N}{h^2} - \frac{\beta}{h^2} + c_N\varphi_N = f_N - \frac{h^2}{12}\varphi^{(4)}(x_N + \theta_N h).$$

The system of equations given above may be written in the matrix form

$$A_h\varphi_h = b_h + \varepsilon_h(\varphi),$$

by setting

$$A_h = \frac{1}{h^2}\begin{pmatrix} 2+c_1h^2 & -1 & & & \\ -1 & 2+c_2h^2 & -1 & & \\ & \ddots & \ddots & \ddots & \\ & & -1 & 2+c_{N-1}h^2 & -1 \\ & & & -1 & 2+c_Nh^2 \end{pmatrix},$$

$$\varphi_h = \begin{pmatrix} \varphi_1 \\ \varphi_2 \\ \vdots \\ \varphi_{N-1} \\ \varphi_N \end{pmatrix}, \quad b_h = \begin{pmatrix} f_1 + \alpha/h^2 \\ f_2 \\ \vdots \\ f_{N-1} \\ f_N + \beta/h^2 \end{pmatrix}, \quad \varepsilon_h(\varphi) = -\frac{h^2}{12}\begin{pmatrix} \varphi^{(4)}(x_1 + \theta_1 h) \\ \varphi^{(4)}(x_2 + \theta_2 h) \\ \vdots \\ \varphi^{(4)}(x_{N-1} + \theta_{N-1} h) \\ \varphi^{(4)}(x_N + \theta_N h) \end{pmatrix}.$$

The method relies upon the following heuristic observation. Since the vector $\varepsilon_h(\varphi)$ is 'smaller', the finer the mesh-size h (because of the factor h^2), one is naturally led to *neglect* it and to define the following *discrete problem*, associated with the boundary-value problem under consideration and corresponding to a mesh-size h: *find a vector $u_h \in \mathbb{R}^N$ which is a solution of the matrix equation*

$$A_h u_h = b_h.$$

Remark

Naturally, the statement that the vector $\varepsilon_h(\varphi)$ becomes 'small' as the mesh-size h tends to zero needs to be made more precise, since it depends essentially on the vector norm chosen. More specifically, supposing the function φ to be four times continuously differentiable over the interval $[0, 1]$, one obtains, depending on the norm chosen,

$$\|\varepsilon_h(\varphi)\|_1 = O(h), \quad \|\varepsilon_h(\varphi)\|_2 = O(h^{3/2}), \quad \|\varepsilon_h(\varphi)\|_\infty = O(h^2),$$

the differences in the asymptotic estimates arising, of course, from the variation with h in the *number* of components in the vector $\varepsilon_h(\varphi)$.   □

It is essential to note that the elements of the matrix A_h, as well as the components of the vector b_h, are *directly calculable from the data, which is not the case for* $\varepsilon_h(\varphi)$ *in the right-hand side of the system* $A_h\varphi_h = b_h + \varepsilon_h(\varphi)$.

At this stage, we are in a position to explain the terminology 'finite-difference'. In effect, the method comes down to the replacing of the derivative $\varphi'(x_i)$ by one of the 'finite-difference' quotients

$$\frac{\varphi(x_{i+1}) - \varphi(x_i)}{x_{i+1} - x_i} = \frac{\varphi_{i+1} - \varphi_i}{h} \quad \text{or} \quad \frac{\varphi_i - \varphi_{i-1}}{h}.$$

A single iteration of this process then replaces the second derivative $\varphi''(x_i)$ by the 'finite-difference' quotient

$$\frac{1}{h}\left\{\frac{\varphi_{i+1} - \varphi_i}{h} - \frac{\varphi_i - \varphi_{i-1}}{h}\right\} = \frac{\varphi_{i-1} - 2\varphi_i + \varphi_{i+1}}{h^2}.$$

Under suitable hypotheses about the data, our aim is to prove that

(i) *the discrete problem* $A_h u_h = b_h$ *has one and only one solution,* that is, the matrix A_h is invertible,

(ii) *the method converges as the mesh-size h tends to zero,* that is, the vector $u_h - \varphi_h$ tends to zero for a suitable norm.

As regards the question (i), the identity

$$v^{\mathrm{T}} A_h v = \sum_{i=1}^{N} c_i v_i^2 + \frac{1}{h^2}\left\{v_1^2 + v_N^2 + \sum_{i=2}^{N} (v_i - v_{i-1})^2\right\},$$

which is true for every vector $v \in \mathbb{R}^N$, shows that *the symmetric matrix A_h is positive definite if the function c is non-negative.*

As regards the question (ii), it may be pointed out, even at this stage, that if the solution φ of the boundary-value problem satisfies $\varphi^{(4)}(x) = 0$, $0 < x < 1$ (φ being a polynomial of degree $\leqslant 3$), then $u_i = \varphi(x_i)$, $1 \leqslant i \leqslant N$, since the corresponding vector $\varepsilon_h(\varphi)$ is null. This actually exhibits a case where convergence is certainly assured, seeing that we have exactly $u_h = \varphi_h$ for all h!

To prove the result generally, we introduce the following notions. A matrix $A = (a_{ij})$ is said to be *non-negative* if each of its elements a_{ij} is non-negative; in particular, a vector $v = (v_i)$ is said to be *non-negative* if each of its components v_i is non-negative. We will write in shorthand form

$$A = (a_{ij}) \geqslant 0 \quad \text{and} \quad v = (v_i) \geqslant 0.$$

Finally, a real square matrix A is called *monotone* if it is invertible and if the matrix A^{-1} is non-negative.

Theorem 3.1-1
Suppose that c is non-negative. Then the matrix A_h is monotone.

Proof
(i) We begin by establishing a useful characterisation. *A real matrix A of order N is monotone if and only if the inclusion*

$$\{v \in \mathbb{R}^N : Av \geqslant 0\} \subset \{v \in \mathbb{R}^N : v \geqslant 0\}$$

is satisfied. For, if the matrix A is monotone and if the vector Av is non-negative, it follows that

$$v = A^{-1}(Av) \geqslant 0.$$

Conversely, suppose that the inclusion is satisfied. Then

$$Av = 0 \Rightarrow A(\pm v) \geqslant 0 \Rightarrow \pm v \geqslant 0 \Rightarrow v = 0,$$

which shows that the matrix A is invertible. Furthermore, the jth column vector of the matrix A^{-1} is nothing other than the vector $b_j = A^{-1} e_j$, where e_j is the jth basis vector. Since the vector $Ab_j = e_j$ is non-negative, so, too, is b_j, which proves that the matrix A^{-1} is non-negative.

(ii) Because of the preceding, it is enough to show that

$$A_h v \geqslant 0 \Rightarrow v \geqslant 0.$$

Given any such vector v, let $p \in \{1, \ldots, N\}$ be an integer satisfying

$$v_p \leqslant v_i \quad \text{for} \quad i = 1, \ldots, N.$$

We then have

$$0 \leqslant (2 + c_1 h^2) v_1 - v_2 \leqslant (1 + c_1 h^2) v_1 \quad \text{if} \quad p = 1,$$
$$0 \leqslant -v_{p-1} + (2 + c_p h^2) v_p - v_{p-1} \leqslant c_p h^2 v_p, \quad \text{if} \quad 2 \leqslant p \leqslant N-1,$$
$$0 \leqslant -v_{N-1} + (2 + c_N h^2) v_N \leqslant (1 + c_N h^2) v_N \quad \text{if} \quad p = N.$$

which proves that

$$\min_{1 \leqslant i \leqslant N} v_i \geqslant 0 \quad \text{if} \quad c_i > 0, 2 \leqslant i \leqslant N-1.$$

There remains the need to look at the case where one (at least) of the numbers c_i, $2 \leqslant i \leqslant N-1$, is zero. Now, from the preceding, the matrix $(A_h + \alpha I_h)$ (I_h being the identity matrix of order N, $N+1 = 1/h$) is monotone for every $\alpha > 0$. As the matrix A_h is invertible (it was shown earlier that such a matrix is positive definite) and as the elements of the matrices $(A_h + \alpha I_h)^{-1}$ are continuous functions of $\alpha \geqslant 0$, it follows that $A_h^{-1} \geqslant 0$. □

Remark
The monotone character of the matrix A_h may be interpreted as the discrete analogue of the property known as 'the maximum principle', which holds good for the boundary-value problem under consideration: we refer the reader to Exercise 3.1-3. □

We are now in a position to establish the *convergence* of the method for the vector norm $\|\cdot\|_\infty$ and the corresponding subordinate matrix norm. The proof relies, in an essential way, upon the fact that the norms $\|\cdot\|_\infty$ of the *inverses* of the matrices A_h are bounded *independently* of h (cf. part (i) of the proof below). We have there an example of a notion that is fundamental in Numerical Analysis, called the *stability* of a family of methods of approximation.

Theorem 3.1-2
Suppose that the function c is non-negative. If the solution φ of the boundary-value problem satisfies

$$\varphi \in \mathscr{C}^4([0,1]),$$

then we have the bound

$$\max_{1 \le i \le N} |u_i - \varphi(x_i)| = \|u_h - \varphi_h\|_\infty \le \left\{ \frac{1}{96} \sup_{0 \le x \le 1} |\varphi^{(4)}(x)| \right\} h^2.$$

Proof
(i) We first show that

$$\|A_h^{-1}\|_\infty \le \frac{1}{8}.$$

For this reason, we introduce the matrix

$$A_{oh} = \frac{1}{h^2} \begin{pmatrix} 2 & -1 & & & \\ -1 & 2 & -1 & & \\ & \ddots & \ddots & \ddots & \\ & & -1 & 2 & -1 \\ & & & -1 & 2 \end{pmatrix},$$

which is just one particular matrix of type A_h, the one corresponding to the case where the function c is identically zero. By Theorem 3.1-1,

$$A_h^{-1} \ge 0 \quad \text{and} \quad A_{oh}^{-1} \ge 0.$$

The hypothesis that c is non-negative implies that

$$A_h - A_{oh} = \text{diag}(c_i) \ge 0.$$

so that

$$A_{oh}^{-1} - A_h^{-1} = A_{oh}^{-1}(A_h - A_{oh})A_h^{-1} \geqslant 0,$$

and, in turn, using the expression for the matrix norm $\| \cdot \|_\infty$ ($\|(b_{ij})\|_\infty = \max_i \sum_j |b_{ij}|$; cf. Theorem 1.4-2),

$$\| A_h^{-1} \|_\infty \leqslant \| A_{oh}^{-1} \|_\infty,$$

the matrices A_h^{-1} and A_{oh}^{-1} being non-negative.

It is enough, then, to find a bound for the norm $\| A_{oh}^{-1} \|_\infty$. Now it may be observed, on the one hand, that

$$A_{oh}^{-1} \geqslant 0 \Rightarrow \| A_{oh}^{-1} \|_\infty = \| A_{oh}^{-1} e \|_\infty,$$

where e is the vector in $\mathbb{R}^N$ all of whose components are equal to 1 and, on the other, that the vector $A_{oh}^{-1} e$ is nothing other than the solution of the discrete problem associated with the particular boundary value problem

$$\begin{cases} -u''(x) = 1, & 0 < x < 1, \\ u(0) = u(1) = 0. \end{cases}$$

The solution $\psi(x) = x(1 - x)/2$ of this problem has the zero function as its exact third derivative, so that

$$(A_{oh}^{-1} e)_i = \psi(x_i), \quad 1 \leqslant i \leqslant N,$$

exactly, since $\varepsilon_h(\psi) = 0$. Hence,

$$\| A_{oh}^{-1} e \|_\infty = \max_{1 \leqslant i \leqslant N} |\psi(x_i)| \leqslant \sup_{0 \leqslant x \leqslant 1} |\psi(x)| = \tfrac{1}{8},$$

which proves the assertion.

(ii) From the equations $A_h \varphi_h = b_h + \varepsilon_h(\varphi)$ and $A_h u_h = b_h$, we have

$$\varphi_h - u_h = A_h^{-1} \varepsilon_h(\varphi), \quad \text{with } \varepsilon_h(\varphi) = -\frac{h^2}{12}(\varphi^{(4)}(x_i + \theta_i h))_{i=1}^N.$$

Consequently,

$$\| \varphi_h - u_h \|_\infty \leqslant \| A_h^{-1} \|_\infty \| \varepsilon_h(\varphi) \|_\infty,$$

so that it is enough to combine the bound $\| A_h^{-1} \|_\infty \leqslant \tfrac{1}{8}$ with the inequality

$$\| \varepsilon_h(\varphi) \|_\infty \leqslant \frac{h^2}{12} \max_{0 \leqslant x \leqslant 1} |\varphi^{(4)}(x)|$$

in order to obtain the required result. $\square$

The theorem given above, which goes by the name of *Gerschgorin's Theorem*, shows that the finite-difference method applied to the boundary-value problem under consideration leads to a *convergence of order* h^2, in the sense that the *error*, here measured by the $\| \cdot \|_\infty$ norm of the vector $u_h - \varphi(h)$, has the asymptotic expression

$$\| u_h - \varphi_h \|_\infty = O(h^2).$$

This very simple example of a method of approximation opens the way to the introduction of certain ideas which are quite general in the numerical analysis of ordinary and partial differential equations.

(i) In general, the error can only be defined correctly *after* the method in question has itself been analysed. For example, in the problem currently of interest, it was not at all evident *a priori* that the norm $\| \cdot \|_\infty$ was an appropriate norm for the analysis of the asymptotic behaviour of the error. It was the question of *stability*, that is to say, the uniform bound of the norms $\| \cdot \|_\infty$ of the matrices A_h^{-1}, which led us to that choice.

(ii) The proof of the previous theorem shows that convergence depends not only on stability, but also on the asymptotic behaviour of the *consistency error*, here measured by the norm

$$\| \varepsilon_h(\varphi) \|_\infty = \max_{1 \leqslant x \leqslant N} \left| \{ - \varphi''(x_i) + c(x_i)\varphi(x_i) \} - \left\{ \frac{- \varphi(x_{i-1}) + (2 + c(x_i)h^2)\varphi(x_i) - \varphi(x_{i+1})}{h^2} \right\} \right|.$$

The convergence to zero (with h) of this norm simply gives expression to a very natural idea, which is that the finite-difference expression under consideration is truly an approximation of the differential equation in question (at least for functions that are sufficiently smooth; cf. point (iii)).

(iii) The asymptotic behaviour of the error $\| u_h - \varphi_h \|_\infty$ depends on the *smoothness* of the solution (which it is generally possible to decide upon from the smoothness of the data) and the asymptotic behaviour of the consistency error. So, for example, if the solution had been 'only' thrice continuously differentiable, the result would have read $\| u_h - \varphi_h \|_\infty = O(h)$; on this question, it is useful to look at Exercise 3.1-4.

(iv) What makes the relation $\| u_h - \varphi_h \|_\infty = O(h^2)$ even more valuable is the further proof that it is the *best possible* (we have already seen that it is possible for the equation $u_h - \varphi_h = 0$ to hold), in the sense that examples may be found where the solution φ is four times continuously differentiable and an *equality* of the form

$$\| u_h - \varphi_h \|_\infty = Ch^2(1 + \delta(h)), \quad C > 0, \quad \lim_{h \to 0} \delta(h) = 0,$$

can be established; which is actually the case here, as is shown in Exercise 3.1-5.

Table 3.1-1

N	$h = 1/(N+1)$	$\|u_h - \varphi_h\|_\infty$	$(\|u_h - \varphi_h\|_\infty)(N+1)^2$
1	1/2	0.014 33	0.057 32
3	1/4	0.003 657	0.058 51
7	1/8	0.000 919 2	0.058 83
15	1/16	0.000 232 6	0.059 55
31	1/32	0.000 058 52	0.059 92

(v) The asymptotic behaviour of the error can be discovered 'empirically'. Consider, for example, the boundary-value problem

$$\begin{cases} -u''(x) + xu(x) = (1 + 2x - x^2)e^x, \\ u(0) = 1, \quad u(1) = 0, \end{cases}$$

with solution $\varphi(x) = (1 - x)e^x$. The use of the finite-difference method yields the numerical results shown in table 3.1-1.

It appears from the column on the right of the table that the product of the error $\|u_h - \varphi_h\|_\infty$ by h^{-2} 'very soon' behaves like a constant. The same fact can also be observed by plotting the error as a function of h on a graph in logarithmic scale. It is quite striking to see the points align themselves with great precision on a straight line of slope 2. The reader can hardly be recommended too strongly to devise his own examples and to analyse them 'experimentally' in this way.

Exercises

3.1-1. The aim in this exercise is to prove, *precisely with the help of the finite-difference method*, the *existence* of a solution, which is twice continuously differentiable over the interval $[0, 1]$, of the boundary-value problem

$$\begin{cases} -u''(x) + c(x)u(x) = f(x), \quad 0 < x < 1, \\ u(0) = 0, \quad u(1) = 0, \end{cases}$$

where c and f are functions which are continuous over $[0, 1]$, the function c being non-negative over this same interval (the *uniqueness* of the solution can be deduced from Exercise 3.1-3; it is also possible to prove it very simply, using integration by parts). It should be observed that the transformation to non-homogeneous boundary conditions $u(0) = \alpha$, $u(1) = \beta$, is achieved quite simply by substracting an affine function.

(1) Define the function

$$G: (x, \xi) \in [0, 1] \times [0, 1] \to G(x, \xi) = \begin{cases} x(1 - \xi) & \text{if } x \leqslant \xi, \\ \xi(1 - x) & \text{if } \xi < x. \end{cases}$$

Show that a function is a solution of the boundary-value problem if and only if it is a solution of the *integral equation*

$$u(x) = - \int_0^1 G(x, \xi)[c(\xi)u(\xi) - f(\xi)]\,d\xi, \quad 0 \leqslant x \leqslant 1.$$

A main result will be to establish that it is enough for this solution to be merely continuous over the interval [0, 1], in order that this property may entail that it is twice continuously differentiable over the same interval. This simplification plays an *essential* role in what follows.

(2) Show that the vector u_h is a solution of the equation $A_h u_h = b_h$ if and only if its components $u_i, 1 \leqslant i \leqslant N$, satisfy *the summation equation* (which is the discrete analogue of the integral equation of question (1))

$$u_i = - h \sum_{j=1}^N G(ih, jh)\{c(jh)u_j - f(jh)\}, \quad 1 \leqslant i \leqslant N.$$

(3) With the vector u_h, whose components are $u_i, 1 \leqslant i \leqslant N$, and which is a solution of the equation $A_h u_h = b_h$, there is associated the continuous *function* $\theta_h : [0, 1] \to \mathbb{R}$ defined by the following conditions:

$$\begin{cases} \theta_h(0) = \theta_h(1) = 0; \quad \theta_h(ih) = u_i, \quad 1 \leqslant i \leqslant N; \\ \theta_h \text{ is affine over } [ih, (i+1)h], \quad 0 \leqslant i \leqslant N. \end{cases}$$

Prove that the sequence of continuous functions (θ_h) (the parameter h taking the values $h = 1/(N+1)$, N an integer $\geqslant 1$) is *uniformly bounded* and *equicontinuous*, that is to say (respectively), there exists a constant M, *independent* of h, such that

$$\sup_{0 \leqslant x \leqslant 1} |\theta_h(x)| \leqslant M,$$

and, given $\varepsilon > 0$, there exists a number $\delta = \delta(\varepsilon)$, *independent of h*, such that

$$x, y \in [0, 1] \quad \text{and} \quad |x - y| \leqslant \delta \Rightarrow |\theta_h(x) - \theta_h(y)| \leqslant \varepsilon.$$

To establish this last property, use should be made of the discrete analogue of the inequality

$$|\varphi'(x)| \leqslant |\varphi(1) - \varphi(0)| + \tfrac{1}{2} \sup_{0 \leqslant \xi \leqslant 1} |\varphi''(\xi)|, \quad 0 \leqslant x \leqslant 1,$$

which holds for every function $\varphi \in \mathscr{C}^2([0, 1])$.

(4) Deduce from *Ascoli's theorem* (for a proof, see, for example, Dieudonné (1968), p. 143) that there exists a subsequence of the sequence (θ_h) which converges uniformly to a function φ, continuous over the interval [0, 1]. Show that this function φ is a solution of the boundary-value problem and that the 'full' sequence (θ_h) converges uniformly to this function φ.

(5) Deduce from the result of question (2) the expression for the elements of the matrix A_{oh}^{-1}. Deduce from it the exact value of the norm $\| A_{oh}^{-1} \|_\infty$ and thus discover again the bound $\| A_{oh}^{-1} \|_\infty \leqslant \tfrac{1}{8}$ established by other means in the course of the proof of Theorem 3.1-2.

3.1-2. Discuss the existence and uniqueness of a solution of the problem

$$\begin{cases} -u''(x) + \lambda u(x) = \mu, & 0 < x < 1, \\ u(0) = \alpha, & u(1) = \beta, \end{cases}$$

depending on the values of the real parameters λ, μ, α and β.

3.1-3. (1) Let c be a function which is non-negative over the interval $(0, 1)$ and φ a function which is continuous over $[0, 1]$ and twice continuously differentiable over $(0, 1)$, satisfying

$$\begin{cases} -\varphi''(x) + c(x)\varphi(x) \geqslant 0, & 0 < x < 1, \\ \varphi(0) \geqslant 0, & \varphi(1) \geqslant 0. \end{cases}$$

Deduce that the function φ is non-negative over $[0, 1]$ (the first step is to show that the function $\varphi_\varepsilon(x) = \varphi(x) + \frac{1}{2}\varepsilon x(1 - x)$ is non-negative over $[0, 1]$ for all $\varepsilon > 0$).

(2) This property is known as the *maximum principle*. Owing to the monotone character of the matrix A, an analogous property may be found for the finite-difference method, which goes by the name of the *discrete maximum principle*. In effect, one may verify the implication

$$\left. \begin{array}{l} c_i \geqslant 0, \quad 1 \leqslant i \leqslant N, \\[2mm] \dfrac{-u_{i-1} + (2 + c_i h^2)u_i - u_{i+1}}{h^2} \geqslant 0, \quad 1 \leqslant i \leqslant N, \\[2mm] u_0 \geqslant 0, \quad u_{N+1} \geqslant 0, \end{array} \right\} \Rightarrow u_i \geqslant 0, \quad 1 \leqslant i \leqslant N.$$

(3) Deduce from (1) the *uniqueness* of the solution of the boundary-value problem

$$-u''(x) + c(x)u(x) = f(x), \quad u(0) = \alpha, \quad u(1) = \beta,$$

when the function c is non-negative.

3.1-4. Examine the asymptotic behaviour of the error function $\| u_h - \varphi_h \|_\infty$, under the assumption that the solution is three times, or only twice, continuously differentiable over the interval $[0, 1]$.

3.1-5. Given an integer $r \geqslant 1$, the functions $c \geqslant 0$ and f are assumed to be $2r + 2$ times continuously differentiable over the interval $[0, 1]$. Setting $\varphi = \varphi_0$, let φ_p, $1 \leqslant p \leqslant r$, denote the (unique) solutions of the boundary-value problems

$$\begin{cases} -u''(x) + c(x)u(x) = \displaystyle\sum_{k=1}^{p} \frac{2}{(2k+2)!} \varphi_{p-k}^{(2k+2)}(x), & 0 < x < 1, \\ u(0) = u(1) = 0. \end{cases}$$

Show that there exists a constant $C(r, \varphi)$, independent of h, such that

$$\max_{1 \leqslant i \leqslant N} \left| u_i - \sum_{p=0}^{r} h^{2p} \varphi_p(ih) \right| \leqslant C(r, \varphi)h^{2r+2}.$$

Deduce, in particular, that the convergence of order h^2, established in Theorem 3.1-2, is actually the best possible.

3.1-6. This exercise has the objective of examining the special properties of non-negative matrices, especially as regards their spectral radii. The results of questions (1) to (5) constitute the *Perron–Frobenius theorem*.

In questions (1) to (7), $A = (a_{ij})$ denotes a *non-negative* (all its elements a_{ij} are non-negative) and *irreducible* matrix (the definitions of an irreducible and reducible matrix are given in Exercise 1.1-6; the notation $A \geqslant 0$ signifies that the matrix A is non-negative).

(1) Prove that the matrix A has (at least) one positive eigenvalue. For this, it is advantageous to make use of *Brouwer's fixed point theorem* (a proof can be found in Berger & Gostiaux (1972), p. 217): if K is a convex, compact subset of $\mathbb{R}^n$, then every continuous function from K into K possesses (at least) one fixed point.

(2) Show that with every positive eigenvalue of the matrix A may be associated an eigenvector all of whose components are positive.

(3) Let r be a positive eigenvalue of the matrix A and B a (possibly reducible) non-negative matrix such that $A - B \geqslant 0$. Show that $\rho(B) \leqslant r$ and that, moreover, $\rho(B) = r \Rightarrow B = A$.

(4) Deduce that $r = \rho(A)$ and that

$$B \geqslant 0, \quad A - B \geqslant 0 \Rightarrow \rho(B) \leqslant \rho(A).$$

(5) Prove that the eigenvalue $r = \rho(A)$ is simple (it can be shown that $p'_A(r) \neq 0$, by expressing the polynomial p_A in terms of the principal minors of the matrix $A - \lambda I$).

(6) Let $H = \{v = (v_i) \in \mathbb{R}^n : v_i > 0, 1 \leqslant i \leqslant n\}$. Given a vector $v \in H$, prove that

$$\text{either} \quad \min_i \frac{\sum_j a_{ij} v_j}{v_i} < \rho(A) < \max_i \frac{\sum_j a_{ij} v_j}{v_i},$$

$$\text{or} \quad \frac{\sum_j a_{ij} v_j}{v_i} = \rho(A) \quad \text{for } i = 1, 2, \ldots, n.$$

(7) Prove that

$$\sup_{v \in H} \left\{ \min_i \frac{\sum_j a_{ij} v_j}{v_i} \right\} = \rho(A) = \inf_{v \in H} \left\{ \max_i \frac{\sum_j a_{ij} v_j}{v_i} \right\}.$$

(8) Prove that there exists a subordinate matrix norm $\|\cdot\|$ such that $\|A\| = \rho(A)$.

(9) Let A be a non-negative, reducible matrix. Show that the matrix A has an eigenvalue $r \geqslant 0$ such that $r = \rho(A)$ and such that the corresponding eigenvector is non-negative. Finally, show that

$$B \geqslant 0, \quad A - B \geqslant 0 \Rightarrow \rho(B) \leqslant \rho(A).$$

(10) Let $A = (a_{ij})$ and $B = (b_{ij})$ be two matrices of order n such that $0 \leqslant |b_{ij}| \leqslant a_{ij}, 1 \leqslant i, j \leqslant n$. Prove that $\rho(B) \leqslant \rho(A)$.

(11) Let A be a non-negative matrix. Prove the equivalence

$$s > \rho(A) \Leftrightarrow \text{the matrix } sI - A \text{ is monotone.}$$

(12) Let A be a non-negative matrix. Prove the equivalence

$$\left.\begin{array}{l} s > \rho(A), \\ A \text{ is irreducible} \end{array}\right\} \Leftrightarrow \left\{\begin{array}{l} \text{the matrix } sI - A \text{ is monotone and all} \\ \text{the elements of the matrix } (sI - A)^{-1} \text{ are} \\ \text{positive.} \end{array}\right.$$

For this, the first step is to establish that if an irreducible matrix $A \in \mathscr{A}_n(\mathbb{R})$ is non-negative, then all the elements of the matrix $(I + A)^{n-1}$ are positive.

3.1-7. This exercise has the objective of examining the special properties of monotone matrices. Some of the results of the previous exercise are put to use. A matrix $A = (a_{ij}) \in \mathscr{A}_n(\mathbb{R})$ is said to be an *M-matrix* if it is monotone and if, furthermore, all its off-diagonal elements a_{ij}, $i \neq j$, are non-positive.

(1) Let $A = (a_{ij}) \in \mathscr{A}_n(\mathbb{R})$ be a matrix all of whose off-diagonal elements a_{ij}, $i \neq j$, are non-positive. Show that the following properties are equivalent:

 (i) A is an M-matrix;

 (ii) $\lambda \in \mathrm{sp}(A) \Rightarrow \mathrm{Re}\,\lambda > 0$;

 (iii) The diagonal elements a_{ii} are positive and the matrix $I - DA \geqslant 0$, where $D = \mathrm{diag}(a_{ii}^{-1})$, is such that

$$\rho(I - DA) < 1.$$

(2) Let $A = (a_{ij}) \in \mathscr{A}_n(\mathbb{R})$ be an M-matrix and let $B = (b_{ij})$ be a matrix such that

$$b_{ii} = a_{ii}, \quad b_{ij} = a_{ij} \quad \text{or } 0 \text{ if } i \neq j.$$

Show that the matrix B is also an M-matrix.

(3) Let $A = (a_{ij}) \in \mathscr{A}_n(\mathbb{R})$ be a matrix all of whose off-diagonal elements are non-positive. Prove that the following properties are equivalent.

 (i) A is an M-matrix and $(A^{-1})_{ij} > 0$, $1 \leqslant i, j \leqslant n$.

 (ii) A is an irreducible matrix and $\lambda \in \mathrm{Sp}(A) \Rightarrow \mathrm{Re}\,\lambda > 0$.

 (iii) The diagonal elements of the matrix A are all positive and the matrix $I - DA \geqslant 0$, where $D = \mathrm{diag}(a_{ii}^{-1})$, is irreducible and satisfies $\rho(I - DA) < 1$.

(4) Let $A = (a_{ij})$ be an irreducible matrix of order n such that

$$a_{ii} > 0, \quad 1 \leqslant i \leqslant n,$$

$$a_{ij} \leqslant 0, \quad 1 \leqslant i, j \leqslant n, \quad i \neq j,$$

$$|a_{ii}| \geqslant \sum_{j \neq i} |a_{ij}|, \quad 1 \leqslant i \leqslant n,$$

$$|a_{i_0 i_0}| > \sum_{j \neq i_0} |a_{i_0 j}| \text{ for at least one index } i_0 \in \{1, 2, \ldots, n\}.$$

Prove that the matrix A is an M-matrix and that $(A^{-1})_{ij} > 0$, $1 \leqslant i, j \leqslant n$ (the reader is advised to make use of the result of Exercise 1.1-6 (3)).

3.1-8. It is desired to approximate the solution of the problem

$$\begin{cases} -u''(x) + \dfrac{u'(x)}{1+x} = f(x), & 0 < x < 1, \\ u(0) = u(1) = 0, \end{cases}$$

by the finite-difference method. To this end, we approximate (with the usual notation)

$$-\varphi''(x_i) \text{ by } \frac{-\varphi_{i-1} + 2\varphi_i - \varphi_{i+1}}{h^2},$$

$$\varphi'(x_i) \text{ by } \frac{\varphi_{i+1} - \varphi_{i-1}}{2h},$$

for every sufficiently smooth function φ.

(1) Write down the approximate problem in the form of a linear system $A_h u_h = b_h$ and show that the matrix A_h is monotone.

(2) Consider the 'auxiliary' function

$$\theta: x \in [0,1] \to -\frac{(1+x)^2}{2} \log(1+x) + \frac{2}{3}(x^2 + 2x)\log 2.$$

Show that there exists a constant C, independent of h, such that

$$\max_{1 \leqslant i \leqslant N} \left| \frac{-\theta_{i-1} + 2\theta_i - \theta_{i+1}}{h^2} + \frac{\theta_{i+1} - \theta_{i-1}}{2h(1+ih)} - 1 \right| \leqslant Ch^2,$$

where $\theta_i = \theta(x_i)$.

(3) Set $\theta_h = \begin{pmatrix} \theta_1 \\ \theta_2 \\ \vdots \\ \theta_N \end{pmatrix}$. Show that $(A_h \theta_h)_i \geqslant 1 - Ch^2$, $1 \leqslant i \leqslant N$.

(4) Prove that there exists a constant M, independent of h, such that

$$\|A_h^{-1}\|_\infty \leqslant M.$$

(5) Deduce the convergence of the method, stating explicitly what assumptions are made.

3.2 The finite-difference method for a two-dimensional boundary-value problem

Consider the following physical problem. An elastic membrane is fixed round a contour whose projection onto the 'horizontal' plane, which is spanned by the two basis vectors e_1 and e_2, is the boundary Γ of an open, bounded, connected region Ω of this same plane. Suppose that the membrane is subjected to the action of a given vertical force $\tau f(x)\,dx$ per surface element of the horizontal plane, τ being the tension of the membrane (it is convenient to make explicit the physical constant τ in the expression for the force density). The problem can then be posed of the determination of the vertical displacement $u(x)$, $x \in \Omega$, of points of the membrane when subjected to this force (cf. figure 3.2-1).

With the help of certain simplifying hypotheses, it can be shown that the unknown function u must satisfy the partial differential equation

$$-\Delta u(x) = f(x) \quad \text{for} \quad x \in \Omega, \quad \text{where } \Delta = \frac{\partial^2}{\partial x_1^2} + \frac{\partial^2}{\partial x_2^2},$$

which is called *Poisson's equation*, or *Laplace's equation* if $f = 0$. The operator Δ is called the *Laplacian* (here, two-dimensional). We are then led to the problem of finding a function $u: \bar{\Omega} \to \mathbb{R}$ which is the solution of

$$\begin{cases} -\Delta u(x) = f(x), & x \in \Omega, \\ u(x) = g(x), & x \in \Gamma, \end{cases}$$

where the function g, representing the height of the contour of the membrane, is given. Another example of a *boundary-value problem* is provided in this way, the unknown function having to satisfy the *boundary condition* $u = g$ on the boundary of the open region in which the partial differential equation has to be satisfied.

It can be proved (though it is a much more delicate task than in the one-dimensional case) that, if the data f and g, as well as the boundary Γ, are sufficiently smooth, then the problem has a unique solution, continuous in

Figure 3.2-1

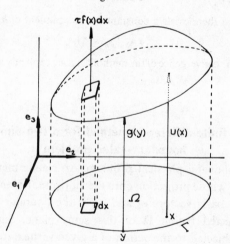

$\bar\Omega$ and twice continuously differentiable in Ω (just to fix ideas). Let φ denote the solution.

As in the one-dimensional case, the *finite-difference method* provides an approximation of the solution at a *finite* number of points of the open region Ω. To obtain which, the first step is to set up a *uniform mesh* of the plane, of fixed mesh-size h in both directions (at times it is useful to consider non-uniform meshes; we shall not do so here), the *nodes* being the points (ih, jh), $i, j \in \mathbb{Z}$. Let Ω_h be the set of nodes of the mesh belonging to Ω, and let Γ_h be the set of points of the plane which fall on the intersection of the boundary Γ and a horizontal and/or vertical line of the mesh. Attention is drawn to the fact that, except in particular cases, there is no reason to expect the points of Γ_h to coincide with the nodes of the mesh! (Cf. figure 3.2-2.)

Figure 3.2-2

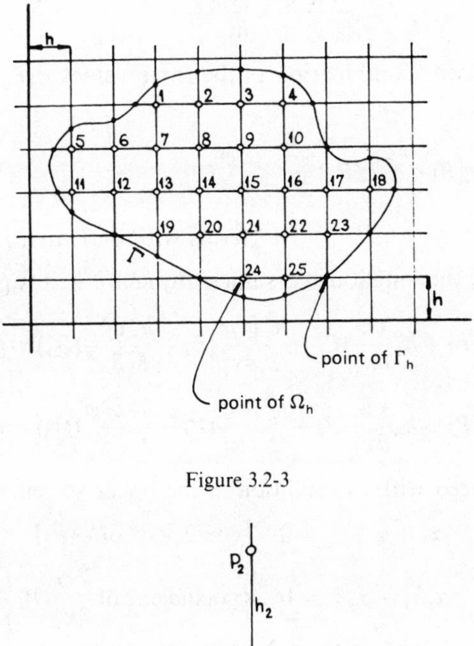

point of Γ_h

point of Ω_h

Figure 3.2-3

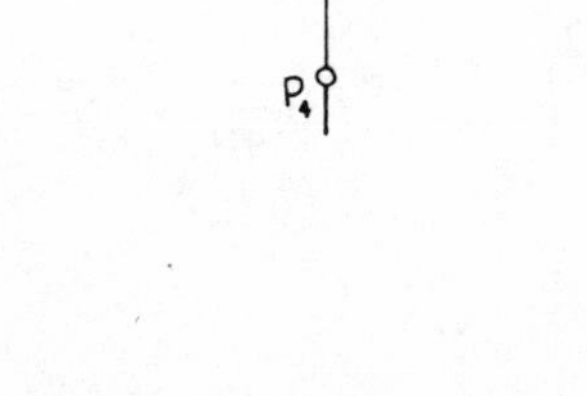

The discrete problem consists in finding an *approximation of the solution at the points of* Ω_h. As a first step, the points are numbered from left to right and from top to bottom, as is indicated in figure 3.2-2 for a particular example. Although *a priori* it may seem that the order in which these points are numbered is of no particular significance, in fact it plays a fundamental role in the *practical* solution of the associated discrete problem. We shall return to this point.

The first step of the method consists in obtaining an approximate expression for the Laplacian at the nodes $P \in \Omega_h$ by replacing the partial derivatives with appropriate 'finite-difference' quotients. Let P be a point of Ω_h, and let P_i, $1 \le i \le 4$, be the four points of the set $\Omega_h \cup \Gamma_h$ which are 'its closest neighbours in the four directions', arranged as in figure 3.2-3. Let h_i denote the distance of the point P from the point P_i.

If, for example, $h_1 = h_3$, then the approximation of $-(\partial^2 \varphi / \partial x_1^2)(P)$ is known, namely (see the previous section)

$$\frac{-\varphi(P_3) + 2\varphi(P) - \varphi(P_1)}{h_1^2}.$$

Otherwise, a linear combination of the three values $\varphi(P_3)$, $\varphi(P)$, $\varphi(P_1)$, is sought such that

$$\alpha_1 \varphi(P_1) + \alpha \varphi(P) + \alpha_3 \varphi(P_3) = -\frac{\partial^2 \varphi}{\partial x_1^2}(P)$$

$$+ \{\text{terms with derivatives of order} \ge 3\}.$$

Supposing that the function φ is sufficiently smooth, it is possible to write

$$\varphi(P_1) = \varphi(P) + h_1 \frac{\partial \varphi}{\partial x_1}(P) + \frac{h_1^2}{2}\frac{\partial^2 \varphi}{\partial x_1^2}(P) + \frac{h_1^3}{6}\frac{\partial^3 \varphi}{\partial x_1^3}(Q_1), \quad Q_1 \in (P, P_1),$$

$$\varphi(P_3) = \varphi(P) - h_3 \frac{\partial \varphi}{\partial x_1}(P) + \frac{h_3^2}{2}\frac{\partial^2 \varphi}{\partial x_1^2}(P) - \frac{h_3^3}{6}\frac{\partial^3 \varphi}{\partial x_1^3}(Q_3), \quad Q_3 \in (P, P_3).$$

We are then faced with the solution of the linear system

$$\alpha_1 + \alpha + \alpha_3 = 0 \quad \{\text{coefficient of } \varphi(P)\},$$

$$\alpha_1 h_1 - \alpha_3 h_3 = 0 \quad \left\{\text{coefficient of } \frac{\partial \varphi}{\partial x_1}(P)\right\},$$

$$\alpha_1 \frac{h_1^2}{2} + \alpha_3 \frac{h_3^2}{2} = -1 \quad \left\{\text{coefficient of } \frac{\partial^2 \varphi}{\partial x_1^2}(P)\right\},$$

with solution

$$\alpha_1 = -\frac{2}{h_1(h_1 + h_3)}, \quad \alpha = \frac{2}{h_1 h_3}, \quad \alpha_3 = -\frac{2}{h_3(h_1 + h_3)}.$$

In this way, we obtain

$$-\frac{\partial^2\varphi}{\partial x_1^2}(P) = -\frac{2}{h_1(h_1+h_3)}\varphi(P_1) + \frac{2}{h_1h_3}\varphi(P) - \frac{2}{h_3(h_1+h_3)}\varphi(P_3)$$
$$-\frac{h_1^2}{3(h_1+h_3)}\frac{\partial^3\varphi}{\partial x_1^3}(Q_1) + \frac{h_3^2}{3(h_1+h_3)}\frac{\partial^3\varphi}{\partial x_1^3}(Q_3),$$

together with an analogous expression for the partial derivative $-(\partial^2\varphi/\partial x_2^2)(P)$. We observe, in passing, that there is no advantage in incorporating the fourth derivatives in the Taylor expansions if $h_1 \neq h_3$, because *in that case* the terms involving the third derivatives do not cancel.

Neglecting, as in the one-dimensional case, terms involving third or higher order derivatives of the function φ, the *Laplacian approximation* is defined as the linear operator Δ_h which, with every function u defined over the set $\Omega_h \cup \Gamma_h$, associates the function $\Delta_h u$ defined over the set Ω_h by

$$\Delta_h u(P) \stackrel{\text{def}}{=} \frac{2}{h_1(h_1+h_3)}u(P_1) + \frac{2}{h_2(h_2+h_4)}u(P_2) + \frac{2}{h_3(h_3+h_1)}u(P_3)$$
$$+ \frac{2}{h_4(h_4+h_2)}u(P_4) - \left\{\frac{2}{h_1h_3} + \frac{2}{h_2h_4}\right\}u(P), \quad P \in \Omega_h,$$

which simplifies to

$$\Delta_h u(P) = \frac{u(P_1) + u(P_2) + u(P_3) + u(P_4) - 4u(P)}{h^2},$$

if $h_1 = h_2 = h_3 = h_4 = h$. For an obvious reason, the formulas given above are often appropriately called *five-point approximations* of the Laplacian operator. Once the Laplacian approximation has been defined, the *discrete problem*, associated with the boundary-value problem being considered and the chosen mesh, consists in finding a function u_h defined over the discrete set $\Omega_h \cup \Gamma_h$ such that

$$\begin{cases} -\Delta_h u_h(P) = f(P), & P \in \Omega_h, \\ u_h(P) = g(P), & P \in \Gamma_h. \end{cases}$$

After choosing a *numbering* of the nodes of Ω_h, the two previous relations may be written as a *linear system*

$$A_h u_h = b_h,$$

the vector u_h having as its components the values $u_h(P)$, $P \in \Omega_h$, arranged in the particular order under consideration. For example, in the case of figure 3.2-2, we obtain the matrix (where the points represent the non-zero elements).

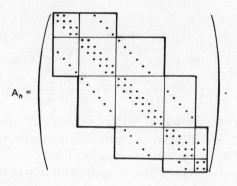

$$A_h =$$

Returning to the general case, let us now go over the 'immediate' properties of the matrix A_h. It is a *sparse matrix*, since each row contains at most five non-zero elements. Moreover, to the chosen numbering of the points of Ω_h there corresponds a partitioning of the matrix A_h which is altogether remarkable, the matrix being *block tridiagonal*. Each block diagonal submatrix corresponds to a subset of equations $- \Delta_h u_h(P) = f(P)$ associated with the points P of a single line of the mesh, let us say the pth, in order to fix ideas. The only non-zero elements associated with such a point P correspond (in the notation of figure 3.2-3) to the points P_1 and P_3 (causing the block diagonal matrices to be *tridiagonal*) and the points P_2 and P_4 (which correspond to a non-zero element of the submatrix $A_{p-1,p}$ and to a non-zero element of the submatrix $A_{p+1,p}$); this explains the block tridiagonal structure.

It is hardly possible to overstress the *practical* significance of this result. As will be seen in Chapter 5, there exist methods for the solution of linear systems which are especially well suited to the treatment of block tridiagonal matrices. Moreover, the reader should make his own effort to convince himself that an 'arbitrary' numbering of the points leads to a matrix with no particular structure (though, of course, still sparse).

Though the matrix A_h has a 'symmetric structure', in the sense that an element $(A_h)_{ij}$ is non-zero if and only if the element $(A_h)_{ji}$ is also non-zero, *it is not generally symmetric*, except when only the second expression for the Laplacian approximation is used; this eventuality corresponds to the case where the set Γ consists of a polygon with its sides parallel to the coordinate axes.

Remark
In contrast, the variational methods of approximation *always* lead to symmetric matrices (see sections 3.4 and 3.5). □

We leave as an exercise the study of the convergence of the method (Exercise 3.2-2); as in the one-dimensional case, it depends in an essential way on the monotone character of the matrix A_h (Exercise 3.2-1).

Exercises

3.2-1. Show that the matrix A_h associated with the discrete Laplacian is monotone. For this, one may obtain guidance from the proof of Theorem 3.1-1.

3.2-2. The aim in this problem is to establish the *convergence* of the finite-difference method applied to the solution of the problem

$$\begin{cases} -\Delta u(x) = f(x), & x \in \Omega, \\ u(x) = g(x), & x \in \Gamma, \end{cases}$$

and to obtain, in this way (at question(6)), an analogue of the result obtained in Theorem 3.1-2 for the one-dimensional case. Let Q be an arbitrary point of Ω_h. Define the function $G_h(P, Q)$, $P \in \Omega_h \cup \Gamma_h$, as the solution of the discrete problem

$$\begin{cases} -\Delta_h u(P) = \begin{cases} 0 & \text{if } P \neq Q, \quad P \in \Omega_h, \\ 1/h^2 & \text{if } P = Q, \quad P \in \Omega_h, \end{cases} \\ u(P) = 0, \quad P \in \Gamma_h. \end{cases}$$

Similarly, let Q be an arbitrary point of Γ_h. Define the function $g_h(P, Q)$, $P \in \Omega_h \cup \Gamma_h$, as the solution of the discrete problem

$$\begin{cases} -\Delta_h u(P) = 0, \quad P \in \Omega_h, \\ u(P) = \begin{cases} 0 & \text{if } P \neq Q, \quad P \in \Gamma_h, \\ -1 & \text{if } P = Q, \quad P \in \Gamma_h. \end{cases} \end{cases}$$

Each of these two problems does, in fact, have a unique solution, by the result of Exercise 3.2-1.

(1) Let u be any function defined over the set $\Omega_h \cup \Gamma_h$. Prove the identity (which should be compared with the summation equation of Exercise 3.1-1, question (2)), for every $P \in \Omega_h \cup \Gamma_h$,

$$u(P) = -h^2 \sum_{Q \in \Omega_h} G_h(P, Q) \Delta_h u(Q) - \sum_{Q \in \Gamma_h} g_h(P, Q) u(Q).$$

(2) Prove the inequality

$$G_h(P, Q) \geq 0 \quad \text{for every } P \in \Omega_h \cup \Gamma_h, \quad Q \in \Omega_h.$$

(3) Prove that there exists a constant C, independent of the mesh-size, such that

$$h^2 \sum_{Q \in \Omega_h} G_h(P, Q) \leq C \quad \text{for every } P \in \Omega_h \cup \Gamma_h.$$

(4) Denote by Ω_h^* the subset of points of Ω_h which correspond to the 'general' expression for the approximate Laplacian, that is, those for which $h_1 \neq h_3$ and/or $h_2 \neq h_4$ (these are necessarily points of Ω_h situated at a distance $< h$ from the

boundary Γ). Prove the inequality

$$\sum_{Q \in \Omega_h^*} G_h(P, Q) \leqslant 1 \quad \text{for every } P \in \Omega_h \cup \Gamma_h.$$

(5) Given a function φ which is four times continuously differentiable over the set $\bar{\Omega}$, let

$$M_\alpha(\varphi) = \max_{x \in \bar{\Omega}} \left\{ \left| \frac{\partial^\alpha \varphi}{\partial x_1^\alpha}(x) \right|, \left| \frac{\partial^\alpha \varphi}{\partial x_2^\alpha}(x) \right| \right\} \quad \text{for} \quad \alpha = 3 \text{ and } 4.$$

Prove the inequalities

$$|\Delta_h \varphi(P) - \Delta \varphi(P)| \leqslant \frac{2M_3(\varphi)}{3} h \quad \text{if} \quad P \in \Omega_h^*,$$

$$|\Delta_h \varphi(P) - \Delta \varphi(P)| \leqslant \frac{M_4(\varphi)}{6} h^2 \quad \text{if} \quad P \in \Omega_h - \Omega_h^*.$$

(6) Assuming the solution φ of the problem to be four times continuously differentiable over the set $\bar{\Omega}$, obtain the bound

$$\max_{P \in \Omega_h} |u_h(P) - \varphi(P)| \leqslant \left(\frac{C}{6} M_4(\varphi) \right) h^2 + \left(\frac{2}{3} M_3(\varphi) \right) h^3.$$

As in the one-dimensional case, deduce a *convergence of order* h^2, in respect of the norm $\| \cdot \|_\infty$, despite the fact that the 'local' consistency error $|\Delta_h \varphi(P) - \Delta \varphi(P)|$ is only of order h for the points of Ω_h^* (question (5)), this effect being compensated by the 'h^2 behaviour' of the function $h^2 G_h(P, Q)$ at these same points (question (4)).

3.3 The finite-difference method for time-dependent boundary-value problems

Consider a rod of length l supported along an axis between the two abscissae 0 and l, in which heat is transmitted by conduction. It is supposed that the heat exchanges with the exterior arise from heat sources along the length of the rod, the quantity of heat provided at the point x and instant t having the density function $f(x, t)$. Lastly, it is supposed that there is no heat exchange by radiation or convection. Then the temperature $u(x, t)$ at the point x and instant t satisfies the following partial differential equation

$$\sigma \frac{\partial u}{\partial t}(x, t) - \gamma \frac{\partial^2 u}{\partial x^2}(x, t) = f(x, t)$$

(σ = linear specific heat; γ = thermal conductivity), which is called the *one-dimensional heat equation*.

If the temperature of the rod $u(x, 0) = u_0(x)$, $0 \leqslant x \leqslant l$, is known at the 'initial' instant $t = 0$ and the ends of the rod are maintained at known temperatures for $t \geqslant 0$, then we are led to *look for a function* $u(x, t)$ *defined for* $0 \leqslant x \leqslant l$ *and* $t \geqslant 0$ *which is the solution of*

$$
\begin{cases}
\sigma \dfrac{\partial u}{\partial t}(x, t) - \gamma \dfrac{\partial^2 u}{\partial x^2}(x, t) = f(x, t), & 0 < x < l, \quad t > 0, \\[2mm]
u(x, 0) = u_0(x), & 0 \leqslant x \leqslant l, \\[2mm]
u(0, t) = \alpha(t) \text{ and } u(l, t) = \beta(t), & t \geqslant 0.
\end{cases}
$$

Supposing, without loss of generality, that $\sigma = \gamma = l = 1$ and that $\alpha(t) = \beta(t) = 0$ for $t \geqslant 0$ (subtract the function $(1 - x)\alpha(t) + x\beta(t)$), we are led to look for a function $u(x, t)$, defined for $0 \leqslant x \leqslant 1$ and $t \geqslant 0$, satisfying

$$
\begin{cases}
\dfrac{\partial u}{\partial t}(x, t) - \dfrac{\partial^2 u}{\partial x^2}(x, t) = f(x, t), & 0 < x < 1, \quad 0 < t, \\[2mm]
u(x, 0) = u_0(x), & 0 \leqslant x \leqslant 1, \\[2mm]
u(0, t) = u(1, t) = 0, & t \geqslant 0.
\end{cases}
$$

This provides yet another example of a *boundary-value problem*, because the equations $u(0, t) = u(1, t) = 0$ for $t \geqslant 0$ are, in effect, *boundary conditions*. But it has to be noted that this problem is further complicated by the *initial condition* $u(x, 0) = u_0(x)$, $0 \leqslant x \leqslant 1$, which is characteristic of *time-dependent problems*, that is to say, problems involving time.

It can be shown that, if the data f and u_0 are sufficiently smooth (and if the function u_0 satisfies the natural condition of compatibility $u_0(0) = u_0(1) = 0$), then the problem has a unique solution, continuous over the set $[0, 1] \times \mathbb{R}_+$ and sufficiently smooth over the set $]0, 1[\times \{t \in \mathbb{R}: t > 0\}$ for the partial differential equation under consideration to make sense. This solution will be denoted by φ.

Figure 3.3-1

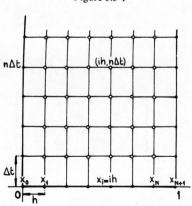

In order to discretise this problem by a *finite-difference method*, we set up a mesh of mesh-size $h = 1/(N + 1)$, N an integer $\geqslant 1$, for the variable x, and a *time step* Δt for the variable t (figure 3.3-1). *Both h and Δt are meant to be free to tend to zero.*

If we denote by u_i^n an approximation (to be determined) of the solution at the point $(ih, n\Delta t)$, the considerations of section 3.1 lead us to approximate

$$-\frac{\partial^2 \varphi}{\partial x^2}(ih, n\Delta t) \quad \text{by} \quad \frac{-u_{i-1}^n + 2u_i^n - u_{i+1}^n}{h^2},$$

and

$$\frac{\partial \varphi}{\partial t}(ih, n\Delta t) \text{ by } \begin{cases} \dfrac{u_i^{n+1} - u_i^n}{\Delta t} & \text{(schema (I)),} \\[2mm] \text{or} \\[2mm] \dfrac{u_i^n - u_i^{n-1}}{\Delta t} & \text{(schema (II)).} \end{cases}$$

The *discrete problem* associated with schema (I) can be reset as follows. Find numbers $u_i^n, 0 \leqslant i \leqslant N + 1$, $n \geqslant 0$, which solve

$$\begin{cases} \dfrac{u_i^{n+1} - u_i^n}{\Delta t} + \left(\dfrac{-u_{i-1}^n + 2u_i^n - u_{i+1}^n}{h^2} \right) = f(ih, n\Delta t) & 1 \leqslant i \leqslant N, \quad n \geqslant 0, \\[3mm] u_i^0 = u_0(ih), \quad 1 \leqslant i \leqslant N, \\[2mm] u_0^n = u_{N+1}^n = 0, \quad n \geqslant 0. \end{cases}$$

If we define the vectors

$$u^n = (u_i^n)_{i=1}^N \in \mathbb{R}^N \quad \text{and} \quad f^n = (f(ih, n\Delta t))_{i=1}^N \in \mathbb{R}^N,$$

the discrete problem can also be set in the following matrix form: find vectors $u^n, n \geqslant 0$, satisfying

$$\begin{cases} \dfrac{u^{n+1} - u^n}{\Delta t} + Au^n = f^n, \quad n \geqslant 0, \\[3mm] u^0 \text{ given,} \end{cases}$$

where the matrix A is of order N and is given by the familiar expression

$$A = \frac{1}{h^2} \begin{pmatrix} 2 & -1 & & & \\ -1 & 2 & -1 & & \\ & \ddots & \ddots & \ddots & \\ & & -1 & 2 & -1 \\ & & & -1 & 2 \end{pmatrix}.$$

Remark

To avoid making the notation cumbersome, we have chosen to suppress the dependence on $h = 1/(N + 1)$ of the vectors u^n, f^n and of the matrix A. However, it is as well not to forget this fact! □

The vector u^{n+1} is then given *explicitly* in terms of the vector u^n by the formula

$$u^{n+1} = (I - \Delta t\, A)u^n + \Delta t\, f^n, \quad n \geq 0.$$

More precisely, each component u_i^{n+1} of the vector u^{n+1} is given explicitly in terms of only the three components u_{i-1}^n, u_i^n, u_{i+1}^n of the vector u^n, which is assumed to be known (recall that $u_0^n = u_{N+1}^n = 0$). Schema (I) is said to be *explicit*.

Schema (II) may be written, in like manner, as

$$\begin{cases} \dfrac{u_i^n - u_i^{n-1}}{\Delta t} + \dfrac{-u_{i-1}^n + 2u_i^n - u_{i+1}^n}{h^2} = f(ih, n\Delta t), \quad 1 \leq i \leq N, \quad n \geq 1, \\[2mm] u_i^0 = u_0(ih), \quad 1 \leq i \leq N, \\[2mm] u_0^n = u_{N+1}^n = 0, \quad n \geq 0, \end{cases}$$

or again, in matrix form, as

$$\begin{cases} \dfrac{u^n - u^{n-1}}{\Delta t} + Au^n = f^n, \quad n \geq 1, \\[2mm] u^0 \quad \text{given.} \end{cases}$$

Unlike the case for schema (I), it is not possible to calculate immediately a component u_i^n of the vector u^n since the equation in which it appears makes use also of the components u_{i-1}^n and u_{i+1}^n of the same vector: thus the vector u^n is calculated in terms of the vector u^{n-1} by solving a *linear system*, namely

$$(I + \Delta t\, A)u^n = u^{n-1} + \Delta t\, f^n,$$

and that is the reason for calling schema (II) *implicit*. The execution of schema (II) thus leads to *the solution of a sequence of linear systems*, with the *same* matrix $I + \Delta t\, A$, which is tridiagonal, symmetric and positive definite (this last property is the result of the positive definite character of the matrix A, established in section 3.1).

We shall not deal here with the question of the convergence of such schemata, referring the reader to specialist works, with references given in the Bibliography and comments. We merely mention the fact that the implicit schema (II) is superior to the explicit schema (I) from the point of view of 'stability'; this superiority is, however, counteracted by an increase in computational effort at each iteration.

We could as readily consider the problem of the determination of the temperature in the interior Ω of a surface in $\mathbb{R}^2$, or of a volume in $\mathbb{R}^3$, with boundary Γ. We would then be led to look for a function $u(x, t)$ defined for $x \in \bar{\Omega}$ and $t \geqslant 0$, that would satisfy

$$
\begin{cases}
\sigma \dfrac{\partial u}{\partial t}(x, t) - \gamma \Delta u(x, t) = f(x, t), & x \in \Omega, \quad t > 0, \\[2mm]
u(x, 0) = u_0(x), & x \in \bar{\Omega} \quad \text{(initial condition)}, \\[2mm]
u(x, t) = g(x, t), & x \in \Gamma, \quad t \geqslant 0 \quad \text{(boundary condition)},
\end{cases}
$$

for the given functions f, u_0 and g. It is just as possible to approximate such time-dependent boundary-value problems using finite-difference methods, with the Laplacian Δ being approximated in the way that was indicated in section 3.2 for the two-dimensional case.

We next consider a homogeneous string of constant cross-section, length l, and stretched between its two ends which are fixed along an axis, one being at the point 0 and the other at the point l. The string is subjected to the action of a transverse force $\tau f(x, t)\,dx$ per element of length dx (τ = the tension in the string). This leads to the investigation of the 'small' transverse displacements of the string in the vertical plane. In other words, a function $u(x, t)$ is sought which is defined for $0 \leqslant x \leqslant l$ and $t \geqslant 0$, and which represents at the point x and instant t the vertical deformation of the string. It can then be shown that the function u has to satisfy the following partial differential equation

$$
\frac{1}{c^2}\frac{\partial^2 u}{\partial t^2}(x, t) - \frac{\partial^2 u}{\partial x^2}(x, t) = f(x, t), \quad 0 < x < l, \quad t > 0,
$$

where $c = \sqrt{(\tau/\varrho)}$ (ϱ being the linear density of the string). It is called the *one-dimensional wave equation*.

The shape of the string at the 'initial' moment $t = 0$ is supposed known, as well as the distribution of the initial velocities along the length of the string; in other words, the functions $u(x, 0) = u_0(x)$ and $(\partial u/\partial t)(x, 0) = u_1(x)$ for $0 \leqslant x \leqslant l$ are taken as known (for example, if the string is simply released from a given starting position, we have $u_1(x) = 0$ for $0 \leqslant x \leqslant l$). Lastly, we need to have $u(0, t) = u(l, t) = 0$ for $t \geqslant 0$, because the ends of the string are supposed fixed.

We are then led to *look for a function $u(x, t)$ defined over $0 \leqslant x \leqslant l$ and $t \geqslant 0$ which solves*

$$
\begin{cases}
\dfrac{1}{c^2}\dfrac{\partial^2 u}{\partial t^2}(x, t) - \dfrac{\partial^2 u}{\partial x^2}(x, t) = f(x, t), & 0 < x < l, \quad t > 0, \\[2mm]
u(0, t) = u(l, t) = 0, & t \geqslant 0 \quad \text{(boundary condition)}, \\[2mm]
u(x, 0) = u_0(x), & 0 \leqslant x \leqslant l \quad \text{(initial condition)}, \\[2mm]
\dfrac{\partial u}{\partial t}(x, 0) = u_1(x) & 0 \leqslant x \leqslant l \quad \text{(initial condition)}.
\end{cases}
$$

This provides yet another example of a *time-dependent boundary-value problem*, now provided with *two* initial conditions. In order to discretise this problem by means of a finite-difference method, we establish the same mesh as that given in figure 3.3-1 (with the assumption $l = 1$). The most 'natural' discretisation consists in finding the numbers $u_i^n, 0 \leqslant i \leqslant N + 1, n \geqslant 0$, which satisfy

$$\begin{cases} \dfrac{1}{c^2} \dfrac{u_i^{n+1} - 2u_i^n + u_i^{n-1}}{\Delta t^2} + \left(\dfrac{-u_{i-1}^n + 2u_i^n - u_{i+1}^n}{h^2} \right) = f(ih, n\Delta t), \\ \qquad\qquad\qquad\qquad\qquad\qquad\qquad 1 \leqslant i \leqslant N, \quad n \geqslant 1, \\ u_i^0 = u_0(ih), \quad 1 \leqslant i \leqslant N, \\ u_i^1 = u_0(ih) + \Delta t u_1(ih), \quad 1 \leqslant i \leqslant N, \\ u_0^n = u_N^n = 0, \quad n \geqslant 0, \end{cases}$$

or, in vector form (with the same notation as before),

$$\begin{cases} \dfrac{1}{c^2} \dfrac{u^{n+1} - 2u^n + u^{n-1}}{\Delta t^2} + Au^n = f^n, \quad n \geqslant 1, \\ u^0, u^1 \quad \text{given.} \end{cases}$$

To the *explicit* schema given above an *implicit* schema is generally preferable, for example

$$\begin{cases} \dfrac{1}{c^2} \dfrac{u^{n+1} - 2u^n + u^{n-1}}{\Delta t^2} + \dfrac{1}{4}(Au^{n+1} + 2Au^n + Au^{n-1}) = f^n, \quad n \geqslant 1, \\ u^0, u^1 \quad \text{given,} \end{cases}$$

whose solution, as it develops, requires *the solution of a sequence of linear systems* with the *same matrix* $I + (c^2\Delta t^2/4)A$, which is once again tridiagonal, symmetric and positive definite.

In the same way, it would be possible to consider the small displacements of a thin membrane clamped round a contour in $\mathbb{R}^3$ whose horizontal projection is the boundary Γ of an open region Ω of the plane, subjected to the action of a vertical force $\tau f(x)\,dx$ per element of projected surface dx (τ being the tension in the membrane). If $x = (x_1, x_2)$ represents any point in the open region Ω, the vertical deformation of the membrane satisfies the *time-dependent boundary-value problem*

$$\begin{cases} \dfrac{1}{c^2} \dfrac{\partial^2 u}{\partial t^2}(x, t) - \Delta u(x, t) = f(x), \quad x \in \Omega, \quad t > 0, \\ u(x, 0) = u_0(x), \quad x \in \bar{\Omega} \quad \text{(initial condition),} \\ \dfrac{\partial u}{\partial t}(x, 0) = u_1(x), \quad x \in \bar{\Omega} \quad \text{(initial condition),} \\ u(x, t) = g(x), \quad x \in \Gamma, \quad t \geqslant 0 \quad \text{(boundary condition),} \end{cases}$$

where c is a physical constant of the problem (as in the case of the string), and the known function g represents the height of the contour of the membrane. The partial differential equation for this problem is called the *two-dimensional wave equation*. This kind of problem can also be approximated by means of finite-difference methods, making use of the Laplacian approximation described in section 3.2.

Exercises

3.3-1. The aim in this exercise is to examine the *convergence* of finite-difference methods, applied to the approximate solution of the heat equation in one dimension (refer to the text, whose notation and definitions are retained).

The schemata considered are particular cases of the following schema. Two integers $k \geqslant 0$ and $l \geqslant 0$ are given which satisfy $k + l \geqslant 1$, and matrices B_j, $-l \leqslant j \leqslant k$, of order N, which *depend* on the quantities $h = 1/(N+1)$ and Δt (however, in order not to overload the notation, this last dependence is not brought out explicitly), the matrices B_k being invertible. The vectors $u^n \in \mathbb{R}^N$, $n \geqslant 0$, are solutions of the schema

$$(S) \qquad \begin{cases} B_k u^{n+k} + \cdots + B_0 u^n + \cdots + B_{-l} u^{n-1} = f^n, & n \geqslant 1, \\ u^0, u^1, \ldots, u^{k+l-1} \text{ given,} \end{cases}$$

which is said to be *explicit* if the matrix B_k is diagonal (in practice, some multiple of the identity matrix) and *implicit* otherwise.

In analysing such schemata, use is made of the two essential notions of *consistency* and *stability*, which we are about to define. In order to justify the introduction of stability, we first establish a fundamental property of the continuous problem.

(1) Let $\varphi(x, t)$, $0 \leqslant x \leqslant 1$, $0 \leqslant t \leqslant T$, $T > 0$ fixed, be a sufficiently smooth solution of the heat equation. Prove that there exist two constants $C_1(T)$ and $C_2(T)$, *independent of the functions* φ, u_0 and f, such that

$$\sup_{0 \leqslant t \leqslant T} \left(\int_0^1 |\varphi(x, t)|^2 \, dx \right)^{1/2} \leqslant C_1(T) \left(\int_0^1 |u_0(x)|^2 \, dx \right)^{1/2}$$
$$+ C_2(T) \sup_{0 \leqslant t \leqslant T} \left(\int_0^1 |f(x, t)|^2 \, dx \right)^{1/2}.$$

Hint. For $t \in [0, T]$, write

$$\int_0^1 |\varphi(x, t)|^2 \, dx = \int_0^1 |\varphi(x, 0)|^2 \, dx + \int_0^1 \left\{ \frac{d}{d\lambda} \int_0^1 |\varphi(x, \lambda)|^2 \, dx \right\} d\lambda.$$

The inequality given above is called the *stability inequality* for the continuous problem. An immediate consequence of it (apart from that of the *uniqueness of the solution*) is the *continuous dependence of the solution on the data* u_0 and f, with respect to norms which will be given explicitly. Likewise, it will be made clear in what precise sense this continuous dependence affects the *continuity of the inverse* of the linear operator associated with the heat equation (however, the existence of this inverse remains a matter to be established by some other means).

(2) In view of the stability inequality given above, we are naturally led to the following definition. The schema (S) is said to be *stable* over the interval $[0, T]$, $T > 0$ fixed, if there exist two constants $D_1(T)$ and $D_2(T)$, *independent of the data u^i*, $0 \leqslant i \leqslant k+l-1$, and f^n, $l \leqslant n \leqslant T/\Delta t$ (and, hence, independent of the vectors u^n, $k+l \leqslant n \leqslant T/\Delta t$, which are solutions of (S)), and *independent of the quantities h and Δt* as these tend to zero (this may hold, however, only if some restrictions are imposed on the way h and Δt tend to zero; cf. question (5)), such that

$$\max_{n\Delta t \leqslant T} \|u^n\|_h \leqslant D_1(T) \max_{0 \leqslant i \leqslant k+l-1} \|u^i\|_h + D_2(T) \max_{n\Delta t \leqslant T} \|f^n\|_h$$

where

$$\|v\|_h \overset{\text{def}}{=} \left(h \sum_{i=1}^{N} |v_i|^2 \right)^{1/2} \quad \text{for} \quad v \in \mathbb{R}^N.$$

Lastly, the schema (S) is said to be *consistent* if, for every sufficiently smooth solution φ of the heat equation,

$$\lim_{\substack{h \to 0 \\ \Delta t \to 0}} \max_{n\Delta t \leqslant T} \|\varepsilon^n(\varphi)\|_h = 0,$$

and

$$\lim_{\substack{h \to 0 \\ \Delta t \to 0}} \max_{0 \leqslant i \leqslant k+l-1} \|\varphi^i - u^i\|_h = 0,$$

where

$$\varepsilon^n(\varphi) \overset{\text{def}}{=} (B_k\varphi^{n+k} + \cdots + B_{-l}\varphi^{n-l}) - f^n,$$

$$\varphi^n \overset{\text{def}}{=} (\varphi(ih), n\Delta t)_{i=1}^N.$$

A schema is said to be *consistent of order p in x and of order q in t* if, for every sufficiently smooth function φ,

$$\lim_{\substack{h \to 0 \\ \Delta t \to 0}} \max_{n\Delta t \leqslant T} \|\varepsilon^n(\varphi)\|_h \leqslant C(T)(h^p + \Delta t^q).$$

Prove that the schemata (I) and (II) are consistent. Find their orders of consistency.

(3) Prove that stability and consistency imply the *convergence* of schema (S), in the following sense

$$\lim_{\substack{h \to 0 \\ \Delta t \to 0}} \max_{n\Delta t \leqslant T} \|\varphi^n - u^n\|_h = 0,$$

with resultant restrictions on the way in which the parameters h and Δt tend to zero simultaneously (refer to question (5) for an example of one such restriction).

(4) Prove that a schema of the form

$$\begin{cases} u^{n+1} = Bu^n + \Delta t f^n, & n \geq 0, \\ u^0 \text{ given}, \end{cases}$$

is stable over the interval $[0, T]$ if and only if there exists a constant $M(T)$ such that

$$\max_{n\Delta t \leq T} \| B^n \|_h \leq M(T)$$

as h and Δt tend to zero, where

$$\| B \|_h = \sup_{v \in \mathbb{R}^N} \frac{\| Bv \|_h}{\| v \|_h}.$$

(5) Deduce that schema (I) is stable over every interval $[0, T]$ if and only if

$$\frac{\Delta t}{h^2} \leq \frac{1}{2}.$$

Because there appears a restriction in the way that the parameters h and Δt tend to zero, schema (I) is said to be *conditionally stable*.

(6) Arguing as in questions (4) and (5), prove that schema (II) is *unconditionally stable*, that is, stable over every interval $[0, T]$, but with no restrictions.

(7) Define the schema

(III) $$\begin{cases} \dfrac{u^{n+1} - u^{n-1}}{2\Delta t} + Au^n = f^n, & n \geq 1, \\ u^0, u^1 \text{ given}. \end{cases}$$

As a choice of the vector u^1, one may use, for example, the vector u^1 found by one of the schemata (I) or (II). Prove that this schema is consistent. Find its order of consistency.

(8) Prove that this schema is never stable (rewrite it as a schema in $\mathbb{R}^{2N}$).

(9) Prove that it is not convergent (find a counterexample).

3.4 Variational approximation of a one-dimensional boundary-value problem

Let us return to the problem considered in section 3.1.

Given two functions $c, f \in \mathscr{C}^0([0, 1])$, find a function $u \in \mathscr{C}^2([0, 1])$ which satisfies

$$\begin{cases} -u''(x) + c(x)u(x) = f(x), & 0 < x < 1, \\ u(0) = u(1) = 0. \end{cases}$$

It is for the sake of convenience that we suppose here that the boundary conditions are homogeneous: it is not a restriction. For, in effect, if they are given in the form $u(0) = \alpha$, $u(1) = \beta$, then it is enough, in order to obtain the homogeneous form, to introduce a change in the unknown function by subtracting the function $\alpha(1 - x) + \beta x$.

We introduce the vector space V consisting of functions which are continuous over the interval $[0, 1]$, take the value 0 at the points 0 and 1 and are once piecewise continuously differentiable over the same interval; that is to say, differentiable at every point of the interval $[0, 1]$ except at a finite number of points x_i of the open interval $]0, 1[$, the derivative coinciding in each open interval between two consecutive points x_i with the restriction of a function which is continuous over the corresponding closed interval (the number and distribution of the points x_i varies with the function under consideration). When equipped with the mapping

$$v \in V \to \|v\|_V = \left(\int_0^1 (|v'|^2 + |v|^2) \, dx \right)^{1/2},$$

the vector space is normed (actually it is a pre-Hilbert space).

To carry out the analysis of the method we have in mind, it is essential to pose the boundary-value problem in another form, called its *variational formulation*. This is the purpose of the result which follows.

Theorem 3.4.1

(1) *If u is a solution of the boundary-value problem, then*

$$a(u, v) = f(v) \quad \text{for every } v \in V,$$

where the bilinear form $a: V \times V \to \mathbb{R}$ and the linear form $f: V \to \mathbb{R}$ are given respectively by the expressions

$$a(u, v) = \int_0^1 (u'v' + cuv) \, dx, \quad f(v) = \int_0^1 fv \, dx,$$

for arbitrary functions $u, v \in V$.

(2) *Suppose that c is non-negative. A function $u \in V$ is the solution of the equations $a(u, v) = (f, v)$ for every $v \in V$ if and only if*

$$J(u) = \inf_{v \in V} J(v), \quad \text{where } J(v) \overset{\text{def}}{=} \tfrac{1}{2} a(v, v) - f(v).$$

Proof

(i) Let v be an arbitrary function in the space V. Denoting by x_i, $1 \leqslant i \leqslant N$, the points, arranged in increasing order, where the derivative of the function v is undefined and setting $0 = x_0$, $1 = x_{N+1}$, it is possible to write

$$\int_0^1 u''(x)v(x)\,dx = \sum_{i=1}^N \int_{x_i}^{x_{i+1}} u''(x)v(x)\,dx$$

$$= \sum_{i=1}^N \left(-\int_{x_i}^{x_{i+1}} u'(x)v'(x)\,dx + [u'(x)v(x)]_{x=x_i}^{x_{i+1}} \right)$$

$$= -\int_0^1 u'(x)v'(x)\,dx,$$

the last equality above resulting from the continuity of the functions u' and v over the interval $[0,1]$ and the relations $v(0) = v(1) = 0$.

(ii) We prove the following inequality, which will be of service time and again in the sequel. If the function c is non-negative, then there exists a constant $\alpha > 0$ such that

$$\alpha \|v\|_V^2 \leqslant a(v,v) \quad \text{for every } v \in V.$$

For this, it is enough to establish that

$$\alpha \|v\|_V^2 \leqslant \int_0^1 |v'|^2\,dx \quad \text{for every } v \in V.$$

Now, for every $x \in [0,1]$, one can write

$$|v(x)| = \left| \int_0^x v'(t)\,dt \right| \leqslant \int_0^1 |v'(t)|\,dt \leqslant \left(\int_0^1 |v'(t)|^2\,dt \right)^{1/2}$$

by the Cauchy–Schwarz inequality for functions. From which it follows that the stated inequality is satisfied with $\alpha = \frac{1}{2}$.

(iii) The characterisation given in point (2) can be proved from the identity (which is immediately verifiable)

$$J(u+v) - J(u) = \{a(u,v) - f(v)\} + \tfrac{1}{2}a(v,v) \quad \text{for every } u, v \in V.$$

For, from this follow, firstly,

$$J(u+v) - J(u) = \tfrac{1}{2}a(v,v) \geqslant \frac{\alpha}{2}\|v\|_V^2 \geqslant 0 \quad \text{for every } v \in V,$$

since $a(u,v) - f(v) = 0$ for every $v \in V$, and, secondly, for fixed $v \in V$, the inequality

$$0 \leqslant J(u+\theta v) - J(u) = \theta\{a(u,v) - f(v)\} + \frac{\theta^2}{2}a(v,v) \quad \text{for every } \theta \in \mathbb{R},$$

which necessarily implies that $a(u,v) = f(v)$. □

Remarks

(1) When the function c is non-negative, the proof given above provides another demonstration of the uniqueness of the solution u (already highlighted, especially in Exercise 3.1-3).

For it follows that

$$v \in V \quad \text{and} \quad a(v, v) = 0 \Rightarrow \|v\|_V = 0 \Rightarrow v = 0,$$

so that

$$v \neq 0 \Rightarrow J(u + v) - J(u) > 0.$$

(2) The expression $a(u, v) - f(v)$ is nothing other than the derivative of the function J at the point u, acting on the 'variation' v. This explains its characterisation as a minimum through the vanishing of the 'first variation' of the function J (these ideas are further developed in Chapter 7). And that is why the relations '$a(u, v) = f(v)$ for every $v \in V$' are called *variational equations*.

(3) Conversely, it is possible to define the boundary-value problem directly by either of the formulations (1) or (2) of Theorem 3.4-1. The first difficulty then to arise consists in proving the existence of a solution. For, in fact, it can only be proved in all generality for a *complete* space; specifically, the completion of the space V (with respect to its norm) which here is the Sobolev space $H_0^1(0, 1)$. While the 'abstract' result relating to existence is relatively easy to prove (cf. Theorem 8.2-3), the study of these completed spaces is somewhat delicate (especially in dimensions ≥ 2, to which these ideas may be generalised).

A second difficulty consists in showing that the solution obtained in this way is sufficiently smooth for it to be also a solution in the 'classical' sense that we have assumed so far. □

The *variational method of approximation* consists in approximating, in the most natural way possible, the variational formulation of a boundary-value problem. For the problem in hand, a *finite-dimensional* subspace V_h of the space V is obtained, and the associated *discrete problem* consists in finding a function $u_h \in V_h$ such that

$$a(u_h, v_h) = f(v_h) \quad \text{for every } v_h \in V_h.$$

We then have the following simple, yet crucial, result relating to the existence, uniqueness and characterisation of the *approximate solution* u_h, and, lastly, its comparison with the 'exact' solution u.

Theorem 3.4-2
Suppose that the function c is non-negative.

(1) *Given a finite-dimensional subspace V_h of the space V, there exists a unique element $u_h \in V_h$ satisfying*

$$a(u_h, v_h) = f(v_h) \quad \text{for every } v_h \in V_h.$$

(2) *This element is also characterised as the unique solution of the problem*

find a $u_h \in V_h$ *such that*

$$J(u_h) = \inf_{v_h \in V_h} J(v_h).$$

(3) *Lastly, there exists a constant C, independent of the subspace* V_h *under consideration and of the solution u, such that*

$$\|u - u_h\|_V \leqslant C \inf_{v_h \in V_h} \|u - v_h\|_V.$$

Proof

(i) Since finding a solution of the discrete problem is equivalent to solving a linear system with square matrix, existence and uniqueness are equivalent properties. We now verify the second:

$$a(u_h, v_h) = 0 \quad \text{for every } v_h \in V_h \Rightarrow a(u_h, u_h) = 0 \Rightarrow u_h = 0,$$

since $a(v, v) = 0 \Rightarrow v = 0$ (cf. part (ii) of the proof of Theorem 3.4-1). The characterisation of point (2) is proved in exactly the same way as the corresponding characterisation in the previous theorem.

(ii) Applying the Cauchy–Schwarz inequality, first for functions and then for vectors in $\mathbb{R}^2$, we obtain the inequality

$$\left| \int_0^1 (u'v' + uv) \, dx \right| \leqslant \left(\int_0^1 |u'|^2 \, dx \right)^{1/2} \left(\int_0^1 |v'|^2 \, dx \right)^{1/2}$$
$$+ \left(\int_0^1 |u|^2 \, dx \right)^{1/2} \left(\int_0^1 |v|^2 \, dx \right)^{1/2}$$
$$\leqslant \|u\|_V \|v\|_V \quad \text{for every } u, v \in V,$$

from which it follows that

$$|a(u, v)| \leqslant M \|u\|_V \|v\|_V,$$

with

$$M = \max \left\{ 1, \sup_{0 \leqslant x \leqslant 1} c(x) \right\}, \quad \text{for every } u, v \in V.$$

From the relations

$$a(u, v) = f(v) \quad \text{for every } v \in V, \quad a(u_h, v_h) = f(v_h) \quad \text{for every } v_h \in V_h,$$

and from the inclusion $V_h \subset V$, it can further be deduced that

$$a(u - u_h, w_h) = 0 \quad \text{for every } w_h \in V_h.$$

Consequently,

$$\alpha \|u - u_h\|_V^2 \leqslant a(u - u_h, u - u_h) = a(u - u_h, u - v_h + v_h - u_h)$$
$$= a(u - u_h, u - v_h)$$
$$\leqslant M \|u - u_h\|_V \|u - v_h\|_V \quad \text{for every } v_h \in V_h,$$

and the stated inequality is obtained with $C = M/\alpha$. $\qquad \square$

Let us now consider *examples* of subspaces of the space V that are used in practice. Given an integer $N \geqslant 1$, we set $h = 1/(N+1)$, and consider a uniform mesh of the interval $[0, 1]$ of mesh-size h, consisting of the nodes $x_i = ih$, $0 \leqslant i \leqslant N+1$. Given also an integer $m \geqslant 0$, we define the space

$$V_h^m = \{v \in \mathscr{C}^m([0, 1]): v(0) = v(1) = 0;$$
$$v|_{[x_i, x_{i+1}]} \in P_{2m+1}([x_i, x_{i+1}]), 0 \leqslant i \leqslant N\},$$

denoting in general by $P_k(I)$ the vector space formed by the restriction to an interval $I \subset \mathbb{R}$ of all polynomials of degree $\leqslant k$, and by $v|_I$ the restriction of the function v to the set I.

We recall that a polynomial of degree $\leqslant 2m+1$ is uniquely determined by its values, as well as the values of its derivatives of order $\leqslant m$, at two distinct points. The space V_h^m is in fact a subspace of the space V. Since the inclusion $V_h^m \subset \mathscr{C}^1([0, 1]) \subset V$ is satisfied for $m \geqslant 1$, it is enough to consider the case $m = 0$. Now since the space V_h^0 consists of functions which are *piecewise linear*, an example being given in figure 3.4-1, the conclusion is immediate.

Let us now look at the *convergence* of the process as the mesh-size h tends to zero. It is instructive to compare this result with that of Theorem 3.1-2; see also Exercise 3.4-2.

Theorem 3.4-3
Suppose that the function c is non-negative. For every integer $m \geqslant 0$, there exists a constant $C(m)$ independent of h such that, if the solution u of the boundary-value problem satisfies

$$u \in \mathscr{C}^{2m+2}([0, 1]),$$

then

$$\|u - u_h\|_V = \left(\int_0^1 \{|(u - u_h)'|^2 + |u - u_h|^2\} \, dx \right)^{1/2}$$

$$\leqslant C(m) \sup_{0 \leqslant x \leqslant 1} |u^{(2m+2)}(x)| h^{2m+1},$$

Figure 3.4-1

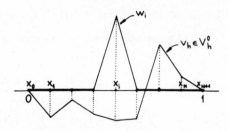

where u_h is the solution of the discrete problem associated with the subspace V_h^m.

Proof

(i) To begin with, we prove a *result on interpolation*, which is of interest in its own right: *let m be a non-negative integer; given a function $u \in \mathscr{C}^{2m+2}([a,b])$, $[a,b]$ being an interval of $\mathbb{R}$ with non-empty interior, we designate by $\Pi u: [a,b] \to \mathbb{R}$ the function defined uniquely by the relations*

$$\begin{cases} \Pi u \in P_{2m+1}([a,b]), \\ (\Pi u)^{(l)}(a) = u^{(l)}(a), \quad (\Pi u)^{(l)}(b) = u^{(l)}(b), \quad 0 \leqslant l \leqslant m. \end{cases}$$

Then there exist two constants $C_0(m)$ and $C_1(m)$, independent of the function u and the interval $[a,b]$, such that

$$\sup_{a \leqslant x \leqslant b} |(u - \Pi u)(x)| \leqslant C_0(m) \sup_{a \leqslant x \leqslant b} |u^{(2m+2)}(x)|(b-a)^{2m+2},$$

$$\sup_{a \leqslant x \leqslant b} |(u - \Pi u)'(x)| \leqslant C_1(m) \sup_{a \leqslant x \leqslant b} |u^{(2m+2)}(x)|(b-a)^{2m+1}.$$

We first remark that it is enough to prove the second inequality, since $(\Pi u(a) = u(a)$ in either case)

$$(u - \Pi u)(x) = \int_a^x (u - \Pi u)'(t)\, dt, \quad a \leqslant x \leqslant b.$$

Since the function $w = u - \Pi u$ satisfies

$$w^{(l)}(a) = w^{(l)}(b) = 0, \quad 0 \leqslant l \leqslant m,$$

successive applications of Rolle's theorem to the functions $w, w', \ldots, w^{(2m)}$ show that there exists a point η satisfying

$$a < \eta < b \quad \text{and} \quad w^{(2m+1)}(\eta) = 0.$$

Given a point $t \in\,]a,b[$, the auxiliary function

$$\psi_t: x \in [a,b] \to \psi_t(x) \overset{\text{def}}{=} w'(x) - \left(\frac{(x-a)(x-b)}{(t-a)(t-b)} \right)^m w'(t)$$

satisfies (the derivatives of the function ψ_t are in respect of the variable x)

$$\psi_t(t) = 0, \quad \psi_t^{(l)}(a) = \psi_t^{(l)}(b) = 0, \quad 0 \leqslant l \leqslant m-1.$$

By further applications of Rolle's theorem, we conclude that there exists a point ξ satisfying

$$a < \xi < b, \quad \text{and} \quad 0 = \psi_t^{(2m)}(\xi) = w^{(2m+1)}(\xi) - \frac{(2m)!}{((t-a)(t-b))^m} w'(t),$$

which allows us to express the derivative $w'(t)$ as

$$w'(t) = \frac{((t-a)(t-b))^m}{(2m)!} w^{(2m+1)}(\xi),$$

and to deduce from this the bound (extended by continuity to the points $t = a$ and b)

$$|w'(t)| \leqslant \frac{1}{(2m)!} \left(\frac{b-a}{2} \right)^{2m} \sup_{a \leqslant \xi \leqslant b} |w^{(2m+1)}(\xi)|, \quad a \leqslant t \leqslant b.$$

Since $w^{(2m+1)}(\eta) = 0$, and since $w^{(2m+2)} = u^{(2m+2)}$ (the function Πu is a polynomial of degree $\leqslant 2m+1$), it is possible to write

$$w^{(2m+1)}(\xi) = \int_{\eta}^{\xi} u^{(2m+2)}(\gamma) \mathrm{d}\gamma,$$

from which follows the bound

$$|w^{(2m+1)}(\xi)| \leqslant (b-a) \sup_{a \leqslant x \leqslant b} |u^{(2m+2)}(x)|, \quad a \leqslant \xi \leqslant b,$$

and the statement is established.

(ii) Denote by $\prod_h u : [0, 1] \to \mathbb{R}$ the function defined (uniquely) by the relations

$$\begin{cases} \prod_h u \in V_h^m, \\ (\prod_h u)^{(l)}(x_i) = u^{(l)}(x_i), & 0 \leqslant l \leqslant m, \quad 0 \leqslant i \leqslant N+1, \end{cases}$$

where u denotes the solution of the boundary-value problem. By 'putting together' the earlier inequalities and by making use of the definition of the norm of the vector space V, we obtain

$$\begin{aligned} \|u - \prod_h u\|_V &= \left(\int_0^1 (|u - \prod_h u)'|^2 + (u - \prod_h u)|^2) \mathrm{d}x \right)^{1/2} \\ &\leqslant \sup_{0 \leqslant x \leqslant 1} |(u - \prod_h u)'(x)| \\ &\quad + \sup_{0 \leqslant x \leqslant 1} |(u - \prod_h u)(x)| \\ &\leqslant (C_0(m) + C_1(m)) \sup_{0 \leqslant x \leqslant 1} |u^{(2m+2)}(x)| h^{2m+1}. \end{aligned}$$

It is then enough to make use of the inequality

$$\|u - u_h\|_V \leqslant C \|u - \prod_h u\|_V,$$

which follows from the bound (3) of Theorem 3.4-2. $\qquad\square$

To conclude, we give some indications of the *practical use* of the method. When $m = 0$, we choose as a basis of the space V_h^0 the set of functions

w_i, $1 \leqslant i \leqslant N$, defined uniquely by the relations

$$w_i \in V_h^0, \quad w_i(x_j) = \delta_{ij}, \quad 1 \leqslant j \leqslant N.$$

The search for the approximate solution

$$u_h = \sum_{j=1}^{M} u_j w_j$$

of the discrete problem is equivalent to the solution of the *linear system*

$$\sum_{j=1}^{M} a(w_j, w_i) u_j = f(w_i), \quad 1 \leqslant i \leqslant M,$$

where

$$a(w_j, w_i) = \int_0^1 \{w_j' w_i' + c w_j w_i\} \, dx, \quad f(w_i) = \int_0^1 f w_i \, dx.$$

The matrix $(a(w_j, w_i))$ is *tridiagonal*, the support of each basis function w_i being the union of the two intervals $[x_{i-1}, x_i]$ and $[x_i, x_{i+1}]$, *symmetric* (since $a(u, v) = a(v, u)$ for every $u, v \in V$) and *positive definite*, since

$$\sum_{i,j=1}^{M} u_i a(w_j, w_i) u_j = a\left(\sum_{i=1}^{M} u_i w_i, \sum_{j=1}^{M} u_j w_j \right) \geqslant \alpha \left\| \sum_{i=1}^{M} u_i w_i \right\|_V^2.$$

It should be noted that, while the matrix of the linear system is symmetric and positive definite for every other choice of basis, and even for every other choice of finite-dimensional subspace V_h, its *tridiagonal character is linked essentially to the choice of basis*.

By way of an example, we write down the linear system which corresponds to the case where the functions c and f are constant:

$$\frac{1}{h} \begin{pmatrix} 2 + \frac{2c}{3} h^2 & -1 + \frac{c}{6} h^2 & & & \\ -1 + \frac{c}{6} h^2 & 2 + \frac{2c}{3} h^2 & -1 + \frac{c}{6} h^2 & & \\ & \ddots & \ddots & \ddots & \\ & & -1 + \frac{c}{6} h^2 & 2 + \frac{2c}{3} h^2 & -1 + \frac{c}{6} h^2 \\ & & & -1 + \frac{c}{6} h^2 & 2 + \frac{2c}{3} h^2 \end{pmatrix} \begin{pmatrix} u_1 \\ u_2 \\ \vdots \\ u_{N-1} \\ u_N \end{pmatrix} = h \begin{pmatrix} f \\ f \\ \vdots \\ f \\ f \end{pmatrix}.$$

The analogy with the linear system obtained through the use of the finite-difference method should not escape notice (section 3.1); an analogy all the more 'troubling' because the unknowns u_i represent precisely the values of the approximate solution at the nodes x_i. The functions v_h of the space V_h^0 in fact satisfy the identity

$$v_h = \sum_{i=1}^{M} v_h(x_i) w_i.$$

When $m = 1$, we choose as a basis of the space V_h^1 the set of functions w_i^0, $1 \leqslant i \leqslant N$, and w_i^1, $0 \leqslant i \leqslant N + 1$, defined uniquely by the relations

$$w_i^0 \in V_h^1, \quad w_i^0(x_j) = \delta_{ij}, \quad 1 \leqslant j \leqslant N, \quad (w_i^0)'(x_j) = 0, \quad 0 \leqslant j \leqslant N + 1,$$
$$w_i^1 \in V_h^1, \quad w_i^1(x_j) = 0, \quad 1 \leqslant j \leqslant N, \quad (w_i^1)'(x_j) = \delta_{ij}, \quad 0 \leqslant j \leqslant N + 1.$$

Here again, the components of the expression of a function of V_h^1 in terms of the basis have a remarkable significance, since the identity

$$v_h = \sum_{i=1}^{N} v_h(x_i) w_i^0 + \sum_{i=0}^{N+1} v_h'(x_i) w_i^1$$

is valid for every function $v_h \in V_h^1$.

Since the basis functions have 'small' support (cf. figure 3.4-2), one would expect a *band-matrix structure*, provided, as ever, that the basis functions are ordered in some 'natural' order, here

$$w_0^1, w_1^0, w_1^1, w_2^0, w_2^1, \ldots, w_N^0, w_N^1, w_{N+1}^1.$$

And, indeed, the matrix corresponding to the case $c = 0$ (the bilinear form reducing to $a(u, v) = \int_0^1 u'v' \, dx$) has the form

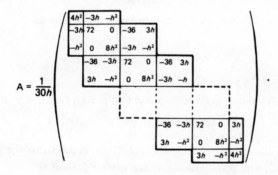

$$A = \frac{1}{30h} \begin{pmatrix} \end{pmatrix}$$

Figure 3.4-2

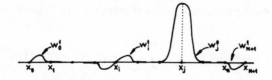

This, then, is another example of a *block tridiagonal matrix*, a characteristic which, together with the properties of symmetry and positive definiteness, facilitates the solution of the associated linear systems, as has already been observed.

Had the basis functions been arranged in the order (which, after all, is just as 'natural' as the preceding)

$$w_1^0, w_2^0, \ldots, w_N^0, w_0^1, w_1^1, w_2^1, \ldots, w_N^1, w_{N+1}^1,$$

one would have obtained the matrix

$$A = \frac{1}{30h} \begin{pmatrix} \begin{array}{ccccc|ccccc} 72 & -36 & & & & -3h & 0 & 3h & & \\ -36 & 72 & -36 & & & & -3h & 0 & 3h & \\ & \ddots & \ddots & \ddots & & & & \ddots & \ddots & \ddots \\ & & -36 & 72 & & & & & -3h & 0 & 3h \\ \hline -3h & & & & 4h^2 & -h^2 & & & \\ 0 & -3h & & & -h^2 & 8h^2 & -h^2 & & \\ 3h & 0 & & & & \ddots & \ddots & \ddots & \\ & 3h & \ddots & -3h & & & & -h^2 & 8h^2 & -h^2 \\ & & \ddots & 0 & & & & & -h^2 & 4h^2 \\ & & & 3h & & & & & \end{array} \end{pmatrix} \begin{matrix} \left. \begin{matrix} \\ \\ \\ \\ \end{matrix} \right\} N \text{ rows} \\ \\ \left. \begin{matrix} \\ \\ \\ \\ \\ \end{matrix} \right\} N + 2 \text{ rows} \end{matrix}$$

which has lost completely the block tridiagonal structure!

Exercises

3.4-1. For what values of the real constant γ is the function

$$v \to \left(\int_0^1 (|v'|^2 + \gamma |v|^2) \, dx \right)^{1/2}$$

a norm over the vector space V?

3.4-2. (1) Prove that there exists a constant C such that

$$\sup_{0 \leqslant x \leqslant 1} |v(x)| \leqslant C \|v\| \quad \text{for every } v \in V.$$

(2) Deduce from this inequality and Theorem 3.4-3 that the approximate solution u_h, with corresponding vector space V_h^0, satisfies

$$\sup_{0 \leqslant x \leqslant 1} |(u - u_h)(x)| = O(h).$$

Show, by the application of a more 'detailed' analysis, that, in fact,

$$\sup_{0 \leqslant x \leqslant 1} |(u - u_h)(x)| = O(h^2).$$

3.4-3. Employing the same notation as in the proof of Theorem 3.4-3, prove the bounds

$$\sup_{0 \leqslant x \leqslant 1} |(u - \Pi u)^{(k)}(x)| \leqslant \frac{h^{2m+2-k}}{(2m-2k)! \, k! 2^{2m-2k}} \sup_{0 \leqslant x \leqslant 1} |u^{(2m+2)}(x)|, \quad 0 \leqslant k \leqslant m.$$

3.4-4. Assuming the same conditions as in Theorem 3.1-2, let $\tilde{u}_h$ denote the function in the space V_h^0 which satisfies $\tilde{u}_h(x_i) = u_i$, $1 \leqslant i \leqslant N$. Establish a bound, as a function of a suitable power of h, for the expression $\sup_{0 \leqslant x \leqslant 1} |(u - \tilde{u}_h)(x)|$. Compare this with the result of Exercise 3.4-2.

3.4-5. Apply the method of variational approximation to the problem

$$\begin{cases} -u''(x) + xu(x) = (-x^2 + 2x + 1)e^x + x^2 - x, & 0 < x < 1, \\ u(0) = u(1) = 0. \end{cases}$$

With regard to the spaces V_h^0 and V_h^1 and the values $h = 1/2^m$, $1 \leqslant m \leqslant 5$, calculate the numbers

$$\max_{1 \leqslant i \leqslant N} |u(x_i) - u_h(x_i)| \quad \text{and} \quad \sup_{0 \leqslant x \leqslant 1} |u(x) - u_h(x)|.$$

Compare these with the numerical results given in section 3.1 (in Chapters 4 and 5 may be found whatever guidance is needed as regards the choice of methods for the actual solution of the linear systems obtained).

3.4-6. It was seen in the text that the variational formulation of the boundary-value problem

$$\begin{cases} -u''(x) + c(x)u(x) = f(x), & c \geqslant 0, \\ u(0) = u(1) = 0 \end{cases}$$

consists in finding a function u such that

$$u \in V \quad \text{and} \quad J(u) = \inf_{v \in V} J(v),$$

where

$$J(v) = \frac{1}{2} \int_0^1 (|v'(x)|^2 + c(x)|v(x)|^2) \, dx$$

$$- \int_0^1 f(x)v(x) \, dx$$

and V is the space of functions which are piecewise continuously differentiable over the interval $[0, 1]$, taking the value 0 at 0 and 1. This formulation suggests a procedure of approximation, which is *different* from the variational approximation and which simply consists in replacing the integrals which appear in the functional J by the Riemann sum associated with the uniform mesh of mesh-size h of the interval $[0, 1]$, each derivative $v'(x_i)$ being, furthermore, approximated by the 'finite-difference' quotient $(v_{i+1} - v_i)/h$ (an exact resumption is made of the notation of section 3.1). We are thus led to the definition of the *approximate functional*

$$J_h : v = (v_i) \in \mathbb{R}^N \to J_h(v) = \frac{h}{2} \sum_{i=1}^N \left(\left[\frac{v_{i+1} - v_i}{h} \right]^2 + c_i v_i^2 \right) - h \sum_{i=1}^N f_i v_i,$$

with the convention, quite naturally, that $v_{N+1} = 0$ in the expression given above.

(1) Prove that there exists a unique vector $u_h \in \mathbb{R}^N$ such that

$$J_h(u_h) = \inf_{v \in \mathbb{R}^N} J_h(v).$$

(2) Prove that this vector u_h coincides with the solution of the matrix equation $A_h u_h = b_h$ obtained in section 3.1 through the application of the finite-difference method.

3.5 Variational approximation of a two-dimensional boundary-value problem

We shall now describe the application of this method to the problem of a membrane considered in section 3.2, following a procedure similar to that of the previous section, but omitting any proofs (the interested reader will find these by consulting the references indicated in the Bibliography and comments). It is recalled that this problem consists in finding a solution $u: \bar{\Omega} \to \mathbb{R}$ of the boundary-value problem

$$\begin{cases} -\Delta u(x) = f(x), & x \in \Omega, \\ u(x) = 0, & x \in \Gamma, \end{cases}$$

assuming, for the sake of simplicity, homogeneous boundary conditions. It can then be shown (effectively after an integration by parts, suitably adapted; compare with Theorem 3.4-1) that the solution of this problem satisfies

$$a(u, v) = f(v) \quad \text{for every } v \in V,$$

where

$$a(u, v) \overset{\text{def}}{=} \int_\Omega \left\{ \frac{\partial u}{\partial x_1} \frac{\partial v}{\partial x_1} + \frac{\partial u}{\partial x_2} \frac{\partial v}{\partial x_2} \right\} dx, \quad f(v) \overset{\text{def}}{=} \int_\Omega fv \, dx,$$

and V is a suitable space of 'functions' which are defined over the set Ω and zero (in a particular sense) on its boundary Γ (the space in question is the Sobolev space $H_0^1(\Omega)$). Equivalently, the solution satisfies

$$J(u) = \inf_{v \in V} J(v), \quad \text{where } J(v) = \tfrac{1}{2} a(v, v) - f(v).$$

Remark

From the point of view of mechanics, the function J is nothing other than the *energy* of the physical system comprising the membrane, while the *variational equations* '$a(u, v) = f(v)$ for every $v \in V$' are a translation of the *principle of virtual work*.   □

This *variational formulation* of the boundary-value problem being considered makes possible the definition of the *method of variational approximation*, also known as the *Galerkin method*, or the *Ritz method*. Given a finite-dimensional subspace V_h of the space V, the *discrete solution* is the (unique) solution of the following *discrete problem*: find $u_h \in V_h$ such that

$$a(u_h, v_h) = f(v_h) \quad \text{for every } v_h \in V_h,$$

or equivalently, such that

$$J(u_h) = \inf_{v_h \in V_h} J(v_h).$$

We show that, once again, this comes down to the solution of a linear system. Let $(w_i)_{i=1}^M$ be a basis of the vector space V_h. Writing

$$u_h = \sum_{i=1}^M u_i w_i,$$

the vector $u = (u_i)_{i=1}^M$ turns out to be the solution of the *linear system*

$$Au = b, \quad \text{where } A = (a(w_j, w_i)), \quad b = (f(w_i)).$$

In this way, a matrix is obtained which is *symmetric* and *positive definite*.

A very important particular case of the method of variational approximation is the *finite element method*, which we shall describe briefly for a *very simple example*. Suppose that the boundary Γ is a *polygon*, thus making possible a *triangulation* of the set $\bar{\Omega}$, as is indicated in figure 3.5-1.

The 'simplest' subspace V_h associated with such a triangulation consists of *piecewise affine functions*, i.e., functions that are affine over each triangle, continuous over $\bar{\Omega}$ and zero on Γ. Numbering the points as in figure 3.5-1, a basis of the space V_h is established by associating with every vertex i

Figure 3.5-1

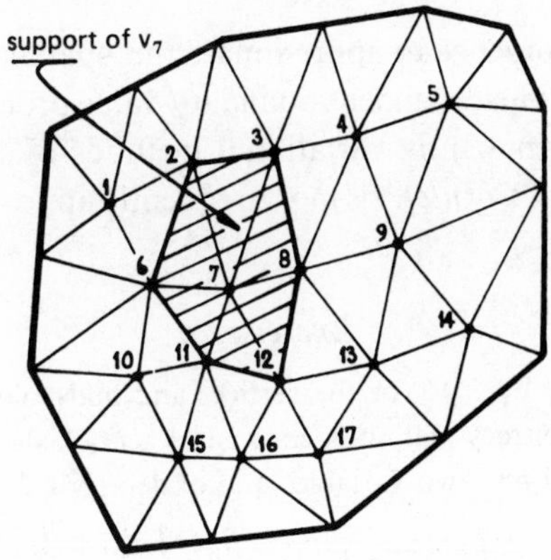

of the triangulation lying in the open set Ω that function of V_h which takes the value 1 at the vertex i and 0 at the other vertices. With this numbering, the corresponding matrix $A = (a(w_j, w_i))$ has the appearance shown below; the dots represent the non-zero elements.

In fact, it can be verified that the non-zero elements $(A)_{ij}$ correspond either to the case $i = j$, or to the case where the vertices i and j belong to the same triangle, since these are the only two cases where the intersection of the supports of the associated basis functions has non-zero measure. This, again, is an example of a *block tridiagonal* matrix.

This very simple example brings out the three essential characteristics of finite-element methods:

(i) the existence of a *triangulation*;
(ii) the construction of a space V_h whose functions are '*piecewise polynomials*' (cf. Exercise 3.5-1 for another example);
(iii) the existence of a 'canonical' basis of the space V_h whose elements have '*small*' *supports*. As a result of this property, the matrix of the associated linear system is a *band matrix*.

Indications regarding the convergence of the method are to be found in Exercise 3.5-2.

Remark

While it is current practice to approximate the 'spatial partial derivatives' $(\partial^2 u/\partial x^2, \Delta u, \ldots)$ of time-dependent boundary-value problems (such as those considered in section 3.3) by variational methods of approximation, the 'time derivative' $(\partial u/\partial t, \partial^2 u/\partial t^2)$ is most frequently approximated by a finite-difference method. □

Exercises

3.5-1. (1) Let $a_i, b_i, 1 \leq i \leq 3$, be the vertices and mid-points, respectively, of a triangle. Given arbitrary real numbers $\alpha_i, \beta_i, 1 \leq i \leq 3$, show that there exists a unique polynomial p in two variables and of degree ≤ 2 such that

$$p(a_i) = \alpha_i, \quad p(b_i) = \beta_i, \quad 1 \leq i \leq 3.$$

(2) Deduce from this the definition of a space V_h associated with a triangulation similar to that of figure 3.5-1, the functions of this space being continuous over the set $\bar{\Omega}$, zero on the boundary Γ and their restriction to each triangle of the triangulation being a polynomial of degree $\leqslant 2$ in two variables. Describe, in particular, the 'canonical' basis of this space.

(3) Does there exist a numbering of the vertices and mid-points of the triangles for which the matrix of the associated linear system is block tridiagonal?

3.5-2. (1) Let T be a triangle with vertices a_i, $1 \leqslant i \leqslant 3$. Given a function $u: T \to \mathbb{R}$ which is twice continuously differentiable over the triangle T, show that there exists a unique function $\Pi u: T \to \mathbb{R}$ which satisfies

$$\begin{cases} \Pi u \text{ is affine } (\Pi u(x) = c_0 + c_1 x_1 + c_2 x_2 \text{ for } x \in T), \\ \Pi u(a_i) = u(a_i), \quad 1 \leqslant i \leqslant 3. \end{cases}$$

(2) Prove that there exists a constant C_0, independent of the function u, such that

$$\sup_{x \in T} |(u - \Pi u)(x)| \leqslant C_0 \sup_{\substack{x \in T \\ \alpha, \beta = 1, 2}} \left| \frac{\partial^2 u}{\partial x_\alpha \partial x_\beta}(x) \right| h_T^2,$$

where h_T is the size of the largest side of the triangle T.

(3) Prove that there exists a constant C_1, independent of the function u, such that

$$\sup_{x \in T} \left| \frac{\partial}{\partial x_\alpha}(u - \Pi u)(x) \right| \leqslant C_1 \sup_{\substack{x \in T \\ \alpha, \beta = 1, 2}} \left| \frac{\partial^2 u}{\partial x_\alpha \partial x_\beta}(x) \right| \frac{h_T^2}{\rho_T}, \quad \alpha = 1, 2,$$

where ρ_T is the diameter of the inscribed circle of T.

Hint. For these first three questions, it is a considerable help to make use of the barycentric co-ordinates $\lambda_1, \lambda_2, \lambda_3$ of a point with co-ordinates x_1 and x_2, which are defined as the solution of the linear system

$$\begin{pmatrix} a_{11} & a_{12} & a_{13} \\ a_{21} & a_{22} & a_{23} \\ 1 & 1 & 1 \end{pmatrix} \begin{pmatrix} \lambda_1 \\ \lambda_2 \\ \lambda_3 \end{pmatrix} = \begin{pmatrix} x_1 \\ x_2 \\ 1 \end{pmatrix},$$

a_{1i} and a_{2i} denoting the co-ordinates of the vertex a_i.

(4) Given the fact that the bound (3) of Theorem 3.4-2 continues to be valid for the norm

$$\|v\|_V \stackrel{\text{def}}{=} \left[\int_\Omega \left(\left| \frac{\partial v}{\partial x_1} \right|^2 + \left| \frac{\partial v}{\partial x_2} \right|^2 + |v|^2 \right) dx \right]^{1/2},$$

prove that there exists a constant C, independent of the subspace V_h and the solution u, assumed to be twice continuously differentiable over the set $\bar{\Omega}$, such that

$$\|u - u_h\|_V \leqslant C \sup_{\substack{x \in \Omega \\ \alpha, \beta = 1, 2}} \left| \frac{\partial^2 u}{\partial x_\alpha \partial x_\beta}(x) \right| \max_{T \in \mathcal{T}_h} \frac{h_T^2}{\rho_T}.$$

where $\mathcal{T}_h$ denotes the set of triangles of each triangulation. One cannot fail to notice the similarity between the results given above and those of Theorem 3.4-3.

3.6 Eigenvalue problems

Returning to the problem of the small displacements of a string (considered in section 3.3), it is possible, *in the absence of any force* $(f = 0)$, to look for *stationary solutions*, or waves, that is to say, types of motion in which the solution $U(x, t)$ is the product of a function of x by a function of t:

$$U(x, t) = u(x)v(t).$$

Of course, solutions identically zero are discarded. The one-dimensional wave equation then becomes

$$u''(x)v(t) = \frac{1}{c^2} u(x)v''(t),$$

which leads to two differential equations for the functions u and v:

$$-u''(x) = \lambda u(x) \quad \text{and} \quad -v''(t) = \lambda c^2 v(t),$$

λ being a constant *to be determined*. Since

$$U(0, t) = U(l, t) = 0 \quad \text{for all } t$$

(the ends of the string being fixed), we are led to *look for real numbers λ and functions u not identically zero* such that

$$\begin{cases} -u''(x) = \lambda u(x), & 0 \leqslant x \leqslant l, \\ u(0) = u(1) = 0, \end{cases}$$

which constitutes an *eigenvalue problem for the second derivative operator*. Here it is easy to see that the problem given above has as its only solutions

$$\lambda_k = \frac{k^2 \pi^2}{l^2}, \quad \varphi_k(x) = C_k \sin \frac{k\pi x}{l}, \quad k = 1, 2, \ldots.$$

(C_k being an arbitrary, non-zero constant), so that the corresponding solutions for the function v are

$$v_k(t) = C_{1k} \cos \frac{k\pi ct}{l} + C_{2k} \sin \frac{k\pi ct}{l}$$

(C_{1k} and C_{2k} being arbitrary constants). In this way, we obtain the solutions

$$U_k(x, t) = \sin \frac{k\pi x}{l} \left(C_{1k} \cos \frac{k\pi ct}{l} + C_{2k} \sin \frac{k\pi ct}{l} \right), \quad k = 1, 2, \ldots.$$

The same solutions are arrived at by seeking *a priori* solutions which are of the form

$$U(x, t) = u(x)e^{i\mu t}, \quad \mu \text{ a constant to be determined.}$$

It is then found that the only possible values of μ are

$$\mu_k = \frac{k^2\pi^2 c}{l^2} = \lambda_k c, \quad k = 1, 2, \ldots$$

We observe that the numbers $2\pi/\mu_k$ are nothing other than the *periods* of the stationary solutions.

It is likewise possible to be interested in the same problem for a membrane, by looking for solutions which are of the type

$$U(x, t) = u(x)v(t),$$

again assuming $f = 0$. The same analysis leads to a *search for real numbers λ and functions u not identically zero* such that

$$\begin{cases} -\Delta u(x) = \lambda u(x), & x \in \Omega, \\ u(x) = 0, & x \in \Gamma. \end{cases}$$

This is an *eigenvalue problem for the operator* $-\Delta$. Every solution λ is called an *eigenvalue* and the corresponding function u, not identically zero, is called an *eigenfunction*.

Such problems can be approximated as well by finite-difference methods as by variational methods of approximation. For example, the application of the former method to the problem of the stationary solutions for a string of unit length leads to the search for vectors with components u_i, $1 \leqslant i \leqslant N$ (the approximations of the values of the eigenfunctions at the nodes of the mesh), and of numbers λ_h (the approximations of the eigenvalues) such that

$$\frac{1}{h^2} \begin{pmatrix} 2 & -1 & & & \\ -1 & 2 & -1 & & \\ & \ddots & \ddots & \ddots & \\ & & -1 & 2 & -1 \\ & & & -1 & 2 \end{pmatrix} \begin{pmatrix} u_1 \\ u_2 \\ \vdots \\ u_{N-1} \\ u_N \end{pmatrix} = \lambda_h \begin{pmatrix} u_1 \\ u_2 \\ \vdots \\ u_{N-1} \\ u_N \end{pmatrix}.$$

It is easy to verify that the eigenvalues of this matrix are the numbers

$$\lambda_{kh} = \frac{4}{h^2} \sin^2 \frac{k\pi}{2(N+1)}, \quad 1 \leqslant k \leqslant N,$$

associated with the eigenvectors (C_k being non-zero arbitrary constants)

$$u_k = (C_k u_i^k)_{i=1}^N, \quad \text{where } u_i^k = \sin \frac{k\pi i}{N+1}, \quad 1 \leqslant i \leqslant N.$$

It is then possible to obtain the bound (for $N \geqslant k$)

$$0 < \lambda_k - \lambda_{kh} = k^2\pi^2 - \frac{4}{h^2}\sin^2\frac{k\pi}{2(N+1)} \leqslant \Gamma_k h^2,$$

where the constant Γ_k depends on the integer k but not on the mesh-size h, as well as the equations

$$\varphi_k(x_i) - u_i^k = 0, \quad 1 \leqslant i \leqslant N,$$

which establishes the convergence of the process, *for fixed k*, in this particular case (evidently, the equalities found here are altogether exceptional!).

Similarly, the membrane problem leads to a search for the solutions (u_h, λ_h) of the equation

$$A_h u_h = \lambda_h u_h,$$

the matrix A_h being of the same type as that in section 3.2. In each of the two cases, we have to do with an *eigenvalue problem*, whose matrix is real, symmetric and positive definite in the first case, but not necessarily symmetric in the second case. Obviously, the convergence of the process is more difficult to establish here than in the first example!

The *method of variational approximation* applied to such problems bears the name of the *Rayleigh–Ritz* method. It gives rise, once again, to eigenvalue problems, but, generally, through the mediation of *two* matrices. In order to fix ideas, let us consider one or other of the boundary-value problems treated in this section. With arguments such as those employed in sections 3.4 and 3.5 (and with the same notation), it can be shown that the *variational formulation* of such a problem consists in finding real numbers λ and functions $u \in V$ not identically zero such that

$$a(u, v) = \lambda(u, v) \quad \text{for every } v \in V,$$

where

$$a(u, v) = \int_0^1 u'v' \, \mathrm{d}x, \quad (u, v) = \int_0^1 uv \, \mathrm{d}x,$$

for the first problem, and

$$a(u, v) = \int_\Omega \left(\frac{\partial u}{\partial x_1} \frac{\partial v}{\partial x_1} + \frac{\partial u}{\partial x_2} \frac{\partial v}{\partial x_2} \right) \mathrm{d}x, \quad (u, v) = \int_\Omega uv \, \mathrm{d}x,$$

for the second.

Given a finite-dimensional subspace V_h of the space V, the associated *discrete problem* consists in finding numbers λ and functions $u_h \in V_h$ which satisfy

$$a(u_h, v_h) = \lambda(u_h, v_h) \quad \text{for every } v_h \in V_h.$$

Given a basis $(w_i)_{i=1}^M$ of the space V_h, we set

$$u_h = \sum_{i=1}^M u_i w_i.$$

The vector $u = (u_i)_{i=1}^M$ has to be the solution of the matrix equation

$$Au = \lambda Bu, \quad \text{where } A = (a(w_j, w_i)) \quad \text{and} \quad B = (w_j, w_i).$$

This is a *generalised eigenvalue problem*, since the matrix B is not diagonal in general. For example, the discrete problem corresponding to the first problem (in one dimension) and to the subspace V_h^0 introduced in section 3.4 can be written as

$$\frac{1}{h} \begin{pmatrix} 2 & -1 & & & \\ -1 & 2 & -1 & & \\ & \ddots & \ddots & \ddots & \\ & & -1 & 2 & -1 \\ & & & -1 & 2 \end{pmatrix} \begin{pmatrix} u_1 \\ u_2 \\ \vdots \\ u_{N-1} \\ u_N \end{pmatrix} = \lambda \frac{h}{6} \begin{pmatrix} 4 & 1 & & & \\ 1 & 4 & 1 & & \\ & \ddots & \ddots & \ddots & \\ & & 1 & 4 & 1 \\ & & & 1 & 4 \end{pmatrix} \begin{pmatrix} u_1 \\ u_2 \\ \vdots \\ u_{N-1} \\ u_N \end{pmatrix}.$$

It is easy to see that the symmetric matrix B is always positive definite, since the functions w_i are linearly independent. It would thus seem desirable to replace this problem by the equivalent eigenvalue problem of the familiar kind

$$B^{-1}Au = \lambda u.$$

But, in so doing, *symmetry is destroyed* (the matrix $B^{-1}A$ is in general no longer symmetric). Now it is advantageous to maintain the symmetry of the matrices under consideration, whenever this is possible, as was indicated in the previous chapter; and that is why the first step is to carry out a *Cholesky factorisation* of the matrix B. As will be shown later (Theorem 4.4-1), it is possible to write every symmetric, positive definite matrix B as the product

$$B = CC^T.$$

where the matrix C is lower triangular. Writing the problem as

$$C^{-1}A(C^T)^{-1}(C^Tu) = \lambda C^Tu,$$

we are now led to the usual eigenvalue problem for the *symmetric* matrix $C^{-1}A(C^T)^{-1}$, the required vectors u being subsequently obtained from the eigenvectors of the matrix $C^{-1}A(C^T)^{-1}$ by solving linear systems with triangular matrices.

Finally, we consider the calculation of the periods of 'small' displacements, in the neighbourhood of a state of equilibrium, of a mechanical system having a finite number n of degrees of freedom. The differential equation of such a system takes the form

$$\mathbf{A}x''(t) + \mathbf{B}x'(t) + \mathbf{C}x(t) = 0,$$

where $x(t)$ is a vector function of the time t, whose components are the degrees of freedom $x_i(t)$, $1 \leqslant i \leqslant n$, of the system, and A, B and C are real matrices of order n. The matrix A is the matrix of 'kinetic energy' and, on this account, is symmetric and positive definite. The matrix C corresponds to forces which are functions of position. Lastly, the matrix B is the matrix of 'damping forces'; it corresponds to forces which are functions of velocity.

Solutions of the differential equation given above are sought which have the form

$$x(t) = e^{\mu t}u,$$

where u is a non-zero vector of $\mathbb{R}^n$ which is time-independent and μ is a constant, real or complex. We are thus led to *look for scalars μ and corresponding non-zero vectors u satisfying*

$$(\mu^2 \mathbf{A} + \mu \mathbf{B} + \mathbf{C})u = 0.$$

With each solution μ of the problem is associated a solution with period $T = 2\pi (\mathrm{Im}\,\mu)^{-1}$, where $\mathrm{Im}(\mu)$ denotes the imaginary part of the complex number μ. Yet again, we have another example of a *generalised eigenvalue problem*.

If a scalar μ is to be a solution, it is necessary and sufficient that the matrix $\mu^2 \mathbf{A} + \mu \mathbf{B} + \mathbf{C}$ be singular, that is

$$\det(\mu^2 \mathbf{A} + \mu \mathbf{B} + \mathbf{C}) \equiv \mu^{2n} \det(\mathbf{A}) + \sum_{k=0}^{2n-1} \alpha_k \mu^k = 0.$$

It thus appears that the problem can have at most $2n$ solutions (whether distinct or not) and exactly $2n$ solutions if and only if the matrix A is invertible, which is in fact the case for the problem we are considering. But while this observation allows us to foresee the possibility of the existence of $2n$ 'generalised eigenvalues', it gives no guidance as to the *practical* solution of the problem. As might be expected, the line to take is to try to obtain the usual eigenvalue problem with a matrix of order $2n$, to be determined. On this topic, one can follow the indications given in Exercise 3.6-1.

Exercise

3.6-1. (1) Let D and E be two matrices of order n. Establish the identity

$$\det(\mu^2 \mathbf{I} + \mu \mathbf{D} + \mathbf{E}) = \det(\mu \mathbf{I} - \mathscr{A})$$

for every scalar μ, where the matrix $\mathscr{A}$ of order $2n$ is given by

$$\mathscr{A} = \left(\left[\begin{array}{c|c} 0 & E \\ \hline -I & -D \end{array} \right] \right).$$

(2) Deduce a correspondence between the solutions (μ, u), $u \neq 0$, of the generalised eigenvalue problem $(\mu^2 I + \mu D + E)u = 0$ and the solutions (λ, p), $p \neq 0$, of the eigenvalue problem $\mathscr{A}p = \lambda p$.

3.7 Interpolation and approximation problems

Let x_i be a finite number of distinct points of $\mathbb{R}$. To each of the points x_i there corresponds a number $c_i \in \mathbb{R}$, which may be (by way of example) either an experimental value or the value $u(x_j)$ of a known function. It is then possible to set one of the following problems.

(i) Find a curve of given type which passes through the points (x_i, c_i). That provides an *interpolation* problem (cf. figure 3.7-1(a)), the function represented by the curve being called the *interpolating function*.

(ii) Find a curve of given type which, *in a sense yet to be made precise*, approximates the values c_i at the points x_i; this provides an *approximation* problem (cf. figure 3.7-1(b)). In each case, 'a curve of given type' is understood to imply a curve that is 'very easy to calculate'; *in practice*, this means a polynomial or piecewise polynomial function (in the sense understood below).

To solve the first problem, the most 'immediate' idea is to find a polynomial of degree $\leqslant m$ which passes through the points x_i, $m + 1$ in number. There then exists a unique $p_m u$, called the *interpolating polynomial*. This procedure, however, is generally to be discouraged, since it introduces undesirable oscillations, which are due to analytical properties of poly-

Figure 3.7-1

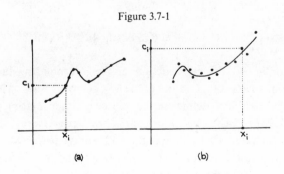

(a)　　　　　　　　(b)

nomials. It is possible, in actual fact, to construct examples of functions u (which are actually very smooth) over an interval $[a, b]$ for which

$$\lim_{m \to \infty} \sup_{a \leqslant x \leqslant b} |(u - p_m u)(x)| = + \infty,$$

even though the number of points x_i in the interval increases indefinitely (cf. Exercise 3.7-4)!

That is why it is preferred to solve interpolation problems with the use of *piecewise polynomial interpolation*, that is to say, with functions whose restrictions to each interval $[x_i, x_{i+1}]$ are polynomials which 'join together' appropriately at the points x_i. In order to fix ideas, let us suppose that the points x_i are the nodes of a uniform mesh, of mesh-size h, of the interval $[0, 1]$:

$$x_i = ih, \quad 0 \leqslant i \leqslant N + 1, \quad \text{with } h = \frac{1}{N+1}.$$

This hypothesis in no way restricts generality; it aims only to simplify the presentation.

Given an integer $m \geqslant 0$, it is possible to define uniquely an interpolating function $\Pi_h u$ by the conditions

$$\begin{cases} \Pi_h u \in \mathscr{C}^m([0, 1]), \\ \Pi_h u|_{[x_i, x_{i+1}]} \in P_{2m+1}([x_i, x_{i+1}]), \quad 0 \leqslant i \leqslant N, \\ (\Pi_h u)^{(l)}(x_i) = u^{(l)}(x_i), \quad 0 \leqslant l \leqslant m, \quad 0 \leqslant i \leqslant N + 1. \end{cases}$$

We can recognise here a type of function already used in the variational approximation of two-point boundary-value problems (section 3.4). Their construction, in terms of known 'canonical' bases, of which we have given examples for $m = 0$ and 1, is immediate. Moreover, it follows from the proof of Theorem 3.4-3 that the process is *convergent*, as h tends to zero (at least, for smooth functions), since we established there the bounds

$$\sup_{0 \leqslant x \leqslant 1} |(u - \Pi_h u)(x)| \leqslant C_0(m) \sup_{0 \leqslant x \leqslant 1} |u^{2m+2}(x)| h^{2m+2},$$

$$\sup_{0 \leqslant x \leqslant 1} |(u - \Pi_h u)'(x)| \leqslant C_1(m) \sup_{0 \leqslant x \leqslant 1} |u^{2m+2}(x)| h^{2m+1}.$$

While, for $m = 1$, the construction of the interpolating function $\Pi_h u$ requires a knowledge of the values of the first derivatives $u'(x_i)$, it is remarkable that it is possible to define another piecewise interpolating polynomial of degree 3 *without using the values of the first derivatives* $u'(x_i)$ (except at the ends of the interval). In so doing, an advantage is gained, at the same time, over the smoothness of the interpolating function, which can, in fact, be made twice continuously differentiable over the interval $[0, 1]$, while the function $\Pi_h u$ for $m = 1$ is only once continuously differentiable. More precisely, we establish the following result on existence.

Theorem 3.7-1
There exists a unique function $s_h u$: $[0, 1] \to \mathbb{R}$ with the following properties

$$\begin{cases} s_h u \in \mathscr{C}^2([0, 1]), \\ s_h u|_{[x_i, x_{i+1}]} \in P_3[x_i, x_{i+1}], \quad 0 \leqslant i \leqslant N, \\ s_h u(x_i) = u(x_i), \quad 0 \leqslant i \leqslant N + 1, \\ (s_h u)'(0) = u'(0), \quad (s_h u)'(1) = u'(1). \end{cases}$$

Proof
Let $p_i, 1 \leqslant i \leqslant N$, be arbitrary real parameters. There exists a unique function $w : [0, 1] \to \mathbb{R}$ with the following properties (such a function depends both on the given function u and on the parameters p_i):

$$\begin{cases} w \in \mathscr{C}^1([0, 1]). \\ w|_{[x_i, x_{i+1}]} \in P_3([x_i, x_{i+1}]), \quad 0 \leqslant i \leqslant N, \\ w(x_i) = u(x_i), \quad 0 \leqslant i \leqslant N + 1, \\ w'(0) = u'(0), \quad w'(1) = u'(1), \quad w'(x_i) = p_i, \quad 1 \leqslant i \leqslant N. \end{cases}$$

To prove the theorem, we shall establish that there exists a unique choice of the parameters p_i for which the function w is twice continuously differentiable over the interval $[0, 1]$. For this, it is enough to write

$$\lim_{x \to x_i^-} w''(x) = \lim_{x \to x_i^+} w''(x), \quad 1 \leqslant i \leqslant N.$$

This requires an expression for the second derivatives of a polynomial of degree $\leqslant 3$ at the ends of an interval of length h, the polynomial itself being determined by its values, as well as those of its first derivative, at the ends of this same interval. Now a simple calculation shows that every polynomial r of degree $\leqslant 3$ over the interval $[x_i, x_{i+1}]$ may be written as

$$r(x) = r(x_i)\left(\frac{2\xi^3 - 3h\xi^2 + h^3}{h^3}\right) + r(x_{i+1})\left(\frac{-2\xi^3 + 3h\xi^2}{h^3}\right)$$
$$+ r'(x_i)\left(\frac{\xi^3 - 2h\xi^2 + h^2\xi}{h^2}\right) + r'(x_{i+1})\left(\frac{\xi^3 - h\xi^2}{h^2}\right),$$
$$\xi = x - x_i,$$

from which follow easily the relations

$$r''(x_i) = \frac{6}{h^2}\{r(x_{i+1}) - r(x_i)\} - \frac{2}{h}\{r'(x_{i+1}) + 2r'(x_i)\},$$
$$r''(x_{i+1}) = \frac{6}{h^2}\{r(x_i) - r(x_{i+1})\} + \frac{2}{h}\{r'(x_i) + 2r'(x_{i+1})\}.$$

Accordingly, the continuity of the second derivative of the function w at the nodes translates itself to the equalities

$$4p_1 + p_2 = \frac{3}{h}\{u(x_2) - u(0)\} - u'(0),$$

$$p_{i-1} + 4p_i + p_{i+1} = \frac{3}{h}\{u(x_{i+1}) - u(x_{i-1})\}, \quad 2 \leqslant i \leqslant N - 1,$$

$$p_{N-1} + 4p_N = \frac{3}{h}\{u(1) - u(x_{N-1})\} - u'(1).$$

The symmetric, tridiagonal matrix of order N

$$A = \begin{pmatrix} 4 & 1 & & & \\ 1 & 4 & 1 & & \\ & \ddots & \ddots & \ddots & \\ & & 1 & 4 & 1 \\ & & & 1 & 4 \end{pmatrix},$$

being invertible (by an argument similar, for example, to that used for the matrix A_h of section 3.1), the assertion follows.   □

The function $s_h u$ defined in this way is called the *cubic spline* interpolating the function u. In contrast to the piecewise interpolating polynomials considered earlier, its determination requires the *solution of a linear system*, as was shown by the preceding proof; there, again, we have another source of linear systems.

Remarks
(1) More generally, it is possible to define interpolating splines of higher degree. However, cubic splines are the ones most frequently used.

(2) It is remarkable that the bounds for the error are identical with those obtained for the cubic piecewise interpolating polynomials; in this regard, there are some indications to be found in Exercise 3.7-1.

(3) In contrast to the preceding interpolation, this interpolation is no longer 'local', in the sense that the interpolating function over a given interval $[x_j, x_{j+1}]$ depends on *all* the values $u(x_i)$. This is the very reason why it turned out to be necessary to solve a linear system.   □

We now go on to the second class of problem introduced at the beginning of this section. Given m distinct points of $\mathbb{R}$, x_i, $1 \leqslant i \leqslant m$, together with their corresponding numerical values c_i, it may turn out to be too costly, or simply undesirable for various reasons (the data may be too scattered, or obviously spoilt by error, etc.), to try to interpolate a curve *exactly* through the points (x_i, c_i). In that case, the interpolation problem is replaced by an *approximation problem*. Taking a finite-dimensional space, whose dimension n is in general 'much smaller' than the number m of points x_i, we look for a function U of this space which approximates 'as well as possible' the given values, since there is no longer any question of satisfying exactly the equalities $U(x_i) = c_i$, $1 \leqslant i \leqslant m$.

More precisely, let $(w_j)_{j=1}^n$ be a set of n real, linearly independent functions, defined over a set which contains the points x_i. The problem then consists in finding a function

$$U = \sum_{j=1}^{n} u_j w_j,$$

for which the equalities $U(x_i) = c_i$, $1 \leqslant i \leqslant m$, are approximated 'as well as can be'. The most commonly used method for attaining this approximation consists in approximating the equalities $U(x_i) = c_i$, $1 \leqslant i \leqslant m$, *in terms of least squares*. A function $U = \sum_{j=1}^n u_j w_j$ is sought which minimises the number

$$\sum_{i=1}^{m} \left| \sum_{j=1}^{n} v_j w_j(x_i) - c_i \right|^2, \quad \text{as the vector } (v_j)_{j=1}^n \text{ runs through } \mathbb{R}^n.$$

The *decisive* advantage of a least squares approximation is its *linearity*. To see this, observe that it may also be expressed as follows:

find a u such that

$$u \in \mathbb{R}^n \quad \text{and} \quad \| Bu - c \|_{2,m} = \inf_{v \in \mathbb{R}^n} \| Bv - c \|_{2,m},$$

where $\| \cdot \|_{2,m}$ denotes the Euclidean norm in $\mathbb{R}^m$, $B = (b_{ij})$ is a matrix of type (m,n) with elements $b_{ij} = w_j(x_i)$, and $c = (c_i)_{i=1}^m \in \mathbb{R}^m$. Now a simple calculation (whose justification will be given in section 7.4; it is nothing other than a Taylor's expansion) shows that, if u and w are any two vectors in $\mathbb{R}^n$,

$$\| B(u + w) - c \|_{2,m}^2 = \| Bu - c \|_{2,m}^2 + 2(B^{\mathrm{T}}Bu - B^{\mathrm{T}}c, w)_n + \| Bw \|_{2,m}^2,$$

where $(\cdot, \cdot)_n$ denotes the scalar product in $\mathbb{R}^n$.

It is then clear that *a vector in $\mathbb{R}^n$ is a solution of the problem if and only if it is a solution of the linear system*

$$B^{\mathrm{T}}Bu = B^{\mathrm{T}}c,$$

whose n equations are called the *normal equations*, associated with the least squares approximation problem under consideration.

Suppose that the matrix B has rank n; this is so, for example, if $w_j(x) = x^{j-1}$, $1 \leqslant j \leqslant n$. The symmetric matrix $B^{\mathrm{T}}B$ is then positive definite (it is always non-negative definite), since

$$B^T B v = 0 \Rightarrow v^T B^T B v = \| B v \|_{2,m}^2 = 0 \Rightarrow B v = 0 \Rightarrow v = 0.$$

Hence, in this case, the normal equations have a unique solution.

Remarks

(1) By misuse of language (even though very suggestive), the vector u obtained above is sometimes said to be the *least squares solution of the linear system* $Bu = c$.

(2) It will be shown in section 8.1 that the normal equations *always* have at least one solution. The reader is invited even now to try to establish this result; it is an excellent exercise!

(3) Of course, if $m = n$, and if the matrix B is invertible, the solution of the normal equations coincides with that of the linear system $Bu = c$.

(4) Quite evidently, the least squares approximation is not the only conceivable one. One could, for example, try to minimise any of the norms $\| Bv - c \|_p, 1 \leqslant p \leqslant \infty$. However, in so doing, one loses, for $p \neq 2$, the linearity of the function $c \rightarrow u$ (which, by the way, is not so easy to prove), and hence the simplicity of the calculations.   □

Exercises

3.7-1. The aim in this exercise is to prove the convergence of the interpolation which uses cubic splines. The notation is the same as in Theorem 3.7-1. The constants C, C_0, C_1 and C_2, which are used to settle the matter of existence, are independent of the mesh-size h as well as of the function u.

(1) Define the matrix of order $N + 1$

$$B_h = \begin{pmatrix} 2 & 1 & & & & & \\ 1 & 4 & 1 & & & & \\ & 1 & 4 & 1 & & & \\ & & \ddots & \ddots & & & \\ & & & 1 & 4 & 1 & \\ & & & & 1 & 4 & 1 \\ & & & & & 1 & 2 \end{pmatrix},$$

and the vector $q_h \in \mathbb{R}^{N+2}$ whose components have the values $(s_h u)''(x_i), 0 \leqslant i \leqslant N + 1$. Calculate the components of the vector $B_h q_h$ in terms of the values $u(x_i)$, $0 \leqslant i \leqslant N + 1, u'(0), u'(1)$.

(2) In questions (2), (3) and (4), the function u is assumed to be four times continuously differentiable over the interval $[0, 1]$. Let $q \in \mathbb{R}^{N+2}$ be a vector whose components take the values $u''(x_i), 0 \leqslant i \leqslant N + 1$. Show that there exists a constant C such that

$$\| B_h(q - q_h) \|_\infty \leqslant C \sup_{0 \leqslant x \leqslant 1} |u^{(4)}(x)| h^2.$$

(3) Using the result of Exercise 1.4-4 and that of the first part of the proof of Theorem 3.4-3 for $m = 0$, show that there exists a constant C_2 such that

$$\sup_{0 \leqslant x \leqslant 1} |(u - s_h u)''(x)| \leqslant C_2 \sup_{0 \leqslant x \leqslant 1} |u^{(4)}(x)| h^2.$$

Hint. Introduce, over each interval $[x_i, x_{i+1}]$, the affine function which takes the values $u''(x_i)$ and $u''(x_{i+1})$ at the end points of the interval.

(4) Deduce the existence of two constants C_1 and C_0 such that

$$\sup_{0 \leqslant x \leqslant 1} |(u - s_h u)'(x)| \leqslant C_1 \sup_{0 \leqslant x \leqslant 1} |u^{(4)}(x)| h^3,$$

$$\sup_{0 \leqslant x \leqslant 1} |(u - s_h u)(x)| \leqslant C_0 \sup_{0 \leqslant x \leqslant 1} |u^{(4)}(x)| h^4.$$

Compare this result with the bounds obtained for the piecewise cubic interpolation polynomial.

(5) In questions (5), (6) and (7), the function u is assumed to be twice continuously differentiable over the interval $[0, 1]$. Establish the relation

$$\int_0^1 (u''(x))^2 dx = \int_0^1 ((u - s_h u)''(x))^2 dx + \int_0^1 (s_h''u(x))^2 dx.$$

(6) Define the set

$$U = \{v \in \mathscr{C}^2([0, 1]): v(x_i) = u(x_i), 0 \leqslant i \leqslant N + 1; v'(0) = u'(0), v'(1) = u'(1)\}.$$

Prove the relation

$$\int_0^1 (s_h''u(x))^2 dx = \inf_{v \in U} \int_0^1 (v''(x))^2 dx.$$

(7) Define the vector space

$$S_h = \{r_h \in \mathscr{C}^2([0, 1]): r_h|_{[x_i, x_{i+1}]} \in P_3([x_i, x_{i+1}]), 0 \leqslant i \leqslant N\}.$$

Prove the relation

$$\int_0^1 ((u - s_h u)''(x))^2 dx = \inf_{r_h \in S_h} \int_0^1 ((u - r_h)''(x))^2 dx.$$

(8) Is it possible to interpret the relations established in questions (5), (6) and (7) in terms of the projection operator (cf. section 8.1)?

3.7-2. Define the space

$$V_h = \{v \in \mathscr{C}^2([0, 1]): v(0) = v(1) = 0; v|_{[x_i, x_{i+1}]} \in P_{2m+1}([x_i, x_{i+1}]), 0 \leqslant i \leqslant N\}.$$

(1) What is the dimension of this space?

(2) With the end in mind of constructing a variational approximation of a two-point boundary-value problem, show that there exists a basis of V_h whose functions have 'small' support, that is, one which is the union of a number, which is independent of h and as small as possible, of intervals $[x_i, x_{i+1}]$.

3.7-3. Let u be a function which is continuous over the interval $[0, 1]$.

(1) Prove that there exists a unique polynomial q_n of degree $\leqslant n-1$ such that

$$\|u - q_n\| = \inf\{\|u - p\| : p \in P_{n-1}([0,1])\},$$

where $\|v\| = (\int_0^1 |v(x)|^2 dx)^{1/2}$.

(2) Setting $q_n(x) = \sum_{i=1}^n u_i x^{i-1}$, prove that the vector $u = (u_i)_{i=1}^n$ is the solution of a linear system of the form $Au = b$, where A is the *Hilbert matrix* (cf. Exercise 2.2-8; such a system, then, is very ill-conditioned!).

3.7-4. Let $x_i, 0 \leqslant i \leqslant m$, be distinct points of $\mathbb{R}$. With a real function u, defined over a subset of $\mathbb{R}$ which contains the points x_i, $0 \leqslant i \leqslant m$, we associate the *interpolation polynomial Pu of the function u at the points* $x_i, 0 \leqslant i \leqslant m$, which is, by definition, a polynomial (with real coefficients) of degree $\leqslant m$, satisfying

$$Pu(x_i) = u(x_i), \quad 0 \leqslant i \leqslant m.$$

The aim now is to examine, firstly, the practical details of the construction of such a polynomial and, secondly, the convergence or divergence of this interpolation procedure, as the number of points x_i tends towards infinity.

(1) Prove the existence and uniqueness of the polynomial Pu.

(2) For what follows, assume that $m \geqslant 1$. Let x_λ and x_μ be two distinct points of the set $\bigcup_{i=0}^m \{x_i\}$. Denote by p_λ the unique polynomial of degree $\leqslant m-1$ which satisfies

$$p_\lambda(x_i) = u(x_i), \quad 0 \leqslant i \leqslant m, i \neq \lambda,$$

and denote by p_μ the unique polynomial of degree $\leqslant m-1$ which satisfies

$$p_\mu(x_i) = u(x_i) \quad 0 \leqslant i \leqslant m, i \neq \mu.$$

Prove that

$$Pu(x) = \frac{(x_\lambda - x)p_\lambda(x) - (x_\mu - x)p_\mu(x)}{x_\lambda - x_\mu}.$$

(3) Define the sequence of polynomials

$$p_i^l, 1 \leqslant l \leqslant i \leqslant m,$$

by means of the recurrence relations

$$p_i^1(x) = \frac{(x_i - x)u(x_0) - (x_0 - x)u(x_i)}{x_i - x_0}, \quad 1 \leqslant i \leqslant m,$$

$$p_i^{j+1}(x) = \frac{(x_i - x)p_j^j(x) - (x_j - x)p_i^j(x)}{x_i - x_j}, \quad 2 \leqslant j+1 \leqslant i \leqslant m.$$

Prove that

$$Pu = p_m^m.$$

(4) Assume that the function u is $m + 1$ times differentiable over an interval I containing the points $x_i, 0 \leqslant i \leqslant m$. Given any point x of the interval I, show that

there exists (at least) one point $\xi \in I$ such that

$$u(x) - Pu(x) = \frac{1}{(m+1)!} \prod_{i=0}^{m} (x - x_i) u^{(m+1)}(\xi).$$

(5) Give an example of a function u, other than a polynomial, defined over a compact interval $I = [a, b]$ of $\mathbb{R}$, with $a < b$, for which the expression $\sup_{x \in I} |u(x) - Pu(x)|$ tends towards 0 as the number of points x_i tends towards infinity.

(6) Given $n + 1$ distinct points $x_j \geqslant 0$, $0 \leqslant j \leqslant n$, we denote by Pw the polynomial which interpolates the particular function

$$w(x) = \frac{1}{1 + x^2}$$

at the points x_j and $-x_j$, $0 \leqslant j \leqslant n$ (hence, at $2n + 1$ points, or $2n + 2$ points, depending on whether one of the points x_j is zero or not). Prove that the degree of the polynomial Pw is even and that, in either case, it is at most $2n$.

(7) Prove that

$$w(x) - Pw(x) = w(x) \prod_{j=0}^{n} \frac{x_j^2 - x^2}{x_j^2 + 1} \quad \text{for every } x \in \mathbb{R}.$$

(8) Let ξ and a be two numbers satisfying

$$0 \leqslant \xi \leqslant a, 0 < a.$$

Given a number $h > 0$, we denote by x_j, $0 \leqslant j \leqslant n_h$, all the distinct points of the interval $[0, a]$ which are of the form $\xi + (k + \frac{1}{2})h$, $k \in \mathbb{Z}$.

Prove that, as h tends to zero, the expression

$$h \log \prod_{j=0}^{n_h} \left| \frac{x_j^2 - \xi^2}{x_j^2 + 1} \right|$$

has a limit, which should be calculated.

(9) Given numbers ξ, a and h satisfying the inequalities

$$0 \leqslant \xi \leqslant a, 0 < a, 0 < h,$$

we denote by $P_h^{(\xi, a)} w$ the polynomial which interpolates the function w at the points x_j and $-x_j$, $0 \leqslant j \leqslant n_h$, these points being defined as in question (8). Prove that, for certain values of the numbers ξ and a,

$$\lim_{h \to 0} |w(\xi) - P_h^{(\xi, a)} w(\xi)| = 0,$$

while, for other values of the numbers ξ and a,

$$\lim_{h \to 0} |w(\xi) - P_h^{(\xi, a)} w(\xi)| = +\infty.$$

Direct methods for the solution of linear systems

Introduction

The problem which we are about to consider is that of the numerical solution of a linear system $Au = b$ whose matrix A is invertible.

The guiding principle behind the *direct methods* studied in this chapter comes down to the determination of an invertible matrix M such that *the matrix MA is upper triangular* (Theorem 4.2-1); this corresponds to the process of *elimination*. It then remains to solve the linear system

$$MAu = Mb,$$

by the method of *back-substitution*, described in section 4.1 (in *actual* calculations, one does not derive explicitly the matrix M, but only the matrix MA and the vector Mb).

This principle is central to *Gaussian elimination* for linear systems with 'general' matrices and to the *Cholesky method* for linear systems with matrices that are symmetric and positive definite; these we describe in sections 4.2 and 4.4 respectively. We shall study, in passing, the calculation of the *determinant* of square matrices, the particular case (very important in applications) of *point* or *block tridiagonal matrices*, as well as *Jordan elimination*, which may be considered as a variant of Gaussian elimination. It is particularly well suited to the calculation of the *inverse* of a given matrix. We insist, in this regard, on the fact that the *calculation of the inverse of a matrix is unnecessary for the solution of a linear system*; this point will be underlined in section 4.1.

The matrix formulation of Gaussian elimination is *the LU factorisation of a matrix* (Theorem 4.3-1). This *very important* result, notably for its many and varied applications in the Numerical Analysis of Matrices, shows that every invertible matrix, with the possible incorporation of row permutations, may be written as the product of a *lower triangular matrix* L by an *upper triangular matrix* U. This factorisation is somewhat simpler in the case of *symmetric, positive definite matrices*; it is then the *Cholesky factorisation* (Theorem 4.4-1), which is central to Cholesky's method, mentioned earlier.

While the methods of Gauss and Cholesky rest on the factorisation of the matrix of a linear system into the product of a lower triangular matrix by an upper triangular one, *Householder's method* is associated with the factorisation of a matrix (not necessarily symmetric) into the product of an *orthogonal* matrix Q by an upper triangular matrix R. This method is described in section 4.5, where it is shown that such a QR *factorisation* can be carried out in an ingenious manner, with the help of certain 'elementary' orthogonal matrices, called *Householder matrices*.

We highlight the fact that the QR factorisation of a matrix will appear as an essential stage in the 'QR algorithm' for the calculation of the eigenvalues of a general matrix (section 6.3) and that Householder matrices will be used once again for the reduction to tridiagonal form of a symmetric matrix (section 6.2).

4.1 Two remarks concerning the solution of linear systems

Contrary to the conclusion that might be reached by a cursory analysis, *the solution of a linear system* $Au = b$, *with A an invertible matrix, is not obtained by calculating the matrix* A^{-1} *and then calculating the vector* $A^{-1}b$ *(the first remark)*. Calculating the matrix A^{-1} is effectively equivalent to solving the n linear systems (where n is the order of the matrix A)

$$Au_j = e_j, \ 1 \leqslant j \leqslant n,$$

where e_j is the jth basis vector of $\mathbb{K}^n$. In other words, by such a method one would be replacing the solution of *one* linear system (the problem presented) by the solution of n linear systems, followed by the multiplication of the matrix A^{-1} by the vector b!

The methods which we shall study are based on the following obvious *second remark. If the invertible matrix* A *is upper triangular* (or lower triangular), *then the numerical solution of a linear system* $Au = b$ *is immediate*; in fact, it can be written as

$$a_{11}u_1 + \cdots + a_{1,n-1}u_{n-1} + a_{1n}u_n = b_1,$$
$$\ddots \qquad \qquad \vdots$$
$$a_{n-1,n-1}u_{n-1} + a_{n-1,n}u_n = b_{n-1},$$
$$a_{nn}u_n = b_n,$$

and, since $a_{11}a_{22}\cdots a_{nn} = \det(A) \neq 0$, the system is solved by calculating u_n

from the last equation, then u_{n-1} from the last but one, etc., giving

$$u_n = a_{nn}^{-1} b_n,$$
$$u_{n-1} = a_{n-1,n-1}^{-1} (b_{n-1} - a_{n-1,n} u_n),$$
$$\vdots$$
$$u_1 = a_{11}^{-1} (b_1 - a_{12} u_2 - \cdots - a_{1,n-1} u_{n-1} - a_{1n} u_n).$$

Since each component u_i appears as a linear function of $b_i, b_{i+1}, \ldots, b_n$, this shows, in passing, that *the inverse of a triangular matrix is a triangular matrix of the same type (upper or lower)*.

The method of calculation given above, called *back-substitution*, requires a total of

$$\begin{cases} 1 + 2 + \cdots + (n-1) = \dfrac{n(n-1)}{2} \text{ additions,} \\[2mm] 1 + 2 + \cdots + (n-1) = \dfrac{n(n-1)}{2} \text{ multiplications,} \\[2mm] n \text{ divisions.} \end{cases}$$

Back-substitution can be extended to *block triangular matrices*. So, for example, the solution of the linear system

$$\begin{pmatrix} A_{11} & A_{12} & A_{13} \\ & A_{22} & A_{23} \\ & & A_{33} \end{pmatrix} \begin{pmatrix} u_1 \\ u_2 \\ u_3 \end{pmatrix} = \begin{pmatrix} b_1 \\ b_2 \\ b_3 \end{pmatrix}$$

comes down to the solution of the sequence of linear systems

$$A_{33} u_3 = b_3,$$
$$A_{22} u_2 = b_2 - A_{23} u_3,$$
$$A_{11} u_1 = b_1 - A_{12} u_2 - A_{13} u_3.$$

Naturally, this presupposes a knowledge of how to solve the linear systems whose matrices are the diagonal submatrices A_{ii}.

In studying the methods for the solution of linear systems (direct as well as iterative), we shall frequently use matrix notation. In particular, we shall consider 'intermediate' linear systems, $Cu = d$, say, which we shall 'solve' and write as $u = C^{-1} d$, in apparent contradiction to the first remark made earlier.

However, *this is no more than a convenient way of writing*, and on closer inspection, it will appear that the matrix C is point or block triangular, so that the solution of the system $Cu = d$ is by way of back-substitution, *without the explicit calculation of the matrix* C^{-1}. That is why it is sometimes said, *with evident misuse of language*, that such matrices C are 'easy to invert'.

Exercise

4.1-1. This exercise gives the main lines of a direct method for the calculation of the inverse of a given matrix, called the *rank reduction method*.

(1) Prove that every matrix of rank 1 may be written in the form uv^T, u and v being two column vectors.

(2) Prove that every invertible matrix of order n is the sum of n matrices of, rank 1.

(3) Let B be an invertible matrix and u and v two column vectors such that $B + uv^T$ is invertible. Prove that

$$(B + uv^T)^{-1} = B^{-1} - \frac{B^{-1}uv^T B^{-1}}{1 + v^T B^{-1} u}.$$

(4) Let A be a matrix of the form

$$A = D + \sum_{i=1}^{n} u_i v_i^T,$$

where D is an invertible diagonal matrix and the column vectors u_i and v_i are such that the matrices $(D + \sum_{i=1}^{l} u_i v_i^T)$, $1 \leqslant l \leqslant n$, are all invertible. Set

$$C_l = \left(D + \sum_{i=1}^{l} u_i v_i^T \right)^{-1},$$

so that

$$C_{l+1} = C_l - \frac{C_l u_{l+1} v_{l+1}^T C_l}{1 + v_{l+1}^T C_l u_{l+1}}.$$

Deduce a direct method for the calculation of the matrix A^{-1}.

4.2 Gaussian elimination

Gaussian elimination is a general method for the solution of a linear system

$$Au = b, \text{ with A an invertible matrix.}$$

It consists of three stages:

(i) one step, called *elimination* (of the unknowns, in succession), is equivalent to the determination of an invertible matrix M such that the matrix MA is *upper triangular*;

(ii) the simultaneous calculation of the vector Mb;

(iii) the solution of the linear system

$$MAu = Mb,$$

with upper triangular matrix, by the method of *back-substitution* described in the previous section.

Remark

In practice, *the matrix M is not calculated explicitly*, but only the matrix MA and the vector Mb. The introduction of the matrix M is just for *convenience of expression*. □

We now describe *the first stage, that of elimination*. At least one of the elements a_{i1}, $1 \leqslant i \leqslant n$, of the first column of the matrix $A = (a_{ij})$ is non-zero, otherwise the matrix would be singular. One of the *non-zero* elements of the first column of the matrix A is then *chosen* (how this non-zero element is chosen *in practice* will be examined later; for the moment, the actual choice is not an important matter) and is called the first *pivot* of the elimination.

Next, *the row containing the pivot is permuted with the first row*. In matrix notation, this amounts to pre-multiplying the matrix A by a particular permutation matrix P. For it may be verified that the interchange of the i_0th and the i_1th rows of a matrix is equivalent to its pre-multiplication by the *transposition matrix* (assuming, in order to fix ideas, that $i_0 < i_1$):

$$T(i_0, i_1) = \begin{pmatrix} 1 & & & & & & & & & \\ & 1 & \vdots & & & \vdots & & & & \\ & & 0 & \cdots & 1 & \cdots & & & & \\ & & & 1 & & & & & & \\ & & & & 1 & & & & & \\ & & & & & 1 & & & & \\ & & 1 & \cdots & 0 & \cdots & & & & \\ & & & & & 1 & & & \\ & & & & & & & 1 & \\ & & & & & & & & 1 \end{pmatrix} \begin{matrix} \\ \\ \longleftarrow i_0 \\ \\ \\ \\ \longleftarrow i_1 \\ \\ \\ \\ \end{matrix}$$

with arrows $\downarrow$ over columns i_0 and i_1.

It should be observed, in passing, that
$$\det(T(i_0, i_1)) = -1.$$

We then set
$$P = \begin{cases} I \text{ if } a_{11} \text{ is the pivot (so that } \det(P) = 1), \\ T(1, i) \text{ if } a_{i1}, i \neq 1, \text{ is the pivot (so that } \det(P) = -1), \end{cases}$$
and the matrix
$$PA = (\alpha_{ij})$$

is obtained in this way, and satisfies

$$\alpha_{11} \neq 0,$$

by construction. *Using appropriate linear combinations of the first and the remaining rows of the matrix* PA, *we then reduce to zero all the elements in the first column of the matrix* PA *situated below the diagonal*, the first row remaining unchanged.

In matrix notation, this amounts to pre-multiplying the matrix PA by the matrix

$$E = \begin{pmatrix} 1 & & & & \\ -\pi^{-1}\alpha_{21} & 1 & & & \\ -\pi^{-1}\alpha_{31} & & 1 & & \\ \vdots & & & \ddots & \\ -\pi^{-1}\alpha_{n1} & & & & 1 \end{pmatrix}, \quad \pi = \alpha_{11},$$

so that the matrix

$$B = EPA$$

is of the form

$$B = \begin{pmatrix} \alpha_{11} & \alpha_{12} & \alpha_{13} & \cdots & \alpha_{1n} \\ & b_{22} & b_{23} & \cdots & b_{2n} \\ & b_{32} & b_{33} & \cdots & b_{3n} \\ & \vdots & \vdots & & \vdots \\ & b_{n2} & b_{n3} & \cdots & b_{nn} \end{pmatrix}.$$

Remark
It is hardly possible to overstress the simplicity of the operations carried out (row interchanges, linear combinations of rows), somewhat 'concealed' by the matrix notation. It would be very *naïf* to think that the matrices P and E are actually worked out and that the products PA, EPA, ... are then formed. In this respect, the detailed numerical example given further on should be instructive. □

Since

$$\det(E) = 1,$$

it follows that

$$\det(B) = \det(E)\det(P)\det(A) = \pm \det(A),$$

depending on whether there has or has not been an interchange of rows. It also proves, in passing, that the matrix B is itself invertible. Hence, at least one of the elements b_{i2}, $2 \leqslant i \leqslant n$, is non-zero, so that, naturally enough, the second step of the elimination consists in carrying out the same operations as before, but only on the submatrix (b_{ij}), $2 \leqslant i,j \leqslant n$, with the first row left unaltered; and so on ...

From here on we set

$$A = A_1 = (a_{ij}^1), \quad P = P_1, \quad P_1 A_1 = (\alpha_{ij}^1),$$
$$E = E_1, \quad B = A_2 = E_1 P_1 A_1 = (a_{ij}^2),$$

and obtain, step by step, as a *result of the $(k-1)$th step of the elimination*, $2 \leqslant k \leqslant n$, the matrix

$$A_k = E_{k-1} P_{k-1} \cdots E_2 P_2 E_1 P_1 A_1,$$

which is of the from

$$A_k = \begin{pmatrix} a_{11}^k & a_{12}^k & \cdots & & a_{1n}^k \\ & a_{22}^k & \cdots & & a_{2n}^k \\ & & \ddots & & \vdots \\ & & a_{kk}^k & \cdots & a_{kn}^k \\ & & \vdots & & \vdots \\ & & a_{nk}^k & \cdots & a_{nn}^k \end{pmatrix} = \begin{pmatrix} \alpha_{11}^1 & \alpha_{12}^1 & \cdots & & \alpha_{1n}^1 \\ & \alpha_{22}^2 & \cdots & & \alpha_{2n}^2 \\ & & \ddots & & \\ & & a_{kk}^k & \cdots & a_{kn}^k \\ & & \vdots & & \vdots \\ & & a_{nk}^k & \cdots & a_{nn}^k \end{pmatrix},$$

the second expression making obvious the fact that *the first i rows remain unaltered after the ith step*.

We next describe the *kth step of the elimination*. Since $\det(A_k) = \pm \det(A)$, the matrix A_k is invertible, so that at least one of the elements $a_{ik}^k, k \leqslant i \leqslant n$, is non-zero. One of these *non-zero* elements is chosen as *pivot* and the row containing the pivot is interchanged with the *kth* row of the matrix A_k. In matrix notation, this amounts to pre-multiplying the matrix A_k by a matrix P_k which is either the identity matrix or a transposition matrix, so that, by construction, the element α_{kk}^k of the matrix

$$P_k A_k = (\alpha_{ij}^k)$$

is non-zero.

The elimination step is equivalent to the premultiplication of the matrix $P_k A_k$ by the matrix

$$E_k = \begin{pmatrix} 1 \\ & \ddots \\ & & 1 \\ & & -\pi_k^{-1}\alpha_{k+1,k}^k & 1 \\ & & \vdots & & \ddots \\ & & -\pi_k^{-1}\alpha_{nk}^k & & & 1 \end{pmatrix}, \quad \pi_k = \alpha_{kk}^k,$$

an operation which leaves unaltered the first k rows of the matrix $P_k A_k$.

After the $(n-1)$th step, *the matrix*

$$A_n = E_{n-1} P_{n-1} \cdots E_2 P_2 E_1 P_1 A$$

is upper triangular; so that an invertible matrix has been found,

$$M = E_{n-1} P_{n-1} \cdots E_2 P_2 E_1 P_1,$$

such that the matrix MA is upper triangular; which was, we recall, the objective 'in matrix terms' of the process of elimination. We observe that

$$\det(M) = \begin{cases} +1 \text{ if an even number of row interchanges was used,} \\ -1 \text{ if an odd number of row interchanges was used,} \end{cases}$$

so that, as a 'by-product' of the process of elimination, *a particularly fast method is obtained for calculating the determinant of a matrix* A, which is just, up to a sign, the *product of the pivots,* since

$$\det(A) = \frac{\det(A_n)}{\det(M)} = \pm \alpha_{11}^1 \alpha_{22}^2 \cdots \alpha_{nn}^n,$$

with the convention that $\alpha_{nn}^n = (A_n)_{nn}$ is the *n*th *pivot.*

The matrix form of the process of elimination is contained in the following result.

Theorem 4.2-1
Let A *be a square matrix, invertible or not. There exists (at least) one invertible matrix* M *such that the matrix* MA *is upper triangular.*

Proof
The result has been proved for the case where the matrix A is invertible. Now the matrix A is singular if and only if the elements $a_{ik}^k, k \leqslant i \leqslant n$, of the matrix A_k are zero for at least one value of the integer k. But then the matrix A_k is already of the form A_{k+1}: it is then enough to set $P_k = E_k = I$. $\quad\square$

We give now a *numerical example* of the use of Gaussian elimination for the solution of a linear system. The pivots are enclosed in square brackets and the linear combinations of rows employed are indicated in the margin; the notation is that used earlier.

$$
\begin{array}{ll}
(\text{I}_1) & [5]u_1 + 2u_2 + u_3 = 12 \\
(\text{II}_1) & 5u_1 - 6u_2 + 2u_3 = -1 \\
(\text{III}_1) & -4u_1 + 2u_2 + u_3 = 3
\end{array}
$$

$$
\begin{array}{ll}
(\text{I}_1) = (\text{I}_2) & 5u_1 + 2u_2 + u_3 = 12 \\
-(\text{I}_1) + (\text{II}_1) = (\text{II}_2) & [-8]u_2 + u_3 = -13 \\
\tfrac{4}{5}(\text{I}_1) + (\text{III}_1) = (\text{III}_2) & \tfrac{18}{5}u_2 + \tfrac{9}{5}u_3 = \tfrac{63}{5}
\end{array}
$$

$$
\begin{array}{ll}
(\text{I}_2) = (\text{I}_3) & 5u_1 + 2u_2 + u_3 = 12 \\
(\text{II}_2) = (\text{II}_3) & -8u_2 + u_3 = -13 \\
\tfrac{9}{20}(\text{II}_2) + (\text{III}_2) = (\text{III}_3) & [\tfrac{9}{4}]u_3 = \tfrac{27}{4}
\end{array}
$$

$$A = A_1 = \begin{pmatrix} 5 & 2 & 1 \\ 5 & -6 & 2 \\ -4 & 2 & 1 \end{pmatrix},$$

$$P_1 = I, \quad E_1 = \begin{pmatrix} 1 & & \\ -1 & 1 & \\ \frac{4}{5} & & 1 \end{pmatrix}, \qquad A_2 = \begin{pmatrix} 5 & 2 & 1 \\ & -8 & 1 \\ & \frac{18}{5} & \frac{9}{5} \end{pmatrix},$$

$$P_2 = I, \quad E_2 = \begin{pmatrix} 1 & & \\ & 1 & \\ & \frac{9}{20} & 1 \end{pmatrix}, \qquad A_3 = \begin{pmatrix} 5 & 2 & 1 \\ & -8 & 1 \\ & & \frac{9}{4} \end{pmatrix}.$$

Since the matrix A_3 is triangular, the solution of the final system is immediate, by means of the method of back-substitution, giving the result

$$u_3 = 3, \quad u_2 = 2, \quad u_1 = 1.$$

Furthermore,

$$\det(A) = a_{11}^1 a_{22}^2 a_{33}^3 = 5 \times (-8) \times \tfrac{9}{4} = -90.$$

Remark

Evidently it is possible to *calculate* the matrix

$$M = E_2 E_1 = \begin{pmatrix} 1 & & \\ -1 & 1 & \\ \frac{7}{20} & \frac{9}{20} & 1 \end{pmatrix},$$

which makes it possible to write

$$MA = A_3, \text{ that is } \begin{pmatrix} 1 & & \\ -1 & 1 & \\ \frac{7}{20} & \frac{9}{20} & 1 \end{pmatrix} \begin{pmatrix} 5 & 2 & 1 \\ 5 & -6 & 2 \\ -4 & 2 & 1 \end{pmatrix} = \begin{pmatrix} 5 & 2 & 1 \\ & -8 & 1 \\ & & \frac{9}{4} \end{pmatrix},$$

but, in general, *an explicit expression of the matrix M is of no practical use.* Rather, as we shall see, it is the *matrix* M^{-1} which holds interest; it is found, moreover, that this matrix M^{-1} may be written down *immediately* from the elements of the matrices E_k (as long as there have been no row interchanges), while, to calculate the matrix M, it is necessary to multiply the matrices E_k.   ☐

We give now some details about the *choice of pivot* at each step of the elimination. Obviously, if, at the beginning of the kth step, the element a_{kk}^k of the matrix A_k is non-zero, there is no *theoretical* reason why it may not be chosen as pivot (so that $P_k = I$).

However, *because of rounding errors, this way of proceeding is, in certain cases, to be totally discouraged.* In this regard, the following numerical example (taken from Forsythe and Moler (1967), page 35) is both spectacular and instructive. Let us assume that floating point arithmetic is used, with a three digit mantissa, and in the decimal system (in order to fix ideas and, above all, to facilitate calculation...); in other words, *the data and the intermediate results of all calculations are rounded to three significant digits* (refer back to the discussion of section 2.1). Consider the 'exact' system

$$(I_1)10^{-4}u_1 + u_2 = 1,$$
$$(II_1) \qquad u_1 + u_2 = 2,$$

With 'exact' solution

$$u_1 = 1.00010... \approx 1, \quad u_2 = 0.99990... \approx 1.$$

It is possible to take the number $a_{11} = 10^{-4}$ as pivot, since it is not zero, leading to the following procedure of elimination

$$(I_1) = (I_2) \quad 10^{-4}u_1 + u_2 = 1,$$
$$-10^4(I_1) + (II_1) = (II_2) \quad -9990u_2 = -9990,$$

since the numbers $-10^4 + 1 = -9999$ and $-10^4 + 2 = -9998$ are both rounded to the same number -9990. The 'solution' found in this way,

$$u_2 = 1, \quad u_1 = 0,$$

is very far from the true solution!

On the other hand, if we begin by *interchanging* the two equations, that is to say, if the pivot is the element $a_{21} = 1$, then we are led to the following calculations

$$(I'_1) \qquad [1]u_1 + u_2 = 2,$$
$$(II'_1) \quad 10^{-4}u_1 + u_2 = 1,$$
$$(I'_1) = (I'_2) \qquad u_1 + u_2 = 2,$$
$$-10^{-4}(I'_1) + (II'_1) = (II'_2) \qquad 0.999u_2 = 0.999,$$

since the numbers $-10^{-4} + 1 = 0.9999$ and $-10^{-4} + 2 = 0.9998$ are both rounded to the same number 0.999. The corresponding 'solution'

$$u_2 = 1, \quad u_1 = 1,$$

is now very satisfactory!

This example shows that rounding errors with disastrous effects arise as a result of *division by pivots which are 'too small'*. That is why, *in practice*, one of the following two strategies is used at the beginning of the kth step, $1 \leqslant k \leqslant n - 1$, of the elimination process.

Strategy of partial pivoting. The pivot is any one of the elements

$a_{ik}^k, k \leqslant i \leqslant n$, satisfying

$$|a_{ik}^k| = \max_{k \leqslant p \leqslant n} |a_{pk}^k|.$$

Strategy of complete pivoting. The pivot is any one of the elements $a_{ij}^k, k \leqslant i, j \leqslant n$, satisfying

$$|a_{ij}^k| = \max_{k \leqslant p,q \leqslant n} |a_{pq}^k|.$$

If the pivot chosen by this strategy is not in the kth column, it is then also necessary to carry out *column interchanges* (in addition to row interchanges). This strategy is then similar to, but not identical with, the process of elimination which we have described, since such an operation is equivalent to an additional *post-multiplication* of the matrix A_k by a transposition matrix (on this point, some further indication is given in Exercise 4.3–4).

While referring readers who are interested in this type of question to the more specialised texts cited in the Bibliography and comments, we will confine ourselves just to this observation. *A pivotal strategy* (partial or complete) *is indispensable if it is desired to avoid rounding errors which are too large in the course of Gaussian elimination applied to linear systems with 'general' matrices. On the other hand, there are particular cases where there is no advantage in employing a pivotal strategy*; this is notably so for linear systems with *matrices which are symmetric and positive definite* (see especially Wilkinson (1965), page 220), which we shall study in section 4.4.

Finally, let us count the number of elementary operations required in Gaussian elimination.

(i) *Elimination.* In advancing from the matrix A_k to the matrix A_{k+1}, $1 \leqslant k \leqslant n-1$, $n-k$ divisions are carried out and $(n-k+1)(n-k) = (n-k)^2 + n - k$ additions and multiplications, or a total of

$$\begin{cases} (n-1)^2 + (n-2)^2 + \cdots + 1^2 + (n-1) + (n-2) + \cdots + 1 \\ \quad = \dfrac{n^3 - n}{3} \text{ additions,} \\[2mm] (n-1)^2 + (n-2)^2 + \cdots + 1^2 + (n-1) + (n-2) + \cdots + 1 \\ \quad = \dfrac{n^3 - n}{3} \text{ multiplications,} \\[2mm] (n-) + (n-2) + \cdots + 1 = \dfrac{n(n-1)}{2} \text{ divisions.} \end{cases}$$

(ii) *In advancing from the vector* $E_{k-1}P_{k-1} \cdots E_1 P_1 b$ *to the vector*

$E_k P_k \cdots E_1 P_1 b$, $n - k$ additions and multiplications are carried out, or a total of

$$\begin{cases} (n-1) + (n-2) + \cdots + 1 = \dfrac{n(n-1)}{2}\, \text{additions,} \\[2mm] (n-1) + (n-2) + \cdots + 1 = \dfrac{n(n-1)}{2}\, \text{multiplications.} \end{cases}$$

(iii) The *back-substitution* step requires (cf. section 4.1)

$$\begin{cases} \dfrac{n(n-1)}{2} & \text{additions,} \\[3mm] \dfrac{n(n-1)}{2} & \text{multiplications,} \\[3mm] n & \text{divisions.} \end{cases}$$

In the final count, then, *Gaussian elimination requires operations of the order of*

$$\begin{cases} \dfrac{n^3}{3}\, \text{additions,} \\[3mm] \dfrac{n^3}{3}\, \text{multiplications,} \\[3mm] \dfrac{n^2}{2}\, \text{divisions.} \end{cases}$$

Remark

In estimating the total time required for the calculation (we recall, in this regard, the orders of magnitude given in section 2.1 for the relative times required for each of the elementary operations), it is necessary to take also into consideration the time needed to search for a pivot, *especially when using the strategy of complete pivoting.* □

It was thought at one time that the function $2n^3/3$ was a lower bound for the leading term of the total number of elementary operations required for the solution of a general linear system by a *direct method* (no matter which). Although the question is still not completely settled at the present time, nevertheless it does seem that the power 3 is not far from its optimal value, and *it is precisely this observation which justifies the use of Gaussian elimination whenever the matrix of the system is a 'general' one.* See, on this question, Pan (1984).

It is very instructive to compare the number of elementary operations needed for Gaussian elimination with the number of elementary operations required in applying *Cramer's rule*:

$$u_i = \frac{\det(B_i)}{\det(A)}, \text{ where } B_i = \begin{pmatrix} a_{11} & \cdots & a_{1,i-1} & b_1 & a_{1,i+1} & \cdots & a_{1n} \\ a_{21} & \cdots & a_{2,i-1} & b_2 & a_{2,i-1} & \cdots & a_{2n} \\ \vdots & & \vdots & \vdots & \vdots & & \vdots \\ a_{n1} & \cdots & a_{n,i-1} & b_n & a_{n,i+1} & \cdots & a_{nn} \end{pmatrix},$$

for which it is necessary to evaluate $n + 1$ determinants and to carry out n divisions. Now the 'brute-force' calculation of a determinant requires $n! - 1$ additions and $(n - 1)n!$ multiplications, so that the use of Cramer's rule requires something of the order of

$$\begin{cases} (n + 1)! & \text{additions,} \\ (n + 2)! & \text{multiplications,} \\ n & \text{divisions.} \end{cases}$$

For $n = 10$, for example, this gives approximately

$$\begin{cases} 700 \text{ operations for Gaussian elimination,} \\ 400\,000\,000 \text{ operations for Cramer's rule.} \\ \text{No comment}\ldots \end{cases}$$

It should be remembered that Gaussian elimination is the method most frequently used to solve linear systems whose matrices do not have any special properties. In particular, this method is employed for systems *with full matrices*.

A method which at root is closely related to Gaussian elimination is *Jordan elimination*. Instead of looking for a matrix M such that MA is upper triangular (A being a given invertible matrix), *we look for a matrix $\tilde{M}$ such that the matrix $\tilde{M}A$ is diagonal*. The result at the $(k - 1)$th step is then a matrix of the form (compare this with the process in Gaussian elimination)

$$\tilde{A}_k = \tilde{E}_{k-1}\tilde{P}_{k-1}\cdots\tilde{E}_2\tilde{P}_2\tilde{E}_1\tilde{P}_1 A = \begin{pmatrix} a_{11}^k & & & a_{1k}^k & \cdots & a_{1r}^k \\ & a_{22}^k & & a_{2k}^k & \cdots & a_{2n}^k \\ & & \ddots & \vdots & & \vdots \\ & & & a_{kk}^k & \cdots & a_{kn}^k \\ & & & \vdots & & \vdots \\ & & & a_{nk}^k & \cdots & a_{nn}^k \end{pmatrix}.$$

As in Gaussian elimination, one of the elements $a_{ik}^k, k \leqslant i \leqslant n$, must be non-zero (the matrix A being invertible) and is chosen as *pivot*, which amounts to pre-multiplying the matrix $\tilde{A}_k$ by a permutation matrix $\tilde{P}_k$ such that the element α_{kk}^k (that is to say, the pivot chosen) of the matrix

$(\alpha_{ij}^k) = \tilde{P}_k \tilde{A}_k$ is non-zero. The matrix $\tilde{E}_k$ then has the form

$$\tilde{E}_k = \begin{pmatrix} 1 & & & -\pi_k^{-1}\alpha_{1k}^k & & \\ & \ddots & & \vdots & & \\ & & 1 & -\pi_k^{-1}\alpha_{k-1,k}^k & & \\ & & & 1 & & \\ & & & -\pi_k^{-1}\alpha_{k+1,k}^k & 1 & \\ & & & \vdots & & \ddots \\ & & & -\pi_k^{-1}\alpha_{nk}^k & & 1 \end{pmatrix}, \quad \pi_k = \alpha_{kk}^k,$$

so that after the $(n-1)$th step the matrix

$$\tilde{A}_n = \tilde{M}A, \quad \text{with} \quad \tilde{M} = \tilde{E}_{n-1}\tilde{P}_{n-1}\cdots\tilde{E}_2\tilde{P}_2\tilde{E}_1\tilde{P}_1,$$

is diagonal.

Jordan elimination is used to calculate the inverse of a given matrix. The following n linear systems are solved *simultaneously*–

$$Au_j = e_j, \quad 1 \leqslant j \leqslant n$$

–by applying to each right-hand side e_j the transformations (interchanges and linear combinations of rows) represented by the matrices $\tilde{P}_k$ and $\tilde{E}_k$, thus obtaining

$$\tilde{A}_n u_j = \tilde{M}e_j, \quad 1 \leqslant j \leqslant n,$$

and each of these systems can be solved immediately, since the matrix $\tilde{A}_n$ is diagonal.

Exercises

4.2-1. Calculate the number of elementary operations which are required for the calculation of the determinant of a matrix by the method indicated in the text (the product of the pivots). Compare it with that for the 'direct' calculation for $n = 10$.

4.2-2. Denote by $\tilde{A}_k$ the square matrix of order $n - k + 1$ consisting of the elements a_{ij}^k, $k \leqslant i,j \leqslant n$, of the matrix $A_k = (a_{ij}^k)$ obtained as the result of the $(k-1)$th step of the elimination process (we retain all details of notation adopted in the text).

Assume that the matrix $A = A_1$ is *symmetric and positive definite*.

(1) Denoting the Euclidean scalar product by $(\cdot,\cdot)$ and by $v' \in \mathbb{R}^{n-k}$ the vector consisting of the last $n - k$ components of a general vector $v = (v_i)_{i=k}^n \in \mathbb{R}^{n-k+1}$, establish the identity

$$(\tilde{A}_k v, v) = (\tilde{A}_{k+1} v', v') + \frac{1}{a_{kk}^k}\left| a_{kk}^k v_k + \sum_{i=k+1}^{n} a_{ik}^k v_i \right|^2.$$

(2) Prove that each matrix $\tilde{A}_k$ is symmetric and positive definite.

(3) Establish the following inequalities:

$$0 < a_{ii}^{k+1} \leqslant a_{ii}^{k}, \quad k+1 \leqslant i \leqslant n,$$

$$\max_{k+1 \leqslant i \leqslant n} a_{ii}^{k+1} = \max_{k+1 \leqslant i,j \leqslant n} |a_{ij}^{k+1}| \leqslant \max_{k \leqslant i,j \leqslant n} |a_{ij}^{k}| = \max_{k \leqslant i \leqslant n} a_{ii}^{k}.$$

Remark. The last inequalities are of *considerable* practical importance, since they show that *overflow* does not occur in the computer in the course of the calculations.

4.2-3. How many elementary operations are required to calculate the inverse of a matrix of order n using Jordan elimination (account should be taken of the special form of the right-hand sides)? Compare this with the number of operations required to calculate the square of a matrix.

4.2-4. Calculate the condition numbers, $\text{cond}_p(\cdot)$ for $p = 1, 2, \infty$, of the 'exact' matrices obtained at the second step of Gaussian elimination applied to the solution of the linear system

$$10^{-4} u_1 + u_2 = 1,$$
$$u_1 + u_2 = 2,$$

according as the two equations are initially interchanged, or not. Compare with the 'numerical' results found in the text, p. 133.

4.3 The LU factorisation of a matrix

Suppose that there is no particular concern as to the effect of rounding errors in Gaussian elimination and that, as a result, no attempt is made to apply any particular pivotal strategy. In other words, if the element a_{11}^{1} of the matrix $A_1 = A$ is non-zero, then it is chosen as pivot; if the element a_{22}^{2} of the matrix $A_2 = E_1 A_1$ (here $P_1 = I$) is non-zero, then it is chosen as pivot; and so on.... If it is possible to choose from step to step $P_1 = P_2 = \cdots = P_{n-1} = I$ (the pivots being, as a result, the elements a_{kk}^{k} of the matrices A_k), then the matrix

$$M = E_{n-1} \cdots E_2 E_1,$$

which is the product of lower triangular matrices, is itself lower triangular (immediately verifiable), as well as the matrix

$$L \stackrel{\text{def}}{=} M^{-1}$$

(cf. section 4.1), and *the matrix* A *can be written as the product* A = LU *of a lower triangular matrix* L *by an upper triangular matrix* U. Summing up,

$$P_1 = P_2 = \cdots = P_{n-1} = I \Rightarrow A = LU, \quad \text{with} \quad \begin{cases} L = (E_{n-1} \cdots E_2 E_1)^{-1}, \\ U = (E_{n-1} \cdots E_2 E_1)A. \end{cases}$$

At this stage, we know how to calculate the matrix U, since

$$U = A_n = \begin{pmatrix} a_{11}^1 & a_{12}^1 & \cdots & a_{1n}^1 \\ & a_{22}^2 & \cdots & a_{2n}^2 \\ & & \ddots & \vdots \\ & & & a_{nn}^n \end{pmatrix},$$

where (recall)

$$(a_{ij}^k) = E_{k-1} \cdots E_2 E_1 A = A_k, \quad 1 \leqslant k \leqslant n-1.$$

It is an altogether remarkable fact that the matrix L *is calculable immediately from the matrices* $E_k, 1 \leqslant k \leqslant n-1$. Since these actually have the form

$$E_k = \begin{pmatrix} 1 \\ & \ddots \\ & & 1 \\ & & -l_{k+1,k} & 1 \\ & & \vdots & & \ddots \\ & & -l_{nk} & & & 1 \end{pmatrix}, \quad \text{with } l_{ik} \stackrel{\text{def}}{=} \frac{a_{ik}^k}{a_{kk}^k}, \quad k+1 \leqslant i \leqslant n,$$

it is easy to verify that (note the change of sign in the elements l_{ik}!)

$$(E_k)^{-1} = \begin{pmatrix} 1 \\ & \ddots \\ & & 1 \\ & & l_{k+1,k} & 1 \\ & & \vdots & & \ddots \\ & & l_{nk} & & & 1 \end{pmatrix},$$

$$L = (E_1)^{-1}(E_2)^{-1} \cdots (E_{n-1})^{-1} = \begin{pmatrix} 1 \\ l_{21} & 1 \\ l_{31} & l_{32} & 1 \\ \vdots & \vdots & \vdots & \ddots \\ l_{n1} & l_{n2} & l_{n3} & \cdots & 1 \end{pmatrix},$$

Thus, in the numerical example of section 4.2 (where there were no row interchanges), we found that

$$E_1 = \begin{pmatrix} 1 \\ -1 & 1 \\ \frac{4}{5} & & 1 \end{pmatrix}, \quad E_2 = \begin{pmatrix} 1 \\ & 1 \\ & \frac{9}{20} & 1 \end{pmatrix}, \quad A_3 = \begin{pmatrix} 5 & 2 & 1 \\ & -8 & 1 \\ & & \frac{9}{4} \end{pmatrix},$$

enabling us to set

$$A = (E_2 E_1)^{-1} A_3 = \begin{pmatrix} 1 & & \\ 1 & 1 & \\ -\frac{4}{5} & -\frac{9}{20} & 1 \end{pmatrix} \begin{pmatrix} 5 & 2 & 1 \\ & -8 & 1 \\ & & \frac{9}{4} \end{pmatrix} = LU.$$

However, care is needed! While the matrix L is immediately obtainable from the matrices E_k, the same is not true of the matrix $M = L^{-1} = E_{n-1} \ldots E_2 E_1$, for which there is no simple expression for its elements in terms of those of the matrices E_k. However, as we have already remarked, there is no need to calculate the matrix M....

It remains to give reasons which are *sufficient* to ensure that there is no need to employ row interchanges in the implementation of Gaussian elimination, that is to say, in the 'factorisation' of a matrix into the product of a lower triangular matrix L by an upper triangular matrix U. It will even be shown that such an LU *factorisation* is unique if the diagonal elements of the matrix L are chosen to have the value 1 (this is precisely the value employed in the previous construction), thus making possible an unambiguous definition of such a factorisation (see, further, Exercise 4.3–1).

Theorem 4.3-1 (LU *factorisation of a matrix*)
Let $A = (a_{ij})$ *be a square matrix of order n such that the n diagonal submatrices*

$$\Delta_k = \begin{pmatrix} a_{11} & \cdots & a_{1k} \\ \vdots & & \vdots \\ a_{k1} & \cdots & a_{kk} \end{pmatrix}, \quad 1 \leqslant k \leqslant n,$$

are invertible.

Then there exists a lower triangular matrix $L = (l_{ij})$ *with* $l_{ii} = 1, 1 \leqslant i \leqslant n$, *and an upper triangular matrix* U *such that*

$$A = LU.$$

Moreover, this factorisation is unique.

Proof
Since $a_{11} \neq 0$, it is possible to choose $P_1 = I$. Suppose that it has been possible to choose

$$P_1 = P_2 = \cdots = P_{k-1} = I,$$

so that the equality

$$(E_{k-1} \cdots E_2 E_1) A = A_k$$

may be written as

$$
\begin{pmatrix}
1 & & & & & \\
\times & 1 & & & & \\
\vdots & \vdots & \cdots & & & \\
\times & \times & \cdots & 1 & & \\
\vdots & \vdots & & & \cdots & \\
\times & \times & . & . & . & 1
\end{pmatrix}
\begin{pmatrix}
a_{11} & \cdots & a_{1k} & \cdots & \times \\
\vdots & (\Delta_k) & \vdots & & . \\
a_{k1} & \cdots & a_{kk} & & \\
\vdots & & & & . \\
\times & . & . & . & \times
\end{pmatrix}
$$

$$
= \begin{pmatrix}
a_{11}^1 & \cdots & a_{1k}^1 & \cdots & \times \\
& \ddots & \vdots & & . \\
& & a_{kk}^k & & . \\
& & \vdots & & \\
& & \times & \cdots & \times
\end{pmatrix}.
$$

Using the rules for the block multiplication of matrices and keeping in mind the particular form of the matrices, we have

$$\det(\Delta_k) = a_{11}^1 \cdots a_{kk}^k.$$

Since $\det(\Delta_k) \neq 0$ by hypothesis, it follows that the element a_{kk}^k is not zero. Hence it can be chosen as pivot, which is equivalent to the choice $P_k = I$.

The *existence* of an LU factorisation with the properties stated in the theorem is, therefore, established upon setting

$$A = (E_{n-1} \cdots E_2 E_1)^{-1}(E_{n-1} \cdots E_2 E_1 A) \overset{\text{def}}{=} LU.$$

We now prove the *uniqueness*. From the existence of two such factorisations,

$$A = L_1 U_1 = L_2 U_2,$$

it follows that

$$
L_2^{-1} L_1 = \begin{pmatrix}
1 & & & \\
\times & 1 & & \\
\vdots & \vdots & \ddots & \\
\times & \times & \cdots & 1
\end{pmatrix}
= \begin{pmatrix}
\times & \times & \cdots & \times \\
& \times & \cdots & \times \\
& & \ddots & \vdots \\
& & & \times
\end{pmatrix}
= U_2 U_1^{-1}.
$$

Now this equality of matrices is only possible if $L_2^{-1} L_1 = U_2 U_1^{-1} = I$, or $L_1 = L_2$ and $U_1 = U_2$. $\qquad \square$

An important case in which the conditions for the application of the previous theorem are met is that in which the matrix A is *symmetric and positive definite*; but, as we shall see, advantage may be taken of the symmetry in order slightly to modify (and simplify) the LU factorisation. That is why this case is treated separately in the following section.

A matter of *major importance* follows upon the existence of an LU factorisation. If there are *many* linear systems, with the *same* matrix A, to be solved, it is enough to store the two matrices L and U once they have been calculated 'for the first time', that is to say, after the solution of the 'first' linear system. *Thereafter, each linear system* $Av = b$ *is solved by solving the two linear systems with triangular matrices,*

$$Lw = b, \quad \text{then} \quad Uv = w.$$

Remark

If the sufficient condition of Theorem 4.3-1 is not satisfied, it is possible, nevertheless, to bring it about by *preliminary* row permutations of the matrix (cf. Exercise 4.3-4). From this point of view, then, the LU factorisation of invertible matrices is always possible. □

Finally, we look at the case of *tridiagonal matrices*, which allow a particularly simple LU factorisation. It should be observed, in passing, that the *band structure of the matrix is conserved*. This is a particular example of a general feature (cf. Exercise 4.3-5). The case of *block* tridiagonal matrices is considered in Exercise 4.3-2.

Theorem 4.3-2

Let

$$A = \begin{pmatrix} b_1 & c_1 & & & & \\ a_2 & b_2 & c_2 & & & \\ & \ddots & \ddots & \ddots & & \\ & & a_{n-1} & b_{n-1} & c_{n-1} \\ & & & a_n & b_n \end{pmatrix}$$

be a tridiagonal matrix. Define the sequence

$$\delta_0 = 1, \quad \delta_1 = b_1, \quad \delta_k = b_k \delta_{k-1} - a_k c_{k-1} \delta_{k-2}, \quad 2 \leqslant k \leqslant n.$$

Then

$$\delta_k = \det(\Delta_k), \quad \text{where } \Delta_k = \begin{pmatrix} b_1 & c_1 & & & \\ a_2 & b_2 & c_2 & & \\ & \ddots & \ddots & \ddots & \\ & & a_{k-1} & b_{k-1} & c_{k-1} \\ & & & a_k & b_k \end{pmatrix}, \quad 1 \leqslant k \leqslant n,$$

and, if the numbers δ_k, $1 \leqslant k \leqslant n$, are all non-zero, the LU factorisation of the *matrix A is*

$$
A = LU = \begin{pmatrix} 1 & & & & & \\ a_2\dfrac{\delta_0}{\delta_1} & 1 & & & & \\ & a_3\dfrac{\delta_1}{\delta_2} & & & & \\ & & \ddots & & & \\ & & & \ddots & 1 & \\ & & & & a_n\dfrac{\delta_{n-2}}{\delta_{n-1}} & 1 \end{pmatrix} \begin{pmatrix} \dfrac{\delta_1}{\delta_0} & c_1 & & & \\ & \dfrac{\delta_2}{\delta_1} & c_2 & & \\ & & \dfrac{\delta_3}{\delta_2} & \ddots & \\ & & & \ddots & c_{n-1} \\ & & & & \dfrac{\delta_n}{\delta_{n-1}} \end{pmatrix}
$$

Proof

Firstly, it may be shown (by induction) that $\delta_k = \det(\Delta_k)$, by expanding this determinant by its last row.

If, then, $\delta_k \neq 0$, $1 \leqslant k \leqslant n$, Theorem 4.3-1 guarantees the existence and uniqueness of the LU factorisation, so that it is enough to *verify* that the factorisation given does in fact equal A. Now a calculation of the matrix LU shows that

$$(LU)_{k,k+1} = c_k, \quad 1 \leqslant k \leqslant n-1,$$

$$(LU)_{11} = \frac{\delta_1}{\delta_0} = b_1, \quad (LU)_{kk} = \frac{a_k c_{k-1}\delta_{k-2} + \delta_k}{\delta_{k-1}}, \quad 2 \leqslant k \leqslant n,$$

$$(LU)_{k,k-1} = a_k, \quad 2 \leqslant k \leqslant n,$$
$$(LU)_{kl} = 0 \quad \text{for } |k - l| \geqslant 2.$$

which, keeping in mind the recurrence formula established for the determinants δ_k, proves the assertion. $\qquad\square$

From the theorem given above, we derive a *particularly simple process for the solution of the linear system* $Av = d$ *whose matrix A is tridiagonal, provided, of course, that the determinants* δ_k, $1 \leqslant k \leqslant n$, *are non-zero*. To do this, it is convenient to 'interpose' the matrix $\Delta = \text{diag}(\delta_i/\delta_{i-1})$ in the LU factorisation of the matrix A, giving the factorisation

$$
A = (L\Delta)(\Delta^{-1}U) = \begin{pmatrix} \dfrac{c_1}{z_1} & & & & \\ a_2 & \dfrac{c_2}{z_2} & & & \\ & \ddots & \ddots & & \\ & & a_n & \dfrac{c_n}{z_n} \end{pmatrix} \begin{pmatrix} 1 & z_1 & & & \\ & 1 & \ddots & & \\ & & \ddots & z_{n-1} \\ & & & 1 \end{pmatrix},
$$

with

$$z_1 = \frac{c_1}{b_1}, \ z_k = c_k \frac{\delta_{k-1}}{\delta_k}, \quad 2 \leqslant k \leqslant n.$$

The solution of the linear system $Av = d$ is then obtained by using in succession the three sequences

$$\begin{cases} z_1 = \dfrac{c_1}{b_1}, \quad z_k = \dfrac{c_k}{b_k - a_k z_{k-1}}, \quad k = 2, 3, \ldots (n, -1) \\[2ex] w_1 = \dfrac{d_1}{b_1}, \quad w_k = \dfrac{z_k}{c_k}(d_k - a_k w_{k-1}) = \dfrac{d_k - a_k w_{k-1}}{b_k - a_k z_{k-1}}, \quad k = 2, 3, \ldots, n, \\[2ex] v_n = w_n, \quad v_k = w_k - z_k v_{k+1}, \quad k = n-1, n-2, \ldots, 1, \end{cases}$$

these steps being equivalent respectively to the recurrence relation $\delta_0 = 1$, $\delta_1 = b_1$, $\delta_k = b_k \delta_{k-1} - a_k c_{k-1} \delta_{k-2}$, $2 \leqslant k \leqslant n$, of Theorem 4.3-2, to the solution of the linear system $L\Delta w = d$, and, finally, to the solution of the linear system $\Delta^{-1} Uv = w$.

This method requires

$$\begin{cases} 3(n-1) & \text{additions,} \\ 3(n-1) & \text{multiplications,} \\ 2n & \text{divisions,} \end{cases}$$

or $8n - 6$ operations altogether, which constitutes a *considerable reduction* in comparison with the number of operations (of the order of $2n^3/3$) required in Gaussian elimination, for a general matrix.

Exercises

4.3-1. Prove that the result of Theorem 4.3-1 remains true if the matrix A is singular, provided that the $n-1$ diagonal submatrices Δ_k, $1 \leqslant k \leqslant n-1$, are invertible. Prove that the element u_{nn} of the matrix $U = (u_{ij})$ is, in that case, zero.

4.3-2. (1) State an equivalent of Theorem 4.3-2 for block tridiagonal matrices.
(2) Consider the linear system whose matrix is block tridiagonal

$$\begin{pmatrix} B_1 & C_1 & & & \\ A_2 & B_2 & C_2 & & \\ & \ddots & \ddots & \ddots & \\ & & & A_N & B_N \end{pmatrix} \begin{pmatrix} V_1 \\ V_2 \\ \vdots \\ V_N \end{pmatrix} = \begin{pmatrix} D_1 \\ D_2 \\ \vdots \\ D_N \end{pmatrix}.$$

With the provision of certain hypotheses, which should be stated explicitly, show that the solution of this linear system may be obtained by constructing successively the three sequences (the first of matrices, the other two of vectors)

$$\begin{cases} Z_1 = B_1^{-1}C_1, \quad Z_k = (B_k - A_k Z_{k-1})^{-1}C_k, \quad k = 2, 3, \ldots, N, \\ W_1 = B_1^{-1}D_1, \quad W_k = (B_k - A_k Z_{k-1})^{-1}(D_k - A_k W_{k-1}), \quad k = 2, 3, \ldots, N, \\ V_N = W_N, \quad V_k = W_k - Z_k V_{k+1}, \quad k = N-1, N-2, \ldots, 1. \end{cases}$$

4.3-3. (1) Find the LU factorisation of the matrices

$$A = \begin{pmatrix} 2 & -1 & 4 & 0 \\ 4 & -1 & 5 & 1 \\ -2 & 2 & -2 & 3 \\ 0 & 3 & -9 & 4 \end{pmatrix}, \quad B = \begin{pmatrix} 3 & -2 & 6 & -5 \\ 24 & -12 & 41 & -39 \\ -27 & 18 & -62 & 54 \\ 9 & 14 & 15 & -47 \end{pmatrix}.$$

(2) Calculate the determinants of the matrices A and B.

4.3-4. (1) Show that partial pivoting in Gaussian elimination applied to a matrix A is equivalent to the LU factorisation of a matrix PA, where P is a permutation matrix.

(2) Show that complete pivoting is equivalent to the LU factorisation of a matrix PAQ, where P and Q are two permutation matrices.

4.3-5. Show that LU *factorisation preserves the band structure of band matrices*, in the following sense (setting $L = (l_{ij})$ and $U = (u_{ij})$):

$$a_{ij} = 0 \quad \text{for } |i - j| \geqslant p \Rightarrow \begin{cases} l_{ij} = 0 & \text{for } i - j \geqslant p, \\ u_{ij} = 0 & \text{for } j - i \geqslant p. \end{cases}$$

4.3-6. It was seen in Chapter 3 that the approximation of the solution of a one-dimensional boundary-value problem leads, under certain conditions, to the solution of equations of the form

$$\begin{cases} a_i v_{i-1} + b_i v_i + c_i v_{i+1} = d_i, \quad 1 \leqslant i \leqslant N, \\ u_0 = u_{N+1} = 0, \end{cases}$$

the numbers a_i, b_i and c_i satisfying the inequalities

$$a_i < 0, \quad c_i < 0, \quad -(a_i + c_i) < b_i, \quad 1 \leqslant i \leqslant N.$$

(1) Show that the numbers z_k which appear in the recurrence relations (with the notation adopted in the text)

$$v_k = w_k - z_k v_{k+1}, \quad k = N-1, N-2, \ldots, 1,$$

satisfy the inequalities

$$-1 < z_k < 0.$$

(2) Prove that if the approximation method is convergent, in the sense that

$$\lim_{h \to 0} \max_{1 \leqslant i \leqslant N} |v_i - \varphi(ih)| = 0, \quad h = \frac{1}{N+1},$$

the function φ being the solution of the boundary-value problem, then the numbers $\max_{1 \leqslant k \leqslant N} |w_k|$ remain bounded independently of N.

Remark. The two properties established above are very important *in practice,* as they show that *overflow* does not occur in the computer in the course of the calculations.

4.3-7. Let p be a non-negative integer. Define the square matrix $A(p)$, of arbitrary order n, by the relations

$$(A(p))_{ij} = \binom{p+j-1}{i-1}, \quad 1 \leqslant i, j \leqslant n.$$

(1) Write down the LU factorisation of this matrix. Deduce that $\det(A(p)) = 1$ (this matrix has already been considered in Exercise 2.2-11, where the aim was to calculate the inverse matrix, whose elements are also integers).

(2) Calculate the elements of the matrices L^{-1} and U^{-1}.

4.3-8. Given a matrix $A = (a_{ij})$ of order n and integers i_l and j_l satisfying

$$1 \leqslant i_1 < i_2 < \cdots < i_p \leqslant n, \quad 1 \leqslant j_1 < j_2 < \cdots < j_p \leqslant n,$$

let $\begin{bmatrix} i_1, i_2, \ldots, i_p \\ j_1, j_2, \ldots, j_p \end{bmatrix}_A$ denote the determinant of the submatrix

$$\begin{pmatrix} a_{i_1 j_1} & a_{i_1 j_2} & \cdots & a_{i_1 j_p} \\ a_{i_2 j_1} & a_{i_2 j_2} & \cdots & a_{i_2 j_p} \\ \vdots & \vdots & & \vdots \\ a_{i_p j_1} & a_{i_p j_2} & \cdots & a_{i_p j_p} \end{pmatrix}.$$

Let $A = (a_{ij})$ be an invertible matrix which admits the factorisation $A = LU$. Defining the matrices $D = \text{diag}((U)_{ii})$ and $V = D^{-1}U$, we may write

$$A = LDV,$$

with

     $L = (l_{ij})$ a lower triangular matrix, $l_{ii} = 1$,

     $D = \text{diag}(d_i)$,

     $V = (v_{ij})$ an upper triangular matrix, $v_{ii} = 1$.

Show that the elements of the matrices L, D and V may be expressed as

$$l_{ij} = \frac{\begin{bmatrix} 1, 2, \ldots, (i-1), i \\ 1, 2, \ldots, (i-1), j \end{bmatrix}_A}{\begin{bmatrix} 1, 2, \ldots, i \\ 1, 2, \ldots, i \end{bmatrix}_A} \quad \text{for} \quad i \geqslant j,$$

$$d_i = \frac{\begin{bmatrix} 1, 2, \ldots, i \\ 1, 2, \ldots, i \end{bmatrix}_A}{\begin{bmatrix} 1, 2, \ldots, (i-1) \\ 1, 2, \ldots, (i-1) \end{bmatrix}_A},$$

$$v_{ij} = \frac{\begin{bmatrix} 1, 2, \ldots, (j-1), i \\ 1, 2, \ldots, (j-1), j \end{bmatrix}_A}{\begin{bmatrix} 1, 2, \ldots, j \\ 1, 2, \ldots, j \end{bmatrix}_A} \quad \text{for} \quad j \geqslant i.$$

Deduce another proof of the uniqueness of the LU factorisation.

4.4 The Cholesky factorisation and method

We shall establish that a *symmetric, positive definite* matrix satisfies the conditions of Theorem 4.3-1; so that such a matrix has a unique LU factorisation. But there is more. It turns out, in fact, that there is an analogous factorisation, but one which is much more simple, because *only one* matrix is involved. More precisely, it will be shown in the theorem given below that it is possible to write such a matrix as

$$A = BB^T, \text{ with B } a \text{ real lower triangular matrix;}$$

this equality constitutes a *Cholesky factorisation* of the matrix A.

Theorem 4.4-1 (Cholesky factorisation of a matrix).

If A is a symmetric, positive definite matrix, then there exists (at least) one real lower triangular matrix B such that

$$A = BB^T$$

Moreover, it is possible to require that the diagonal elements of the matrix B should all be positive; the corresponding factorisation $A = BB^T$ is then unique.

Proof

Let Δ_k denote the (symmetric) submatrices with elements a_{ij}, $1 \leqslant i,j \leqslant k$, of the matrix $A = (a_{ij})$. If $w = (w_i)_{i=1}^k$ is any vector of $\mathbb{R}^k$, one can write $w^T \Delta_k w = v^T A v$, with $v = (v_i)_{i=1}^n \in \mathbb{R}^n$, $v_i = w_i$ for $1 \leqslant i \leqslant k$, and $v_i = 0$ for $k+1 \leqslant i \leqslant n$, which shows that the n submatrices Δ_k are positive definite and, hence, invertible.

Applying Theorem 4.3-1, it is possible to write

$$A = LU = \begin{pmatrix} 1 & & & \\ \times & 1 & & \\ \vdots & \vdots & \ddots & \\ \times & \times & \cdots & 1 \end{pmatrix} \begin{pmatrix} u_{11} & \times & \cdots & \times \\ & u_{22} & \cdots & \times \\ & & \ddots & \vdots \\ & & & u_{nn} \end{pmatrix},$$

and the numbers u_{ii} are all positive since

$$\prod_{i=1}^k u_{ii} = \det(\Delta_k) > 0, \quad 1 \leqslant k \leqslant n.$$

If we 'interpose' the (real) matrix $\Lambda = \text{diag}(\sqrt{u_{ii}})$ into this LU factorisation, we obtain

$$A = (L\Lambda)(\Lambda^{-1}U)$$

$$= \begin{pmatrix} \sqrt{u_{11}} & & & \\ \times & \sqrt{u_{22}} & & \\ \vdots & \vdots & \ddots & \\ \times & \times & \cdots & \sqrt{u_{nn}} \end{pmatrix} \begin{pmatrix} \sqrt{u_{11}} & \times & \cdots & \times \\ & \sqrt{u_{22}} & \cdots & \times \\ & & \ddots & \vdots \\ & & & \sqrt{u_{nn}} \end{pmatrix}.$$

Setting

$$B = L\Lambda, \quad C = \Lambda^{-1}U,$$

the symmetry of the matrix A implies that $BC = C^T B^T$, or, in other words,

$$C(B^T)^{-1} = \begin{pmatrix} 1 & \times & \cdots & \times \\ & 1 & \cdots & \times \\ & & \ddots & \vdots \\ & & & 1 \end{pmatrix} = \begin{pmatrix} 1 & & & \\ \times & 1 & & \\ \vdots & \vdots & \ddots & \\ \times & \times & \cdots & 1 \end{pmatrix} = B^{-1}C^T.$$

Now this matrix equality is only possible if $C(B^T)^{-1} = B^{-1}C^T = I$, that is to say, if $C = B^T$, which proves the *existence* of (at least) one Cholesky factorisation.

We now prove the *uniqueness* of such a factorisation $A = BB^T$, when the diagonal elements b_{ii} of the lower triangular matrix $B = (b_{ij})$ are all positive, which was, in fact, the case in the preceding construction. The factorisation

$$A = (B\Delta^{-1})(\Delta B^T), \quad \text{where } \Delta \overset{\text{def}}{=} \text{diag}(b_{ii}),$$

is just the LU factorisation of the matrix A, with $L = B\Delta^{-1}$, $U = \Delta B^T$. Since such a factorisation is unique,

$$A = B_1 B_1^T = B_2 B_2^T \Rightarrow B_1 \Delta_1^{-1} = B_2 \Delta_2^{-1} \quad \text{and} \quad \Delta_1 B_1^T = \Delta_2 B_2^T,$$

setting $\Delta_\alpha = \text{diag}((B_\alpha)_{ii})$, $\alpha = 1, 2$. The equality of the diagonal elements of the matrices $\Delta_1 B_1^T$ and $\Delta_2 B_2^T$ may be expressed as

$$(B_1)_{ii}^2 = (B_2)_{ii}^2, \quad 1 \leqslant i \leqslant n,$$

which proves that $\Delta_1 = \Delta_2$, *since we have made* the hypothesis that $(B_\alpha)_{ii} > 0$, $\alpha = 1, 2$. From this it follows that $B_1 = B_2$. $\qquad \square$

Remarks

(1) It is possible to present the previous result as a *necessary and sufficient* condition. If a matrix A can be written as $A = BB^T$, with B invertible (it is of no great consequence that the matrix B is triangular), then it is clearly symmetric and, furthermore, it is positive definite, since

$$v^T A v = (Bv)^T Bv > 0 \quad \text{if } v \neq 0.$$

(2) The Cholesky factorisation is sometimes called the 'LL^T factorisation' (L for 'lower'), but we shall avoid this terminology, since the 'new' matrix L, in general, *does not coincide* with the matrix L of the LU factorisation of the same matrix. $\qquad \square$

In practice, we proceed as follows. We set, *a priori*,

$$\mathbf{B} = \begin{pmatrix} b_{11} & & & \\ b_{21} & b_{22} & & \\ \vdots & \vdots & \ddots & \\ b_{n1} & b_{n2} & \cdots & b_{nn} \end{pmatrix}.$$

From the equality $\mathbf{A} = \mathbf{B}\mathbf{B}^{\mathrm{T}}$, there follow the equations

$$a_{ij} = (\mathbf{B}\mathbf{B}^{\mathrm{T}})_{ij} = \sum_{k=1}^{n} b_{ik}b_{jk} = \sum_{k=1}^{\min\{i,j\}} b_{ik}b_{jk}, \quad 1 \leqslant i,j \leqslant n,$$

since $b_{pq} = 0$ if $1 \leqslant p < q \leqslant n$. The matrix A being symmetric, it is enough that the relations given above be satisfied for $i \leqslant j$, for example; that is to say, that the elements b_{ij} of the matrix B satisfy the relations

$$a_{ij} = \sum_{k=1}^{i} b_{ik}b_{jk}, \quad 1 \leqslant i \leqslant j \leqslant n.$$

Setting $i = 1$, there follow

$$(j = 1) \quad a_{11} = (b_{11})^2, \quad \text{so that} \quad b_{11} = \sqrt{a_{11}},$$
$$(j = 2) \quad a_{12} = b_{11}b_{21}, \quad \text{so that} \quad b_{21} = a_{12}/b_{11},$$
$$\vdots \qquad\qquad\qquad\qquad\qquad\qquad \vdots$$
$$(j = n) \quad a_{1n} = b_{11}b_{n1}, \quad \text{so that} \quad b_{n1} = a_{1n}/b_{11}.$$

This determines the first column of B. Step by step, the ith column of the matrix B is determined from the relations

$$(j = i)\ a_{ii} = b_{i1}b_{i1} + \cdots + b_{ii}b_{ii}, \qquad \text{so that } b_{ii} = \sqrt{\left[a_{ii} - \sum_{k=1}^{i-1} (b_{ik})^2 \right]},$$

$$(j = i + 1)\ a_{i,i+1} = b_{i1}b_{i+1,1} + \cdots + b_{ii}b_{i+1,i},$$

$$\vdots$$

$$\text{so that } b_{i+1,i} = \frac{a_{i,i+1} - \sum_{k=1}^{i-1} b_{ik}b_{i+1,k}}{b_{ii}},$$

$$(j = n)\ a_{in} = b_{i1}b_{n1} + \cdots + b_{ii}b_{ni}, \qquad \text{so that } b_{ni} = \frac{a_{in} - \sum_{k=1}^{i-1} b_{ik}b_{nk}}{b_{ii}},$$

the first i-1 columns having been determined.

It follows from Theorem 4.4-1 that it is possible to choose all the elements b_{ii} to be positive. This choice ensures that *all the quantities* $a_{ii}, \ldots,$ $a_{ii} - \sum_{k=1}^{i-1} (b_{ik})^2, \ldots$ *which appear under the root sign are positive* (a fact which was by no means evident *a priori*).

The *Cholesky method*, for the solution of a linear system $Au = b$ with a matrix A that is symmetric and positive definite, consists in calculating the Cholesky factorisation $A = BB^T$ of the matrix and then solving successively the two linear systems with triangular matrices,

$$Bw = b \quad \text{and} \quad B^Tu = w.$$

We now count the number of elementary operations involved in the course of applying this method.

(i) *Factorisation*. The calculation of the matrix B through the formulae given above requires

$$n \text{ root extractions,}$$

$$\begin{cases} (n-1) + (n-2) + \cdots + 1 = \dfrac{n(n-1)}{2} \quad \text{divisions,} \\[2mm] (n-1) + 2(n-2) + \cdots + (n-2)2 + (n-1) = \dfrac{n^3 - n}{6} \quad \text{additions,} \\[2mm] (n-1) + 2(n-2) + \cdots + (n-2)2 + (n-) = \dfrac{n^3 - n}{6} \quad \text{multiplications.} \end{cases}$$

(ii) *Solution of the two linear systems* $Bw = b$ *and* $B^Tu = w$. It has already been seen that these solutions require

$$\begin{cases} n(n-1) \text{ additions,} \\ n(n-1) \text{ multiplications,} \\ 2n \text{ divisions.} \end{cases}$$

Altogether, *the Cholesky method requires of the order of*

$$\begin{cases} \dfrac{n^3}{6} \quad \text{additions,} \\[2mm] \dfrac{n^3}{6} \quad \text{multiplications,} \\[2mm] \dfrac{n^2}{2} \quad \text{divisions,} \\[2mm] n \text{ root extractions,} \end{cases}$$

which compares very favourably with the $n^3/3$ additions, $n^3/3$ multiplications and $n^2/2$ divisions of Gaussian elimination. *It is, therefore, advan-*

tageous to use the Cholesky method rather than Gaussian elimination for the solution of a linear system with a matrix which is symmetric and positive definite.

Lastly, it should be observed that the calculation of the determinant of a symmetric, positive definite matrix comes at once from the Cholesky factorisation $A = BB^T$, since

$$\det(A) = (b_{11}b_{22} \cdots b_{nn})^2.$$

Exercises

4.4-1. What is the Cholesky factorisation of the matrix

$$A = \begin{pmatrix} 1 & 2 & 3 & 4 \\ 2 & 5 & 1 & 10 \\ 3 & 1 & 35 & 5 \\ 4 & 10 & 5 & 45 \end{pmatrix}?$$

4.4-2. (1) Let $A = (a_{ij})$ be a symmetric matrix of order n such that the determinants of the matrices Δ_k with elements a_{ij}, $1 \leqslant i, j \leqslant k$, are positive for $k = 1, 2, \ldots, n$. Show that the matrix A is positive definite (the converse was established in the course of the proof of Theorem 4.4-1).

(2) Deduce from this a procedure, based on the Cholesky factorisation, which enables one to decide whether a symmetric matrix is positive definite or not (assuming that all calculations are carried out without any rounding errors).

4.4-3. How many distinct Cholesky factorisations BB^T are there of a symmetric, positive definite matrix (when, of course, it is not assumed that the diagonal elements of the matrix B are necessarily positive)?

4.4-4. Repeat the analysis of the text for Hermitian matrices.

4.4-5. (1) Let A be an invertible, symmetric matrix having an LU factorisation. Show that it is possible to express A in the form $A = B\tilde{B}^T$, where B is a lower triangular matrix and $\tilde{B}$ is a matrix each of whose columns is equal either to the corresponding column of B or to the corresponding column of B with the sign changed.

(2) As a numerical application, try this out on

$$A = \begin{pmatrix} 1 & 2 & 1 & 1 \\ 2 & 3 & 4 & 3 \\ 1 & 4 & -4 & 0 \\ 1 & 3 & 0 & 0 \end{pmatrix}.$$

4.4-6. Let A be a full, invertible matrix. In solving the linear system $A^2u = b$, is it more advantageous to calculate the matrix A^2 or to solve the two linear systems $Av = b$ and $Au = v$, in turn, making use of one and the same LU factorisation? Is the conclusion different if the matrix A is symmetric and positive definite?

4.4-7. Show *directly*, in terms of the formulae giving the elements b_{ij} of the matrix B, that *the Cholesky factorisation* $A = BB^T$ *preserves the band structure of band matrices*, in the sense that (compare with Exercise 4.3-5)

$$a_{ij} = 0 \quad \text{for} \quad |i-j| \geqslant p \Rightarrow b_{ij} = 0 \quad \text{for} \quad i-j \geqslant p.$$

4.5 The QR factorisation of a matrix and Householder's method

A matrix which is of the form

$$H(v) = I - 2\frac{vv^*}{v^*v}, \ v \text{ being a } non\text{-}zero \text{ vector of } \mathbb{C}^n,$$

is called a *Householder matrix*. We shall further adopt the *convention* of regarding *the identity matrix* as a Householder matrix, in order to simplify the presentation of a number of statements (but, of course, this convention represents a slight misuse of language).

Such matrices (whose geometric interpretation is given in Exercise 4.5-1) are both *unitary* and *Hermitian* (this is immediately verifiable); interest in them within the Numerical Analysis of Matrices derives from the following result.

Theorem 4.5-1

Let $a = (a_i)_{i=1}^n$ be a vector in $\mathbb{C}^n$ such that $\sum_{i=2}^n |a_i| > 0$. There exist two Householder matrices H such that the last $n-1$ components of the vector Ha are zero. More precisely, let $\alpha \in \mathbb{R}$ be such that $a_1 = e^{i\alpha}|a_1|$. Then,

$$H(a + \|a\|_2 e^{i\alpha} e_1)a = -\|a\|_2 e_1 \text{ and } H(a - \|a\|_2 e^{i\alpha} e_1)a = \|a\|_2 e_1,$$

where e_1 represents the first basis vector of $\mathbb{C}^n$.

Proof

We observe at once that the condition $\sum_{i=2}^n |a_i| > 0$ implies that the vectors $a \pm \|a\|_2 e_1$ are certainly not zero (a condition that is necessary to the definition of the corresponding Householder matrices). Setting $\|\cdot\|_2 = \|\cdot\|$, in order to make the notation less cumbersome, *we verify* that the Householder matrices put forward are suitable to the purpose (for a more 'natural' approach, see Exercise 4.5-2) if $a_1 > 0$ (in order to fix ideas; the calculations are analogous in other cases):

$$H(a \pm \|a\|e_1)a = a - \frac{2(a \pm \|a\|e_1)(a^* \pm \|a\|e_1^*)a}{(a^* \pm \|a\|e_1^*)(a \pm \|a\|e_1)}.$$

Now,

$$(a \pm \|a\|e_1)(a^* \pm \|a\|e_1^*)a = \|a\|(\|a\| \pm a_1)(a \pm \|a\|e_1),$$
$$(a^* \pm \|a\|e_1^*)(a \pm \|a\|e_1) = 2\|a\|(\|a\| \pm a_1). \qquad \square$$

Remark

If $\sum_{i=2}^{n}|a_i| = 0$, we still have (recall that the identity matrix is a particular Householder matrix)

$$Ia = \|a\|_2 e_1 \quad \text{if} \quad a_1 \geqslant 0,$$
$$H(a - \|a\|_2 e_1)a = \|a\|_2 e_1 \quad \text{if} \quad a_1 < 0.$$

Hence, *if the vector a is real, the first component of the vector Ha may be chosen to be non-negative.* Since the interest behind this remark does not appear until the proof of Theorem 4.5-2, it may be ignored *for the time being.* $\qquad \square$

In practice, we proceed as follows. We calculate the norm $\|a\|_2$ (it is the only root extraction which is necessary), then the vector $v = a \pm \|a\|_2 e^{i\alpha}e_1$, and then the number

$$\frac{v^*v}{2} = \|a\|_2(\|a\|_2 \pm |a_1|).$$

If b is a vector of $\mathbb{C}^n$, the calculation of the vector Hb is effected by calculating first the scalar product v^*b and then the vector

$$Hb = b - \frac{v^*b}{v^*v/2}v.$$

In the real case, the *choice of sign* (in front of the vector $\|a\|_2 e_1$) is guided by the presence of the expression v^*v in the denominator. In order to avoid division by numbers which are too 'small' (an eventuality which can have disastrous consequences, as was seen in section 4.2 in the context of Gaussian elimination), we choose $v = a + \|a\|_2 e_1$ if $a_1 > 0$ and $v = -\|a\|_2 e_1$ if $a_1 < 0$.

Householder's method for the solution of a linear system $Au = b$ is equivalent to finding $n - 1$ Householder matrices $H_1, H_2, \ldots, H_{n-1}$ such that the matrix $H_{n-1} \ldots H_2 H_1 A$ is upper triangular. It then remains to solve the system

$$H_{n-1} \cdots H_2 H_1 Au = H_{n-1} \cdots H_2 H_1 b$$

by back-substitution.

Setting $A = A_1$, each matrix

$$A_k = H_{k-1} \cdots H_2 H_1 A, \quad k \geqslant 1,$$

appears in the form

$$
A_k =
\begin{pmatrix}
\times & \times & \times & \times & \times & \times & \times & \times \\
 & \times & \times & \times & \times & \times & \times & \times \\
 & & \times & \times & \times & \times & \times & \times \\
 & & & \boxed{\times} & \times & \times & \times & \times \\
 & & & \times & \times & \times & \times & \times \\
 & & & \times & \times & \times & \times & \times \\
 & a_k & & \times & \times & \times & \times & \times \\
 & & & \times & \times & \times & \times & \times
\end{pmatrix}
\quad \leftarrow k\text{th row}
$$

$$\uparrow$$
$$k\text{th column}$$

Remark

The distribution of zeros is, therefore, the same as that in the matrix found after the $(k-1)$th step of Gaussian elimination, but the transition from the matrix A_k to the matrix A_{k+1} is different, as we shall see. Furthermore, it should be noted that, in contrast to Gaussian elimination, only the first $k-1$ rows of the matrix A_k remain unchanged in the matrix A_{k+1}. □

We denote by a_k the column vector of $\mathbb{C}^{n-k+1}$ whose components are the elements a_{ik}^k, $k \leqslant i \leqslant n$, of the matrix $A_k = (a_{ij}^k)$. If $\sum_{i=k+1}^{n} |a_{ik}| > 0$, there exists, in accordance with Theorem 4.5-1, a vector $\tilde{v}_k \in \mathbb{C}^{n-k+1}$ such that all the components, except the first, of the vector $H(\tilde{v}_k)a_k \in \mathbb{C}^{n-k+1}$ are zero.

Let v_k denote the vector of $\mathbb{C}^n$ whose first $k-1$ components are zero and the last $n-k+1$ are those of the vector $\tilde{v}_k$. Then, the matrix

$$
H_k = \left(\begin{array}{c|c} I_{k-1} & 0 \\ \hline 0 & H(\tilde{v}_k) \end{array} \right), \quad I_{k-1} = \text{the identity matrix of order } k-1,
$$

is just the Householder matrix $H(v_k)$ and the matrix $A_{k+1} = H_k A_k = (a_{ij}^{k+1})$ is such that $a_{ik}^{k+1} = 0$ for $i = k+1, \ldots, n$. Obviously, if $a_{ik}^k = 0$ for $i = k+1$, $\ldots, n$, then the matrix A_k is already of the form A_{k+1}, and it is enough to set $H_k = I$. We continue in this way until we reach the matrix

$$
A_n = H_{n-1} \cdots H_2 H_1 A,
$$

which is upper triangular by construction.

It should be observed in passing that the condition number of the matrix of the system is not altered since (Theorem 2.2-3)

$$
\text{cond}_2(A_n) = \text{cond}_2(A);
$$

this constitutes an advantage, from the point of view of 'numerical stability', of Householder's method over Gaussian elimination; however, this is achieved at the cost of nearly double the number of elementary operations (cf. Exercise 4.5-3).

Householder's method permits a simple evaluation of the *determinant* of the matrix A. In fact, since the determinant of a Householder matrix other than the identity matrix is equal to -1 (cf. Exercise 4.5-1),

$$\det(A) = \pm a_{11}^1 a_{22}^2 \cdots a_{nn}^n,$$

depending on the parity of the number of Householder matrices (other than the identity matrix) encountered.

The matrix interpretation of Householder's method leads to a remarkable result about the factorisation of square matrices.

Theorem 4.5-2 (*The* QR *factorisation of a matrix*).
Given a matrix A of order n, there exists a unitary matrix Q and an upper triangular matrix R such that

$$A = QR.$$

Moreover, it is possible to ensure that all the diagonal elements of the matrix R are non-negative. If the matrix A is invertible, the corresponding factorisation A = QR is then unique.

Proof
Observing that the existence of the Householder matrices $H_1, H_2, \ldots, H_{n-1}$ is in fact independent of the possibly singular nature of the matrix A, it has already been established that every square matrix A may be written as

$$A = (H_{n-1} \cdots H_2 H_1)^{-1} A_n,$$

the matrix

$$R \overset{\text{def}}{=} A_n$$

being upper triangular. The matrix

$$Q \overset{\text{def}}{=} (H_{n-1} \cdots H_2 H_1)^{-1} = H_1 H_2 \cdots H_{n-1}$$

being unitary (we recall that Householder matrices H_k satisfy $H_k^{-1} = H_k^* = H_k$), *the existence* of (at least) one QR factorisation is established, the matrix Q being the product of the $n-1$ Householder matrices.

Once the factorisation A = QR has been obtained, let $\alpha_j \in \mathbb{R}$ be numbers which satisfy $(R)_{jj} = e^{i\alpha_j} |(R)_{jj}|$, $1 \leqslant j \leqslant n$, and let $D = \text{diag}(e^{i\alpha_j})$. Then the matrix $\tilde{Q} = QD$ is also unitary; the matrix $\tilde{R} = D^{-1}R$ is also upper triangular and, furthermore, the diagonal elements are all non-negative. Hence, the existence of (at least) one factorisation $A = \tilde{Q}\tilde{R}$ with $(\tilde{R})_{jj} \geqslant 0$ has been established.

If the matrix A is invertible, there exists at least one factorisation $A = QR$ such that $(R)_{ii} > 0$, $1 \leqslant i \leqslant n$. Let us now show the uniqueness of this factorisation. From the equalities

$$A = Q_1 R_1 = Q_2 R_2,$$

it follows that

$$Q_2^* Q_1 = R_2 R_1^{-1} \overset{\text{def}}{=} \Delta.$$

Consequently,

$$\Delta^* \Delta = Q_1^* Q_2 Q_2^* Q_1 = I$$

is precisely a Cholesky factorisation of the identity matrix, since the matrix Δ is upper triangular (being the product of the upper triangular matrices R_2 and R_1^{-1}). Since

$$(R_1)_{ii} > 0 \quad \text{and} \quad (R_2)_{ii} > 0 \Rightarrow (\Delta)_{ii} = \frac{(R_2)_{ii}}{(R_1)_{ii}} > 0,$$

the uniqueness of the Cholesky factorisation (established for the real case in Theorem 4.4-1; the extension to the complex case is immediate) implies that $\Delta = I$. $\qquad\qquad\square$

Remarks

(1) If the matrix A is real, the Householder matrices $H_1, H_2, \ldots, H_{n-1}$ are all real (cf. Theorem 4.5-1); it follows that the matrix Q, and hence also the matrix $R = Q^{-1}A$, is real, the matrix Q being, in addition, orthogonal.

(2) If the restriction $(R)_{ii} > 0$, $1 \leqslant i \leqslant n$, is no longer imposed, it is easy to see in what precise way two factorisations of the same matrix A may be different. Let

$$A = Q_1 R_1 = Q_2 R_2.$$

It follows from the second equality that the unitary matrix $Q_2^* Q_1$ is equal to the upper triangular matrix $R_2 R_1^{-1}$. But since every unitary triangular matrix is diagonal (if Δ is such a matrix, write $(\Delta^* \Delta)_{ij} = 0$ for $i \neq j$), it follows that

$$Q_2^* Q_1 = R_2 R_1^{-1} = \text{diag}(d_i), \quad \text{with } |d_i| = 1. \qquad\square$$

Finally, we indicate a remarkable *interpretation of the QR factorisation* of an invertible matrix A. Denoting by $a_1, a_2, \ldots, a_n$ and $q_1, q_2, \ldots, q_n$ the column vectors of the matrices A and Q respectively, the equation $A = QR$ may also be written as

$$\begin{cases} a_1 = r_{11} q_1, \\ a_2 = r_{12} q_1 + r_{22} q_2, \\ \vdots \\ a_n = r_{1n} q_1 + r_{2n} q_2 + \cdots + r_{nn} q_n, \end{cases}$$

while denoting by r_{ij} the elements of the upper triangular matrix R. Now, since the vectors q_i form an orthonormal set (this is another way of expressing the unitary character of the matrix Q), the equations given above are equivalent to the *Gram–Schmidt orthogonalisation process*. It might seem desirable to put them to use for a 'direct' construction of the matrices Q and R (some indications on the subject are given in Exercise 4.5-4), but this method, although 'natural', is to be avoided, as a general rule, because it leads to the sometimes disastrous propagation of rounding errors. The method based on the use of Householder matrices is indubitably preferable.

Exercises

4.5-1. (1) Let v be a real vector which satisfies $v^\mathsf{T}v = 1$. Show that the Householder matrix $H(v) = I - 2vv^\mathsf{T}$ represents the identity transformation when restricted to the vector subspace spanned by the vectors which are orthogonal to the vector v. Deduce that $\det(H(v)) = -1$.

(2) Prove that every orthogonal matrix is the product of n Householder matrices. Deduce a geometrical interpretation of orthogonal matrices.

(3) Apply the results obtained above to the matrix

$$\begin{pmatrix} \cos\theta & \sin\theta \\ -\sin\theta & \cos\theta \end{pmatrix}.$$

4.5-2. Let a and b be two given real vectors satisfying $a^\mathsf{T}a = b^\mathsf{T}b$ and $a \neq b$. Show that the only real solutions of the equations

$$(I - 2uu^\mathsf{T})a = b, \quad u^\mathsf{T}u = 1,$$

are $u = \pm(c^\mathsf{T}c)^{-1/2} c$, where $c = b - a$. We recover, in this way, the Householder matrices obtained in Theorem 4.5-1.

4.5-3. Show that the number of elementary operations which are used in Householder's method (the operations being carried out in the way indicated after the proof of Theorem 4.5-1) is of the order of

$$\begin{cases} n & \text{root extractions,} \\[2mm] \dfrac{2n^3}{3} & \text{additions,} \\[3mm] \dfrac{2n^3}{3} & \text{multiplications,} \\[3mm] \dfrac{n^2}{2} & \text{divisions.} \end{cases}$$

Compare with Gaussian elimination.

4.5-4. Assuming the matrix A to be invertible, show that the relations

$$a_i = r_{1i}q_1 + r_{2i}q_2 + \cdots + r_{ii}q_i, \quad 1 \leqslant i \leqslant n,$$

enable us to calculate, step by step, the numbers r_{ij} and the vectors q_j in the following order: r_{11}, q_1, r_{12}, r_{22}, q_2, etc., and that it is, in fact, possible so to arrange matters as to ensure that all the elements r_{ii} are positive and that the vectors q_i form an orthonormal set. Lastly, calculate the number of elementary operations required by this method, which is 'numerically unstable', as we have already underlined.

4.5-5. Prove that the functions $A \to Q$ and $A \to R$, with $(R)_{ii} > 0$, defined unambiguously over the set of invertible matrices (Theorem 4.5-2), are continuous over this set.

4.5-6. Let $A = (a_{ij})$ be an invertible matrix which admits the factorisation $A = QR$, with $(R)_{ii} > 0$. Defining the matrices $D = \mathrm{diag}\,((R)_{ii})$ and $S = D^{-1}R$, we may write

$$A = QDS,$$

with

Q a unitary matrix,

$D = \mathrm{diag}\,(d_i)$, $d_i > 0$,

$S = (s_{ij})$ an upper triangular matrix, $s_{ii} = 1$.

Using the same notation as in Exercise 4.3-8, show that

$$(d_i)^2 = \frac{\begin{bmatrix} 1,2,\ldots,i \\ 1,2,\ldots,i \end{bmatrix}_{A^*A}}{\begin{bmatrix} 1,2,\ldots,(i-1) \\ 1,2,\ldots,(i-1) \end{bmatrix}_{A^*A}},$$

$$s_{ij} = \frac{\begin{bmatrix} 1,2,\ldots,(i-1),i \\ 1,2,\ldots(i-1),j \end{bmatrix}_{A^*A}}{\begin{bmatrix} 1,2,\ldots,i \\ 1,2,\ldots,i \end{bmatrix}_{A^*A}}, \quad \text{for } j \geqslant i.$$

Iterative methods for the solution of linear systems

Introduction

This chapter gives only a brief look at iterative methods for the solution of linear systems; and we have limited ourselves to the most basic methods. These appear in the form

$$u_{k+1} = Bu_k + c, \quad k \geqslant 0, \quad u_0 \text{ an arbitrary vector,}$$

the matrix B and the vector c being constructed from the data of a linear system $Au = b$ (the matrix B depends only on the matrix A).

We begin by establishing in section 5.1 general results on *convergence* (sufficient conditions which ensure that $\lim_{k \to \infty} u_k = u$) and on the *comparison* (evaluation of the 'rates of convergence') of iterative methods (cf. Theorems 5.1-1 and 5.1-2; see, also, Theorem 5.3-1). Then, in section 5.2, certain methods are described which are much in use at present, *the point* or *block Jacobi, Gauss–Seidel and relaxation methods* (the Gauss–Seidel method being a particular case of the relaxation method).

Finally, in section 5.3, some exact results are given on *convergence* and on the *comparison of rates of convergence*, applicable to the methods described above, notably in the case of linear systems whose matrix A is *symmetric, positive definite and block tridiagonal* (Theorem 5.3-6). This particular case is very important, since it is encountered in methods of approximation of elliptic boundary-value problems such as those reviewed in Chapter 3.

5.1 General results on iterative methods

Given an invertible matrix A and a vector b, we would like to find the solution u of the linear system

$$Au = b.$$

Suppose that a matrix B and a vector c have been found (for example, by one of the methods described in the following section) such that the matrix $I - B$ is invertible and such that the unique solution of the linear system

$$u = Bu + c$$

is also the solution of $Au = b$. The form of the system $u = Bu + c$ suggests the definition of an *iterative method* for the solution of the linear system $Au = b$. Choosing an arbitrary *initial vector* u_0, the sequence of vectors $(u_k)_{k \geqslant 0}$ is defined by

$$u_{k+1} = Bu_k + c, \; k \geqslant 0.$$

We then say (a definition which is entirely natural!) that the iterative method is *convergent* if

$$\lim_{k \to \infty} u_k = u \quad \text{for } \textit{every} \text{ initial vector } u_0.$$

The following result gives the fundamental criterion for the *convergence* of iterative methods. It will be observed that it relies only on the matrix B, which is called the *matrix of the iterative method* being considered.

Theorem 5.1-1
The following statements are equivalent.

(1) *The iterative method is convergent;*
(2) $\varrho(B) < 1$;
(3) $\| B \| < 1$ *for at least one matrix norm* $\| \cdot \|$.

Proof
To say that the method is convergent amounts to saying that

$$\lim_{k \to \infty} e_k = 0 \quad \text{for every vector } e_0 = u_0 - u,$$

where

$$e_k \overset{\text{def}}{=} u_k - u = B^k e_0, \quad k \geqslant 0,$$

is the kth *error vector*. Accordingly, the equivalences we seek follow from Theorem 1.5-1. □

Remark
The iterative method defined above is precisely the *method of successive approximations*, also known as *Picard's method*, for finding a fixed point of the function

$$f : v \in \mathbb{K}^n \to f(v) = Bv + c \in \mathbb{K}^n,$$

which is a *contraction mapping* when property (3) is satisfied. The proof of Theorem 1.4-3 shows, in effect, that the matrix norm, in whose respect $\| B \| < 1$, may be chosen to be subordinate to a vector norm $\| \cdot \|$ in such a way that

$$\| f(v) - f(u) \| \leqslant \| B \| \, \| v - u \|.$$ □

How do we go about choosing one from among a *number* of *convergent* iterative methods for the solution of one and the *same* linear system Au = b? To fix ideas, suppose that the matrix B is normal. Then (Theorem 1.4-2)

$$\| B^k e_0 \|_2 \leqslant \| B^k \|_2 \| e_0 \|_2 = (\varrho(B))^k \| e_0 \|_2,$$

and this inequality is the best possible, in the sense that, for every integer $k \geqslant 0$, there exists a vector $e_0(k) \neq 0$ for which equality is attained. In the case of normal matrices, then, the smaller the number $\varrho(B)$, the faster the method is, since

$$\sup_{\| e_0 \|_2 = 1} \| B^k e_0 \|_2^{1/k} = \varrho(B) \quad \text{for every } k \geqslant 0.$$

In the general case (for any matrix B and any vector norm), the conclusion is analogous: *asymptotically, the error vector* $e_k = B^k e_0$ *behaves 'at worst' like* $(\varrho(B))^k$. This is stated more precisely in the result which follows.

Theorem 5.1-2

(1) *Let* $\| \cdot \|$ *be any vector norm, and let u be such that*

$$u = Bu + c.$$

Consider the iterative method

$$u_{k+1} = Bu_k + c, \quad k \geqslant 0.$$

Then

$$\lim_{k \to \infty} \left\{ \sup_{\| u_0 - u \| = 1} \| u_k - u \|^{1/k} \right\} = \varrho(B).$$

(2) *Let* $\| \cdot \|$ *be any vector norm, and let u be such that*

$$u = Bu + c = \tilde{B}u + \tilde{c}.$$

Consider the iterative methods

$$\tilde{u}_{k+1} = \tilde{B}\tilde{u}_k + \tilde{c}, \quad k \geqslant 0, \quad u_{k+1} = Bu_k + c, \quad k \geqslant 0,$$

with

$$\rho(B) < \rho(\tilde{B}), \quad u_0 = \tilde{u}_0.$$

Then, for any number $\varepsilon > 0$, there exists an integer $l = l(\varepsilon)$ such that

$$k \geqslant l \Rightarrow \sup_{\| u_0 - u \| = 1} \left\{ \frac{\| \tilde{u}_k - u \|}{\| u_k - u \|} \right\}^{1/k} \geqslant \frac{\varrho(\tilde{B})}{\varrho(B) + \varepsilon}.$$

Proof

Let $\| \cdot \|$ be the subordinate matrix norm. For every integer k, one can write

$$(\varrho(B))^k = \varrho(B^k) \leqslant \| B^k \| = \sup_{\| e_0 \| = 1} \| B^k e_0 \|,$$

so that

$$\varrho(B) \leqslant \sup_{\|e_0\|=1} \| B^k e_0 \|^{1/k} = \| B^k \|^{1/k},$$

and the statement (1) follows from Theorem 1.5-2. By the same result, given $\varepsilon > 0$, there exists an integer $l = l(\varepsilon)$ such that

$$k \geqslant l \Rightarrow \sup_{\|e_0\|=1} \| B^k e_0 \|^{1/k} \leqslant (\varrho(B) + \varepsilon).$$

Moreover, for every integer $k \geqslant l$, there exists a vector $e_0 = e_0(k)$ such that

$$\|e_0\| = 1 \quad \text{and} \quad \| \tilde{B}^k e_0 \|^{1/k} = \| \tilde{B}^k \|^{1/k} \geqslant \varrho(\tilde{B}),$$

and the statement (2) is proved. $\qquad\qquad\qquad\qquad\qquad\qquad\qquad\square$

The investigation, therefore, of iterative methods rests on the solution of the following two problems:

(1) *given an iterative method with matrix* B, *determine whether the method is convergent*, that is to say, whether $\varrho(B) < 1$, or, equivalently, whether there exists a matrix norm such that $\| B \| < 1$ (Theorem 5.1-1);

(2) *given two convergent iterative methods, compare them.* The iterative method which is 'faster' is that whose matrix has the smaller spectral radius (Theorem 5.1-2).

Exercises

5.1-1. What can be said of an iterative method whose matrix has zero as the spectral radius?

5.1-2. The aim in this exercise is to study a procedure currently in use for *accelerating the convergence* of an iterative method, making use of *Chebyshev polynomials*. In what follows, A denotes a symmetric, positive definite matrix with eigenvalues $0 < \lambda_1 \leqslant \cdots \leqslant \lambda_n$. In order to approximate the solution of the linear system $Au = b$, we define the iterative method

$$u_{k+1} = u_k + \alpha_k(b - Au_k),$$

the sequence of numbers $\alpha_k > 0$ being adjusted 'as best may be'. Set $e_k = u_k - u$.

(1) Prove that there exists a polynomial $q_k \colon \mathscr{A}_n(\mathbb{R}) \to \mathscr{A}_n(\mathbb{R})$ such that

$$e_k = q_k(A)e_0.$$

What are the zeros of the polynomial $q_k \colon \mathbb{R} \to \mathbb{R}$?

(2) Prove the inequality

$$\| q_k(A) \|_2 \leqslant \max_{\lambda_1 \leqslant t \leqslant \lambda_n} | q_k(t) |.$$

(3) For the case where $\alpha_k = \alpha > 0$ independently of the interger k, what conditions does α have to satisfy in order to ensure convergence? Determine the optimal value of α.

(4) Given a non-negative integer k, the kth *Chebyshev polynomial* is defined by

$$C_k: x \in \mathbb{R} \to C_k(x) = \begin{cases} \cos(k \cos^{-1} x), & |x| \leqslant 1, \\ \frac{1}{2}\{[x + \sqrt{(x^2 - 1)}]^k + [x - \sqrt{(x^2 - 1)}]^k\}, & |x| > 1. \end{cases}$$

Verify that the function C_k is, in fact, a polynomial of degree k, by establishing, for example, the recurrence relation

$$\begin{cases} C_{k+2}(x) = 2x C_{k+1}(x) - C_k(x), & x \in \mathbb{R}, \ k \geqslant 0, \\ C_0(x) = 1, \ C_1(x) = x, & x \in \mathbb{R}. \end{cases}$$

(5) Prove that the polynomial C_k is a solution of the following minimisation problem. Given any real number a satisfying $|a| > 1$, find a polynomial $p \in P_k$ (P_k being the set of polynomials of degree $\leqslant k$) such that

$$\max_{-1 \leqslant x \leqslant 1} |p(x)| = \inf_{\left\{\substack{q \in P_k \\ q(a) = C_k(a)}\right\}} \max_{-1 \leqslant x \leqslant 1} |q(x)|.$$

(6) Show that the solution of this minimisation problem is unique.

(7) Now let α and β be two real numbers which satisfy $-1 < \alpha \leqslant \beta < 1$. Deduce from the preceding that the problem

find a polynomial $p \in P_k$ such that
$$\max_{\alpha \leqslant x \leqslant \beta} |p(x)| = \inf_{\left\{\substack{q \in P_k \\ q(1) = 1}\right\}} \max_{\alpha \leqslant x \leqslant \beta} |q(x)|$$

has a unique solution (which should be given explicitly).

(8) For fixed integer k, what is the best way of choosing the numbers α_l so as to minimise $\|q_k(A)\|_2$?

(9) For 'large' values of the integer k and of the ratio $(\lambda_n + \lambda_1)/(\lambda_n - \lambda_1)$, compare the convergence of the iterative method 'with constant α' of question (3) with that of the 'optimal' iterative method of question (8).

5.2 Description of the methods of Jacobi, Gauss–Seidel and relaxation

The methods which we are about to describe are particular cases of the following method. Given a linear system $Au = b$, it is supposed that it is possible to express the invertible matrix A by way of a regular splitting[†]

$$A = M - N,$$

where M is an *invertible* matrix which *'it is easy to invert'*, that is to say, *in practice*, it is diagonal or triangular, or block diagonal or triangular. Obviously, when we say that it is easy to invert the matrix M, *we do not wish to imply* that the matrix M^{-1} is actually calculated. What there is to calculate is the *solution of the linear system whose matrix is* M.

[†]This usage is different from that in other works

We then have the following equivalences:

$$Au = b \Leftrightarrow Mu = Nu + b \Leftrightarrow u = M^{-1}Nu + M^{-1}b.$$

The last equation is of the form $u = Bu + c$, so that there is associated with it the iterative method (in accordance with the considerations of the previous section)

$$u_{k+1} = M^{-1}Nu_k + M^{-1}b, \quad k \geqslant 0, \quad u_0 \quad \text{an arbitrary vector,}$$

whose matrix is

$$B = M^{-1}N = I - M^{-1}A,$$

which shows, by the way, that the matrix $I - B = M^{-1}A$ is invertible. *In practice*, we are led to the solution of the sequence of linear systems

$$Mu_{k+1} = Nu_k + b, \quad k \geqslant 0.$$

In the case of the first two methods (that of Jacobi and Gauss–Seidel) the regular splitting $A = M - N$ of the matrix $A = (a_{ij})$ is such that

$$(M)_{ij} = a_{ij} \quad \text{or} \quad 0, \quad \text{depending} \quad \text{on the values of the pair } (i, j);$$

this will be represented, with a (slight) misuse of notation, by the equality

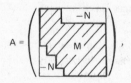

in order to show clearly that the matrices M and N are 'disjoint'. It appears, in this way, in a manner that is visually *very* striking, that the corresponding iterative method, broadly speaking, consists in 'inverting only the part M of A'.

Intuitively, it would seem that, the more M consists of the non-zero elements of the matrix A, the better the method should be (this will be a useful point in the heuristic comparison of the methods of Jacobi and Gauss–Seidel introduced later on), since, in considering the 'limiting' case where $M = A$ (so that $N = 0$), the *exact* solution is found just by solving the *first* equation of the iterative method. There are other regular splittings $A = M - N$ which are not of this type, in that the matrices M and *N* 'overlap one another', as is the case in the relaxation method, for example, which is described further on.

We start with *the Jacobi method*. Let $A = (a_{ij})$ be a matrix of order n such that

$$a_{ii} \neq 0, \quad 1 \leqslant i \leqslant n,$$

and let

$$A = \begin{pmatrix} a_{11} & a_{12} & a_{13} & \cdot & \cdot & \cdot & a_{1n} \\ a_{21} & a_{22} & a_{23} & \cdot & \cdot & \cdot & a_{2n} \\ a_{31} & a_{32} & a_{33} & & -F & & a_{3n} \\ \cdot & \cdot & & D & & & \vdots \\ \cdot & \cdot & -E & & & & \\ \cdot & \cdot & & & & & a_{n-1,n} \\ a_{n1} & a_{n2} & a_{n3} & \cdots & a_{n,n-1} & & a_{nn} \end{pmatrix} = D - E - F,$$

which will be called the *point splitting* $A = D - E - F$ of the matrix A.

Remarks

(1) Obviously enough, the expression given above once again constitutes a notational misuse (but one which is suggestive!), the letters D, E and F representing matrices of order n given by respectively

$$(D)_{ij} = a_{ij}\delta_{ij},$$
$$(-E)_{ij} = a_{ij} \quad \text{if} \quad i > j, \quad 0 \quad \text{otherwise},$$
$$(-F)_{ij} = a_{ij} \quad \text{if} \quad i < j, \quad 0 \quad \text{otherwise}.$$

(2) In contrast to the use of the matrices M and N, which 'vary' from one method to the next, here the regular splitting $A = D - E - F$ should be considered *fixed*, at least as regards the 'point' methods which we are considering at present. $\qquad\square$

The first regular splitting $A = M - N$ to be considered is the 'simplest possible', since the matrix M is equal to the matrix $D = \text{diag}(a_{ii})$; which, as a matter of fact, is the reason why we have assumed that $a_{ii} \neq 0, 1 \leqslant i \leqslant n$. We have, then, the equivalences

$$Au = b \Leftrightarrow Du = (E + F)u + b \Leftrightarrow u = D^{-1}(E + F)u + D^{-1}b,$$

which leads to the (*point*) *Jacobi iterative method*

$$Du_{k+1} = (E + F)u_k + b \Leftrightarrow u_{k+1} = D^{-1}(E + F)u_k + D^{-1}b, \, k \geqslant 0.$$

Consequently, the matrix of this iterative method is

$$J = D^{-1}(E + F) = I - D^{-1}A,$$

and is called the (*point*) *Jacobi matrix*.

The actual calculations appear as follows, setting $u_k = (u_i^k)_{i=1}^n$ and $u_{k+1} = (u_i^{k+1})_{i=1}^n$:

$$
\begin{aligned}
a_{11}[u_1^{k+1}] &= & -a_{12}u_2^k - a_{13}u_3^k \cdots & -a_{1,n-1}u_{n-1}^k & -a_{1n}u_n^k & +b_1 \\
a_{22}[u_2^{k+1}] &= -a_{21}u_1^k & -a_{23}u_3^k & -a_{2,n-1}u_{n-1}^k & -a_{2n}u_n^k & +b_2 \\
&\vdots & \vdots & & \vdots & \vdots \\
a_{n-1,n-1}[u_{n-1}^{k+1}] &= -a_{n-1,1}u_1^k \cdots & -a_{n-1,n-2}u_{n-2}^k & & -a_{n-1,n}u_n^k & +b_{n-1} \\
a_{nn}[u_n^{k+1}] &= -a_{n1}u_1^k & -a_{n2}u_2^k \cdots & -a_{n,n-1}u_{n-1}^k & & +b_n,
\end{aligned}
$$

where we have enclosed in square brackets the components which are calculated in succession. In order to calculate, then, the component u_i^{k+1} of the vector u_{k+1}, $n-1$ components of the vector u_k are used, *which have, therefore, to be kept in memory throughout the whole of the calculation of the vector u_{k+1}*. In other words, one iteration of the method immobilises $2n$ memory locations of the computer, n for the n components of the vector u_k and n for the n components of the vector u_{k+1}.

Furthermore, it would seem that the method could probably be improved by making 'better' use of quantities already calculated. So, for example, in calculating the second component u_2^{k+1} of the vector u_{k+1}, why not use the 'latest' value u_1^{k+1} rather than the 'old' value u_1^k, and so on? The implementation of this remark then leads to the replacing of the previous system by the following:

$$
\begin{aligned}
a_{11}[u_1^{k+1}] &\simeq & -a_{12}u_2^k - a_{13}u_3^k \cdots & -a_{1,n-1}u_{n-1}^k & -a_{1n}u_n^k & +b_1, \\
a_{22}[u_2^{k+1}] &= -a_{21}u_1^{k+1} & -a_{23}u_3^k \cdots & -a_{2,n-1}u_{n-1}^k & -a_{2n}u_n^k & +b_2, \\
&\vdots & \vdots & & \vdots & \vdots \\
a_{n-1,n-1}[u_{n-1}^{k+1}] &= -a_{n-1,1}u_1^{k+1} \cdots & -a_{n-1,n-2}u_{n-2}^{k+1} & & -a_{n-1,n}u_n^k & +b_{n-1} \\
a_{nn}[u_n^{k+1}] &= -a_{n1}u_1^{k+1} & -a_{n2}u_2^{k+1} \cdots & -a_{n,n-1}u_{n-1}^{k+1} & & +b_n.
\end{aligned}
$$

This defines a new iterative method, which is called the *(point) Gauss–Seidel iterative method*, and is written, in matrix notation, as

$$
Du_{k+1} = Eu_{k+1} + Fu_k + b \Leftrightarrow u_{k+1} = (D-E)^{-1}Fu_k + (D-E)^{-1}b, \quad k \geqslant 0.
$$

The matrix $D - E$ is invertible, since $a_{ii} \neq 0, 1 \leqslant i \leqslant n$. The matrix

$$
\mathscr{L}_1 = (D-E)^{-1}F
$$

is called the *(point) Gauss–Seidel matrix*; the reason for the notation $\mathscr{L}_1$ will appear later.

One of the advantages of this method in relation to that of Jacobi lies in the fact that, in order to calculate a component u_i^{k+1} of the vector u_{k+1}, one only needs the $n-1$ components $u_{i+1}^k, \ldots, u_n^k, u_1^{k+1} \ldots, u_{i-1}^{k+1}$. Consequently, one iteration of the method immobilises only n memory locations of the computer, *instead of the 2n memory locations required by the method of Jacobi.* That is a *decisive* advantage for 'large systems'.

Besides, the heuristic considerations set out at the beginning of this section suggest an eventual superiority of this second method, to the extent that a 'greater number' of elements of the matrix are taken into account by the matrix M. This is just what we shall establish in a particular case (Theorem 5.3-4), but the same cannot always be concluded in the general case (cf. Exercise 5.2-2).

We have just seen that the Gauss–Seidel method consists in solving the linear system

$$(D - E)u_{k+1} = Fu_k + b.$$

If the Gauss–Seidel method converges, one could hope to introduce a *real parameter* $\omega \neq 0$, in such a way that the preceding linear system is replaced by

$$\left\{ \frac{D}{\omega} - E \right\} u_{k+1} = \text{some function of } (u_k, \omega).$$

In other words, 'part of the matrix D is shunted over to the matrix N'. In this way, a method is obtained which in fact converges for $\omega = 1$ and, with some luck, one might expect to attain a higher rate of convergence for $\omega \neq 1$, because of the continuity of the spectral radius of the matrix of the iterative method as a function of ω (cf. Exercise 5.2-1), at least in the neighbourhood of $\omega = 1$ (the unfavourable case being that in which $\omega = 1$ is a minimum of this same spectral radius). This form of heuristic reasoning will, in fact, have its confirmation in Theorem 5.3-5, for a particular case.

The regular splitting $A = M - N$ corresponding to the choice $M = (D/\omega) - E$ being

$$A = \left\{ \frac{D}{\omega} - E \right\} - \left\{ \frac{1-\omega}{\omega} D + F \right\},$$

one iteration of the iterative method associated with this splitting may be written as

$$\left\{ \frac{D}{\omega} - E \right\} u_{k+1} = \left\{ \frac{1-\omega}{\omega} D + F \right\} u_k + b, \quad k \geq 0.$$

This is the (*point*) *relaxation iterative method*, which is defined only for *nonzero* values of the *relaxation parameter* ω. The matrix of this method is the (*point*) *relaxation matrix*

$$\mathcal{L}_\omega = \left\{ \frac{D}{\omega} - E \right\}^{-1} \left\{ \frac{1-\omega}{\omega} D + F \right\} = (D - \omega E)^{-1} \{(1-\omega)D + \omega F\},$$

and, as it should, it coincides with the Gauss–Seidel matrix for $\omega = 1$ (it is this which explains the notation $\mathscr{L}_1$ for the latter). If this iterative method is employed with $\omega > 1$, or $\omega < 1$, it is called the *over-relaxation*, or the *under-relaxation*, method respectively.

The study of this method, therefore, consists in determining (if they exist) in turn:

(i) an *interval* $I \subset \mathbb{R}$ (not containing the origin) such that

$$\omega \in I \Rightarrow \varrho(\mathscr{L}_\omega) < 1;$$

(ii) an *optimal relaxation parameter* $\omega_0 \in I$, such that

$$\varrho(\mathscr{L}_{\omega_0}) = \inf_{\omega \in I} \varrho(\mathscr{L}_\omega).$$

It is essential to note that, contrary to appearances (the matrix $\mathscr{L}_\omega$ has a distinctly more 'complicated' look than the matrix $\mathscr{L}_1$), one iteration of the relaxation method is (once the parameter ω has been determined) entirely analogous to one iteration of the Gauss–Seidel method, as regards the calculations which have to be made. In either case, a linear system with lower triangular matrix, $D - E$ and $(D/\omega) - E$ respectively, has to be solved (the equations $a_{ii} \neq 0, 1 \leqslant i \leqslant n$, are equivalent, as before, to the invertibility of the matrix $(D/\omega) - E$, $\omega \neq 0$). More precisely, one iteration corresponds to the solution of the following equations:

$$a_{11}[u_1^{k+1}] = a_{11}u_1^k - \omega\{a_{11}u_1^k + a_{12}u_2^k + a_{13}u_3^k + \cdots + a_{1n}u_n^k - b_1\},$$
$$a_{22}[u_2^{k+1}] = a_{22}u_2^k - \omega\{a_{21}u_1^{k+1} + a_{22}u_2^k + a_{23}u_3^k + \cdots + a_{2n}u_n^k - b_2\},$$
$$\vdots$$
$$a_{nn}[u_n^{k+1}] = a_{nn}u_n^k - \omega\{a_{n1}u_1^{k+1} + \cdots + a_{n,n-1}u_{n-1}^{k+1} + a_{nn}u_n^k - b_n\}.$$

Remark

While one can expect, for certain values of the relaxation parameter, a more rapid convergence (hence less time spent in calculations) than for the Gauss–Seidel method, it should not be forgotten to take into account the time devoted to a preliminary estimation of the relaxation parameter (the optimal one, for example), whenever it is desired to compare the efficiency of the two methods. An example of an optimal value of the parameter is given in Theorem 5.3-6. $\qquad\qquad\square$

It has been said that the preceding methods are *point* methods. It is possible, in actual fact, to define analogous *block* methods. Suppose that the matrix A of the linear system is partitioned into blocks, and set, with a useful misuse (already referred to) of notation,

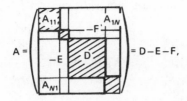

$$A = \begin{bmatrix} A_{11} & & A_{1N} \\ -E & D & -F \\ A_{N1} & & \end{bmatrix} = D - E - F,$$

which we shall call a *block splitting* $A = D - E - F$ of the matrix A.

Assuming, once again, that the matrix D *is invertible* or, equivalently, that the diagonal submatrices A_{ii}, $1 \leqslant i \leqslant N$, are invertible, the *block Jacobi, Gauss–Seidel and relaxation iterative methods and matrices* (tacitly understood, *associated with the block splitting* $A = D - E - F$ given above) are defined by the *same* formulae as before, the letters D, E and F representing this time the 'new' matrices appearing in the block splitting.

According to the heuristic considerations of the beginning of this section, it seems *a priori* that the block methods ought to be better (more rapid) than the corresponding point methods. However, it should be noted that, in each iteration, *there are N linear systems which have to be solved* whose matrices are the submatrices A_{ii}, $1 \leqslant i \leqslant N$. Hence, a block method is to be used in preference to the corresponding point method, only if the increase in the time taken for one iteration, due to the need to solve these N linear systems, is sufficiently compensated by the acceleration of convergence.

Let us write down, by way of example, the *linear systems* corresponding to one iteration of the block Gauss–Seidel method for $N = 4$ (observe that

Table 5.2-1

Name of the method	Regular splitting $A = M - N$ (point or block)	Matrix $M^{-1}N$ of the iterative method	Description of one iteration
Jacobi	$A = D - (E + F)$	$J = D^{-1}(E + F) = I - D^{-1}A$	$Du_{k+1} = (E + F)u_k + b$
Gauss–Seidel	$A = (D - E) - F$	$\mathscr{L}_1 = (D - E)^{-1}F$	$(D - E)u_{k+1} = Fu_k + b$
relaxation	$A = \left\{ \dfrac{D}{\omega} - E \right\}$ $\quad - \left\{ \dfrac{(1 - \omega)}{\omega}D + F \right\}, \quad \omega \neq 0$	$\mathscr{L}_\omega = \left\{ \dfrac{D}{\omega} - E \right\}^{-1}$ $\quad \cdot \left\{ \dfrac{(1 - \omega)}{\omega}D + F \right\}$	$\left\{ \dfrac{D}{\omega} - E \right\} u_{k+1}$ $\quad = \left\{ \dfrac{(1 - \omega)}{\omega}D + F \right\} u_k + b$

the quantities u_i^{k+1} and b_i are now *vectors*).

$$
\begin{aligned}
A_{11}[u_1^{k+1}] = & & -A_{12}u_2^k & \quad -A_{13}u_3^k & \quad -A_{14}u_4^k + b_1, \\
A_{22}[u_2^{k+1}] = & -A_{21}u_1^{k+1} & & \quad -A_{23}u_3^k & \quad -A_{24}u_4^k + b_2, \\
A_{33}[u_3^{k+1}] = & -A_{31}u_1^{k+1} & -A_{32}u_2^{k+1} & & \quad -A_{34}u_4^k + b_3, \\
A_{44}[u_4^{k+1}] = & -A_{41}u_1^{k+1} & -A_{42}u_2^{k+1} & \quad -A_{43}u_3^{k+1} & \quad + b_4.
\end{aligned}
$$

Table 5.2-1 sets out the chief characteristics of the methods (point or block) which have just been described.

Exercises

5.2-1. Prove that, for $\omega \neq 0$, the spectral radius of the matrix $\mathscr{L}_\omega$ is a continuous function of ω.

5.2-2. The aim in this exercise is to prove (by means of examples) that, in general, nothing can be said, by way of comparison, of two iterative methods.
   (1) Let

$$
A = \begin{pmatrix} 1 & 2 & -2 \\ 1 & 1 & 1 \\ 2 & 2 & 1 \end{pmatrix}.
$$

Show that $\rho(J) < 1 < \rho(\mathscr{L}_1)$.
   (2) Let

$$
A = \begin{pmatrix} 2 & -1 & 1 \\ 2 & & 2 & 2 \\ -1 & -1 & 2 \end{pmatrix}.
$$

Show that

$$
\rho(\mathscr{L}_1) < 1 < \rho(J).
$$

5.2-3. Let $A = (a_{ij})$ be a strictly diagonally dominant matrix of order n:

$$
|a_{ii}| > \sum_{j \neq i} |a_{ij}|, \quad 1 \leqslant i \leqslant n.
$$

(1) Prove that the point Jacobi method converges.
(2) Prove that the point Gauss-Seidel method converges.
Remark. For analogous results, see Exercise 5.3-7.

5.2-4. Consider the linear system

$$
\begin{pmatrix} 2 & -1 & & \\ -1 & 2 & -1 & \\ & -1 & 2 & -1 \\ & & -1 & 2 \end{pmatrix} \begin{pmatrix} u_1 \\ u_2 \\ u_3 \\ u_4 \end{pmatrix} = \begin{pmatrix} 19 \\ 19 \\ -3 \\ -12 \end{pmatrix}.
$$

(1) Calculate the exact solution using Gaussian elimination.
(2) Calculate the vectors $u_k = (u_i^k)_{i=1}^4, k \leqslant 10$, obtained by the iterative methods of Jacobi, Gauss-Seidel and relaxation for $\omega = 1.1, 1.2, \ldots, 1.9$, taking $u_0 = 0$ as the initial vector.

5.3 Convergence of the Jacobi, Gauss–Seidel and relaxation methods

In order to fix ideas, the results of this section are enunciated for the complex case. We begin by establishing a sufficient condition (which is also necessary; see Exercise 5.3-1) for the convergence of an iterative method associated with *any* regular splitting $A = M - N$ of a *Hermitian and positive definite* matrix. The result will then be applied to the relaxation method.

Theorem 5.3-1
Let A *be a Hermitian, positive definite matrix, and let*

$$A = M - N, \ M \ \textit{an invertible matrix.}$$

If the Hermitian matrix $M^* + N$ *is positive definite, then*

$$\varrho(M^{-1}N) < 1.$$

Proof
The matrix $M^* + N$ is indeed Hermitian, since $(A = A^*)$

$$M^* + N = A^* + N^* + N = A + N + N^* = M + N^*.$$

We shall establish the inequality $\| M^{-1}N \| < 1$, where $\| \cdot \|$ denotes the matrix norm subordinate to the function

$$\| \cdot \| : v \in \mathbb{C}^n \to \| v \| = (v^*Av)^{1/2},$$

which here is a vector norm, since the Hermitian matrix A is positive definite (it is at this point that the hypothesis is used). Since

$$\| M^{-1}N \| = \| I - M^{-1}A \| = \sup \{ \| v - M^{-1}Av \| : \| v \| = 1 \},$$

we are led to evaluate the expression $\| v - w \|$, where $w = M^{-1}Av$ and $\| v \| = 1$:

$$\begin{aligned}
\| v - w \|^2 &= 1 - v^*Aw - w^*Av + w^*Aw \\
&= 1 - w^*M^*w - w^*Mw + w^*Aw \\
&= 1 - w^*(M^* + N)w.
\end{aligned}$$

Consequently,

$$\| v \| = 1 \Rightarrow \| v - M^{-1}Av \| < 1,$$

since, the matrix $M^* + N$ being positive definite by hypothesis,

$$v \neq 0 \Rightarrow w = M^{-1}Av \neq 0 \Rightarrow w^*(M^* + N)w > 0.$$

The function

$$v \in \mathbb{C}^n \to \| v - M^{-1}Av \| \in \mathbb{R}$$

is continuous (by the composition of continuous functions) over the *compact set* $\{ v \in \mathbb{C}^n : \| v \| = 1 \}$, and so attains its upper bound over this same compact set. This proves the theorem. $\square$

For the rest of this section, we associate with a given, invertible matrix A of order n a (block) splitting $A = D - E - F$ as defined in the previous section. The matrix D is assumed to be *invertible* and consists of the diagonal elements of the matrix A (point methods) or the diagonal submatrices of the matrix A (block methods), the matrix E is lower triangular and the matrix F is upper triangular (point or block, as the case may be). The iterative methods considered are then defined in terms of the associated regular splittings $A = M - N$, as was indicated in the previous section (refer to table 5.2-1).

Theorem 5.3-2 provides a simple criterion for the convergence of the relaxation method (and, as a particular case, that of Gauss–Seidel). It goes by the name of the *Ostrowski–Reich* Theorem.

Theorem 5.3-2 (*Sufficient condition for the convergence of the relaxation method*)

If the matrix A is Hermitian and positive definite, the point or block relaxation method converges if $0 < \omega < 2$.

Proof
The regular splitting $A = M - N$ associated with the relaxation method is

$$A = M - N = \left\{ \frac{D}{\omega} - E \right\} - \left\{ \frac{1 - \omega}{\omega} D + F \right\},$$

so that

$$M^* + N = \frac{D}{\omega} - E^* + \frac{1 - \omega}{\omega} D + F = \frac{2 - \omega}{\omega} D,$$

since $D = D^*$ and $E^* = F$ (ω is a non-zero real number). The matrix D is itself positive definite; in fact, denoting by $A_{ii}, 1 \leqslant i \leqslant N$, the diagonal submatrices, of order n_i, which appear in the block decomposition of the matrix A, we have

$$\text{sp}(D) = \bigcup_{i=1}^{N} \text{sp}(A_{ii}),$$

and, on the other hand, the matrix A being positive definite,

$$v_i \in \mathbb{C}^{n_i} - \{0\} \Rightarrow v_i^* A_{ii} v_i = V_i^* A V_i > 0,$$

the vectors $V_i \in \mathbb{C}^n$ being obtained by extending with zeros the components of the vectors v_i.

The matrix $M^* + N$ is then positive definite if and only if $0 < \omega < 2$. It then suffices to apply the previous theorem.   □

We note, in passing, that the *variational methods of approximation and, in certain cases, the finite-difference methods, for elliptic boundary-value problems*, lead to linear systems whose matrices are *symmetric and positive definite* (see Chapter 3).

Remark

The application of Theorem 5.3-1 to the Jacobi method does not *in general* provide a condition which can easily be 'exploited'. However, here is a particular case to which it does apply. If

$$
A = \begin{pmatrix}
2 & -1 & & & \\
-1 & 2 & -1 & & \\
& \ddots & \ddots & \ddots & \\
& & -1 & 2 & -1 \\
& & & -1 & 2
\end{pmatrix}
$$

(it has already been verified in section 3.1 that this matrix is positive definite), a simple calculation shows that, for the splitting $A = M - N$ associated with the point Jacobi method,

$$
v^T(M^T + N)v = v_1^2 + v_n^2 + \sum_{i=1}^{n-1} (v_i + v_{i+1})^2 > 0 \quad \text{if} \quad v \neq 0.
$$

We shall meet this result again as a consequence of Theorem 5.3-6. □

We shall now establish that, *independently of any hypothesis*, the condition $\omega \in]0, 2[$ is also *necessary* for the convergence of the relaxation method.

Theorem 5.3-3 (*Necessary condition for the convergence of the relaxation method*)

The spectral radius of the point or block relaxation matrix always satisfies the inequality

$$
\varrho(\mathcal{L}_\omega) \geq |\omega - 1|, \omega \neq 0.
$$

Consequently, the point or block relaxation method can only converge if $\omega \in]0, 2[$.

Proof

Observe that

$$
\prod_{i=1}^n \lambda_i(\mathcal{L}_\omega) = \det(\mathcal{L}_\omega) = \frac{\det\left(\dfrac{1-\omega}{\omega}D + F\right)}{\det\left(\dfrac{D}{\omega} - E\right)} = (1 - \omega)^n,
$$

taking account of the particular structure of the matrices D, E and F. Consequently,

$$\varrho(\mathscr{L}_\omega) \geqslant \left| \prod_{i=1}^n \lambda_i(\mathscr{L}_\omega) \right|^{1/n} = |1 - \omega|,$$

with equality if and only if all the eigenvalues have the same modulus $|1 - \omega|$.   □

We shall now go on to see that the fact of the existence of a *point or block tridiagonal* structure of the matrix A allows, *independently of any other hypothesis*, a very precise comparison of the spectral radii of the Jacobi and relaxation matrices, as well in the case of convergence as in that of divergence. The case $\omega \neq 1$ being technically more difficult than the case $\omega = 1$, we begin with a comparison of just the two spectral radii $\rho(J)$ and $\rho(\mathscr{L}_1)$.

Theorem 5.3-4 (*Comparison of the Jacobi and Gauss–Seidel methods*)
Let A be a block tridiagonal matrix. Then the spectral radii of the corresponding block Jacobi and Gauss–Seidel matrices are related by the equation

$$\varrho(\mathscr{L}_1) = (\varrho(J))^2,$$

so that the two methods converge or diverge simultaneously; if they converge, the Gauss–Seidel method converges more rapidly than the Jacobi method.

Proof
(i) We begin with a *preliminary result*. Let $A(\mu)$ (μ a non-zero scalar) be a block tridiagonal matrix of the form

$$A(\mu) = \begin{pmatrix} B_1 & \mu^{-1}C_1 & & & \\ \mu A_2 & B_2 & \mu^{-1}C_2 & & \\ & \ddots & \ddots & \ddots & \\ & & \mu A_{N-1} & B_{N-1} & \mu^{-1}C_{N-1} \\ & & & \mu A_N & B_N \end{pmatrix}.$$

Then

$$\det(A(\mu)) = \det(A(1)) \quad \text{for every} \quad \mu \neq 0.$$

For, if we introduce the diagonal matrix

$$Q(\mu) = \begin{pmatrix} \mu I_1 & & & & \\ & \mu^2 I_2 & & & \\ & & \ddots & & \\ & & & \mu^{N-1}I_{N-1} & \\ & & & & \mu^N I_N \end{pmatrix},$$

each matrix I_j being the identity matrix of the same order as that of the matrix B_j, it can easily be verified that

$$A(\mu) = Q(\mu)A(1)\{Q(\mu)\}^{-1},$$

which proves the required result.

(ii) The eigenvalues of the Jacobi matrix $J = D^{-1}(E + F)$ are the zeros of the characteristic polynomial

$$p_J(\lambda) = \det(D^{-1}(E + F) - \lambda I),$$

and are, therefore, also the zeros of the polynomial

$$q_J(\lambda) \overset{\text{def}}{=} \det(\lambda D - E - F) = \det(-D)p_J(\lambda).$$

Similarly, the eigenvalues of the Gauss–Seidel matrix $\mathscr{L}_1 = (D - E)^{-1}F$ are the zeros of the characteristic polynomial

$$p_{\mathscr{L}_1}(\lambda) = \det((D - E)^{-1}F - \lambda I),$$

and are, therefore, also the zeros of the polynomial

$$q_{\mathscr{L}_1}(\lambda) \overset{\text{def}}{=} \det(\lambda D - \lambda E - F) = \det(E - D)p_{\mathscr{L}_1}(\lambda).$$

Keeping in mind the block tridiagonal structure of the matrix A, an application of the preliminary result (with $\mu = \lambda \neq 0$) shows that

$$q_{\mathscr{L}_1}(\lambda^2) = \det(\lambda^2 D - \lambda^2 E - F) = \det(\lambda^2 D - \lambda E - \lambda F) = \lambda^n q_J(\lambda),$$

for every $\lambda \in \mathbb{C}$, this equation being also valid for $\lambda = 0$ by continuity. From this functional relationship, we deduce the implications

$$\beta \neq 0 \quad \text{and} \quad \beta \in \text{sp}(\mathscr{L}_1) \Rightarrow \{\beta^{1/2}, -\beta^{1/2}\} \in \text{sp}(J),$$
$$\{\alpha \in \text{sp}(J) \Leftrightarrow -\alpha \in \text{sp}(J)\} \Rightarrow \alpha^2 \in \text{sp}(\mathscr{L}_1),$$

denoting by $\beta^{1/2}$ either of the two roots of the complex number β. This concludes the proof. $\qquad\qquad\square$

Remark

The proof given above, then, establishes a bijection between the non-zero eigenvalues of the matrix $\mathscr{L}_1$ and the corresponding pairs of non-zero eigenvalues of the matrix J. $\qquad\qquad\square$

Theorem 5.3-5 (*Comparison of the Jacobi and relaxation methods*)
Let A be a block tridiagonal matrix, such that all the eigenvalues of the corresponding block Jacobi matrix are real.

Then the block Jacobi method and the block relaxation method for $0 < \omega < 2$ converge or diverge simultaneously; if they converge, the function $\omega \in]0, 2[\to \rho(\mathscr{L}_\omega)$ has the form indicated in figure 5.3-1.

Proof

(i) Arguing as in the previous proof, we begin by establishing a *functional relationship between the polynomial*

$$q_J(\lambda) = \det(\lambda D - E - F) = \det(-D)p_J(\lambda)$$

and the polynomial

$$q_{\mathscr{L}_\omega}(\lambda) \overset{\text{def}}{=} \det\left(\frac{\lambda + \omega - 1}{\omega}D - \lambda E - F\right) = \det\left(E - \frac{D}{\omega}\right)p_{\mathscr{L}_\omega}(\lambda),$$

where

$$p_{\mathscr{L}_\omega}(\lambda) = \det\left\{\left(\frac{D}{\omega} - E\right)^{-1}\left(\frac{1 - \omega}{\omega}D + F\right) - \lambda I\right\}$$

denotes the characteristic polynomial of the matrix $\mathscr{L}_\omega$. The block tridiagonal structure of the matrix A allows us to write (applying the 'preliminary result' established in the proof of Theorem 5.3-4)

$$q_{\mathscr{L}_\omega}(\lambda^2) = \det\left(\frac{\lambda^2 + \omega - 1}{\omega} - \lambda^2 E - F\right) = \det\left(\frac{\lambda^2 + \omega - 1}{\omega} - \lambda E - \lambda F\right)$$

$$= \lambda^n q_J\left(\frac{\lambda^2 + \omega - 1}{\lambda \omega}\right),$$

for every $\lambda \in \mathbb{C} - \{0\}$. Consequently,

$$\beta \neq 0 \text{ and } \beta \in \text{sp}(\mathscr{L}_\omega) \Rightarrow \left\{\frac{\beta + \omega - 1}{\beta^{1/2}\omega}, -\frac{\beta + \omega - 1}{\beta^{1/2}\omega}\right\} \subset \text{sp}(J),$$

$$\{\alpha \in \text{sp}(J) \Leftrightarrow -\alpha \in \text{sp}(J)\} \Rightarrow \{\mu_+(\alpha, \omega), \mu_-(\alpha, \omega)\} \subset \text{sp}(\mathscr{L}_\omega),$$

Figure 5.3-1

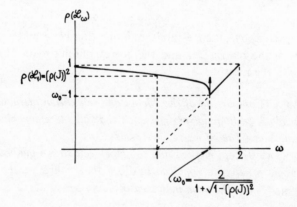

where

$$\mu_+(\alpha, \omega) = \tfrac{1}{2}(\alpha^2\omega^2 - 2\omega + 2) + \frac{\alpha\omega}{2}(\alpha^2\omega^2 - 4\omega + 4)^{1/2},$$

$$\mu_-(\alpha, \omega) = \tfrac{1}{2}(\alpha^2\omega^2 - 2\omega + 2) - \frac{\alpha\omega}{2}(\alpha^2\omega^2 - 4\omega + 4)^{1/2},$$

are the *squares* of the two roots of the quadratic (*in* λ)

$$\lambda^2 - \lambda\alpha\omega + (\omega - 1) = 0 \Leftrightarrow \frac{\lambda^2 + \omega - 1}{\lambda\omega} = \alpha \quad \text{for} \quad \lambda \neq 0.$$

It should be noted that the functional relationship established above is not valid for $\lambda = 0$, so that the second implication is not valid for $\omega = 1$; however, this was precisely the case considered in the previous theorem.

(ii) We have, therefore,

$$\varrho(\mathscr{L}_\omega) = \max_{\alpha \in \mathrm{sp}(J)} \{\max(|\mu_+(\alpha, \omega)|, |\mu_-(\alpha, \omega)|)\}.$$

Hence, on making the *additional* hypothesis that *all the eigenvalues of the matrix* J *are real*, we are led to an *examination of the function*

$$M : (\alpha, \omega) \in \mathbb{R}_+ \times \,]0, 2[\, \to M(\alpha, \omega) \overset{\text{def}}{=} \max\{|\mu_+(\alpha, \omega)|, |\mu_-(\alpha, \omega)|\}.$$

There is, in fact, no need to look at the function M for $\alpha < 0$ since $M(-\alpha, \omega) = M(\alpha, \omega)$, or for $\omega \notin \,]0, 2[$, this latter case having been settled by Theorem 5.3-3.

(iii) *Let us suppose, to start with, that* $0 \leqslant \alpha < 1$. For $\alpha = 0$, we see at once that

$$M(0, \omega) = |\omega - 1|.$$

For $0 < \alpha < 1$, the quadratic $\omega \to \alpha^2\omega^2 - 4\omega + 4$ has two real roots $\omega_0(\alpha)$ and $\omega_1(\alpha)$ satisfying

$$1 < \omega_0(\alpha) = \frac{2}{1 + \sqrt{(1 - \alpha^2)}} < 2 < \omega_1(\alpha).$$

If, then, $\omega_0(\alpha) < \omega < 2$, the complex numbers $\mu_+(\alpha, \omega)$ and $\mu_-(\alpha, \omega)$ are conjugate. As these are the squares of the roots of the quadratic $\lambda \to \lambda^2 - \lambda\alpha\omega + (\omega - 1)$, the product of the roots being $\omega - 1$, it follows that

$$1 < \omega_0(\alpha) < \omega < 2 \Rightarrow M(\alpha, \omega) = |\mu_+(\alpha, \omega)| = |\mu_-(\alpha, \omega)| = \omega - 1.$$

Next, suppose that $0 < \omega \leqslant \omega_0(\alpha)$. Then it may be readily verified that

$$M(\alpha, \omega) = \mu_+(\alpha, \omega) = v^2(\alpha, \omega),$$

where we have set

$$2v(\alpha, \omega) = \alpha\omega + (\alpha^2\omega^2 - 4\omega + 4)^{1/2}.$$

Consequently,

$$\frac{\partial M}{\partial \omega} = 2v \frac{\partial v}{\partial \omega} = 2v \left\{ \frac{v\alpha - 1}{2v - \alpha\omega} \right\} \quad \text{for } 0 < \omega < \omega_0(\alpha).$$

Since

$$0 < \alpha < 1 \text{ and } 0 < \omega < 2 \Rightarrow v^2 = \tfrac{1}{2}(\alpha^2\omega^2 - 2\omega + 2) + \frac{\alpha\omega}{2}(\alpha^2\omega^2 - 4\omega + 4)^{1/2}$$

$$< \tfrac{1}{2}(\omega^2 - 2\omega + 2) + \frac{\omega}{2}(2 - \omega) = 1,$$

$$\lim_{\omega \to 0^+} v(\alpha, \omega) = 1, \quad \lim_{\omega \to \omega_0(\alpha)^-} v(\alpha, \omega) = \frac{\alpha}{2}\omega_0(\alpha) = \{\omega_0(\alpha) - 1\}^{1/2} > 0,$$

we have, on the one hand,

$$\begin{cases} 2v - \alpha\omega = (\alpha^2\omega^2 - 4\omega + 4)^{1/2} > 0 \quad \text{for} \quad 0 < \omega < \omega_0(\alpha), \\ \lim_{\omega \to \omega_0(\alpha)^-} (2v - \alpha\omega) = 0, \quad \lim_{\omega \to 0^+} (2v - \alpha\omega) = 2, \end{cases}$$

and, on the other,

$$\begin{cases} 2v(v\alpha - 1) < 0 \quad \text{for} \quad 0 < \omega < \omega_0(\alpha), \\ \lim_{\omega \to \omega_0(\alpha)^-} \{2v(v\alpha - 1)\} < 0, \quad \lim_{\omega \to 0^+} \{2v(v\alpha - 1)\} < 0. \end{cases}$$

Lastly, as the function

$$\alpha > 0 \to M(\alpha, \omega), \quad \text{for} \quad \omega > 0 \quad \text{fixed},$$

is strictly increasing (refer to the expression for the function v), we have all the elements necessary to trace the family of curves $M(\alpha, \omega)$ for $0 \leqslant \alpha < 1$ and $0 < \omega < 2$ (cf. figure 5.3-2).

Observe, in particular, that

$$\varrho(J) < 1 \Rightarrow \varrho(\mathscr{L}_\omega) = M(\varrho(J), \omega) < 1 \quad \text{for} \quad 0 < \omega < 2,$$

Figure 5.3-2

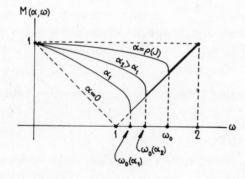

and that

$$\varrho(\mathscr{L}_{\omega_0}) = \inf_{0 < \omega < 2} \varrho(\mathscr{L}_\omega) = \omega_0 - 1,$$

with

$$\omega_0 = \omega(\varrho(\mathrm{J})) = \frac{2}{1 + \sqrt{[1 - (\varrho(\mathrm{J}))^2]}}.$$

(iv) *Lastly, we suppose that* $\alpha \geqslant 1$. Then the quadratic $\omega \to \alpha^2\omega^2 - 4\omega + 4$ is always non-negative, and

$$\mu_+(\alpha, \omega) = \tfrac{1}{2}(\alpha^2\omega^2 - 2\omega + 2) + \frac{\alpha\omega}{2}(\alpha^2\omega^2 - 4\omega + 4)^{1/2}$$

$$\geqslant \tfrac{1}{2}(\omega^2 - 2\omega + 2) + \frac{\omega}{2}(2 - \omega) = 1 \quad \text{for} \quad 0 < \omega < 2,$$

which shows, in particular, that

$$\varrho(\mathrm{J}) \geqslant 1 \Rightarrow \varrho(\mathscr{L}_\omega) \geqslant 1 \quad \text{for } 0 < \omega < 2. \qquad \square$$

Putting together the results of Theorems 5.3-2 and 5.3-5, we obtain a result which is doubly of interest. First, it is an example of a situation where the capability is available of *comparing very precisely the spectral radii of the three matrices* J, $\mathscr{L}_1$, $\mathscr{L}_{\omega_0}$; secondly, this particular case is actually encountered in the *discretisation of boundary-value problems* by variational methods of approximation or by finite-difference methods (Chapter 3).

Theorem 5.3-6
Let A *be a Hermitian, positive definite, block tridiagonal matrix.*

Then the block Jacobi and Gauss–Seidel methods and the relaxation method for $0 < \omega < 2$ *converge, the function* $\omega \in \,]0, 2[\to \varrho(\mathscr{L}_\omega)$ *having the form indicated in figure 5.3-1. There exists, in particular, a unique optimal relaxation parameter*

$$\omega_0 = \frac{2}{1 + \sqrt{[1 - (\varrho(\mathrm{J}))^2]}},$$

such that

$$\varrho(\mathscr{L}_{\omega_0}) = \inf_{0 < \omega < 2} \varrho(\mathscr{L}_\omega) = \omega_0 - 1 < \varrho(\mathscr{L}_1) = (\varrho(\mathrm{J}))^2 < \varrho(\mathrm{J})$$

if $\varrho(\mathrm{J}) > 0$; *if* $\varrho(\mathrm{J}) = 0$, $\omega_0 = 1$ *and* $\varrho(\mathscr{L}_1) = \varrho(\mathrm{J}) = 0$.

Proof
In order to be able to apply Theorem 5.3-5, it is necessary to verify in particular that the eigenvalues of the matrix $\mathrm{J} = \mathrm{D}^{-1}(\mathrm{E} + \mathrm{F})$ are real. Now

$$\mathrm{D}^{-1}(\mathrm{E} + \mathrm{F})v = \alpha v \Rightarrow (\mathrm{E} + \mathrm{F})v = \alpha \mathrm{D}v$$

$$\Rightarrow \mathrm{A}v = (1 - \alpha)\mathrm{D}v \Rightarrow v^*\mathrm{A}v = (1 - \alpha)v^*\mathrm{D}v.$$

The matrices A and D being Hermitian and positive definite, the numbers v^*Av and v^*Dv are strictly positive if $v \neq 0$, and, therefore,

$$\alpha \in \mathrm{sp}(J) \Rightarrow \alpha \in \mathbb{R}.$$

By Theorem 5.3-2,

$$\varrho(\mathcal{L}_\omega) < 1 \quad \text{for } 0 < \omega < 2.$$

But this is the situation in which the Jacobi and relaxation methods converge simultaneously, and the conclusions follow from Theorem 5.3-5 and its proof. $\qquad \square$

Remarks

(1) If the optimal parameter ω_0 is known only approximately, it is clear that it would be preferable to *overestimate* rather than to underestimate it, since

$$\lim_{\omega \to \omega_0^-} \frac{\mathrm{d}\{\varrho(\mathcal{L}_\omega)\}}{\mathrm{d}\omega} = +\infty, \quad \frac{\mathrm{d}\{\varrho(\mathcal{L}_\omega)\}}{\mathrm{d}\omega} = 1 \quad \text{for } \omega > \omega_0.$$

(2) The results of Theorems 5.3-5 and 5.3-6 are applicable, in particular, to the 'usual' tridiagonal matrices. What holds for the general case holds for the particular.... $\qquad \square$

Exercises

5.3-1. (The converse of Theorem 5.3-1). Let A be an invertible Hermitian matrix, with the splitting $A = M - N$, M being an invertible matrix. Prove that, if the Hermitian matrix $M^* + N$ is positive definite and if $\rho(M^{-1}N) < 1$, then the matrix A is positive definite.

5.3-2. Consider the matrix of order N

$$A = \frac{1}{h^2} \begin{pmatrix} 2 & -1 & & & \\ -1 & 2 & -1 & & \\ & \ddots & \ddots & \ddots & \\ & & -1 & 2 & -1 \\ & & & -1 & 2 \end{pmatrix}, \quad \text{with } h = \frac{1}{N+1},$$

obtained, notably, in the course of the discretisation of the problem $-u''(x) = f(x)$, $0 < x < 1$, $u(0)$ and $u(1)$ given (cf. section 3.1).

(1) Calculate the eigenvalues of the point Jacobi matrix J.

(2) Deduce expressions for the spectral radii $\rho(J), \rho(\mathcal{L}_1), \rho(\mathcal{L}_{\omega_0})$ as functions of h (all the matrices are point matrices).

(3) Put each of these quantities into the form $\rho(\cdot) = 1 - f(h)$ and evaluate the leading term of $f(h)$ as h tends to zero.

(4) Answer questions (1), (2) and (3) for the point *and* block Jacobi, Gauss–Seidel and relaxation methods associated with the matrix which corresponds to

the discretisation, using the finite-difference method, of the problem

$$\begin{cases} -\Delta u(x) = f(x), & x \in \Omega, \\ u(x) = 0, & x \in \Gamma, \end{cases}$$

where $\bar{\Omega}$ is the unit square $[0, 1] \times [0, 1]$ of the plane and Γ its boundary.

5.3-3. (1) Let H be a positive definite Hermitian matrix (or just non-negative definite). Show that, for every positive number r, the matrix $rI + H$ is invertible, the matrix $(rI - H)(rI + H)^{-1}$ is Hermitian and, finally,

$$\|(rI - H)(rI + H)^{-1}\|_2 < 1 \quad \text{(or just } \leqslant 1, \text{ respectively)}.$$

(2) Let H_1 and H_2 be two Hermitian matrices, positive definite and non-negative definite, respectively. Given an arbitrary vector u_0, the sequence of vectors (u_k) is defined by

$$(H_1 + rI)u_{k+1/2} = (rI - H_2)u_k + b,$$
$$(H_2 + rI)u_{k+1} = (rI - H_1)u_{k+1/2} + b, \quad k \geqslant 0,$$

the number $r > 0$ being fixed, as well as the vector b. Express the vector u_{k+1} in the form

$$u_{k+1} = Bu_k + c,$$

and show that $\rho(B) < 1$.

(3) Show that the sequence (u_k) converges and that $\lim\limits_{k \to \infty} u_k$ is the solution of a linear system whose matrix and right-hand side depend only on the matrices H_1 and H_2 and on the vector b.

(4) Show that the linear system which arises from the application of the finite-difference method to the approximation of the problem (cf. section 3.2)

$$-\Delta u(x) = f(x), \quad x \in \Omega, \quad u(x) = g(x), \quad x \in \partial\Omega,$$

where $\bar{\Omega}$ is a *rectangle* with sides parallel to the co-ordinate axes, may be solved by the iterative method given above, with

$$H_i = P_i A P_i^T, \quad P_i \text{ some permutation matrix, } i = 1, 2,$$

the matrix A having the same form as in Exercise 5.3-2.

The iterative method given above is called an *alternating-direction method*. The result of question (4) shows, in effect, that, by 'alternating in the direction of each of the co-ordinate axes', it becomes possible to use *the same algorithms* (for example, Gaussian elimination with tridiagonal matrices) *in the two-dimensional, as in the one-dimensional, case, but only for rectangular domains.*

(5) Compare this iterative method with the methods set out in question (4) of Exercise 5.3-2.

5.3-4. Given a symmetric, positive definite matrix and a point splitting of it into the form $A = D - E - F$, we shall examine an iterative method for the solution of the linear system $Au = b$.

(1) Given an arbitrary vector u_0, the sequence (u_k) is defined by

$$(D - E)u_{k+1/2} = Fu_k + b,$$
$$(D - F)u_{k+1} = Eu_{k+1/2} + b,$$

the vector b being fixed (this is a simple variant of the Gauss–Seidel method, where the two parts $D - E$ and $D - F$ of the matrix A are given similar roles to play). Express the vector u_{k+1} in the form

$$u_{k+1} = Bu_k + c,$$

giving the matrix B and the vector c explicitly.

(2) Prove the convergence of this method. For this, begin by establishing the relation

$$Bp = \lambda p \Rightarrow \lambda D^{-1}Ap + (\lambda - 1)D^{-1}ED^{-1}Fp = 0.$$

(3) Extend the preceding results to cover the case of a block splitting $A = D - E - F$.

5.3-5. Consider the matrix

$$A = \begin{pmatrix} 2+\alpha_1 & -1 & & & \\ -1 & 2+\alpha_2 & -1 & & \\ & \ddots & \ddots & \ddots & \\ & & -1 & 2+\alpha_{n-1} & -1 \\ & & & -1 & 2+\alpha_n \end{pmatrix}, \quad \alpha_i \geqslant 0, \quad 1 \leqslant i \leqslant n,$$

and the splitting $A = M_\beta - N_\beta$, where

$$N_\beta = \text{diag}\,(\beta - \alpha_i), \quad \beta \text{ some non-negative parameter.}$$

Discuss the convergence of the iterative method associated with this splitting into the form $M - N$, depending on the values taken by the parameter β (dealing, in particular, with the questions of the existence of an interval $I \subset \mathbb{R}$ such that $\rho(M_\beta^{-1}N_\beta) < 1$ for $\beta \in I$ and of the existence of an optimal parameter).

5.3-6. This exercise makes use of the results established in Exercises 3.1-6 and 3.1-7.

(1) Let A be a monotone matrix and $A = M - N$ a regular splitting of it, where M is also a monotone matrix and N is a non-negative matrix. Prove that

$$\rho(M^{-1}N) = \frac{\rho(A^{-1}N)}{1 + \rho(A^{-1}N)},$$

which proves at once the convergence of the iterative method corresponding to the splitting $A = M - N$.

(2) Let A be an M-matrix (the definition is given in Exercise 3.1-7) and $A = M - N$ a regular splitting of it, the matrix M being obtained from the matrix A by replacing by zero some of the off-diagonal elements of the matrix A. Show that the matrix M is invertible and that $\rho(M^{-1}N) < 1$.

(3) Let A be a monotone matrix, such that all the elements of the matrix A^{-1}

are positive, having the two regular splittings

$$A = M_1 - N_1 = M_2 - N_2,$$

where M_1 and M_2 are monotone matrices and

$$N_1 \geqslant 0, \quad N_2 - N_1 \geqslant 0, \quad N_2 \neq N_1.$$

Show that

$$0 \leqslant \rho(M_1^{-1}N_1) < \rho(M_3^{-1}N_2) < 1,$$

thus allowing a comparison of the iterative methods associated with the splittings $A = M_1 - N_1$ and $A = M_2 - N_2$.

(4) Apply these results to the Jacobi and Gauss–Seidel methods used for the solution of the linear system $A_h u_h = b_h$ obtained in section 3.1 through the use of the finite-difference method.

(5) It was seen in section 1.1 that a partitioning of a matrix of order n is associated with a direct sum decomposition

$$\mathbb{K}^n = V_1 \oplus V_2 \oplus \cdots \oplus V_N \quad \text{where } V_i = \mathbb{K}^{n_i}.$$

A partitioning of the same matrix is said to be *finer* if there corresponds to it a decomposition

$$\mathbb{K}^n = V'_1 \oplus V'_2 \oplus \cdots \oplus V'_{N'} \quad \text{where } V'_j = \mathbb{K}^{n'_j},$$

each of the subspaces V'_j being a subspace of one of the subspaces V_i.

Let A be an irreducible M-matrix, J and $\mathcal{L}_1$ the Jacobi and Gauss–Seidel matrices associated with a partitioning of the matrix A and J' and $\mathcal{L}'_1$ the Jacobi and Gauss–Seidel matrices associated with a finer partitioning. Show that

$$0 < \rho(J) < \rho(J') < 1 \quad \text{if } E + F \neq E' + F',$$
$$0 < \rho(\mathcal{L}_1) < \rho(\mathcal{L}'_1) < 1 \quad \text{if } F \neq F'.$$

(6) Apply these results to the point or block Jacobi and Gauss–Seidel methods, used for the solution of the linear system $A_h u_h = b_h$ which was obtained in section 3.1 through the use of the finite-difference method.

5.3-7. Let $A = (a_{ij})$ be an irreducible matrix of order n which satisfies

$$|a_{ii}| \geqslant \sum_{j \neq i} |a_{ij}|, \quad 1 \leqslant i \leqslant n,$$
$$|a_{i_0 i_0}| > \sum_{j \neq i_0} |a_{i_0 j}| \text{ for at least one index } i_0 \in \{1, 2, \ldots, n\}.$$

Prove that the point Jacobi and Gauss–Seidel methods associated with such a matrix are well-defined and convergent (it is a great help to make use of some of the results of Exercises 1.1-6, 3.1-6, 3.1-7 and 5.3-6).

5.3-8. In order to solve the linear system $Au = b$, we define the sequence of regular splittings

$$A = M_k - N_k, M_k \text{ an invertible matrix}, k \geqslant 0,$$

with which is associated the iterative method

$$u_{k+1} = M_k^{-1}N_k u_k + M_k^{-1}b, \quad k \geqslant 0, u_0 \text{ an arbitrary vector.}$$

(1) Prove that the condition

$$\rho(M_k^{-1}N_k) < 1 \text{ for every } k \geqslant 0$$

is not enough, in general, to guarantee convergence.

(2) We now make the 'stronger' hypothesis that there exists a number δ such that

$$\delta > 0 \quad \text{and} \quad \rho(M_k^{-1}N_k) \leqslant 1 - \delta \quad \text{for every } k \geqslant 0.$$

Does the iterative method converge?

(3) From here on, we restrict attention to matrices M_k of the form

$$M_k = \frac{1}{\mu_k}M,$$

where M is an invertible matrix given once for all and μ_k a real non-zero number. Prove the relation

$$Au_k - b = \left\{ \prod_{l=0}^{k-1} (I - \mu_l AM^{-1}) \right\}(Au_0 - b).$$

(4) Suppose, in what follows, that all the eigenvalues of the matrix AM^{-1} are positive (cf. question (7) for an example where this hypothesis is satisfied). Prove that, if these eigenvalues are known exactly (an assumption evidently not realised in practice!), there exists a choice of numbers $\mu_0, \mu_1, \ldots, \mu_{n-1}$ such that

$$\rho\left(\prod_{l=0}^{n-1} (I - \mu_l AM^{-1})\right) = 0.$$

Prove that this circumstance corresponds to a *direct method* for the solution of the linear system $Au = b$.

(5) Suppose that the matrix AM^{-1} is Hermitian and that an interval is known which contains the eigenvalues (a more realistic hypothesis than the preceding one!), of the form

$$\text{sp}(AM^{-1}) \subset [\alpha, \beta] \quad \text{with} \quad 0 < \alpha \leqslant \beta < +\infty.$$

Taking the integer $k \geqslant 1$ as fixed, prove that an optimal choice (it needs to be stated in what precise sense) of the matrices $M_k = (1/\mu_k)M$ corresponds to the solution of the problem: find numbers $\mu_l \neq 0, 0 \leqslant l \leqslant k - 1$ such that

$$\sup_{\alpha \leqslant \lambda \leqslant \beta} \prod_{l=0}^{k-1} (1 - \mu_l \lambda) = \inf_{\substack{\rho_l \neq 0 \\ 0 \leqslant l \leqslant k-1}} \sup_{\alpha \leqslant \lambda \leqslant \beta} \prod_{l=0}^{k-1} (1 - \rho_l \lambda).$$

(6) Solve the problem given above by means of *Chebyshev polynomials* (cf. Exercise 5.1-2).

(7) Prove that if the matrices A and M are symmetric and positive definite, the eigenvalues of the matrix AM^{-1} are all positive.

5.3-9. A matrix A is said to be *skew-Hermitian* if $A^* = -A$.

(1) Prove that the eigenvalues of the associated (point or block) Jacobi matrix are purely imaginary.

(2) Let A be a block tridiagonal skew-Hermitian matrix. Prove a result concerning the convergence of the associated block relaxation method, adopting a procedure similar to that used in establishing Theorem 5.3-6. It should be verified, in particular, that the corresponding optimal parameter ω_0 satisfies $0 < \omega_0 < 1$; hence, it is an *under-relaxation* method.

Methods for the calculation of eigenvalues and eigenvectors

Introduction

In order to calculate approximations to the set of eigenvalues of a matrix A, an idea commonly exploited consists in constructing a sequence of matrices $(P_k)_{k \geqslant 1}$ such that the matrices $P_k^{-1}AP_k$ 'converge' (in a sense yet to be made precise; in fact, it is not always a case of true convergence) to a matrix whose eigenvalues are known, that is to say, a diagonal or triangular matrix.

This idea underlies the *Jacobi method* for *symmetric* matrices (cf. section 6.1), where the matrices P_k are the product of 'elementary' orthogonal matrices, which are very simple to construct. It is then possible to show that (Theorem 6.1-2)

$$\lim_{k \to \infty} P_k^{-1}AP_k = \text{diag}(\lambda_i),$$

where the numbers λ_i are the eigenvalues of the matrix A (up to a permutation). If these are all distinct, it can be shown that the column vectors of the matrices P_k are, in addition, approximations of the eigenvectors of the matrix A (Theorem 6.1-3).

For *general* matrices, the remarkable QR *algorithm*, whose basic execution is described in section 6.3, derives from the same idea. Making use, in each iteration of the method, of the QR factorisation of matrices (cf. section 4.5), a sequence of matrices (P_k) is constructed such that, with the aid of certain hypotheses (cf. Theorem 6.3-1),

$$\lim_{k \to \infty} (P_k^{-1}AP_k)_{ij} = 0 \quad \text{for} \quad j < i,$$

$$\lim_{k \to \infty} (P_k^{-1}AP_k)_{ii} = \lambda_i,$$

the scalars λ_i being the eigenvalues of the matrix A, nothing being said about the convergence of the elements $(P_k^{-1}AP_k)_{ij}, i < j$; as a result, we are dealing here with a 'pseudo-convergence' of the sequence $(P_k^{-1}AP_k)$. There will also be found in section 6.3 some additional remarks about the convergence of the method and about its practical implementation (initial reduction to Hessenberg form, incorporating shifts into the QR algorithm, etc.).

It will be noted that there are no methods for the calculation of the set of eigenvalues of a *general matrix* which make use of the *characteristic polynomial*. In fact, it is *exactly the opposite* which happens: in order to find the roots of a polynomial of high degree, it is current practice to apply the QR algorithm to its companion matrix (which is already in exact Hessenberg form; cf. section 2.1)!

Yet other methods allow the calculation of certain *chosen* eigenvalues. This is notably the case with the *Givens–Householder method* which we shall study in section 6.2, and which is applicable to *symmetric* matrices. One begins by reducing such a matrix to *tridiagonal* form by means of *Householder matrices* (introduced in section 4.5); then, an ingenious device of *Givens* makes possible the approximate calculation, but to any desired degree of accuracy, of an eigenvalue at a *given position* in the ordered spectrum of such a matrix.

Finally, in section 6.4, a procedure which is widely used is described for the calculation of an eigenvector corresponding to a particular eigenvalue, known with some degree of precision. It is known as the *inverse power method*, and, in Theorem 6.4-1, sufficient conditions for its convergence are given.

6.1 The Jacobi method

This method is used when *all the eigenvalues* and (possibly) *all the eigenvectors* of a *symmetric* matrix are being sought. It is well suited for use with *full* matrices.

A symmetric matrix A is diagonalisable. There exists an orthogonal matrix O such that (Theorem 1.2-1)

$$O^T A O = \text{diag}(\lambda_1, \lambda_2, \ldots, \lambda_n),$$

where the numbers λ_i are the eigenvalues, with multiplicities included, of the matrix A. It is recalled that the column vectors of the matrix O form an orthonormal set of eigenvectors, the ith column vector being an eigenvector associated with the eigenvalue λ_i.

Starting with the matrix $A = A_1$, the Jacobi method consists in constructing a sequence $(\Omega_k)_{k \geqslant 1}$ of 'elementary' orthogonal matrices (whose form, an extremely simple one, is given in the following theorem), so chosen that the sequence of matrices (also symmetric)

$$A_{k+1} = \Omega_k^T A_k \Omega_k = (\Omega_1 \Omega_2 \cdots \Omega_k)^T A (\Omega_1 \Omega_2 \cdots \Omega_k), \quad k \geqslant 1,$$

converges to the matrix diag (λ_i), up to a permutation of the indices (Theorem 6.1-2). One might then expect (at least in certain cases; cf. Theorem 6.1-3) the sequence of orthogonal matrices

$$O_k = \Omega_1 \Omega_2 \cdots \Omega_k$$

to converge to an orthogonal matrix whose columns form an orthonormal set of eigenvectors of the matrix A.

The aim behind each transformation

$$A_k \to A_{k+1} = \Omega_k^T A_k \Omega_k, \quad k \geqslant 1,$$

is the annihilation of two off-diagonal elements in symmetrical positions, say $(A_k)_{pq}$ and $(A_k)_{qp}$, of the matrix A_k, by means of a very simple process, which we shall describe and analyse. In order to make the notation less cumbersome, we set 'for the time being'

$$A_k = A = (a_{ij}), \quad A_{k+1} = B = (b_{ij}), \quad \Omega_k = \Omega.$$

The *actual* choice of the pair (p, q) will be discussed later.

Theorem 6.1-1

Let p and q be two integers satisfying $1 \leqslant p < q \leqslant n$, and θ a real number, with which is associated the orthogonal matrix

$$
\Omega =
\begin{matrix}
& & p & & q & \\
& & \downarrow & & \downarrow & \\
\end{matrix}
\begin{pmatrix}
1 & & & & & & & & & \\
 & 1 & & & & & & & & \\
 & & 1 & & & & & & & \\
 & & & \cos\theta & & & \sin\theta & & & \leftarrow p \\
 & & & & 1 & & & & & \\
 & & & & & 1 & & & & \\
 & & & & & & 1 & & & \\
 & & & -\sin\theta & & & \cos\theta & & & \leftarrow q \\
 & & & & & & & & 1 & \\
 & & & & & & & & & 1 \\
\end{pmatrix}
$$

(1) *If $A = (a_{ij})$ is a symmetric matrix, the matrix*

$$B = \Omega^T A \Omega = (b_{ij}),$$

also symmetric, satisfies

$$\sum_{i,j=1}^{n} b_{ij}^2 = \sum_{i,j=1}^{n} a_{ij}^2.$$

(2) *If* $a_{pq} \neq 0$, *there exists a unique value of the number* θ *belonging to the set* $]-\pi/4, 0[\cup]0, \pi/4[$ *such that*

$$b_{pq} = 0;$$

it is the unique solution, from the same set, of the equation

$$\cot 2\theta = \frac{a_{qq} - a_{pp}}{2a_{pq}}.$$

With the number θ *chosen in this way,*

$$\sum_{i=1}^{n} b_{ii}^2 = \sum_{i=1}^{n} a_{ii}^2 + 2a_{pq}^2.$$

Proof

(1) It is easy to verify that the matrix Ω is orthogonal. The norm $\| \cdot \|_E$ being invariant under orthogonal transformations (Theorem 1.4-4),

$$\sum_{i,j} a_{ij}^2 = \|A\|_E^2 = \|O^T A O\|_E^2 = \sum_{i,j} b_{ij}^2.$$

(2) Next, we observe that the transformation which bears on the elements with indices (p, p), (p, q), (q, p), (q, q) is written as

$$\begin{pmatrix} b_{pp} & b_{pq} \\ b_{qp} & b_{qq} \end{pmatrix} = \begin{pmatrix} \cos\theta & -\sin\theta \\ \sin\theta & \cos\theta \end{pmatrix} \begin{pmatrix} a_{pp} & a_{pq} \\ a_{qp} & a_{qq} \end{pmatrix} \begin{pmatrix} \cos\theta & \sin\theta \\ -\sin\theta & \cos\theta \end{pmatrix},$$

and the same argument as that used in (1) shows that

$$a_{pp}^2 + a_{qq}^2 + 2a_{pq}^2 = b_{pp}^2 + b_{qq}^2 + 2b_{pq}^2$$

for every value of the number θ. Since

$$b_{pq} = b_{qp} = a_{pq} \cos 2\theta + \frac{a_{pp} - a_{qq}}{2} \sin 2\theta,$$

the choice of the number θ indicated in the statement entails

$$b_{pq} = b_{qp} = 0, \quad \text{so that } a_{pp}^2 + a_{qq}^2 + 2a_{pq}^2 = b_{pp}^2 + b_{qq}^2.$$

Moreover, since $a_{ii} = b_{ii}$ for $i \neq p$ and $i \neq q$, the proof is complete. $\quad\square$

Remarks

(1) The matrix Ω represents a rotation through an angle θ in the plane of the pth and qth basis vectors, which is another way of seeing that it is orthogonal.

(2) It may be observed, too, that only the pth and qth rows and columns of the matrix A are modified in the transformation $A \to B = \Omega^T A \Omega$. More precisely, for every value of the angle θ,

$$\begin{cases} b_{ij} = a_{ij} & \text{if } i \neq p,q \text{ and } j \neq p,q, \\ b_{pi} = a_{pi} \cos\theta - a_{qi} \sin\theta & \text{if } i \neq p,q, \\ b_{qi} = a_{pi} \sin\theta + a_{qi} \cos\theta & \text{if } i \neq p,q, \\ b_{pp} = a_{pp} \cos^2\theta + a_{qq} \sin^2\theta - a_{pq} \sin 2\theta \\ b_{qq} = a_{pp} \sin^2\theta + a_{qq} \cos^2\theta + a_{pq} \sin 2\theta, \\ b_{pq} = b_{qp} = a_{pq} \cos 2\theta + \dfrac{a_{pp} - a_{qq}}{2} \sin 2\theta. \end{cases}$$

(3) Because of relationships which exist among trigonometric functions, the elements of the matrix B are determined, despite appearances, by *algebraic equations* in terms of the elements of the matrix A. The following quantities are calculated consecutively:

$$\varkappa = \frac{a_{qq} - a_{pp}}{2a_{pq}} (= \cot 2\theta);$$

$$t = \begin{cases} \text{the root smallest in modulus of the quadratic } t^2 + 2\varkappa t - 1 = 0 \\ \left(\text{since } t = \tan\theta \text{ and } |\theta| \leqslant \dfrac{\pi}{4}\right), & \text{if } \varkappa \neq 0, \\ 1, & \text{if } \varkappa = 0; \end{cases}$$

$$c = \frac{1}{\sqrt{(1+t^2)}} (= \cos\theta);$$

$$s = \frac{t}{\sqrt{(1+t^2)}} (= \sin\theta).$$

Hence, the formulae giving the elements of the matrix B can be re-written as follows (verify):

$$\begin{aligned} b_{pi} &= ca_{pi} - sa_{qi}, & i \neq p,q, \\ b_{qi} &= ca_{qi} + sa_{pi}, & i \neq p,q, \\ b_{pp} &= a_{pp} - ta_{pq}, \\ b_{qq} &= a_{qq} + ta_{pq}, \end{aligned} \qquad \square$$

We describe now one step of the Jacobi method. The matrix $A_k = (a_{ij}^k)$ having been obtained, a pair (p,q), $p \neq q$, is chosen for which the element a_{pq}^k is not zero; then a matrix Ω_k is constructed, as was the matrix Ω of Theorem 6.1-1, the angle θ_k being chosen from the set $]-\pi/4, 0[\cup]0, \pi/4[$ in such a way that

$$\cot 2\theta_k = \frac{a_{qq}^k - a_{pp}^k}{2a_{pq}^k},$$

and we then set

$$A_{k+1} = \Omega_k^T A_k \Omega_k = (a_{ij}^{k+1}).$$

Three strategies for the choice of the pair (p, q) may be distinguished.

(i) *The classical Jacobi method.* Any pair is chosen for which

$$|a_{pq}^k| = \max_{i \neq j} |a_{ij}^k|.$$

Of course, the pairs (p, q) chosen in this way *vary with the integer k*. It is simply in order to avoid a cumbersome notation that we take no steps to bring out this dependence.

(ii) *The cyclic Jacobi method.* Because the search for the off-diagonal elements largest in modulus takes quite a long time in the classical Jacobi method, *all* the off-diagonal elements are annihilated in turn through a *cyclic sweep*, always the same; for example, the pairs (p, q) are chosen in the following order:

$$(1, 2), (1, 3), \ldots, (1, n); \quad (2, 3), \ldots, (2, n); \quad \ldots; (n - 1, n).$$

Obviously, if one of the elements 'within the sweep' is already zero, we move on to the next (in matrix terms, this is equivalent to the choice $\theta_k = 0$, or $\Omega_k = I$).

(iii) *The threshold Jacobi method.* Here we advance as in the cyclic Jacobi method, but the annihilation of any off-diagonal elements which are smaller in modulus than a certain *threshold*, which is reduced with each sweep, is omitted. That is because there appears to be no advantage in annihilating any off-diagonal elements which are already sufficiently small in absolute value, while there are yet others which are of a much greater order of magnitude.

But *be aware*! Whatever the strategy chosen, it is quite evident that the elements annihilated at a given step can be replaced subsequently by non-zero elements. Otherwise, one would obtain a reduction to a diagonal matrix in a finite number of iterations, which is impossible, as was indicated earlier in section 2.1.

We shall examine the convergence of the Jacobi method limiting ourselves to the simplest case (that of the classical method) without looking for error estimates. Readers who are interested in further information on the convergence of the classical and cyclic Jacobi methods can profitably refer to the articles Henrici (1958), Henrici & Zimmermann (1968), van Kempen (1966a, b). At this stage, we observe that the convergence is made possible by the fact that *the sum of the squares of all the elements of each matrix* A_k *remains constant, so that in each step the sum of the squares of the diagonal elements is increased by the sum of the squares of the two elements annihilated* (this is the import of Theorem 6.1-1). There are, then, good grounds for hoping that the matrices A_k might converge to a diagonal matrix, and that this diagonal matrix might turn out to be precisely the matrix $\text{diag}(\lambda_i)$, up to a permutation of the indices; this is just what is proved in the theorem which follows (the approximation of the eigenvectors is examined separately).

In order to avoid trivial situations, we shall assume in what follows that $\max_{i \neq j} |a_{ij}^k| > 0$ for every integer $k \geqslant 1$. We recall that $\mathfrak{S}_n$ denotes the set of permutations of the set of integers $\{1, 2, \ldots, n\}$.

Theorem 6.1-2 (Convergence of the eigenvalues for the classical Jacobi method)

The sequence $(A_k)_{k \geqslant 1}$ of matrices obtained by the classical Jacobi method is convergent, and

$$\lim_{k \to \infty} A_k = \text{diag}(\lambda_{\sigma(i)})$$

for some suitable permutation $\sigma \in \mathfrak{S}_n$.

Proof

(i) We begin by proving a lemma, which plays a key role in this proof as well as in that of the following theorem. *Let X be a finite-dimensional normed vector space, and (x_k) a bounded sequence in X, having only a finite number of cluster points, and such that $\lim_{k \to \infty} \|x_{k+1} - x_k\| = 0$. Then the sequence (x_k) is convergent* (of course, to one of its cluster points...).

For, if $a_\mu, 1 \leqslant \mu \leqslant M$, are the cluster points of the sequence, then, for every $\varepsilon > 0$, there exists an integer $l(\varepsilon)$ such that

$$k \geqslant l(\varepsilon) \Rightarrow x_k \in \bigcup_{\mu=1}^{M} B(a_\mu; \varepsilon),$$

letting $B(a; \varrho) = \{x \in X : \|x - a\| < \varrho\}$. Otherwise, there would exist a subsequence $(x_{k'})$ such that

$$x_{k'} \notin \left\{ \bigcup_{\mu=1}^{M} B(a_\mu; \varepsilon) \right\} \quad \text{for every } k' \geqslant 1.$$

As this subsequence is still a bounded sequence of a space of finite dimension, there would exist another subsequence $(x_{k''})$ converging to a point $x'' \notin \{ \bigcup_{\mu=1}^{M} B(a_\mu; \varepsilon) \}$. But then x'' would be a cluster point of the sequence (x_k) distinct from the points a_μ, thus contradicting the hypothesis.

The particular choice

$$\varepsilon_0 = \tfrac{1}{3} \min_{\mu \neq \mu'} \|a_\mu - a_{\mu'}\| > 0$$

shows the existence of an integer l_0 such that

$$k \geqslant l_0 \Rightarrow \begin{cases} x_k \in \bigcup_{\mu=1}^{M} B(a_\mu; \varepsilon_0), \\ \| x_{k+1} - x_k \| \leqslant \varepsilon_0. \end{cases}$$

Consequently,

$$x_k \in B(a_\mu; \varepsilon_0) \Rightarrow x_{k+1} \in B(a_\mu; \varepsilon_0) \quad \text{for every } k \geqslant l_0,$$

and the lemma is proved.

(ii) Returning to the proof proper of the convergence of the classical Jacobi method, we set, for every integer $k \geqslant 1$,

$$A_k = (a_{ij}^k) = D_k + B_k, \quad \text{with} \quad D_k \overset{\text{def}}{=} \text{diag}\,(a_{ii}^k),$$

and we show, as the first step, that $\lim_{k \to \infty} B_k = 0$. The numbers

$$\varepsilon_k = \sum_{i \neq j} |a_{ij}^k|^2 = \| B_k \|_{\text{E}}^2, \quad k \geqslant 1,$$

satisfy, on the one hand (cf. Theorem 6.1-1),

$$\varepsilon_{k+1} = \varepsilon_k - 2|a_{pq}^k|^2,$$

and, on the other (by definition of the strategy adopted in the classical Jacobi method),

$$\varepsilon_k \leqslant n(n-1)|a_{pq}^k|^2,$$

since there are $n(n-1)$ off-diagonal elements. Combining these equations, we get

$$\varepsilon_{k+1} \leqslant \left(1 - \frac{2}{n(n-1)} \right) \varepsilon_k,$$

which shows that $\lim_{k \to \infty} \varepsilon_k = 0$.

(iii) We show that *the sequence* (D_k) *has only a finite number of cluster points which are necessarily of the form* $\text{diag}(\lambda_{\sigma(i)}), \sigma \in \mathfrak{S}_n$. We recall that $\lambda_1, \lambda_2, \ldots, \lambda_n$ denote the n eigenvalues of the matrix A, assumed to be arranged, once for all, in some fixed, but arbitrary, order.

If $(D_{k'})$ is a subsequence which converges to a matrix D, then we also have

$$\lim_{k' \to \infty} A_{k'} = D, \quad \text{since} \quad A_{k'} = D_{k'} + B_{k'} \quad \text{and} \quad \lim_{k' \to \infty} B_{k'} = 0,$$

so that (consider the coefficients of the characteristic polynomial)

$$\det (D - \lambda I) = \lim_{k' \to \infty} \det (A_{k'} - \lambda I) \quad \text{for every } \lambda \in \mathbb{C}.$$

Since

$$\det (A_{k'} - \lambda I) = \det (A - \lambda I) \quad \text{for every } \lambda \in \mathbb{C},$$

the matrices $A_{k'}$ and A being similar, it follows that the matrices A and

$D = \lim_{k' \to \infty} D_{k'}$ have the same characteristic polynomial and, therefore, the same eigenvalues, multiplicities included.

Since D is a diagonal matrix (it is the limit of a sequence of diagonal matrices), there exists a permutation $\sigma \in \mathfrak{S}_n$ such that

$$D = \text{diag}\,(\lambda_{\sigma(i)}).$$

(iv) *We show that* $\lim_{k \to \infty} (D_{k+1} - D_k) = 0$. It may be verified that

$$a_{ii}^{k+1} - a_{ii}^k = \begin{cases} 0 & \text{if} \quad i \neq p, q, \\ -\tan \theta_k a_{pq}^k & \text{if} \quad i = p, \\ \tan \theta_k a_{pq}^k & \text{if} \quad i = q. \end{cases}$$

Since

$$|\theta_k| \leqslant \frac{\pi}{4} \quad \text{and} \quad |a_{pq}^k| \leqslant \|B_k\|_E \quad \text{with} \quad \lim_{k \to \infty} B_k = 0,$$

the conclusion follows.

(v) *Conclusion.* The sequence (D_k) is bounded, since

$$\|D_k\|_E \leqslant \|A_k\|_E = \|A\|_E$$

(cf. Theorem 6.1-1). Hence, by the lemma (part (i) of the proof), the sequence (D_k) converges, and its limit, which is one of the cluster points, is necessarily of the form $\text{diag}\,(\lambda_{\sigma(i)})$, $\sigma \in \mathfrak{S}_n$ (cf. (iii)). Since $A_k = D_k + B_k$, it follows that the sequence (A_k) is also convergent, and $\lim_{k \to \infty} A_k = \lim_{k \to \infty} D_k$. $\qquad \square$

We move on now to the convergence of the eigenvectors. We recall the relations

$$A_{k+1} = \Omega_k^T A_k \Omega_k = O_k^T A O_k, \quad \text{where} \quad O_k = \Omega_1 \Omega_2 \cdots \Omega_k.$$

Theorem 6.1-3 (*Convergence of the eigenvectors for the classical Jacobi method*)

Suppose that all the eigenvalues of the matrix A are distinct.

Then the sequence $(O_k)_{k \geqslant 1}$ of matrices constructed in the classical Jacobi method converges to an orthogonal matrix whose column vectors form an orthonormal set of eigenvectors of the matrix A.

Proof

The lemma established in (i) of the preceding proof will once again play a crucial role.

(i) We show that *the sequence (O_k) has only a finite number of cluster points, which are necessarily of the form*

$$\left(\boxed{\pm P_{\sigma(1)}} \, \boxed{\pm P_{\sigma(2)}} \cdots \boxed{\pm P_{\sigma(n)}} \right), \quad \sigma \in \mathfrak{S}_n,$$

where $p_1, p_2, \ldots, p_n$ denote the column vectors of the orthogonal matrix O which appears in the relation

$$O^T A O = \mathrm{diag}(\lambda_1, \lambda_2, \ldots, \lambda_n).$$

Let $(O_{k'})$ be a subsequence of the sequence (O_k) converging to an (orthogonal) matrix O'. By the previous theorem, there exists a permutation $\sigma \in \mathfrak{S}_n$ such that

$$\mathrm{diag}(\lambda_{\sigma(i)}) = \lim_{k' \to \infty} A_{k'} = \lim_{k' \to \infty} (O_{k'}^T A O_{k'}) = (O')^T A O',$$

and the assertion is proved. It should be noted that the hypothesis requiring all the eigenvalues of the matrix A to be *distinct* is critical in reaching the conclusion that there exists a finite number of cluster points.

(ii) *We show that* $\lim_{k \to \infty} (O_{k+1} - O_k) = 0$. By construction, the angle θ_k satisfies

$$\tan 2\theta_k = \frac{2a_{pq}^k}{a_{qq}^k - a_{pp}^k}, \quad |\theta_k| \leqslant \frac{\pi}{4}.$$

Using the previous theorem and (once again) the fact that all the eigenvalues of the matrix A are *distinct*, it follows that there exists an integer l such that

$$k \geqslant l \Rightarrow |a_{qq}^k - a_{pp}^k| \geqslant \tfrac{1}{2} \min_{i \neq j} |\lambda_i - \lambda_j| > 0$$

(since the pairs (p, q) vary with the integer k, it is not possible to affirm that the sequences (a_{pp}^k) and (a_{qq}^k) converge). Furthermore, since $\lim_{k \to \infty} a_{pq}^k = 0$, it is, therefore, established that

$$\lim_{k \to 0} \theta_k = 0, \quad \text{so that} \quad \lim_{k \to \infty} \Omega_k = I$$

(refer to the expression for the matrix Ω_k as a function of the angle θ_k). Finally, as

$$O_{k+1} - O_k = O_k(\Omega_{k+1} - I),$$

the assertion follows, since *the sequence* (O_k) *is bounded* (remember that the norm $\| \cdot \|_2$ of an orthogonal matrix is equal to 1). We have, therefore, at our disposal all the elements necessary for the application of the lemma, so ending the proof. □

The proofs of Theorems 6.1-2 and 6.1-3 given above are due to Michel Crouzeix.

Exercises

6.1-1. Repeat the analysis of the text for Hermitian matrices.

6.1-2. This exercise presents the method of Corbato which, in the course of the classical Jacobi method, helps to accelerate the search for a pair (p, q) satisfying $|a_{pq}^k| = \max_{i \neq j} |a_{ij}^k|$.

(1) Introduce the vectors a_k and b_k, with components

$$a_i^k = \max_{j > 1} |a_{ij}^k| = |a_{ij^k(i)}^k|, \quad 1 \leqslant i \leqslant n,$$
$$b_i^k = j^k(i), \quad 1 \leqslant i \leqslant n,$$

respectively. Say how the vectors a_{k+1} and b_{k+1} are constructed in terms of the vectors a_k and b_k.

(2) Deduce a procedure for determining, in terms of the vectors a_k and b_k, a pair (p, q) satisfying

$$|a_{pq}^{k+1}| = \max_{i \neq j} |a_{ij}^{k+1}|.$$

6.2 The Givens–Householder method

This is a method which is particularly well suited to finding *some particular eigenvalues* of a *symmetric* matrix; for example, all the eigenvalues located in an interval given in advance, or all the eigenvalues at given positions (assuming they are set out in increasing order), etc. Furthermore, this method permits the approximation of each eigenvalue with variable precision, according to the position of the eigenvalue under consideration, and at the user's choice. On the other hand, this method does not provide the corresponding eigenvectors, which would need to be obtained by other means (cf. section 6.4).

The *Givens–Householder* method consists of two stages.

(i) Given a *symmetric* matrix A, an *orthogonal* matrix P of particular type is determined such that the symmetric matrix P^TAP is *tridiagonal*. This stage, which requires only a *finite* number of elementary operations, constitutes *Householder's method for the reduction of a symmetric matrix to tridiagonal form*.

(ii) One is then faced with the *calculation of the eigenvalues of a symmetric tridiagonal matrix*, which is achieved by *Givens' method*, also known as the *method of bisection*.

We begin with a description of *Householder's method*. Given a symmetric matrix $A = A_1$, $n - 2$ orthogonal matrices $H_1, H_2, \ldots, H_{n-2}$ are determined, step by step, such that the matrices

$$A_k = H_{k-1}^T A_{k-1} H_{k-1}$$
$$= (H_1 H_2 \cdots H_{k-1})^T A (H_1 H_2 \cdots H_{k-1}), \quad 1 \leqslant k \leqslant n - 2,$$

have the form

$$A_k = \begin{pmatrix} \times & \times \\ \times & \times & \times \\ & \times & \times & \times \\ & & \times & \times & \times \\ & & & \times & \times & \times & \times & \times & \times & \times & \overset{\nearrow a_k^T}{} \\ & & & & & \times & \times & \times & \times & \times & \times \\ & & & & & \times & \times & \times & \times & \times & \times \\ & & & & & \times & \times & \times & \times & \times & \times \\ & & & & a_k \nearrow & \times & \times & \times & \times & \times & \times \\ & & & & & \times & \times & \times & \times & \times & \times \end{pmatrix} \;\rightarrow k\text{th row}$$

$\downarrow$
kth column

Consequently, the matrix (also symmetric)

$$A_{n-1} = (H_1 H_2 \cdots H_{n-2})^T A (H_1 H_2 \cdots H_{n-2})$$

is tridiagonal, on the one hand, and similar to the given matrix A, on the other, the required orthogonal matrix P being the product $P = H_1 H_2 \cdots H_{n-2}$.

In the present method, every transformation $A_k \rightarrow A_{k+1} = H_k^T A_k H_k$ is carried out with the help of a matrix of form

$$H_k = \left(\begin{array}{c|c} I_k & 0 \\ \hline 0 & \tilde{H}_k \end{array} \right),$$

I_k being the identity matrix of order k. Consequently, the properties of the block multiplication of matrices show that the matrix $H_k^T A_k H_k$ has the form

$$H_k^T A_k H_k = \begin{pmatrix} \times & \times \\ \times & \times & \times \\ & \times & \times & \times \\ & & \times & \times & \times \\ & & & \times & \times & \times & \times & \times & \times & \times & \overset{\nearrow a_k^T \tilde{H}_k}{} \\ & & & & & \times & \times & \times & \times & \times & \times \\ & & & & & \times & \times & \times & \times & \times & \times \\ & & & & \tilde{H}_k^T a_k \nearrow & \times & \times & \times & \times & \times & \times \\ & & & & & \times & \times & \times & \times & \times & \times \\ & & & & & \times & \times & \times & \times & \times & \times \end{pmatrix} \;\rightarrow k\text{th row}$$

$\downarrow$
kth column

where a_k denotes the column vector of $\mathbb{R}^{n-k}$ whose components are the elements a_{ik}^k, $k+1 \leqslant i \leqslant n$, of the matrix $A_k = (a_{ij}^k)$, the elements a_{ij}^k, $1 \leqslant i, j \leqslant k$, of the matrix A_k remaining unaltered. *It is enough, therefore, to arrange matters so that the vector $\tilde{H}_k^T a_k \in \mathbb{R}^{n-k}$ has all its components, other than the first, equal to zero*, the matrix $\tilde{H}_k$ being orthogonal and 'as simple as possible to construct'.

This aim will be attained very simply with the help of the *Householder matrices* introduced in section 4.5 which, if it is remembered, are both symmetric and orthogonal, and have the particular form

$$H(v) = 1 - 2\frac{vv^T}{v^T v}, \qquad v \text{ a non-zero vector.}$$

If $\sum_{i=k+2}^{n} |a_{ik}^k| > 0$, it follows from Theorem 4.5-1 that there exists a vector $\tilde{v}_k \in \mathbb{R}^{n-k}$ such that the vector $H(\tilde{v}_k) a_k \in \mathbb{R}^{n-k}$ has all its components except the first equal to zero. We then set $\tilde{H} = H(\tilde{v}_k)$. Naturally, if $a_{ik}^k = 0$ for $i = k+2, \ldots, n$, we choose $\tilde{H}_k = I$ (we recall that it has been agreed to consider the identity matrix also as a Householder matrix).

We have now all the elements necessary for the description of the *transition from the matrix A_k to the matrix $A_{k+1} = H_k^T A_k H_k$*. If $a_{ik}^k = 0$ for $i = k+2, \ldots, n$, the matrix A_k is already of the form A_{k+1}, so that it is enough to set $H_k = I$. Otherwise, the matrix H_k is constructed with the help of the matrix $\tilde{H}_k = H(\tilde{v}_k)$, as was indicated earlier.

It is, moreover, permissible to consider the matrix H_k itself as a Householder matrix:

$$H_k = H(v_k) = I - 2\frac{v_k v_k^T}{v_k^T v_k},$$

the vector $v_k \in \mathbb{R}^n$ having its first k components equal to zero, the last $n-k$ being those of the vector $\tilde{v}_k \in \mathbb{R}^{n-k}$.

From these considerations and from Theorem 4.5-1, it follows that there exist two possible choices of the vector v_k, namely

$$v_k = \begin{pmatrix} 0 \\ \vdots \\ 0 \\ a_{k+1,k}^k \pm \left(\sum_{i=k+1}^{n} |a_{ik}^k|^2 \right)^{1/2} \\ a_{k+2,k}^k \\ \vdots \\ a_{nk}^k \end{pmatrix},$$

the *actual* choice of the sign in the expression for the $(k + 1)$th component being the same as *that of the element* $a_{k+1,k}^k$ (in order to avoid denominators which are too 'small').

Once the vector v_k has been determined, the elements a_{ij}^{k+1}, $k + 1 \leqslant i, j \leqslant n$, of the matrix $A_{k+1} = (a_{ij}^{k+1})$ are obtained as follows. We calculate in succession the vectors

$$w_k = (v_k^T v_k)^{-1/2} v_k,$$
$$q_k = 2(I - w_k w_k^T) A_k w_k,$$

with components (w_i^k) and (q_i^k), respectively. Then the matrix A_{k+1} takes the form

$$A_{k+1} = A_k - w_k q_k^T - q_k w_k^T,$$

so that

$$a_{ij}^{k+1} = a_{ij}^k - w_i^k q_j^k - q_i^k w_j^k, \quad k + 1 \leqslant i, j \leqslant n.$$

Remark

The condition number $\mathrm{cond}_2(A)$ is not altered by this method for

$$\mathrm{cond}_2(A) = \mathrm{cond}_2(A_k) \quad \text{for} \quad k = 1, \ldots, n - 1$$

since it is, in fact, invariant under orthogonal transformations (Theorem 2.2-3); that is one of the advantages of the method, from the point of view of 'numerical stability'. □

Note, in passing, the following result, which states the applicability of Householder's method.

Theorem 6.2-1
Given a symmetric matrix A, *there exists a symmetric matrix* P, *the product of* $n - 2$ *Householder matrices, such that the matrix* $P^T A P$ *is tridiagonal.* □

Another efficient reduction of a symmetric (or Hermitian) matrix to tridiagonal form is achieved by the *Lanczos method* (this method can also be extended to non-Hermitian matrices; it is then called *Arnoldi's method*). The method relies on the fact that n orthonormal vectors $q_1, q_2, \ldots, q_n$ can be constructed using a three-term recurrence formula and n matrix–vector multiplications only, in such a way that, if Q denotes the matrix whose column vectors are the q_i, $1 \leqslant i \leqslant n$, the matrix $Q^T A Q$ is tridiagonal. Details on this popular method may be found in Chatelin (1983, 1987), Lascaux & Théodor (1987), Parlett (1980), and especially in the book Cullum & Willoughby (1985), which is entirely devoted to this method.

Next, we move on to a description of *Givens' method* for finding the eigenvalues of a symmetric tridiagonal matrix:

$$B = \begin{pmatrix} b_1 & c_1 & & & \\ c_1 & b_2 & c_2 & & \\ & \ddots & \ddots & \ddots & \\ & & c_{n-2} & b_{n-1} & c_{n-1} \\ & & & c_{n-1} & b_n \end{pmatrix}.$$

We observe, to begin with, that if one of the elements c_i is zero, then the matrix B decomposes into two tridiagonal submatrices of the same type. Hence, it may be supposed, without loss of generality, that

$$c_i \neq 0, \quad 1 \leqslant i \leqslant n-1.$$

We begin by establishing that the roots of the characteristic polynomials of the submatrices

$$B_i = \begin{pmatrix} b_1 & c_1 & & & \\ c_1 & b_2 & c_2 & & \\ & \ddots & \ddots & \ddots & \\ & & c_{i-2} & b_{i-1} & c_{i-1} \\ & & & c_{i-1} & b_i \end{pmatrix}, \quad 1 \leqslant i \leqslant n,$$

have truly remarkable properties of 'interlacing', as is illustrated in figure 6.2-1.

Figure 6.2-1

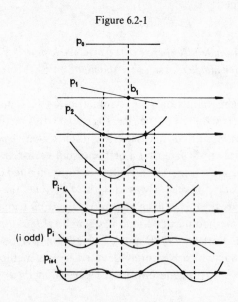

Theorem 6.2-2

The polynomials $p_i(\lambda)$, $\lambda \in \mathbb{R}$, defined for $i = 0, 1, \ldots, n$ by the recurrence formulae

$$p_0(\lambda) = 1,$$
$$p_1(\lambda) = b_1 - \lambda,$$
$$p_i(\lambda) = (b_i - \lambda)p_{i-1}(\lambda) - c_{i-1}^2 p_{i-2}(\lambda), \quad 2 \leqslant i \leqslant n,$$

have the following properties:

(1) The polynomial p_i is the characteristic polynomial of the matrix B_i, $1 \leqslant i \leqslant n$;

(2) $\lim_{\lambda \to -\infty} p_i(\lambda) = +\infty$, $1 \leqslant i \leqslant n$;

(3) $p_i(\lambda_0) = 0 \Rightarrow p_{i-1}(\lambda_0)p_{i+1}(\lambda_0) < 0$, $1 \leqslant i \leqslant n - 1$;

(4) The polynomial p_i has i distinct real roots, which separate the $i + 1$ roots of the polynomial p_{i+1}, $1 \leqslant i \leqslant n - 1$, as is indicated in figure 6.2-1.

Proof

(1) This property is verified by expanding the determinant of the matrix $B_i - \lambda I$ by its last row (or column).

(2) It is easy to see by induction that the term of highest degree of the polynomial p_i is $(-1)^i \lambda^i$.

(3) Suppose that $p_i(\lambda_0) = 0$ for some integer satisfying $1 \leqslant i \leqslant n - 1$. From the recurrence relation, it follows that

$$p_{i+1}(\lambda_0) = -c_i^2 p_{i-1}(\lambda_0),$$

which implies (with the assumption $c_i \neq 0$)

$$\begin{cases} \text{either} & p_{i-1}(\lambda_0)p_{i+1}(\lambda_0) < 0, \\ \text{or} & p_{i-1}(\lambda_0) = p_i(\lambda_0) = p_{i+1}(\lambda_0). \end{cases}$$

In the case of the second alternative, the recurrence relation (used, once again, in conjunction with the hypothesis that $c_i \neq 0$, $1 \leqslant i \leqslant n - 1$) shows that

$$p_i(\lambda_0) = p_{i-1}(\lambda_0) = \cdots = p_1(\lambda_0) = p_0(\lambda_0) = 0,$$

which is impossible, since $p_0(\lambda) = 1$.

(4) This property comes as a consequence of properties (2) and (3). $\square$

A sequence of polynomials satisfying conditions (2), (3) and (4) of the preceding theorem is called a *Sturm sequence*. Givens' method relies upon an altogether remarkable property of these sequences, which is the fact that they allow an immediate calculation of the number of roots $< \mu$, for any given $\mu \in \mathbb{R}$, for each of the polynomials in the sequence. This is what is proved in the result which follows.

Theorem 6.2-3

Let i be an integer satisfying $1 \leqslant i \leqslant n$. Given a number $\mu \in \mathbb{R}$, set

$$\operatorname{sgn} p_i(\mu) = \begin{cases} sign \ of \ p_i(\mu) & if \quad p_i(\mu) \neq 0, \\ sign \ of \ p_{i-1}(\mu) & if \quad p_i(\mu) = 0. \end{cases}$$

Then the number $N(i; \mu)$ of changes of sign among consecutive elements of the ordered set

$$E(i; \mu) = \{ +, \operatorname{sgn} p_1(\mu), \operatorname{sgn} p_2(\mu), \ldots, \operatorname{sgn} p_i(\mu) \}$$

is equal to the number of roots of the polynomial p_i which are $< \mu$.

Proof

To begin with, we verify that the expression $\operatorname{sgn} p_i(\lambda)$ is defined unambiguously in every case; this is so because $p_i(\lambda) = 0 \Rightarrow p_{i-1}(\lambda) \neq 0$ by the previous theorem. Now, since

$$\mu \leqslant b_1 \Rightarrow E(1; \mu) = \{ +, + \} \Rightarrow N(1; \mu) = 0,$$
$$b_1 < \mu \Rightarrow E(1; \mu) = \{ +, - \} \Rightarrow N(1; \mu) = 1,$$

the assertion holds good for $i = 1$.

Suppose that the property is true for the integers $1, 2, \ldots, i$; we show that it is also true for the integer $i + 1$. Calling $\lambda_1^i, \ldots, \lambda_i^i$ and $\lambda_1^{i+1}, \ldots, \lambda_{i+1}^{i+1}$ the roots, arranged in increasing order, of the polynomials p_i and p_{i+1} respectively, we have

$$\lambda_1^i < \cdots < \lambda_{N(i;\mu)}^i < \mu \leqslant \lambda_{N(i;\mu)+1}^i < \cdots < \lambda_i^i,$$

by definition of the integer $N(i; \mu)$, and, furthermore,

$$\lambda_{N(i;\mu)}^i < \lambda_{N(i;\mu)+1}^{i+1} < \lambda_{N(i;\mu)+1}^i,$$

by the preceding theorem. It is enough, therefore, to examine the various possible cases (cf. figure 6.2-1).

$$\lambda_{N(i;\mu)}^i < \mu \leqslant \lambda_{N(i;\mu)+1}^{i+1} \Rightarrow \operatorname{sgn} p_{i+1}(\mu) = \operatorname{sgn} p_i(\mu)$$
$$\Rightarrow N(i+1; \mu) = N(i; \mu);$$

$$\lambda_{N(i;\mu)+1}^{i+1} < \mu < \lambda_{N(i;\mu)+1}^i \Rightarrow \operatorname{sgn} p_{i+1}(\mu) = -\operatorname{sgn} p_i(\mu)$$
$$\Rightarrow N(i+1; \mu) = N(i; \mu) + 1;$$

$$\lambda_{N(i;\mu)+1}^i = \mu \Rightarrow \operatorname{sgn} p_i(\mu) = \operatorname{sgn} p_{i-1}(\mu) = -\operatorname{sgn} p_{i+1}(\mu)$$
$$\Rightarrow N(i+1; \mu) = N(i; \mu) + 1$$

(in the last implication, property (3) of the preceding theorem is used in an essential way). $\qquad \square$

Remark

Let $M(i; \mu)$ be the number of consecutive pairs with the same sign ($\{+, +\}$ or $\{-, -\}$) found in the ordered set $E(i; \mu)$; for example, if

$$E(10; \mu) = \{+, +, -, +, -, -, +, -, -, -, +\},$$

$M(10; \mu) = 4$, while $N(10; \mu) = 6$. It is easy to verify that

$$M(i; \mu) + N(i; \mu) = i, \quad \text{for} \quad i = 1, \ldots, n,$$

so that the number $M(i; \mu)$ is equal to the number of roots of the polynomial p_i which are $\geqslant \mu$. $\qquad \square$

The result of Theorem 6.2-3 allows the approximation *to any desired degree of accuracy* of the eigenvalues of the matrix $B = B_n$ and even the *direct calculation of an eigenvalue at a given position*. For example, suppose that we would like to approximate the ith eigenvalue $\lambda_i = \lambda_i^n$ of the matrix B, the integer i being fixed (it is supposed, as before, that the eigenvalues $\lambda_1, \ldots, \lambda_n$ are arranged in increasing order; we recall that they are all distinct according to Theorem 6.2-2).

The first step is to determine an interval $[a_0, b_0]$ in which the required eigenvalue is sure to be found; *for example,* $-a_0 = b_0 = \|B\|_1$ or $\|B\|_\infty$ (Theorem 1.4-3) may be chosen, or even Gerschgorin's theorem may be used (Exercise 1.1-5). Next, the mid-point of the segment $[a_0, b_0]$ is denoted by c_0 and the integer $N(n; c_0)$ is calculated (the notation is that of Theorem 6.2-3; recall that p_n is just the characteristic polynomial of the matrix B); then

$$\text{either} \quad N(n; c_0) \geqslant i \quad \text{and} \quad \lambda_i \in [a_0, c_0[,$$
$$\text{or} \quad N(n; c_0) < i \quad \text{and} \quad \lambda_i \in [c_0, b_0],$$

so that an interval $[a_1, b_1]$ is acquired, half the length of the previous one, in which the eigenvalue λ_i is situated.

In this way, step by step, a sequence $[a_k, b_k]$, $k \geqslant 0$, of boxed intervals is determined such that

$$\lambda_i \in [a_k, b_k] \quad \text{and} \quad b_k - a_k = 2^{-k}(b_0 - a_0), \quad k \geqslant 0,$$

so that it is possible, in effect, to bracket the ith eigenvalue λ_i with arbitrary (theoretically) precision.

Exercises

6.2-1. Consider the following *direct method* for the solution of a linear system $Au = b$ whose matrix A is symmetric: (i) by Householder's method, an orthogonal matrix P is determined such that the matrix $B = P^T A P$ is tridiagonal;

(ii) the linear system $Bv = P^T b$ is solved by the method of section 4.3; (iii) $u = Pv$ is calculated.

Count the corresponding number of elementary operations and compare it with that of the Cholesky method (section 4.4).

6.2-2. Reduce the following matrix into tridiagonal form–

$$\begin{pmatrix} 120 & 80 & 40 & -16 \\ 80 & 120 & 16 & -40 \\ 40 & 16 & 120 & -80 \\ -16 & -40 & -80 & 120 \end{pmatrix}$$

–and then calculate its eigenvalues.

6.3 The QR algorithm

The QR algorithm, due to J.C.F. Francis and V.N. Kublanovskaya, is the one most frequently used for the calculation of the set of eigenvalues of a *general matrix*, notably one which is not symmetric. Of course, it is applicable *a fortiori* to symmetric matrices, for which it is at least as efficient as the Jacobi method.

We describe the algorithm below in a form *purposely simplified* and for which a result about convergence will be established (Theorem 6.3-1) which is not the most general, but whose proof has the advantage of being simple. Thereafter, some complementary points will be made, especially about the practical implementation of the algorithm.

Let, then, $A = A_1$ be a general square matrix; its QR factorisation (Theorem 4.5-2) is written as $A_1 = Q_1 R_1$, then the matrix $A_2 = R_1 Q_1$ is formed; next the QR factorisation of the matrix A_2 is written as $A_2 = Q_2 R_2$ and the matrix $A_3 = R_2 Q_2$ is formed, and so on:

$$A_1 \rightarrow A_2 \begin{cases} \text{QR factorisation: } A = A_1 = Q_1 R_1; \\ \qquad\quad \text{set: } A_2 = R_1 Q_1; \end{cases}$$

$$\vdots$$

$$A_k \rightarrow A_{k+1} \begin{cases} \text{QR factorisation: } A_k = Q_k R_k; \\ \qquad\quad \text{set: } A_{k+1} = R_k Q_k; \end{cases}$$

$$\vdots$$

In this way, a sequence of matrices A_k is obtained which are all *similar to the matrix* A, since

$$A_2 = R_1 Q_1 = Q_1^* A Q_1,$$

$$\vdots$$

$$A_{k+1} = R_k Q_k = Q_k^* A_k Q_k = \cdots = (Q_1 Q_2 \cdots Q_k)^* A (Q_1 Q_2 \cdots Q_k),$$

$$\vdots$$

Under hypotheses which are somewhat restrictive, we shall establish in the theorem given below that the matrices A_k 'become' upper triangular, in the sense that $\lim_{k \to \infty}(A_k)_{ij} = 0$ for $j < i$, while the diagonal elements of the matrices A_k converge to the eigenvalues of the matrix A. However, care is needed! It is not possible to say anything about the possible convergence of the elements $(A_k)_{ij}$ for $i < j$ *and hence about the sequence* (A_k). Even so, this observation is of no significance from the practical point of view, to the extent that the objective pursued, that is to say, an approximation of the eigenvalues, is in fact attained.

We recall that in the QR factorisation of a matrix (Theorem 4.5-2), the matrix Q is unitary and the matrix R is upper triangular, and that, if the matrix A = QR is real, the matrices Q and R are also real, the matrix Q being thereby orthogonal.

Theorem 6.3-1 (*Convergence of the QR algorithm*)
Suppose that the matrix A is invertible and that all its eigenvalues are distinct in modulus. Then, there exists (at least) one invertible matrix P such that

$$A = P\Lambda P^{-1}, \quad with \ \Lambda = \mathrm{diag}(\lambda_1, \lambda_2, \ldots, \lambda_n),$$

and

$$|\lambda_1| > |\lambda_2| > \cdots > |\lambda_n| > 0.$$

Suppose that the matrix P^{-1} *has an LU factorisation (cf. section 4.3). Then the sequence of matrices* $(A_k)_{k \geqslant 1}$ *is such that*

$$\lim_{k \to \infty} (A_k)_{ii} = \lambda_i, \quad 1 \leqslant i \leqslant n,$$

$$\lim_{k \to \infty} (A_k)_{ij} = 0, \quad 1 \leqslant j < i \leqslant n.$$

Proof
(i) *Outline of the proof*. It has already been seen that

$$A_{k+1} = \mathcal{Q}_k^* A \mathcal{Q}_k, \quad with \ \mathcal{Q}_k \overset{\mathrm{def}}{=} Q_1 Q_2 \cdots Q_k.$$

The aim, then, is to examine the *asymptotic behaviour of the matrices* $\mathcal{Q}_k$. To this end, *it is convenient to consider the successive powers* A^k, $k \geqslant 1$, *of the matrix* A, for which a particular QR factorisation is already available in the form

$$A^k = \mathcal{Q}_k \mathcal{R}_k, \quad with \ \mathcal{R}_k = R_k \cdots R_2 R_1,$$

since

$$A^k = Q_1(R_1 Q_1)(R_1 \cdots Q_1)(R_1 Q_1)R_1 = Q_1 Q_2 (R_2 Q_2) \cdots (R_2 Q_2) R_2 R_1$$
$$= \cdots = (Q_1 Q_2 \cdots Q_k)(R_k \cdots R_2 R_1).$$

Using the LU factorisation of the matrix P^{-1}, we shall, by different means, obtain a second QR factorisation of the matrices A^k, and the comparison of these two factorisations will lead to an expression for the matrices $\mathcal{Q}_k$ which is particularly useful for their further examination.

(ii) *Another expression for the matrices $\mathcal{Q}_k$.* Setting

$$P = QR \quad \text{and} \quad P^{-1} = LU$$

for the QR and LU factorisations of the matrices P and P^{-1}, respectively, the hypothesis of the invertibility of A makes it possible to write

$$A^k = P\Lambda^k P^{-1} = QR(\Lambda^k L \Lambda^{-k})\Lambda^k U,$$

writing Λ^{-k} for the matrix $\mathrm{diag}(\lambda_i^{-k})$. The reason for introducing the matrix $\Lambda^k L \Lambda^{-k}$ is that its asymptotic behaviour is known. The matrix L being lower triangular with $(L)_{ii} = 1$, $1 \leqslant i \leqslant n$, we have

$$(\Lambda^k L \Lambda^{-k})_{ij} = \begin{cases} 0 & \text{if} \quad i < j, \\ 1 & \text{if} \quad i = j, \\ \left(\dfrac{\lambda_i}{\lambda_j}\right)^k (L)_{ij} & \text{if} \quad i > j, \end{cases}$$

so that

$$\lim_{k \to \infty} (\Lambda^k L \Lambda^{-k}) = I,$$

because of the hypothesis regarding the moduli of the eigenvalues of the matrix A (this is the only time that this hypothesis is used). Setting

$$\Lambda^k L \Lambda^{-k} = I + F_k, \quad \text{with} \lim_{k \to \infty} F_k = 0,$$

one can write

$$R(\Lambda^k L \Lambda^{-k}) = (I + RF_k R^{-1})R.$$

For sufficiently large values of the integer k, the matrices $I + RF_k R^{-1}$ are invertible, since $\lim_{k \to \infty} F_k = 0$ (Theorem 1.4-5); hence they admit a unique QR factorisation of the type (cf. Theorem 4.5-2)

$$I + RF_k R^{-1} = \tilde{Q}_k \tilde{R}_k \quad \text{with} (\tilde{R}_k)_{ii} > 0, \quad 1 \leqslant i \leqslant n.$$

The matrices $\tilde{Q}_k$ being unitary, the sequence $(\tilde{Q}_k)$ is bounded ($\| \tilde{Q}_k \|_2 = 1$; cf. Theorem 1.4-2). Hence it is possible to find a subsequence, say $(\tilde{Q}_{k'})$, which converges to a matrix $\tilde{Q}$, also unitary. As

$$\tilde{R}_{k'} = \tilde{Q}_{k'}^*(I + RF_{k'}R^{-1}),$$

it follows that the subsequence $(\tilde{R}_{k'})$ converges to a matrix $\tilde{R}$, also upper triangular, with $(\tilde{R})_{ii} \geqslant 0, 1 \leqslant i \leqslant n$. Taking the limit of the subsequence under consideration, we obtain

$$I = \tilde{Q}\tilde{R},$$

which means that $(\tilde{R})_{ii} > 0$, $1 \leqslant i \leqslant n$. By the uniqueness of such a QR factorisation (Theorem 4.5-2), it follows that $\tilde{Q} = \tilde{R} = I$.

Since the same argument may be repeated for every subsequence of the sequences $(\tilde{Q}_k)$ and $(\tilde{R}_k)$, the uniqueness of the limit shows that the two 'full' sequences converge:

$$\lim_{k \to \infty} \tilde{Q}_k = I, \quad \lim_{k \to \infty} \tilde{R}_k = I.$$

We note in passing the importance of the existence and the uniqueness of the QR factorisation when $(R)_{ii} > 0$, which permit, on the one hand, the unambiguous definition of the matrices $\tilde{Q}_k$ and $\tilde{R}_k$, and, on the other, the demonstration of the convergence of the two sequences.

In conclusion, we have obtained

$$A^k = (Q\tilde{Q}_k)(\tilde{R}_k R \Lambda^k U) = \mathcal{Q}_k \mathcal{R}_k.$$

Now, the matrix $Q\tilde{Q}_k$ being unitary and the matrix $\tilde{R}_k R \Lambda^k U$ being upper triangular (as the product of matrices of the same type), *the first expression for the matrix A^k is nothing other than a QR factorisation of this same matrix.* By the second remark following the proof of Theorem 4.5-2, *there exists, therefore, for every integer k, a diagonal matrix D_k with $|(D_k)_{ii}| = 1$,* $1 \leqslant i \leqslant n$, such that

$$\mathcal{Q}_k = Q\tilde{Q}_k D_k.$$

(iii) *Asymptotic behaviour of the matrices*

$$A_{k+1} = \mathcal{Q}_k^* A \mathcal{Q}_k.$$

Using the expression for the matrices $\mathcal{Q}_k$ found above and the equality $A = QR\Lambda R^{-1} Q^{-1}$, it follows that

$$A_{k+1} = D_k^* \tilde{Q}_k^* R \Lambda R^{-1} \tilde{Q}_k D_k.$$

Since $\lim_{k \to \infty} \tilde{Q}_k = I$, we conclude that

$$\lim_{k \to \infty} (\tilde{Q}_k^* R \Lambda R^{-1} \tilde{Q}_k) = R \Lambda R^{-1} = \begin{pmatrix} \lambda_1 & \times & \cdots & \times \\ & \lambda_2 & \cdots & \times \\ & & \ddots & \vdots \\ & & & \lambda_n \end{pmatrix},$$

the order of the eigenvalues (decreasing in modulus) being preserved, since the matrix R is upper triangular. Setting

$$\mathcal{D}_k = \tilde{Q}_k^* R \Lambda R^{-1} \tilde{Q}_k,$$

we obtain (the matrices D_k are diagonal)

$$(A_{k+1})_{ij} = \overline{(D_k)}_{ii}(D_k)_{jj}(\mathscr{D}_k)_{ij},$$

so that

$$(A_{k+1})_{ii} = (\mathscr{D}_k)_{ii}, \quad 1 \leqslant i \leqslant n,$$

since $|(D_k)_{ii}| = 1, 1 \leqslant i \leqslant n$. The desired conclusions then follow from the relation $\lim_{k\to\infty}\mathscr{D}_k = R\Lambda R^{-1}$ established above.   □

Remarks

(1) If the matrix A is real, the hypothesis that all the eigenvalues are distinct in modulus implies that they are real.

(2) Observe the surprising character of the hypothesis concerning the existence of an LU factorisation of the matrix P^{-1} (a sufficient condition for the existence of such a factorisation has been given in Theorem 4.3-1). On this subject, see exercise 6.3-1 which, however, 'attenuates' the import of this restriction.

(3) While it is not possible to prove the convergence of the elements $(A_k)_{ij}$ for $i < j$, it follows, nevertheless, from the last part of the proof that their *moduli* converge to the moduli of the corresponding elements of the matrix $R\Lambda R^{-1}$.   □

If several eigenvalues are equal in modulus (for example, multiple eigenvalues, or complex eigenvalues of a real matrix), it can be shown, through the existence of an LU factorisation, but now 'by blocks', of the matrix P^{-1}, that the matrices A_k become 'only' block triangular, each 'limiting' diagonal sub-matrix corresponding to a set of eigenvalues equal in modulus. More precisely, if p denotes the multiplicity of all the eigenvalues of some particular modulus, then there will appear in the matrices A_k a diagonal submatrix of order p, whose elements may not necessarily converge, but whose *eigenvalues* converge to the eigenvalues having the particular modulus under consideration, the 'limiting' diagonal sub-matrices remaining arranged in decreasing order of the moduli of the eigenvalues. If, for example, the matrix A has the eigenvalues $\lambda_i, 1 \leqslant i \leqslant 9$, with $|\lambda_1| = |\lambda_2| = |\lambda_3| > |\lambda_4| > |\lambda_5| = |\lambda_6| = |\lambda_7| = |\lambda_8| = |\lambda_9|$, then the 'limiting' appearance of the matrices A_k is the following:

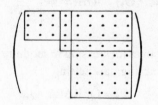

In practice, prior to applying the QR algorithm proper, *the matrix* A *is transformed to upper Hessenberg form* (cf. figure 2.1-1), by the method proposed in Exercise 6.3-2, which is nothing other than the extension to the case of non-symmetric matrices (or non-Hermitian matrices, in the complex case) of Householder's method for the reduction of a symmetric matrix (or Hermitian matrix) to tridiagonal form, described in section 6.2.

The interest behind the preliminary reduction of the matrix A to Hessenberg form derives from the fact that the matrices A_k *constructed by the* QR *algorithm remain in Hessenberg form* (cf. Exercise 6.3-3). This reduces considerably the time required for the corresponding calculation of one iteration of the method.

Lastly, we give a very simple variant of the algorithm, *of universal applicability*, whose effect is to achieve a considerable acceleration of the rate of convergence. In order to fix ideas, let us suppose that all the eigenvalues of the matrix A are real and distinct in modulus; it can then be established (cf. Wilkinson J.H. (1965), p. 491) that the elements $(A_k)_{ij}$ for which $i > j$ converge to zero as $|\lambda_i/\lambda_j|^k$, the eigenvalues of the matrix A being still assumed to be ordered in decreasing order of magnitude, as in Theorem 6.3-1. If, then, the preliminary precaution is taken of reducing the matrix A to Hessenberg form, the convergence to zero of the elements $(A_k)_{i,i-1}$ is consequently found to be governed only by the ratios $|\lambda_i/\lambda_{i-1}|^k$, $2 \leqslant i \leqslant n$. This result, in conjunction with Exercise 6.3-4, shows that *there is great advantage in replacing the matrices* A_k *by 'shifted' matrices of the form*

$$A_k - s_k I, s_k \in \mathbb{R},$$

the number s_k being an approximation of the eigenvalue λ_n which is 'as good as possible', since the elements in position $(n, n-1)$ can then be expected to converge to zero as $|(\lambda_n - s_k)/(\lambda_{n-1} - s_k)|^k$. This observation lies behind the use of the *shifted* QR *algorithm*. At the kth iteration, a QR factorisation of the matrix $A_k - s_k I$ is effected, that is,

$$A_k - s_k I = Q_k R_k,$$

with the 'best available choice' of s_k, that is, obviously enough,

$$s_k = (A_k)_{nn},$$

and we then set

$$A_{k+1} = R_k Q_k + s_k I = Q_k^T A_k Q_k,$$

so that the 'new' matrix A_{k+1} remains similar to the matrix A_k.

Modified in this way, the algorithm converges more rapidly, the 'earliest' convergence to be observed being that of the elements of the last row of the matrix A_k (every effort was made to achieve just that!). Upon deciding that the element $(A_k)_{n,n-1}$ is sufficiently close to zero, which amounts to considering that the element $(A_k)_{nn}$ is a sufficiently close approximation of the eigenvalue λ_n, the implementation of the algorithm is continued using only the submatrix of order $n-1$ consisting of the first $n-1$ rows and columns of the matrix A_k, and so on.

In order to justify the use of shifts, we set ourselves in the particular situation where the eigenvalues were real and all distinct in modulus. It is quite evident that the preceding analysis (which, in any case, is only approximate) does not apply when λ_n and λ_{n-1} are two complex conjugate eigenvalues of a real matrix, because, do what one may, the successive matrices A_k are all real, so that the element $(A_k)_{n,n-1}$ could not tend to zero. In theory, it would be possible to avoid this difficulty by implementing, in succession, two complex conjugate shifts, leading to a real matrix A_{k+2}. However, it would be necessary to introduce complex numbers into the calculations, so that it could happen that, as a result of rounding errors, some of the elements of A_{k+2} 'continued' to have non-zero imaginary parts. For this reason, a particular idea is used, due to J.C.C. Francis, called the *double shift*, which combines the previous two steps without introducing complex numbers.

As G. Strang say, '[the QR algorithm is] one of the most remarkable algorithms in numerical mathematics' (Strang (1976), p. 282). Indeed it is quite remarkable that an algorithm, which is both effective and easy to describe, has resisted, and stoutly continues to resist, a full mathematical analysis, to such an extent that no proof of convergence in the most general case (the matrix not Hermitian; the QR algorithm with shifts) exists at the present, at the same time that no counter-example to convergence exists! By way of example, convergence has been proved ('roughly' in increasing order of generality) in the following cases (the list given below is not exhaustive):

 (i) necessary and sufficient conditions for the convergence of Hessenberg matrices, without shifts (Parlett (1968));
 (ii) symmetric tridiagonal matrices, with appropriate shift strategy (Wilkinson (1968)) (a new proof was proposed in Hoffmann & Parlett (1978));
(iii) normal Hessenberg matrices, with shift strategy (Buurema (1970));
(iv) general Hessenberg matrices, with shift strategy (Lebaud (1970)).

For further material on the practical implementation of the algorithm, which is very useful and often indispensable (for example, the preliminary scaling of the matrix A), see Wilkinson & Reinsch (1971). Moreover, a connection has been discovered between the QR algorithm and dynamical systems; see, for example, Watkins (1984).

Exercises

6.3-1. The same conditions are assumed as those which prevail in Theorem 6.3-1, with the further assumption made that the matrix A is a Hessenberg matrix. Show that it is always possible, in that case, to find a matrix P (such that $A = P\Lambda P^{-1}$) which admits an LU factorisation.

6.3-2. Given a general square matrix $A = A_1$, show that there exist $n - 2$ Householder matrices (cf. section 4.5) such that the matrices

$$A_k = H_{k-1}A_{k-1}H_{k-1} = (H_1H_2\cdots H_{k-1})A(H_1H_2\cdots H_{k-1}), \quad 1 \leqslant k \leqslant n-2,$$

are of the form

$$
A_k = \begin{pmatrix}
\times & \times & \times & \times & \times & \times & \times & \times & \times & \times \\
\times & \times & \times & \times & \times & \times & \times & \times & \times & \times \\
 & \times & \times & \times & \times & \times & \times & \times & \times & \times \\
 & & \times & \times & \times & \times & \times & \times & \times & \times \\
 & & & \times & \times & \times & \times & \times & \times & \times \\
 & & & & \times & \times & \times & \times & \times & \times \\
 & & & & \times & \times & \times & \times & \times & \times \\
 & & & & \times & \times & \times & \times & \times & \times \\
 & & & & \times & \times & \times & \times & \times & \times \\
 & & & & \times & \times & \times & \times & \times & \times
\end{pmatrix} \rightarrow k\text{th row}
$$

$\downarrow$
kth column

and such that the matrix A_{n-1}, firstly, is an upper Hessenberg matrix and, secondly, is similar to A.

Hint. Adapt the various stages proposed in section 6.2 to the case of symmetric (or Hermitian) matrices.

6.3-3. Show that if the matrix $A = A_1$ is a Hessenberg matrix, then the matrices A_k, $k \geqslant 2$, constructed in the course of the QR algorithm are also Hessenberg matrices.

6.3-4. With the same hypotheses as those of Theorem 6.3-1, we set

$$r_1 = \left|\frac{\lambda_2}{\lambda_1}\right|, \quad r_i = \max\left\{\left|\frac{\lambda_i}{\lambda_{i-1}}\right|, \left|\frac{\lambda_{i+1}}{\lambda_i}\right|\right\}$$

for

$$2 \leqslant i \leqslant n-1, \quad r_n = \left|\frac{\lambda_n}{\lambda_{n-1}}\right|.$$

Show that there exists a constant C, independent of the integer k, such that

$$|(A_k)_{ii} - \lambda_i| \leqslant C(r_i)^k, \quad 1 \leqslant i \leqslant n.$$

6.3-5. (Complement to Theorem 6.3-1 when the matrix is singular). Suppose that the matrix $A = A_1$ is singular, that is to say, has rank r, with $1 \leqslant r < n$, and that the first r column vectors of the matrix A are linearly independent. Then it is still possible to apply the QR algorithm to the matrix A_1 (cf. Theorem 4.5-2).

Show that the matrix $A_2 = R_1 Q_1$ (recall that $A = A_1 = Q_1 R_1$), even though it is not uniquely determined, is such that its last $n - r$ rows are zero and that the diagonal submatrix of order r consisting of the first r rows and columns of A_2 is determined uniquely and is invertible (so that, in an actual calculation, the QR algorithm would be applied just to this submatrix).

6.4 Calculation of eigenvectors

When it is intended to calculate *all, or at least a 'large' number, of the eigenvectors* of a given matrix, some of the algorithms described above for the determination of the eigenvalues provide simultaneously the approximations we seek. So, for example, Theorem 6.1-3 established the convergence of the eigenvectors for the classical Jacobi method applied to a symmetric matrix all of whose eigenvalues were distinct.

Likewise, the QR algorithm applied to a general matrix A, which has first been transformed to the Hessenberg form $H = P^{-1} A P$, leads to the construction of matrices

$$A_{k+1} = (Q_1 Q_2 \cdots Q_k)^{-1} H (Q_1 Q_2 \cdots Q_k),$$

which 'become' upper triangular (at least in certain cases; cf. Theorem 6.3-1). If, then, the products $\mathcal{Q}_k = Q_1 Q_2 \cdots Q_k$ are formed, it might be hoped to derive from them approximations of the eigenvectors of the matrix A, from the moment that the matrix A_k is considered to be effectively upper triangular, that is to say, when all the elements $(A_k)_{ij}$, $j < i$, are small enough to be replaceable by zero. But, of course, we cannot lose sight of the fact that considerations of the kind, which are purely intuitive, *in no way* constitute even an incipient justification for the 'method' which is about to be proposed.

We argue, in fact, *as though*, for an appropriate integer k, the matrix

$$\mathcal{Q}_k^{-1} P^{-1} A P \mathcal{Q}_k \overset{\text{def}}{=} T = (t_{ij})$$

were upper triangular, and suppose that the numbers t_{ii}, that is to say, the numbers $(A_{k+1})_{ii}$, were all distinct (which is in fact the case for a sufficiently large value of the integer k, if the hypotheses of Theorem 6.3-1 are satisfied). It is then easily verified that the vector q_i with components q_j^i, $1 \leqslant j \leqslant n$,

where

$$
\begin{cases}
q_1^i = 1, \quad q_2^i = \cdots = q_n^i = 0 \quad \text{if} \quad i = 1, \\
q_{i+1}^i = \cdots = q_n^i = 0, \\
q_i^i = 1, \\
q_j^i = -\dfrac{t_{j,j+1} q_{j+1}^i + \cdots + t_{ji} q_i^i}{(t_{jj} - t_{ii})}, \quad j = (i-1), \ldots, 1, \quad \text{if} \quad i \geqslant 2,
\end{cases}
$$

is an eigenvector of the matrix T corresponding to the eigenvalue t_{ii} of the same matrix. Since the vectors q_i, $1 \leqslant i \leqslant n$, form a basis of eigenvectors of the matrix T, the vectors $P \mathscr{Q}_k q_i$, $1 \leqslant i \leqslant n$, form a basis of eigenvectors of the matrix A.

It may also happen that one is only interested in finding a *single eigenvector* (or, at any rate, a small number of them), corresponding to an eigenvalue of a matrix A of which an *approximation* $\tilde{\lambda}$ is known, found, for example, by one of the methods described at the beginning of this chapter. The method most frequently in use is the *inverse power method*, also called *inverse iteration*. This iterative method is defined as follows:

$$(A - \tilde{\lambda} I) u_{k+1} = u_k, \quad k \geqslant 0, \quad u_0 \text{ an arbitrary non-zero vector.}$$

An immediate observation is that *the method is not defined if $\tilde{\lambda}$ belongs to the spectrum of* A. We shall return to this point.

We shall now state sufficient conditions for the method to provide an approximation of an eigenvector corresponding to the eigenvalue of the matrix A nearest to $\tilde{\lambda}$.

Theorem 6.4-1 (*'Convergence' of inverse iteration*)
Let A be a diagonalisable matrix and λ an eigenvalue, simple or not, of this matrix. Suppose that the number $\tilde{\lambda}$ satisfies

$$\tilde{\lambda} \neq \lambda \quad \text{and} \quad |\tilde{\lambda} - \lambda| < |\tilde{\lambda} - \mu| \quad \text{for every } \mu \in \text{sp}(A) - \{\lambda\},$$

and that the vector u_0 does not belong to the subspace spanned by the vectors corresponding to the eigenvalues which are distinct from λ.

Then, if $\| \cdot \|$ denotes any vector norm,

$$\lim_{k \to \infty} \left(\frac{(\lambda - \tilde{\lambda})^k}{|\lambda - \tilde{\lambda}|^k} \frac{u_k}{\|u_k\|} \right) = q,$$

where q is an eigenvector corresponding to the eigenvalue λ.

Proof
Let μ_i, $1 \leqslant i \leqslant m$, be the eigenvalues of the matrix A distinct from λ, with possible multiplicities included, and let q_i denote the corresponding linearly

independent eigenvectors. The hypotheses make it possible to write

$$u_0 = \tilde{q} + \sum_{i=1}^{m} \alpha_i q_i, \quad \text{with } \tilde{q} \neq 0,$$

the vector $\tilde{q}$ being an eigenvector corresponding to the eigenvalue λ. It follows that (having assumed that $\tilde{\lambda} \notin \mathrm{sp}(A)$)

$$u_k = \frac{1}{(\lambda - \tilde{\lambda})^k} \tilde{q} + \sum_{i=1}^{m} \frac{\alpha_i}{(\mu_i - \tilde{\lambda})^k} q_i.$$

Hence

$$(\lambda - \tilde{\lambda})^k u_k = \tilde{q} + \sum_{i=1}^{m} \alpha_i \left(\frac{\lambda - \tilde{\lambda}}{\mu_i - \tilde{\lambda}} \right)^k q_i,$$

which implies that

$$(\lambda - \tilde{\lambda})^k u_k = \tilde{q} + \delta_k, \quad \text{with } \lim_{k \to \infty} \delta_k = 0,$$

since $|\lambda - \tilde{\lambda}| < |\mu_i - \tilde{\lambda}|$, $1 \leqslant i \leqslant m$, by hypothesis; so that

$$|\lambda - \tilde{\lambda}|^k \|u_k\| = \|\tilde{q}\| + \varepsilon_k, \quad \text{with } \lim_{k \to \infty} \varepsilon_k = 0.$$

Consequently, for sufficiently large values of the integer k (such that $\|u_k\| \neq 0$),

$$\frac{(\lambda - \tilde{\lambda})^k}{|\lambda - \tilde{\lambda}_k|^k} \frac{u_k}{\|u_k\|} = \frac{\tilde{q} + \delta_k}{\|\tilde{q}\| + \varepsilon_k},$$

and the assertion follows from this last relation. $\square$

Remarks

(1) If the inverse power method is used in conjunction with the Givens–Householder method, it is obviously preferable to apply it to the matrix $B = P^T A P$, which is obtained after the first step of the method, rather than to the matrix A; because, in fact, it is much faster to solve linear systems whose matrices are tridiagonal than linear systems whose matrices are full (cf. section 4.3). Once the eigenvector q of the matrix B has been found, it remains to calculate the vector Pq, which is the corresponding eigenvector of the matrix A.

(2) If λ and $\tilde{\lambda}$ are real, then

$$\begin{cases} \text{either} \quad \lim_{k \to \infty} \dfrac{u_k}{\|u_k\|} = q \quad \text{if } \tilde{\lambda} < \lambda, \\[2mm] \text{or} \quad \lim_{k \to \infty} (-1)^k \dfrac{u^k}{\|u_k\|} = q \quad \text{if } \lambda < \tilde{\lambda}, \end{cases}$$

so that the mere inspection of the 'convergence' of the method makes it possible to decide whether $\tilde{\lambda}$ is an approximation of the eigenvalue λ which underestimates it or overestimates it.

(3) The hypothesis $\tilde{q} \neq 0$ is no restriction *in practice* since, as a result of rounding errors, it is probable that some vector u_k will have a 'small' component from the eigenspace corresponding to the eigenvalue λ; from that moment onwards, one is in a situation which makes possible the application of Theorem 6.4-1.

(4) In view of the expression found for the vectors u_k in the preceding proof, it is easy to convince oneself that the sequence $(u_k)_{k \geqslant 0}$ does not converge in general. That is the reason for 'normalising' the vectors u_k.

(5) The essential nature of the condition $\tilde{\lambda} \neq \lambda$, in order that the matrix $A - \lambda I$ may be invertible, has already been remarked upon. On the other hand, it is desirable to have a $\tilde{\lambda}$ which is as close as possible to λ, in order to accelerate the convergence, since the numbers $(\lambda - \tilde{\lambda})/(\mu_i - \lambda)$, $1 \leqslant i \leqslant m$, converge all the more rapidly to zero as $\tilde{\lambda}$ is closer to λ. As a result, there are two opposing tendencies at play. On the one hand, if $\tilde{\lambda}$ is very close to λ, the convergence is accelerated; and, on the other, if $\tilde{\lambda}$ is 'excessively' close to λ, the matrix $A - \tilde{\lambda}I$ is 'nearly' singular (this consideration, bound up as it is with the presence of rounding errors, is obviously a purely 'numerical' one). An excellent discussion of the subject can be found in Peters & Wilkinson (1979). $\qquad\square$

Exercises

6.4-1. Given a square matrix A, the *power method* is an iterative method defined as follows:

$$u_{k+1} = Au_k, \quad k \geqslant 0, \quad u_0 \text{ an arbitrary non-zero vector.}$$

Suppose that the matrix A is diagonalisable and that there exists a unique eigenvalue $\lambda = \lambda_1$ (as regards the multiplicity as well) satisfying $|\lambda| = \rho(A)$.

(1) What can be said about the sequence $(u_k/\|u_k\|)_{k \geqslant 0}$?

(2) Let z be a fixed non-zero vector. Show that, under certain hypotheses, which should be stated explicitly,

$$\lim_{k \to \infty} \frac{z^*Au_k}{z^*u_k} = \lambda_1,$$

and that, hence, in particular,

$$\lim_{k \to \infty} \frac{(Au_k)_i}{(u_k)_i} = \lambda_1, \quad 1 \leqslant i \leqslant n.$$

6.4-2. Show that, in Theorem 6.4-1, it is not mandatory in the following case to make the assumption that the matrix A is diagonalisable: if the eigenvalue λ has multiplicity p, it suffices to be able to associate with it exactly p linearly independent eigenvectors.

A review of differential calculus. Some applications

Introduction

The purpose of the present chapter is to lead the reader, step by step, into the 'heart of the subject' of Optimisation, taking the Differential Calculus as starting point. The various results which are here established will be of constant use in all that follows.

To begin with, we review in section 7.1 the main definitions and fundamental results of the Differential Calculus for normed vector spaces (the differentiability of a composite function, the mean value theorem, the implicit function theorem, Taylor's formulae). We have confined ourselves by choice to first-order and second-order derivatives, as there will be no need to make use of higher order derivatives.

Having recalled (in Theorem 7.2-1) a necessary condition for the existence of a relative extremum, known as *Euler's equation*,

$$J'(u) = 0,$$

for a function J defined over an *open* set, we then examine how this condition needs to be modified when considering functions defined over particular sets which are not open. More precisely, we turn our attention to the following problem. Find conditions which are both *necessary* and *sufficient* for a point of a set U to be a relative extremum of the *restriction* to the set U of a function J defined over a 'larger' set (for example, the entire space or, more generally, some open subset). A first example is that of a *constrained relative extremum*, the set U being of the form

$$U = \{v \in \Omega : \varphi_i(v) = 0, 1 \leqslant i \leqslant m\}, \quad \Omega \text{ an open subset of } \mathrm{R}^n,$$

and the functions $\varphi_i : \Omega \subset \mathbb{R}^n \to \mathbb{R}$ being given. Under suitable hypotheses, we establish the existence of *Lagrange multipliers* λ_i at a constrained relative extremum u, which satisfy (Theorem 7.2-3)

$$J'(u) + \sum_{i=1}^{m} \lambda_i \varphi_i'(u) = 0,$$

this result being itself obtained as the corollary of a general result (Theorem 7.2-2).

It is then shown how the consideration of the *second-order derivatives* makes it possible, on the one hand, to state a second condition, which is *necessary* for the existence of an extremum and, on the other, to give conditions which are *sufficient*, in the particular situation of functions J which are defined over an *open set* Ω of a normed vector space V. If a point u of Ω is a relative minimum, then (Theorem 7.3-1)

$$J''(u)(w, w) \geqslant 0 \quad \text{for every } w \in V.$$

Conversely, if there exists α such that

$$\alpha > 0, \quad J''(u)(w, w) \geqslant \alpha \| w \|^2 \quad \text{for every} \quad w \in V,$$

or else if there exists a neighbourhood B of u such that

$$J''(v)(w, w) \geqslant 0 \quad \text{for every} \quad v \in B, w \in V,$$

then the point u is a relative minimum (Theorem 7.3-2).

It is clear that the necessary condition $J'(u) = 0$ for the existence of a relative extremum at a point u of an *open set* does not apply to move general situations (consider, for example, the function $J(v) = v$ over an interval $[a, b] \subset \mathbb{R}$). The first example of a situation where it is possible to produce a suitable generalisation has to do with sets U defined by the 'constraints' $\varphi_i(v) = 0, 1 \leqslant i \leqslant m$: this is the case, as already indicated, in which Lagrange multipliers are used. A second example, treated in section 7.4, is concerned with sets U which are *convex*. If a function J admits a relative minimum on a convex set U, then (Theorem 7.4-1)

$$J'(u)(v - u) \geqslant 0 \quad \text{for every } v \in U,$$

these conditions, known as *Euler's inequalities*, being also *sufficient* for the existence of a minimum if the function J is itself convex (Theorem 7.4-4).

It should be noted that consideration of the second derivatives, just as consideration of convexity, makes it possible, on the one hand, to distinguish between *minima* and *maxima* (and even, at times, to be able to say more precisely whether, for example, these are *strict* minima, or even whether they are minima over the entire set), and, on the other, to set down *sufficient* conditions for their existence.

Finally, in section 7.5, a family of iterative methods is described which makes possible the approximation of the *zeros* of the derivative of a function J, that is to say, of points u for which $J'(u) = 0$. These are known as *Newton's methods*, of which the 'prototype' consists in defining the sequence

$$x_{k+1} = x_k - \{ f'(x_k) \}^{-1} f(x_k), \quad k \geqslant 0,$$

for the approximation of the zero of a function f defined over an open set. For this method, and its variants, we establish results on convergence and the existence of zeros (Theorems 7.5-1 and 7.5-2). In the case of the search for the zeros of a derivative J', these methods appear in the form (Theorems 7.5-3 and 7.5-4)

$$u_{k+1} = u_k - A_k^{-1}(u_k) J'(u_k), \quad k \geqslant 0,$$

where the linear operators $A_k(u_k)$ are, for example, 'approximations' of the second derivatives $J''(u_k)$.

In order to give a wider generality to the results of this chapter, and because *the proofs are exactly the same*, we have deliberately 'abandoned' $\mathbb{R}^n$ in order to set ourselves in the context of general normed vector spaces. A consequence which stands out is the fact that we are led, time and again, to make use of the concept of a *Banach space*, that is to say, a *complete normed vector space*. The reader who is unfamiliar with this concept will be able, without any inconvenience, to replace throughout 'complete vector space' by $\mathbb{R}^n$, '$A \in \mathscr{L}(X; Y)$' by 'A is a matrix', '$\| \cdot \|_{\mathscr{L}(X;Y)}$' by 'a subordinate matrix norm', etc.

In this chapter, *all the vector spaces considered are real*.

7.1 First and second derivatives of a function

If X and Y are two normed vector spaces, we denote by

$$\mathscr{L}(X; Y), \quad \text{or simply } \mathscr{L}(X) \quad \text{if } X = Y,$$

the vector space consisting of all *continuous linear mappings* from X to Y. It is a normed vector space under the norm

$$\| A \| = \sup_{\substack{x \in X \\ x \neq 0}} \frac{\| Ax \|_Y}{\| x \|_X},$$

and it is a complete vector space if the space Y is complete. In case $X = Y = \mathbb{R}^n$, the norm defined above is just a *subordinate matrix norm* (to a vector norm over $\mathbb{R}^n$). If $Y = \mathbb{R}$, we write, in general,

$$\mathscr{L}(X; \mathbb{R}) = X',$$

and call the space X' the *dual* of X.

In the same way, we denote by $\mathscr{L}_2(X; Y)$ the space of *continuous bilinear mappings* from $X \times X$ to Y, that is, those satisfying

$$B(\alpha_1 x_1 + \alpha_2 x_2, x) = \alpha_1 B(x_1, x) + \alpha_2 B(x_2, x),$$
$$B(x, \alpha_1 x_1 + \alpha_2 x_2) = \alpha_1 B(x, x_1) + \alpha_2 B(x, x_2),$$

for every $x, x_1, x_2 \in X$ and for every $\alpha_1, \alpha_2 \in \mathbb{R}$, and

$$\| B \|_{\mathscr{L}_2(X;Y)} \stackrel{\text{def}}{=} \sup_{\substack{x_1, x_2 \in X \\ x_1 \neq 0, x_2 \neq 0}} \frac{\| B(x_1, x_2) \|_Y}{\| x_1 \|_X \| x_2 \|_X} < +\infty.$$

It is a normed vector space under the function $\| \cdot \|_{\mathscr{L}_2(X;Y)}$ defined above.

Throughout this section, the statement

$$f: \Omega \subset X \to Y$$

will signify systematically that f is a function from an *open set* Ω of a *normed vector space* X into a *normed vector space* Y.

A function $f: \Omega \subset X \to Y$ is said to be *differentiable at a point* $a \in \Omega$ if there exists an element, denoted by $f'(a)$, of the space $\mathscr{L}(X; Y)$ such that one can write

$$f(a + h) = f(a) + f'(a)h + \| h \| \varepsilon(h), \quad \lim_{h \to 0} \varepsilon(h) = 0.$$

It is then easy to show that such an element $f'(a)$ is unique, if it exists, and $f'(a)$ is called the *(first) derivative of the function f at the point a*. If $X = Y = \mathbb{R}$, the notation $f'(a) = (\mathrm{d}f/\mathrm{d}x)(a)$ is also used.

Remark

It is especially in order to ensure the uniqueness of the derivative that it is useful to assume that the domain of definition of the function f is open. □

A function which is differentiable at a point is continuous at that point. Moreover, it should be noted that, if $f: \Omega \subset X \to Y$ is differentiable at $a \in \Omega$, then, for every vector h of X,

$$f'(a)h = \lim_{\theta \to 0} \frac{f(a + \theta h) - f(a)}{\theta}.$$

Remark

Here, as elsewhere, we omit stating explicitly that it is, of course, necessary to restrict the values of θ to those for which the points $a + \theta h$ belong to the domain of definition of the function f; and this with the obvious desire of keeping the notation unencumbered! □

We say that *the function* $f: \Omega \subset X \to Y$ *is differentiable in* Ω if it is differentiable at every point of Ω. It is then possible to define a function

$$f': x \in \Omega \subset X \to f'(x) \in \mathscr{L}(X; Y),$$

which is called the *derivative* *(mapping)*. If the derivative $f':\Omega \subset X \to \mathscr{L}(X;Y)$ is continuous, the function $f:\Omega \subset X \to Y$ is said to be *(once) continuously differentiable in* Ω, and we write

$$f \in \mathscr{C}^1(\Omega).$$

For example, *an affine continuous function*

$$f: x \in X \to f(x) = Cx + d \in Y,$$

where $C \in \mathscr{L}(X;Y)$ and $d \in Y$, is differentiable in X, and

$$f'(a) = C \quad \text{for every } a \in X,$$

since $f(a+h) = f(a) + Ch$. If $B \in \mathscr{L}_2(X;Y)$, the function f defined by

$$f: x \in X \to f(x) = B(x,x) \in Y,$$

is differentiable in X and

$$f'(a)h = B(h,a) + B(a,h),$$

since

$$f(a+h) = f(a) + B(h,a) + B(a,h) + B(h,h),$$
$$\| B(h,a) + B(a,h) \| \leqslant 2\|B\|\,\|a\|\,\|h\|,$$
$$\| B(h,h) \| \leqslant \|B\|\,\|h\|^2.$$

If the bilinear function B is *symmetric*, that is, if

$$B(x,y) = B(y,x) \quad \text{for every } x, y \in X,$$

the preceding formula becomes

$$f'(a)h = 2B(a,h).$$

If $Z = Z_1 \times Z_2 \times \cdots \times Z_p$ is a product of normed vector spaces Z_i (the space Z is naturally equipped with the product topology), we denote by

$$z = \begin{pmatrix} z_1 \\ z_2 \\ \vdots \\ z_p \end{pmatrix}, \quad z_i \in Z_i, \quad 1 \leqslant i \leqslant p,$$

a general element of Z, this notation being the natural generalisation of the matrix notation employed for ordinary vectors. The provision of a function

$$f: \Omega \subset X \to Y = Y_1 \times Y_2 \times \cdots \times Y_m$$

amounts to being provided with *m component functions* $f_i: \Omega \subset X \to Y_i$, in such a way that

$$f(x) = \begin{pmatrix} f_1(x) \\ f_2(x) \\ \vdots \\ f_m(x) \end{pmatrix}, \quad \text{for every } x \in \Omega.$$

It is easy to establish that the *function f is differentiable at a point* $a \in \Omega$ *if and only if each component function is also differentiable*, and that

$$f'(a) = \begin{pmatrix} f'_1(a) \\ f'_2(a) \\ \vdots \\ f'_m(a) \end{pmatrix}, \quad f'_i(a) \in \mathscr{L}(X; Y_i),$$

the space $\mathscr{L}(X; Y)$ being identified in a natural manner with the product of the spaces $\mathscr{L}(X; Y_i)$.

Let us now consider a function

$$f: \Omega \subset X_1 \times X_2 \times \cdots \times X_n \to Y$$

defined on an open subset Ω of a product of normed vector spaces; this being what is sometimes called a 'function of n variables'. Let a be a point of Ω, with components $a_1, a_2, \ldots, a_n$, and let $k \in \{1, 2, \ldots, n\}$ be one of the indices. By definition of the product topology, there exists an open subset $\Omega_k \subset X_k$ such that all the points with components $a_1, \ldots, a_{k-1}, x_k, a_{k+1}, \ldots, a_n$ belong to the open subset Ω whenever the point x_k belongs to the open subset Ω_k. Consequently, one can examine the possible differentiability of the *partial function*

$$f(a_1, \ldots, a_{k-1}, \cdot, a_{k+1}, \ldots, a_n): \Omega_k \subset X_k \to Y.$$

If this function is differentiable at every point $a_k \in \Omega_k$,

$$\partial_k f(a) \in \mathscr{L}(X_k; Y)$$

denotes its derivative, called *the kth partial derivative of the function f at the point a*.

If a function

$$f: \Omega \subset X = X_1 \times X_2 \times \cdots \times X_n \to Y$$

is differentiable at a point $a \in \Omega$, it is easy to establish that *all its partial derivatives exist* and that, moreover,

$$f'(a)h = \sum_{k=1}^{n} \partial_k f(a)h_k, \quad \text{for every } h = \begin{pmatrix} h_1 \\ h_2 \\ \vdots \\ h_n \end{pmatrix} \in X.$$

The converse is not necessarily true. Thus the function

$$f: x = (x_1, x_2) \in \mathbb{R}^2 \to \begin{cases} 0 & \text{if } x_1 x_2 = 0, \\ 1 & \text{otherwise,} \end{cases}$$

possesses two partial derivatives at the point $(0, 0)$ (since the partial functions are here constant), though it is not differentiable at this point, not being continuous there.

Suppose, finally, that

$$X = X_1 \times X_2 \times \cdots \times X_n \quad \text{and} \quad Y = Y_1 \times Y_2 \times \cdots \times Y_m,$$

so that a function $f:\Omega \subset X \to Y$ is determined once the m functions $f_i:\Omega \subset X \to Y_i$ in n variables are given. Then the relation

$$k = f'(a)h, \quad \text{with } h = \begin{pmatrix} h_1 \\ h_2 \\ \vdots \\ h_n \end{pmatrix} \in X \quad \text{and} \quad k = \begin{pmatrix} k_1 \\ k_2 \\ \vdots \\ k_m \end{pmatrix} \in Y,$$

is equivalent to the relations

$$k_i = \sum_{j=1}^{n} \partial_j f_i(a)h_j, \quad 1 \leqslant i \leqslant m.$$

A particular case which is *very important for the sequel* is that where $X = \mathbb{R}^n$ and $Y = \mathbb{R}^m$. Then the previous relations may be written in matrix notation as

$$\begin{pmatrix} k_1 \\ k_2 \\ \vdots \\ k_m \end{pmatrix} = \begin{pmatrix} \partial_1 f_1(a) & \partial_2 f_1(a) & \cdots & \partial_n f_1(a) \\ \partial_1 f_2(a) & \partial_2 f_2(a) & \cdots & \partial_n f_2(a) \\ \vdots & \vdots & & \vdots \\ \partial_1 f_m(a) & \partial_2 f_m(a) & \cdots & \partial_n f_m(a) \end{pmatrix} \begin{pmatrix} h_1 \\ h_2 \\ \vdots \\ h_n \end{pmatrix},$$

the numbers $\partial_j f_i(a)$ being the 'usual' partial derivatives, often denoted by $(\partial f_i / \partial x_j)(a)$. *The matrix $(\partial_j f_i(a))$ then represents the linear transformation* $f'(a) \in \mathscr{L}(\mathbb{R}^n; \mathbb{R}^m)$. That is why it is called, with a certain misuse of language, the *derivative matrix* of the function f at a. If $m = n$, its determinant is called the *Jacobian* of the function f at the point a.

Remark

The space $\mathscr{L}(\mathbb{R})$ being identifiable with $\mathbb{R}$, the partial derivatives $\partial_i f(a)$ of a function $f:\Omega \subset \mathbb{R}^n \to \mathbb{R}$ can, in effect, be considered as real numbers.

It should be observed in passing that the partial derivatives $\partial_i f(a)$ of a function $f:\Omega \subset \mathbb{R}^n \to \mathbb{R}$ satisfy

$$\partial_i f(a) = f'(a)e_i,$$

e_i denoting, as usual, the ith canonical basis vector of $\mathbb{R}^n$.

The following result, which makes possible the calculation of the *derivative of a composite function*, is used constantly.

Theorem 7.1-1

Let $f:\Omega \subset X \to Y$ be a function differentiable at a point $a \in \Omega$ and let $g:\Omega' \subset Y \to Z$ be a function differentiable at the point $b = f(a) \in \Omega'$. Suppose that

$f(\Omega) \subset \Omega'$. *Then the composite function*

$$h = gf : \Omega \subset X \to Z$$

is differentiable at the point $a \in \Omega$ *and*

$$h'(a) = g'(b) f'(a).$$

For the special case of functions

$$f : \Omega \subset \mathbb{R}^n \to \mathbb{R}^m, \quad g : \Omega' \subset \mathbb{R}^m \to \mathbb{R}^l,$$

there corresponds to the composition of the derivatives $g'(b)$ *and* $f'(a)$ *the multiplication of the derivative matrices of the functions concerned:*

$$\begin{pmatrix} \partial_1 h_1(a) & \cdots & \partial_n h_1(a) \\ \vdots & & \vdots \\ \partial_1 h_l(a) & \cdots & \partial_n h_l(a) \end{pmatrix} = \begin{pmatrix} \partial_1 g_1(b) & \cdots & \partial_m g_1(b) \\ \vdots & & \vdots \\ \partial_1 g_l(b) & \cdots & \partial_m g_l(b) \end{pmatrix} \begin{pmatrix} \partial_1 f_1(a) & \cdots & \partial_n f_1(a) \\ \vdots & & \vdots \\ \partial_1 f_m(a) & \cdots & \partial_n f_m(a) \end{pmatrix},$$

which can also be written as

$$\partial_j h_i(a) = \sum_{k=1}^{m} \partial_k g_i(b) \partial_j f_k(a), \quad 1 \leqslant i \leqslant l, \; 1 \leqslant j \leqslant n. \qquad \square$$

By way of completing the review of material concerning first derivatives, we shall state two fundamental results of the Differential Calculus (Theorems 7.1-2 and 7.1-3). The first concerns the extension of the 'usual' mean value theorem: given a real function, continuous over an interval $[a, b]$ and differentiable over the interval $]a, b[$, there exists a point $c \in]a, b[$ such that $f(b) - f(a) = f'(c)(b - a)$. *This formula cannot be generalised just as it stands.* Thus, the function $f : t \in \mathbb{R} \to f(t) = (\cos t, \sin t) \in \mathbb{R}^2$ is such that $f(2\pi) - f(0) = (0, 0)$, while the derivative $f'(t) = (-\sin t, \cos t)$ does not vanish in the interval $]0, 2\pi[$.

Though it is not possible to generalise the formula itself, it is possible, in contrast, to generalise the *bound*

$$|f(b) - f(a)| \leqslant \sup_{t \in]a, b[} |f'(t)| \, |b - a|$$

which follows from it (the case where $\sup_{t \in]a, b[} |f'(t)| = +\infty$ is not excluded; however, it provides no information!). If a and b are two points of a vector space X, the notation

$$[a, b] = \{x = ta + (1 - t)b \in X : \; t \in [0, 1]\},$$
$$]a, b[= \{x = ta + (1 - t)b \in X : \; t \in]0, 1[\}$$

will be used to denote the *closed segment* and the *open segment* respectively, with *end-points* a and b.

Theorem 7.1-2 (*Mean value theorem*)

Let $f : \Omega \subset X \to Y$ and let a and b be two points of Ω such that the segment $[a, b]$ lies within Ω. Suppose that the function f is continuous at every point of the closed segment $[a, b]$ and differentiable at every point of the open segment $]a, b[$. Then

$$\| f(b) - f(a) \|_Y \leqslant \sup_{x \in]a, b[} \| f'(x) \|_{\mathscr{L}(X;Y)} \| b - a \|_X. \qquad \square$$

Several consequences follow upon this result, notably the following. If $X = X_1 \times X_2 \times \cdots \times X_n$ is a product of normed vector spaces, then *the function f is continuously differentiable in Ω if and only if*

(i) *its partial derivatives $\partial_k f(x)$, $1 \leqslant k \leqslant n$, exist at every point $x \in \Omega$, and*

(ii) *the partial derivatives $\partial_k f : x \in \Omega \to \partial_k f(x) \in \mathscr{L}(X_k; Y)$ are continuous in Ω.*

With the help of the mean value theorem (and of the fixed point theorem), another fundamental result of the Differential Calculus can also be proved (one which will, moreover, be used in an essential manner from the next section onwards). This is the *implicit function theorem*, which provides an answer to the following problem. Let X_1, X_2 and Y be three normed vector spaces and let $\varphi : X_1 \times X_2 \to Y$ be a given function. Given a point $b \in Y$, it is possible that, for every element $x_1 \in X_1$, there exists a unique element $x_2 \in X_2$ such that $\varphi(x_1, x_2) = b$, thus defining a function $f : x_1 \in X_1 \to x_2 \in X_2$, called the *implicit function* and satisfying

$$\varphi(x_1, f(x_1)) = b \quad \text{for every } x_1 \in X_1.$$

Of course, *this kind of situation is exceptional*, and for the most part it is illusory to hope to prove the existence of an implicit function over the entire set X_1. On the other hand, it is quite realistic to look for a *local* result of existence. Supposing there exists a particular solution (a_1, a_2) of the equation $\varphi(x_1, x_2) = b$, it can be asked under what conditions it is possible to define x_2 as an implicit function of x_1 in a suitable *neighbourhood* of the point (a_1, a_2). This is the object of the result which follows, where the continuity and differentiability of the implicit function are also examined. It should be observed, in addition, that it is *essential* to have a 'previous' knowledge of some particular solution.

If X and Y are two normed vector spaces, then

$$\text{Isom}(X; Y), \quad \text{or simply} \quad \text{Isom}(X) \quad \text{if } X = Y,$$

denotes the set of functions which are *continuous, linear and bijective* from X onto Y, *with continuous inverse functions*.

Theorem 7.1-3 (*Implicit function theorem*)

Let $\varphi:\Omega \subset X_1 \times X_2 \to Y$ be a function continuously differentiable in Ω and let $(a_1, a_2) \in \Omega$, $b \in Y$ be points such that

$$\varphi(a_1, a_2) = b, \quad \partial_2\varphi(a_1, a_2) \in \text{Isom}(X_2; Y).$$

Suppose that the space X_2 is complete.

Then there exist an open subset $O_1 \subset X_1$, an open subset $O_2 \subset X_2$ and a continuous function, called the implicit function,

$$f: O_1 \subset X_1 \to X_2,$$

such that $(a_1, a_2) \in O_1 \times O_2 \subset \Omega$ and (cf. figure 7.1-1)

$$\{(x_1, x_2) \in O_1 \times O_2 : \varphi(x_1, x_2) = b\} = \{(x_1, x_2) \in O_1 \times X_2 : x_2 = f(x_1)\}.$$

Moreover, the function f is differentiable at the point a_1 and

$$f'(a_1) = -\{\partial_2\varphi(a_1, a_2)\}^{-1}\partial_1\varphi(a_1, a_2). \qquad \square$$

Remarks

(1) Assuming that the differentiability of the implicit function at the point a_1 is established, the derivative may be calculated by an application of Theorem 7.1-1 to the composite function

$$h: x_1 \in O_1 \to \{\varphi(x_1, f(x_1)) - b\} = 0 \in Y.$$

It is enough, in fact, to write

$$0 = h'(a_1) = \partial_1\varphi(a_1, a_2) + \partial_2\varphi(a_1, a_2) f'(a_1).$$

Figure 7.1-1

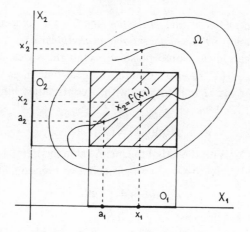

(2) For a given element $x_1 \in O_1$, it could well be that there exist elements $x_2' \in X_2$ such that $x_2' \neq x_2, (x_1, x_2') \in \Omega$ and $\varphi(x_1, x_2') = b$, but $x_2' \notin O_2$ (cf. figure 7.1-1).

(3) In fact, there exists an open set O_1' such that $a_1 \in O_1' \subset O_1$ and such that the implicit function is differentiable at every point of this open set.

(4) The following corollary of the implicit function theorem is often used, called the *inverse function theorem*.

Let $g: \Omega_2 \subset X_2 \to X_1$ be a continuously differentiable function and let $a_1 \in X_1$, $a_2 \in \Omega_2$ be points such that

$$a_1 = g(a_2), \quad g'(a_2) \in \text{Isom}(X_2; X_1).$$

Suppose that the space X_2 is complete.

Then there exist an open subset $O_1 \subset X_1$, an open subset $O_2 \subset X_2$ and a continuous function $f: O_1 \subset X_1 \to X_2$, such that $a_2 \in O_2 \subset \Omega_2$ and

$$\{(x_1, x_2) \in O_1 \times O_2 : x_1 = g(x_2)\} = \{(x_1, x_2) \in O_1 \times X_2 : x_2 = f(x_1)\}.$$

Moreover, the function f is differentiable at a_1 and

$$f'(a_1) = \{g'(a_2)\}^{-1}. \qquad \square$$

We now go on to the notion of the *second derivative* of a function. Let $f: \Omega \subset X \to Y$ be a function which is *differentiable in Ω*. If the derivative

$$f': \Omega \subset X \to \mathscr{L}(X; Y)$$

is differentiable at the point $a \in \Omega$, its derivative, denoted by

$$f''(a) \overset{\text{def}}{=} (f')'(a) \in \mathscr{L}(X; \mathscr{L}(X; Y)),$$

is called the *second derivative of the function f at the point a*, and one says that the *function f is twice differentiable at the point a*.

Using the canonical isomorphism between the space $\mathscr{L}(X; \mathscr{L}(X; Y))$ and the space $\mathscr{L}_2(X; Y)$ of continuous, bilinear mappings from X into Y, the second derivative is identified with a continuous, bilinear mapping from X into Y, and we write

$$f''(a)(h, k) = (f''(a)h)k,$$

for every $h, k \in X$. With the aid of the mean value theorem, it may be shown that the second derivative is a *symmetric* bilinear function, in the sense that

$$f''(a)(h, k) = f''(a)(k, h)$$

for every $h, k \in X$.

The function f is said to be *twice differentiable in Ω* if it is twice differentiable at every point of Ω; it is then possible to define the *second derivative* (*mapping*)

$$f'': \Omega \subset X \to \mathcal{L}_2(X; Y).$$

If this last function is continuous, the function f is said to be *twice continuously differentiable in* Ω, and one writes

$$f \in \mathscr{C}^2(\Omega).$$

As regards the actual *calculation* of the second derivatives, constant use is made of the following result, which allows us to have recourse to the calculation of first derivatives. *Given any two vectors* $h, k \in X$, *the element* $f''(a)(h, k) \in Y$ *is equal to the derivative at the point* $a \in \Omega$ *of the function* $x \in \Omega \to f'(x)k \in Y$, *applied to the vector h.*

Let us consider, for example, a function of the form

$$f : x \in X \to f(x) = B(x, x) + Cx + d \in Y$$

where $B \in \mathcal{L}_2(X; Y)$, $C \in \mathcal{L}(X; Y)$, $d \in Y$. It has already been established that this function is differentiable in X and that

$$f'(x)k = B(k, x) + B(x, k) + Ck,$$

for every $x, k \in X$. For a fixed vector $k \in X$ the function $g : x \in X \to f'(x)k \in Y$ is here affine; it is, therefore, differentiable in X, so that we obtain

$$f''(a)(h, k) = g'(a)h = B(h, k) + B(k, h),$$

applying the result given above.

In the particular case where $X = \mathbb{R}^n$ *and* $Y = \mathbb{R}$, let $h = (h_i)$ and $k = (k_i)$ be two vectors of $\mathbb{R}^n$, with (e_i) as its basis. The second derivative $f''(a)$ being a bilinear function,

$$f''(a)(h, k) = \sum_{i,j} h_i k_j f''(a)(e_i, e_j),$$

where, by what was said above,

$$f''(a)(e_i, e_j) = \partial_i(\partial_j f)(a) = f''(a)(e_j, e_i).$$

Hence the numbers

$$\partial_{ij} f(a) \stackrel{\text{def}}{=} \partial_i(\partial_j f)(a)$$

are just the 'usual' *second partial derivatives*, also denoted by $(\partial^2 f/\partial x_i \partial x_j)(a)$. If $n = 1$, the second derivative $f''(a)$ is also denoted by $(d^2 f/dx^2)(a)$, which, in this case, is just some element of $\mathbb{R}$.

We now review various Taylor formulae. The first formula generalises the definition of the first derivative of a function and the second generalises the mean value theorem (Theorem 7.1-2). As to the third and fourth formulae, they give an expression for the 'remainder', the fourth generalising the formula

$$f(a+h) - f(a) = \int_a^{a+h} f'(\theta)d\theta = \int_0^1 f'(a+th)h\,dt$$

which is well-known for real functions of a real variable. For the reader's convenience, the results are presented in the form of two theorems. Finally, it will be seen that the formulae are given with hypotheses which become increasingly 'stronger'.

Theorem 7.1-4 (*Taylor's formulae for differentiable functions*)
Let $f: \Omega \subset X \to Y$ and let $[a, a+h]$ be any closed segment lying within Ω.
(1) *If f is differentiable at a, then*

$$f(a+h) = f(a) + f'(a)h + \|h\|\varepsilon(h), \quad \lim_{h \to 0} \varepsilon(h) = 0.$$

(2) Mean value theorem: *if $f \in \mathscr{C}^0(\Omega)$ and f is differentiable over* $]a, a+h[$, *then*

$$\|f(a+h) - f(a)\| \leqslant \sup_{x \in]a, a+h[} \|f'(x)\| \|h\|.$$

(3) The Taylor–Maclaurin formula: *if $f = \mathscr{C}^0(\Omega)$, f is differentiable over* $]a, a+h[$ *and $Y = \mathbb{R}$, then*

$$f(a+h) = f(a) + f'(a+\theta h)h, \quad 0 < \theta < 1.$$

(4) Taylor's formula with integral remainder: *if $f \in \mathscr{C}^1(\Omega)$ and Y is a complete vector space, then*

$$f(a+h) = f(a) + \int_0^1 \{f'(a+th)h\}\,dt.$$

Theorem 7.1-5 (*Taylor's formulae for twice differentiable functions*)
Let $f: \Omega \subset X \to Y$ and let $[a, a+h]$ be any closed segment lying within Ω.
(1) The Taylor–Young formula: *if f is differentiable in Ω and twice differentiable at a, then*

$$f(a+h) = f(a) + f'(a)h + \tfrac{1}{2}f''(a)(h, h) + \|h\|^2\varepsilon(h), \quad \lim_{h \to 0} \varepsilon(h) = 0.$$

(2) The generalised mean value theorem: *if $f \in \mathscr{C}^1(\Omega)$ and f is twice differentiable over* $]a, a+h[$, *then*

$$\|f(a+h) - f(a) - f'(a)h\| \leqslant \tfrac{1}{2} \sup_{x \in]a, a+h[} \|f''(x)\| \|h\|^2.$$

(3) The Taylor–Maclaurin formula: *if $f \in \mathscr{C}^1(\Omega)$, f is twice differentiable*

over $]a, a + h[$ *and* $Y = \mathbb{R}$, *then*

$$f(a + h) = f(a) + f'(a)h + \tfrac{1}{2}f''(a + \theta h)(h, h), \quad 0 < \theta < 1.$$

(4) Taylor's formula with integral remainder: *if* $f \in \mathscr{C}^2(\Omega)$ *and* Y *is a complete vector space, then*

$$f(a + h) = f(a) + f'(a)h + \int_0^1 (1 - t)\{f''(a + th)(h, h)\}\, dt.$$

Remarks

(1) While formula (1) of Theorem 7.1-4 is exactly the definition of the first derivative, formula (1) of Theorem 7.1-5 is not equivalent to the definition of the second derivative at a point. On this, see Exercise 7.1-1.

(2) In the generalised mean value theorem (2), the expression $\|f''(x)\|$ denotes, of course, the norm of the element $f''(x)$ of the normed vector space $\mathscr{L}_2(X; Y)$.

(3) It is known that there exists (at least) one number $\theta \in]0, 1[$ such that the Taylor–Maclaurin formulae (3) hold, but, in general, nothing else is known about θ; we repeat, in passing, that it is indispensable to restrict oneself to the case $Y = \mathbb{R}$ in this formula (refer to the example given above).

(4) If the formulae (4) are to make sense, it is necessary to be able to integrate the functions

$$t \in [0, 1] \to \{f'(a + th)h\} \in Y,$$
$$t \in [0, 1] \to \{f''(a + th)(h, h)\} \in Y;$$

this is the reason why the assumption is made that these functions are continuous and the space Y is complete. $\qquad\square$

Finally, we give some *notation* which is special to functions which are of the form

$$f : \Omega \subset \mathbb{R}^n \to \mathbb{R}.$$

At every point a where this function is once, or twice, differentiable, the *vector* $\nabla f(a) \in \mathbb{R}^n$ and the *matrix* $\nabla^2 f(a) \in \mathscr{A}_n(\mathbb{R})$ are introduced, and are defined respectively by the relations

$$f'(a)h = (\nabla f(a), h) \quad \text{for every } h \in \mathbb{R}^n,$$
$$f''(a)(h, k) = (\nabla^2 f(a)h, k) = (h, \nabla^2 f(a)k) \quad \text{for every } h, k \in \mathbb{R}^n,$$

$(\cdot, \cdot)$ denoting, as usual, the Euclidean scalar product in $\mathbb{R}^n$. The vector

$$\nabla f(a) = \begin{pmatrix} \partial_1 f(a) \\ \partial_2 f(a) \\ \vdots \\ \partial_n f(a) \end{pmatrix}$$

is called the *gradient* of the function f at the point a, and the (symmetric) matrix

$$\nabla^2 f(a) = \begin{pmatrix} \partial_{11}f(a) & \cdots & \partial_{1n}f(a) \\ \partial_{21}f(a) & \cdots & \partial_{2n}f(a) \\ \vdots & & \vdots \\ \partial_{n1}f(a) & \cdots & \partial_{nn}f(a) \end{pmatrix}$$

is called the *Hessian* of the function f at the point a.

While the first and second derivatives $f'(a)$ and $f''(a)$ are defined in an 'intrinsic' manner in the spaces $\mathscr{L}(\mathbb{R}^n; \mathbb{R})$ and $\mathscr{L}_2(\mathbb{R}^n; \mathbb{R})$ respectively (the space $\mathscr{L}_2(\mathbb{R}^n; \mathbb{R})$ being here identified with the space $\mathscr{L}(\mathbb{R}^n)$), it should not be forgotten that their gradient and Hessian are *particular* representations, corresponding to the *Euclidean scalar product*: to the choice of another scalar product in $\mathbb{R}^n$ there would correspond another vector and another matrix! It should also be observed that the gradient vector is the *transpose* of the derivative matrix (here a row vector) of the function f at a.

In order to illustrate these considerations, here are three equivalent ways of writing the Taylor–Young formula (for example) for functions $f:\Omega \subset \mathbb{R}^n \to \mathbb{R}$ which are twice differentiable:

$$f(a+h) = f(a) + f'(a)h + \tfrac{1}{2}f''(a)(h,h) + \|h\|_2^2 \varepsilon(h),$$
$$f(a+h) = f(a) + (\nabla f(a), h) + \tfrac{1}{2}(\nabla^2 f(a)h, h) + (h, h)\varepsilon(h),$$
$$f(a+h) = f(a) + (\nabla f(a))^{\mathsf{T}}h + \tfrac{1}{2}h^{\mathsf{T}}\nabla^2 f(a)h + h^{\mathsf{T}}h\varepsilon(h).$$

Exercises

7.1-1. Let $f:\Omega \subset \mathbb{R} \to \mathbb{R}$ be a function and $a \in \Omega$. Suppose that there exist numbers b_0, b_1 and b_2, such that

$$f(a+h) = b_0 + b_1 h + b_2 h^2 + |h|^2 \varepsilon(h), \quad \lim_{h \to 0} \varepsilon(h) = 0.$$

(1) Show that, if the function f is twice differentiable at a, then $b_0 = f(a)$, $b_1 = f'(a), b_2 = 2f''(a)$.

(2) Show that the existence of the relation given above does not necessarily imply that the function f is twice differentiable at a.

7.1-2. (1) Let X be a normed vector space. Show that the function $f: x \in X \to \|x\| \in \mathbb{R}$ is not differentiable at the origin.

(2) Show that, if $X = \mathbb{R}^n$ and $\|\cdot\| = \|\cdot\|_2$, then the function f is differentiable in $\Omega = \mathbb{R}^n - \{0\}$, and that

$$f'(a)h = \frac{\displaystyle\sum_{i=1}^{n} a_i h_i}{\|a\|_2}$$

at every point $a = (a_i) \in \Omega$, for every vector $h = (h_i) \in \mathbb{R}^n$.

(3) Show that if $X = \mathbb{R}^n$ and $\|\cdot\| = \|\cdot\|_\infty$, then the function f is differentiable at a point $a = (a_i) \in \mathbb{R}^n$ if and only if there exists an index $i_0 \in \{1, 2, \ldots, n\}$ such that $|a_{i_0}| > |a_i|$ for every $i \neq i_0$.

7.1-3. Let $\mathscr{R} \subset \mathscr{L}(\mathbb{R}^n)$ be the set of invertible linear functions from $\mathbb{R}^n$ to $\mathbb{R}^n$.

(1) Show that the set $\mathscr{R}$ is open in $\mathscr{L}(\mathbb{R}^n)$.

(2) Show that the function

$$F: A \in \mathscr{R} \subset \mathscr{L}(\mathbb{R}^n) \to F(A) = A^{-1} \in \mathscr{R} \subset \mathscr{L}(\mathbb{R}^n)$$

is continuous over $\mathscr{R}$.

(3) Show that the function F is differentiable in $\mathscr{R}$ and that

$$F'(A)H = -A^{-1}HA^{-1}$$

for every $A \in \mathscr{R}$ and for every $H \in \mathscr{L}(\mathbb{R}^n)$.

(4) Prove that, for every $A \in \mathscr{R}$, the derivative $F'(A)$ is an invertible element of the space $\mathscr{L}(\mathscr{L}(\mathbb{R}^n))$, and calculate its inverse.

(5) Show that the function F is continuously differentiable.

7.1-4. Let $f: \Omega \subset X \to Y$ be a function continuous in Ω, differentiable in $\Omega - \{a\}$, $a \in \Omega$, and such that $\lim_{x \to a} f'(x)$ exists (in $\mathscr{L}(X; Y)$). Show that the function f is differentiable at the point a.

7.1-5. (1) Let $f: \Omega \subset X \to Y$, where Ω is an open, connected set, be a function which is differentiable in Ω. Show that if the derivative is constant, then the function f is affine.

(2) Let $f: \Omega \subset \mathbb{R}^n \to \mathbb{R}$ be a function twice differentiable in an open, connected set Ω and such that $f''(x) = 0$ at every point $x \in \Omega$. Show that f is a polynomial of degree $\leqslant 1$ in the n variables $x_1, x_2, \ldots, x_n$.

7.1-6. Suppose $A \in \mathscr{L}(\mathbb{R}^n)$, Ω is an open subset of $\mathbb{R}^n$, a is a point of Ω and $f: A(\Omega) \subset \mathbb{R}^n \to \mathbb{R}$ is a function twice differentiable at the point Aa. Given the vectors $h, k \in \mathbb{R}^n$, calculate $g''(a)(h, k)$, the function $g: \Omega \to \mathbb{R}$ being defined by $g(x) = f(Ax)$ for $x \in \Omega$.

7.1-7. Let $f: \mathbb{R}^n \to \mathbb{R}$ be a function which is twice continuously differentiable in $\mathbb{R}^n$. Show that there exist constants α and β_i, $1 \leqslant i \leqslant n$, and continuous functions g_{ij}: $\mathbb{R}^n \to \mathbb{R}$, $1 \leqslant i, j \leqslant n$, such that

$$f(x) = \alpha + \sum_{i=1}^{n} \beta_i x_i + \sum_{i,j=1}^{n} g_{ij}(x) x_i x_j \quad \text{for every } x \in \mathbb{R}^n.$$

7.1-8. Let $f: \mathbb{R} \to \mathbb{R}$ be a function which is differentiable in $\mathbb{R}$ and twice differentiable at a point $a \in \mathbb{R}$. Show that the function

$$g: (x, y) \in \mathbb{R}^2 \to g(x, y) = \begin{cases} \dfrac{f(y) - f(x)}{y - x} & \text{if } x \neq y, \\ f'(x) & \text{if } x = y, \end{cases}$$

is differentiable at the point (a, a).

7.1-9. Let $f: \Omega \subset X \to Y$ be a function, a a point of Ω and v a non-zero vector of X.

We define the *directional derivative of the function f at the point a in the direction of the vector v* to be the derivative (if it exists) at the point $t = 0$ of the function

$$t \in I \subset \mathbb{R} \to f(a + tv) \in Y,$$

where I is an open interval containing the origin, chosen in such a way that all the points $a + tv, t \in I$, are also points of Ω. This derivative is denoted by $\partial_v f(a)$ and is, by definition, an element of the space $\mathscr{L}(\mathbb{R}, Y)$. Identifying the spaces $\mathscr{L}(\mathbb{R}; Y)$ and Y, it is also possible to write

$$\partial_v f(a) = \lim_{t \to \infty} \frac{f(a + tv) - f(a)}{t}.$$

(1) Show that, if $f : \Omega \subset X \to Y$ is differentiable at a point $a \in \Omega$, then the function f possesses at a a directional derivative in the direction of every vector $v \in X$, given by

$$\partial_v f(a) = f'(a)v.$$

(2) Considering (for example) the function

$$f(x, y) = \begin{cases} 0 & \text{if} \quad (x, y) = (0, 0), \\ \dfrac{x^5}{(y - x^2)^2 + x^8} & \text{if} \quad (x, y) \neq (0, 0), \end{cases}$$

about the origin, show that the converse of (1) is, in general, false

(3) Let $f : \mathbb{R}^2 \to \mathbb{R}$ be a function having a directional derivative at the point $(0, 0)$ in the direction of every vector. Show that the function f is differentiable at $(0, 0)$ if and only if the point with co-ordinates $(\cos \theta, \sin \theta, \partial_{v(\theta)} f(0, 0))$, where $v(\theta) = (\cos \theta, \sin \theta)$, describes an ellipse in $\mathbb{R}^3$, with centre at the origin, as θ varies from 0 to 2π.

7.1-10. Let Ω_1 and Ω_2 be two open subsets of the finite-dimensional vector spaces X_1 and X_2, respectively, and let $f : \Omega_1 \to \Omega_2$ be a bijection differentiable at a point $a_1 \in \Omega_1$. Show that, if the inverse bijection $f^{-1} : \Omega_2 \to \Omega_1$ is differentiable at the point $a_2 = f(a_1)$, then the spaces X_1 and X_2 have the same dimension.

7.2 Extrema of real functions: Lagrange multipliers

Let $J : W \to \mathbb{R}$ be a function defined over a topological space W. We say that the function J has a *relative minimum* (or a *relative maximum*) at the point $u \in W$ if there exists a neighbourhood O of u such that

$$J(u) \leqslant J(v) \quad (\text{or } J(u) \geqslant J(v)) \quad \text{for every } v \in O.$$

If there is no need to distinguish as between relative maximum and minimum, we say that the function J has a *relative extremum* at the point u.

Remark

It is a useful misuse of language (one to which we shall casually accede) to say that the point u *itself* is a relative minimum, maximum or extremum. □

We begin with a well-known result.

Theorem 7.2-1 (*Necessary condition for a relative extremum*)
Let Ω be an open subset of a normed vector space V and $J : \Omega \subset V \to \mathbb{R}$ a function. If the function J has a relative extremum at a point $u \in \Omega$ and if it is differentiable at this point, then

$$J'(u) = 0.$$

Proof
Let v be any vector of V. The set Ω being open, there exists an open interval I containing 0 such that the function

$$\varphi : t \in I \to \varphi(t) = J(u + tv)$$

is well-defined. Applying Theorem 7.1-1, we find that the function φ is differentiable at $t = 0$, and

$$\varphi'(0) = J'(u)v.$$

In order to fix ideas, suppose that the point u is a relative minimum. Then

$$0 \geqslant \lim_{t \to 0^-} \frac{\varphi(t) - \varphi(0)}{t} = \varphi'(0) = \lim_{t \to 0^+} \frac{\varphi(t) - \varphi(0)}{t} \geqslant 0,$$

which shows that

$$J'(u)v = 0.$$

As the vector v is arbitrary, it follows that $J'(u) = 0$. $\qquad \square$

The relation $J'(u) = 0$ is sometimes called *Euler's equation*.

Remarks
(1) The fact that Ω is *open* is evidently essential. Consider, for example, the function $f(v) = v$ over the interval $[0, 1]$.
(2) If $V = \mathbb{R}^n$, the condition $J'(u) = 0$ is equivalent to the system of equations

$$\begin{cases} \partial_1 J(u_1, \ldots, u_n) = 0, \\ \vdots \\ \partial_n J(u_1, \ldots, u_n) = 0. \end{cases} \qquad \square$$

Now let $J : \Omega \to \mathbb{R}$ be a function defined over a topological space Ω and let U be a *subset* of Ω. The function J is said to have a relative minimum at a point $u \in U$ (or a relative maximum, or a relative extremum) *with respect to the set U* if the restriction of the function J to the set U, endowed with the topology induced by that of Ω, has a relative minimum (or relative maximum, or relative extremum) at u.

The problem of *constrained relative extrema* is an example of the search for such relative extrema, for *particular* sets U and Ω. The set Ω is an *open subset* of a product $V_1 \times V_2$ of two normed vector spaces, and the set U is of the form

$$U = \{(v_1, v_2) \in \Omega: \quad \varphi(v_1, v_2) = 0\},$$

where

$$\varphi: \Omega \subset V_1 \times V_2 \to V_2$$

is a given function. If there is no ambiguity as to the set U in question, we shall simply say that the function J has a *constrained relative extremum* at u.

Observe that the *set U is not open in general* (think of a curve in $\mathbb{R}^2$ when $V_1 = V_2 = \mathbb{R}$; on the contrary, it is *closed* whenever the function φ is continuous). That is why the necessary condition established in Theorem 7.2-1 does not hold in this situation.

Theorem 7.2-2 (*Necessary condition for a constrained relative extremum*)
Let Ω be an open subset of a product $V_1 \times V_2$ of normed vector spaces, the space V_1 being complete, let $\varphi: \Omega \to V_2$ be a function of class $\mathscr{C}^1$ over Ω and let $u = (u_1, u_2)$ be a point of the set

$$U = \{(v_1, v_2) \in \Omega: \quad \varphi(v_1, v_2) = 0\} \subset \Omega$$

at which

$$\partial_2 \varphi(u_1, u_2) \in \text{Isom}(V_2).$$

Let $J: \Omega \to \mathbb{R}$ be a function differentiable at u. If the function J has a relative extremum at u with respect to the set U, then there exists an element $\Lambda(u) \in \mathscr{L}(V_2; \mathbb{R})$ such that

$$J'(u) + \Lambda(u)\varphi'(u) = 0.$$

Proof
The hypotheses which have been made are such as enable us to apply the implicit function theorem (Theorem 7.1-3) in a neighbourhood of the point u. Hence, there exists an open set $O_1 \subset V_1$, an open set $O_2 \subset V_2$ and a continuous function $f: O_1 \to O_2$, such that $(u_1, u_2) \in O_1 \times O_2 \subset \Omega$ and

$$(O_1 \times O_2) \cap U = \{(v_1, v_2) \in O_1 \times O_2: \quad v_2 = f(v_1)\}.$$

Moreover, the implicit function f is differentiable at the point $u_1 \in O_1$ and its derivative can be expressed as

$$f'(u_1) = -\{\partial_2 \varphi(u)\}^{-1} \partial_1 \varphi(u).$$

Hence, the restriction of the function J to the set $(O_1 \times O_2) \cap U$ becomes a function 'of a single variable'

$$G: v_1 \in O_1 \to G(v_1) \overset{\text{def}}{=} J(v_1, f(v_1)) \in \mathbb{R},$$

to which can be applied the necessary condition (Theorem 7.2-1)

$$G'(u_1) = 0;$$

in effect, then, the function G has a relative extremum at the point u_1 (use the definition of the topology induced by $V_1 \times V_2$ onto U) and it is differentiable at this point (Theorem 7.1-1), its derivative being equal to

$$G'(u_1) = \partial_1 J(u) + \partial_2 J(u) f'(u_1)$$
$$= \partial_1 J(u) - \partial_2 J(u) \{\partial_2 \varphi(u)\}^{-1} \partial_1 \varphi(u).$$

Hence, we have, on the one hand,

$$\partial_1 J(u) = \partial_2 J(u) \{\partial_2 \varphi(u)\}^{-1} \partial_1 \varphi(u),$$

and since

$$\partial_2 J(u) = \partial_2 J(u) \{\partial_2 \varphi(u)\}^{-1} \partial_2 \varphi(u),$$

on the other, we obtain the result sought by setting

$$\Lambda(u) = - \partial_2 J(u) \{\partial_2 \varphi(u)\}^{-1}. \qquad \square$$

In practice, the preceding result is frequently used in the following situation. Given two integers m and n satisfying $1 \leqslant m < n$ and functions

$$J: \Omega \subset \mathbb{R}^n \to \mathbb{R} \quad \text{and} \quad \varphi_i: \Omega \subset \mathbb{R}^n \to \mathbb{R}, \quad 1 \leqslant i \leqslant m,$$

defined over the *open set* Ω, *a necessary condition* is sought *for a relative extremum of the function J with respect to the set*

$$U = \{v \in \Omega: \quad \varphi_i(v) = 0, \quad 1 \leqslant i \leqslant m\}.$$

It is clear that this problem is a particular case of the preceding (with V_1 and V_2 identified with the spaces $\mathbb{R}^{n-m}$ and $\mathbb{R}^m$ respectively), so that Theorem 7.2-2 leads to the following result.

Theorem 7.2-3 (*Necessary condition for a constrained relative extremum*)
Let Ω be an open subset of $\mathbb{R}^n$, let $\varphi_i: \Omega \to \mathbb{R}, 1 \leqslant i \leqslant m$, be functions of class $\mathscr{C}^1$ over Ω and let u be a point of the set

$$U = \{v \in \Omega: \quad \varphi_i(v) = 0, \quad 1 \leqslant i \leqslant m\} \subset \Omega,$$

at which the derivatives $\varphi_i'(u) \in \mathscr{L}(\mathbb{R}^n; \mathbb{R}), 1 \leqslant i \leqslant m$, are linearly independent.

Let $J: \Omega \to \mathbb{R}$ be a function differentiable at u. If the function J has a relative extremum at u with respect to the set U, there exist m numbers $\lambda_i(u), 1 \leqslant i \leqslant m$, uniquely defined, such that

$$J'(u) + \lambda_1(u)\varphi_1'(u) + \cdots + \lambda_m(u)\varphi_m'(u) = 0.$$

Proof

The condition of the linear independence of the derivatives $\varphi'_i(u)$ amounts to stating that the matrix with elements $\partial_j\varphi_i(u)$, $1 \leqslant i \leqslant m$, $1 \leqslant j \leqslant n$, has rank m. Suppose (simply in order to fix ideas) that the submatrix with elements $\partial_j\varphi_i(u)$, $1 \leqslant i,j \leqslant m$, is invertible. We then find ourselves exactly in the situation which makes possible the application of the previous theorem, with

$$V_2 = \left\{ \sum_{i=1}^{m} v_i e_i \in \mathbb{R}^m \right\}, \quad V_1 = \left\{ \sum_{i=m+1}^{n} v_i e_i \in \mathbb{R}^{n-m} \right\},$$

$$\varphi: v \in \Omega \subset V_1 \times V_2 \to \varphi(v) = \sum_{i=1}^{n} \varphi_i(v) e_i \in V_2,$$

(e_i) denoting the canonical basis of $\mathbb{R}^n$. Therefore, there exists an element $\Lambda(u)$ of the space $\mathscr{L}(\mathbb{R}^m; \mathbb{R})$ such that $J'(u) + \Lambda(u)\varphi'(u) = 0$; equivalently, there exist m real numbers $\lambda_i(u)$, $1 \leqslant i \leqslant m$, such that $J'(u) + \sum_{i=1}^{m} \lambda_i(u)\varphi'_i(u) = 0$. The uniqueness of the numbers $\lambda_i(u)$ is a result of the linear independence of the derivatives $\varphi'_i(u)$. $\quad\square$

The numbers $\lambda_i(u)$, $1 \leqslant i \leqslant m$, found in the theorem given above are called the *Lagrange multipliers* associated with the 'constrained extremum u' (with the misuse of language to which allusion has already been made).

Let us consider an example. Suppose that the relative extrema of the function

$$J: v = (v_1, v_2) \in \mathbb{R}^2 \to J(v) = -v_2,$$

with respect to the set

$$U = \{(v_1, v_2) \in \mathbb{R}^2 : \varphi(v) \overset{\text{def}}{=} v_1^2 + v_2^2 - 1 = 0\}$$

are to be found (cf. figure 7.2-1(a)). With the help of some geometrical intuition, we first deduce the existence of a Lagrange multiplier (here $m = 1$) at a constrained extremum by heuristic means. Accordingly, if we represent the level curves of the two surfaces in the plane $v_3 = 0$ (cf. figure 7.2-1(b)), it would seem intuitively that a point (u_1, u_2) is a constrained extremum only if the tangents to the curves $\varphi(v_1, v_2) = 0$ and $J(v_1, v_2) - J(u_1, u_2) = 0$ coincide at the point (u_1, u_2). Since the tangents have respectively the equations

$$\begin{cases} (v_1 - u_1)\partial_1\varphi(u) + (v_2 - u_2)\partial_2\varphi(u) = 0, \\ (v_1 - u_1)\partial_1 J(u) + (v_2 - u_2)\partial_2 J(u) = 0, \end{cases}$$

it is enough to express the fact that the coefficients of these equations are proportional, that is to say, that there exists $\lambda \in \mathbb{R}$ such that (the hypothesis given in the theorem of the linear independence of the derivatives $\varphi_i'(u)$ here amounts to verifying that the two partial derivatives $\partial_1\varphi(u)$ and $\partial_2\varphi(u)$ are not simultaneously zero)

$$\begin{cases} \partial_1 J(u) + \lambda\partial_1\varphi(u) = 0, \\ \partial_2 J(u) + \lambda\partial_2\varphi(u) = 0, \end{cases}$$

which is precisely the necessary condition of the theorem! By including the equation $\varphi(u_1, u_2) = 0$, we are then led to a solution of the system

$$\left\{\begin{matrix} 2\lambda u_1 = 0, \\ -1 + 2\lambda u_2 = 0, \\ u_1^2 + u_2^2 - 1 = 0, \end{matrix}\right\}$$

whose only two solutions are

$$\begin{cases} \lambda = \tfrac{1}{2}, & (u_1, u_2) = (0, 1), \\ \lambda' = -\tfrac{1}{2}, & (u_1', u_2') = (0, -1). \end{cases}$$

It turns out, as may readily be verified, that the points $u = (u_1, u_2)$ and $u' = (u_1', u_2')$ obtained in this way are, in fact, extrema; but we must not lose sight of the fact that the conditions given above are only *necessary* conditions; so that we *always* have to end with an investigation of the 'local' behaviour, if we wish to be able to tell whether the points so discovered are, in fact, relative maxima, or minima, of the function J with respect to the set U of the problem.

In order to solve a problem posed in the general form of Theorem 7.2-3, we proceed as follows. We express the fact that the $m + n$ unknowns u_i,

Figure 7.2.1

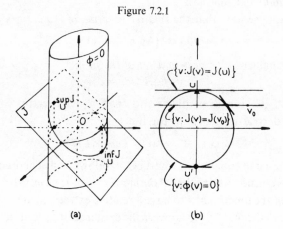

(a)   (b)

$1 \leqslant i \leqslant n$, and λ_j, $1 \leqslant j \leqslant m$, are solutions of the system of $m + n$ equations (in general, non-linear)

$$
\begin{cases}
\partial_1 J(u) + \lambda_1 \partial_1 \varphi_1(u) + \cdots + \lambda_m \partial_1 \varphi_m(u) = 0, \\
\quad \vdots \\
\partial_n J(u) + \lambda_1 \partial_n \varphi_1(u) + \cdots + \lambda_m \partial_n \varphi_m(u) = 0, \\
\varphi_1(u) = 0, \\
\quad \vdots \\
\varphi_m(u) = 0.
\end{cases}
$$

The first n equations may also be written in the matrix form

$$
\begin{pmatrix} \partial_1 J(u) \\ \vdots \\ \partial_n J(u) \end{pmatrix} + \begin{pmatrix} \partial_1 \varphi_1(u) & \cdots & \partial_1 \varphi_m(u) \\ \vdots & & \vdots \\ \partial_n \varphi_1(u) & \cdots & \partial_n \varphi_m(u) \end{pmatrix} \begin{pmatrix} \lambda_1 \\ \vdots \\ \lambda_m \end{pmatrix} = \nabla J(u) + \sum_{i=1}^m \lambda_i \nabla \varphi_i(u) = 0.
$$

Observe, at this point, that it is the *transpose* of the derivative matrix (of the function φ), and *not* the derivative matrix itself, which appears.

To end this section, let us consider an *important example*, to which we shall return in the following chapters. A *quadratic functional over* $\mathbb{R}^n$ is a function of the form

$$
J : v \in \mathbb{R}^n \rightarrow J(v) = \tfrac{1}{2}(Av, v) - (b, v),
$$

where $A \in \mathscr{A}_n(\mathbb{R})$ is a given *symmetric* matrix and $b \in \mathbb{R}^n$ a given vector. Such a function is differentiable in $\mathbb{R}^n$ and

$$
\nabla J(u) = Au - b, \quad \text{for every } u \in \mathbb{R}^n.
$$

This calculation (in which the symmetry of the matrix is used in an essential manner) shows at once that *the solution of linear systems with symmetric matrix is equivalent to the search for points at which the derivative of a quadratic functional vanishes.*

Let us suppose next that the *relative extrema of a quadratic functional*

$$
J(v) = \tfrac{1}{2}(Av, v) - (b, v)
$$

are being sought *with respect to a set of the form*
$$
U = \{ v \in \mathbb{R}^n : Cv = d \},
$$

where $C \in \mathscr{A}_{m,n}(\mathbb{R})$ is a given matrix and $d \in \mathbb{R}^m$ a given vector. Suppose that $m < n$. The derivative matrix of the function

$$
\varphi : v \in \mathbb{R}^n \rightarrow \varphi(v) = Cv - d \in \mathbb{R}^m
$$

being constant and equal to the matrix C, it follows from Theorem 7.2-3 and the previous considerations that, *if the matrix C has rank m*, then a *necessary* condition for the functional J to have a relative extremum at the point $u \in U$ with respect to the set U given above is the existence of a solution $(u, \lambda) \in \mathbb{R}^{n+m}$

of the *linear system*

$$\begin{cases} Au + C^T\lambda = b \\ Cu \qquad = d \end{cases} \Leftrightarrow \left(\begin{array}{c|c} A & C^T \\ \hline C & O \end{array} \right) \left(\begin{array}{c} u \\ \hline \lambda \end{array} \right) = \left(\begin{array}{c} b \\ \hline d \end{array} \right).$$

Remark

The counterpart, then, of taking into consideration the constraint $Cv = d$ is the need to solve a 'larger' linear system than that associated with the case in which there are 'no constraints'. In fact, it is not possible to dispense with the calculation of the vector $\lambda \in \mathbf{R}^m$ even if, as is often the case, one is *only* interested in the possible relative extrema $u \in U$, the unknown λ then making its appearance simply as a necessary 'intermediate' step in the course of the calculations. ☐

Exercises

7.2-1. (1) Find the relative extrema of the function $J(v_1, v_2) = v^2$ with respect to the set

$$U = \{v = (v_1, v_2) \in \mathbb{R}^2: \quad \varphi(v) = 0\}, \quad \text{with} \quad \varphi(v) = v_1^2 + v_2^2 - 1.$$

(2) Find the relative extrema of the function $J(v_1, v_2) = v_1 + (v_2 - 1)^2$ with respect to the set

$$U = \{v = (v_1, v_2) \in \mathbb{R}^2: \varphi(v) = 0\}, \quad \text{with} \quad \varphi(v) = v_1^2.$$

7.2-2. Suppose that the conditions of Theorem 7.2-3 are satisfied, the sets U taking the form

$$U_d = \{v \in \Omega: \varphi(v) = d\},$$

where $\varphi: \Omega \subset \mathbb{R}^n \to \mathbb{R}^m$ is a given function and d a vector which varies in an open subset D of $\mathbb{R}^m$. Suppose that, at each point $d \in D$, the function J has a unique relative extremum u_d with respect to the set U_d, thus defining a function

$$f: d \in D \subset \mathbb{R}^m \to f(d) = J(u_d) \in \mathbb{R}.$$

Assuming that the various functions involved have all the required properties of differentiability, show that the Lagrange multiplier λ_d associated with the constrained extremum u_d may be expressed as $\lambda_d = -\nabla f(d)$.

7.2-3. Let A be a symmetric matrix of order n, B a positive definite symmetric matrix of order n and b a vector of $\mathbb{R}^n$. State a necessary condition for the existence of a relative extremum of the functional

$$J: v \in \mathbb{R}^n \to J(v) = \tfrac{1}{2}(Av, v) - (b, v)$$

in relation to the set

$$U = \{v \in \mathbb{R}^n: (Bv, v) = 1\}.$$

It should be found that, for $b = 0$, a *generalised eigenvalue problem* is thus encountered.

7.3 Extrema of real functions: consideration of the second derivatives

The results which follow are stated for relative *minima*, the consideration of the second derivatives (as the consideration of convexity in the following section) making it actually possible to state the precise nature (maximum or minimum) of the extrema in question. Of course, it would be equally possible to state analogous results for relative *maxima*.

Theorem 7.3-1 (*Necessary condition for a relative minimum*)
Let Ω be an open subset of a normed vector space V and $J: \Omega \subset V \to \mathbb{R}$ a function differentiable in Ω and twice differentiable at the point $u \in \Omega$. If the function J has a relative minimum at u, then

$$J''(u)(w, w) \geqslant 0 \quad \text{for every } w \in V.$$

Proof
Let w be a non-zero vector of V. There exists an open interval $I \subset \mathbb{R}$ containing the origin such that

$$t \in I \Rightarrow u + tw \Rightarrow \Omega \quad \text{and} \quad J(u + tw) \geqslant J(u).$$

The Taylor–Young formula and the relation $J'(u) = 0$ (Theorem 7.2-1) allow us to write

$$0 \leqslant J(u + tw) - J(u) = \frac{t^2}{2}(J''(u)(w, w) + \varepsilon(t)), \quad \lim_{t \to 0} \varepsilon(t) = 0.$$

which proves the assertion (if $J''(u)(w, w)$ had been strictly negative, a similar result would have been true of the difference $J(u + tw) - J(u)$, for t sufficiently small). $\quad\square$

There is no converse of the previous result, as is shown by the example of the function $J(v) = v^3$, $v \in \mathbb{R}$. In order to obtain a sufficient condition for a relative minimum, one is, in effect, led to posit a 'stronger' hypothesis at the point u (case (1) of the theorem which follows), or to assume some analogous property, but one which holds in an entire *neighbourhood* of the point u (case (2)).

We shall also need the following definition. Let $J: W \to \mathbb{R}$ be a function defined over a topological space W. The function J is said to have a *strict relative minimum* (or a *strict relative maximum*) at the point $u \in W$ if there exists a neighbourhood O of the point u such that

$$J(u) < J(v) \text{ (or } J(u) > J(v)) \quad \text{for every } v \in O - \{u\}.$$

Theorem 7.3-2 (*Sufficient conditions for a relative minimum*)

Let Ω be an open subset of a normed vector space V, u a point of Ω and $J:\Omega \subset V \to \mathbb{R}$ a function differentiable in Ω such that $J'(u) = 0$.

(1) *If the function J is twice differentiable at u and if there exists a number α such that*

$$\alpha > 0 \quad and \quad J''(u)(w, w) \geqslant \alpha \|w\|^2 \quad for \; every \; w \in V,$$

then the function J has a strict relative minimum at u.

(2) *If the function J is twice differentiable in Ω and if there exists a ball $B \subset \Omega$ centred at u such that*

$$J''(v)(w, w) \geqslant 0 \quad for \; every \; v \in B, w \in V,$$

then the function J has a relative minimum at u.

Proof

(1) The Taylor–Young formula allows us to write, for every vector w sufficiently small,

$$J(u + w) - J(u) = \tfrac{1}{2}(J''(u)(w, w) + \|w\|^2 \varepsilon(w)),$$
$$\geqslant \tfrac{1}{2}(\alpha - \varepsilon(w))\|w\|^2, \quad \lim_{w \to 0} \varepsilon(w) = 0,$$

which shows that $J(u + w) > J(u)$ for $(u + w) \in B$, where B is a ball centred at u, whose radius r is sufficiently small to ensure that $\|\varepsilon(w)\| < \alpha$ for $\|w\| \leqslant r$.

(2) The Taylor–Maclaurin formula shows that

$$J(u + w) = J(u) + \tfrac{1}{2}J''(v)(w, w) \geqslant J(u), \quad v \in \,]u, u + w[,$$

for every $u + w \in B$. $\qquad\qquad\square$

There are no converses of the two assertions of the preceding theorem (cf. Exercise 7.3-1).

Exercise

7.3-1. (1) Give an example of a twice differentiable function J having a strict minimum at a point u and such that $J''(u)(w, w) = 0$ for at least one vector $w \neq 0$.

(2) Give an example of a twice differentiable function J having a strict minimum at a point u and such that, in every ball B with centre at u, there exists a point $v \in B$ and a vector $w \in V$ satisfying $J''(v)(w, w) < 0$.

7.4 Extrema of real functions: consideration of convexity

A subset of a vector space is said to be *convex*, if, whenever it contains two points u and v, it also contains the closed segment $[u, v]$ which joins them.

For example, *a vector subspace is convex, a ball in a normed vector space is convex* (apply the triangle inequality), *any intersection of convex subsets is convex*.

Theorem 7.4-1 (*Necessary condition for a relative minimum over a convex set*)

Let $J: \Omega \subset \mathbb{R}$ *be a function defined over an open subset* Ω *of a normed vector space V and let U be a convex subset of Ω. If the function J is differentiable at a point $u \in U$ and if it has a relative minimum at u with respect to the set U, then*

$$J'(u)(v - u) \geqslant 0 \quad \text{for every } v \in U.$$

Proof

Let $v = u + w$ be any point of the set U. This set being convex, all points of the form $u + \theta w, 0 \leqslant \theta \leqslant 1$, are also in U. The differentiability of the function J at u allows us to write

$$J(u + \theta w) - J(u) = \theta(J'(u)w + \varepsilon(\theta)), \quad \lim_{\theta \to 0} \varepsilon(\theta) = 0,$$

for every $\theta \in [0, 1]$ (the vector w being fixed). Hence, the number $J'(u)w$ is necessarily non-negative; otherwise, the difference $J(u + \theta w) - J(u)$ would be strictly negative for $\theta > 0$ sufficiently small. $\qquad \square$

Remarks

(1) If the set U is a vector subspace, the preceding condition becomes simply

$$J'(u)w = 0 \quad \text{for every } w \in U.$$

In particular, if $U = V$, we recover the necessary condition $J'(u) = 0$ of Theorem 7.2-1.

(2) Some other special cases are set down in Exercise 7.4-1. $\qquad \square$

A function $J: U \subset V \to \mathbb{R}$ defined over a *convex set* U of a vector space V is said to be *convex over the set* U if

$$u, v \in U \quad \text{and} \quad \theta \in [0, 1]$$

imply

$$J(\theta u + (1 - \theta)v) \leqslant \theta J(u) + (1 - \theta)J(v),$$

and *strictly convex over the set* U if

$$u, v \in U, \quad u \neq v, \quad \text{and} \quad \theta \in \,]0, 1[$$

imply

$$J(\theta u + (1 - \theta)v) < \theta J(u) + (1 - \theta)J(v).$$

For example, *a linear form* $f: V \to \mathbb{R}$ *is convex but not strictly convex*; similarly, *a norm* $\|\cdot\|: V \to \mathbb{R}$ *is convex*. Of course, *if a function is (strictly)* convex over a convex set U, it is also (strictly) convex over every convex subset of U.

A function $G: U \subset V \to \mathbb{R}$ defined over a *convex* subset U of a vector space V is said to be (strictly) concave if the function $- G$ is (strictly) convex.

Before applying the notion of convexity to the extrema of functions, we shall characterise it in terms of the first derivative (Theorem 7.4-2), or the second (Theorem 7.4-3).

Theorem 7.4-2 (Convexity and first derivative)

Let $J: \Omega \subset V \to \mathbb{R}$ be a function differentiable in an open set Ω of a normed vector space V and let U be a convex subset of Ω.

(1) *The function J is convex over U if and only if*

$$J(v) \geqslant J(u) + J'(u)(v - u) \quad \text{for every } u, v \in U.$$

(2) *The function J is strictly convex over U if and only if*

$$J(v) > J(u) + J'(u)(v - u) \quad \text{for every } u, v \in U, \quad u \neq v.$$

Proof

We remark, to begin with, that the geometric interpretation of the conditions given above is clear (figure 7.4-1). They merely state that the function is situated 'above' its tangent plane. Let u and v be two distinct points of U and $\theta \in]0, 1[$. If the function J is convex,

$$J(u + \theta(v - u)) \leqslant (1 - \theta)J(u) + \theta J(v).$$

which can also be written as

$$\frac{J(u + \theta(v - u)) - J(u)}{\theta} \leqslant J(v) - J(u).$$

Consequently,

$$J'(u)(v - u) = \lim_{\theta \to 0} \frac{J(u + \theta(v - u)) - J(u)}{\theta} \leqslant J(v) - J(u).$$

Figure 7.4-1

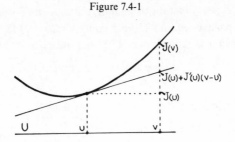

If the function J is strictly convex, the preceding argument does not now lead to the desired conclusion, since it is not possible to affirm that the strict inequality still holds 'in the limit'. Let $\omega \in]0, 1[$, therefore, be a fixed number. Since

$$u + \theta(v - u) = \frac{\omega - \theta}{\omega} u + \frac{\theta}{\omega}(u + \omega(v - u)),$$

it follows that

$$J(u + \theta(v - u)) \leqslant \frac{\omega - \theta}{\omega} J(u) + \frac{\theta}{\omega} J(u + \omega(v - u)), \quad 0 \leqslant \theta \leqslant \omega.$$

Hence, if the function is strictly convex,

$$\frac{J(u + \theta(v - u)) - J(u)}{\theta} \leqslant \frac{J(u + \omega(v - u)) - J(u)}{\omega}$$

$$< J(v) - J(u), \quad 0 < \theta \leqslant \omega,$$

since $\omega < 1$ by hypothesis; hence, strict inequality now holds upon taking the limit.

Conversely, suppose that

$$J(v) \geqslant J(u) + J'(u)(v - u) \quad \text{for every } u, v \in U.$$

Let u and v, then, be two distinct points of U and $\theta \in]0, 1[$; hence, in particular,

$$J(v) \geqslant J(v + \theta(u - v)) - \theta J'(v + \theta(u - v))(u - v).$$
$$J(u) \geqslant J(v + \theta(u - v)) + (1 - \theta)J'(v + \theta(u - v))(u - v),$$

and it is enough to add the two inequalities given above, multiplied respectively by $1 - \theta$ and θ, in order to obtain

$$J(\theta u + (1 - \theta)v) \leqslant \theta J(u) + (1 - \theta)J(v),$$

which establishes the convexity of the function J, or its strict convexity if the inequalities are strict.   □

Theorem 7.4-3 (*Convexity and second derivative*)

Let $J : \Omega \subset V \to \mathbb{R}$ *be a function twice differentiable in an open subset Ω of a normed vector space V and let U be a convex subset of Ω.*

(1) *The function J is convex over U if and only if*

$$J''(u)(v - u, v - u) \geqslant 0 \quad \text{for every } u, v \in U.$$

(2) *If*

$$J''(u)(v - u, v - u) > 0 \quad \text{for every } u, v \in U, \quad u \neq v,$$

then the function J is strictly convex over U.

Proof
Let u and v be two distinct points of U. Applying the Taylor–Maclaurin formula, we obtain

$$J(v) - J(u) - J'(u)(v - u) = \tfrac{1}{2}J''(w)(v - u, v - u)$$

$$= \frac{\varrho^2}{2}J''(w)(v - w, v - w), \quad w \in (u, v),$$

the number $\varrho > 0$ being defined by the equality $v - u = \varrho(v - w)$. The convexity, or strict convexity, of the function then follows upon applying the previous theorem.

To establish the converse of (1), we introduce the auxiliary function

$$G \colon v \in \Omega \to G(v) = J(v) - J'(u)v,$$

for any point $u \in U$, but considered fixed in what follows. A glance at figure 7.4-1 will do much to justify this introduction. If the function J is convex, the function G has a minimum at u *with respect to the set U*; in fact,

$$G(v) - G(u) = J(v) - J(u) - J'(u)(v - u) \geqslant 0, \quad \text{for every } v \in U,$$

by the necessary condition (1) of Theorem 7.4-2. The function G being twice differentiable in Ω and with derivative $G'' = J''$, the Taylor–Young formula gives, for every $v = u + w \in U$ and for every $t \in [0, 1]$,

$$0 \leqslant G(u + tw) - G(u) = \frac{t^2}{2}(J''(u)(w, w) + \varepsilon(t)), \quad \lim_{t \to 0} \varepsilon(t) = 0,$$

since $G'(u) = 0$ by construction.

The usual kind of argument then implies that $J''(u)(w, w) \geqslant 0$. $\qquad \square$

Remarks
(1) It was not possible to apply directly the necessary condition of Theorem 7.3-1, which was established for relative minima with respect to *open* sets.

(2) The example of the strictly convex function $J(v) = v^4$, $v \in \mathbb{R}$, shows that there does not exist *in general* a converse of condition (2).

(3) *In the particular case of a quadratic functional over $\mathbb{R}^n$, the converse does hold.* It follows, in fact, from the expression $J(v) = \tfrac{1}{2}(Av, v) - (b, v)$, $A = A^\mathrm{T}$, that

$$J(v) - J(u) - J'(u)(v - u) = \tfrac{1}{2}(A(v - u), v - u),$$

and Theorem 7.4-2 gives the required conclusion. Gathering together the various results relating to this type of function, we have established that *a quadratic functional over R(of the type given above) is convex if and only* *if the symmetric matrix* A *is non-negative definite, and strictly convex if and only if the matrix* A *is positive definite.*   □

While, up to the present, we have only considered *relative* extrema, convexity is going to enable us to free ourselves of the 'local' character of this property. This is the reason for introducing the following definitions. Let $J: W \to \mathbb{R}$ be a function defined over a set W. We say that a function J has a *minimum* (or a *maximum*) at a point $u \in W$ if

$$J(u) \leqslant J(v) \quad (\text{or } J(u) \geqslant J(v)) \quad \text{for every } v \in W,$$

and a *strict minimum* (or a *strict maximum*) if

$$J(u) < J(v) \quad (\text{or } J(u) > J(v)) \quad \text{for every } v \in W - \{u\}.$$

Lastly, if U is a subset of the set W, we say that the function J has a *minimum* (or a *strict minimum*, etc.) at a point $u \in W$ *with respect to the set U* if the restriction of the function J to the set U has a minimum at u (or a strict minimum, etc.).

The result which follows gathers together a number of properties of *the minima of convex functions* which are of constant use.

Theorem 7.4-4
Let U be a convex subset of a normed vector space V.

(1) If a convex function $J: U \subset V \to \mathbb{R}$ has a relative minimum at a point of U, it has, in fact, a minimum there, that is to say, with respect to the entire set U.

(2) A strictly convex function $J: U \subset V \to \mathbb{R}$ has at most one minimum, and that minimum is strict.

(3) Let $J: \Omega \subset V \to \mathbb{R}$ be a convex function defined over an open subset Ω of V containing U and differentiable at a point $u \in U$. Then the function J has a minimum at u with respect to the set U if and only if

$$J'(u)(v - u) \geqslant 0 \quad \text{for every } v \in U.$$

(4) If the set U is open, the preceding condition is equivalent to Euler's equation $J'(u) = 0$.

Proof
(1) Let $v = u + w$ be any point of the set U. By the convexity of the function J,

$$J(u + \theta w) \leqslant (1 - \theta)J(u) + \theta J(v), \quad 0 \leqslant \theta \leqslant 1,$$

which can also be written as

$$J(u + \theta w) - J(u) \leqslant \theta(J(v) - J(u)), \quad 0 \leqslant \theta \leqslant 1.$$

As the point u is a relative minimum, there exists a number θ_0 such that

$$\theta_0 > 0 \quad \text{and} \quad 0 \leqslant J(u + \theta_0 w) - J(u),$$

which implies that $J(v) \geqslant J(u)$.

(2) If the function is strictly convex, the same reasoning leads to the inequalities

$$0 \leqslant J(u + \theta_0 w) - J(u) < \theta_0 (J(v) - J(u)), \quad \theta_0 > 0, \quad w \neq 0,$$

which establishes the strict nature of the minimum and, hence, at the same time, its uniqueness.

(3) In Theorem 7.4-1, the necessity was shown of the condition $J'(u)(v - u) \geqslant 0$ for every $v \in U$ (without supposing J to be convex). In order to establish that it is sufficient, we observe that

$$J(v) - J(u) \geqslant J'(u)(v - u) \quad \text{for every } v \in U,$$

by Theorem 7.4-2.

(4) The last property is an immediate consequence of property (3). $\quad \Box$

The relations '$J'(u)(v - u) \geqslant 0$ for every $v \in U$' are frequently called *Euler's inequalities*.

The reader interested in the application of the preceding theorems to *quadratic functionals* may refer to Exercise 7.4-2 (it is, in any case, an exercise which is strongly recommended).

As an illustration of the preceding results, we return to the problem of the *least squares solution of a linear system* (cf. Section 3.7). Given a real matrix B of type (m, n) and a vector $c \in \mathbb{R}^m$, we look for a vector $u \in \mathbb{R}^n$ such that

$$\| Bu - c \|_m = \inf_{v \in \mathbb{R}^n} \| Bv - c \|_m,$$

where $\| \cdot \|_m$ denotes the Euclidean norm in $\mathbb{R}^m$. We introduce the quadratic functional

$$
\begin{aligned}
J(v) &\overset{\text{def}}{=} \tfrac{1}{2} \| Bv - c \|_m^2 - \tfrac{1}{2} \| c \|_m^2 \\
&= \tfrac{1}{2} (Bv, Bv)_m - (c, Bv)_m \\
&= \tfrac{1}{2} (B^{\mathrm{T}} Bv, v)_n - (B^{\mathrm{T}} c, v)_n, \quad v \in \mathbb{R}^n,
\end{aligned}
$$

where $(\cdot, \cdot)_m$ and $(\cdot, \cdot)_n$ denote the scalar products in the spaces $\mathbb{R}^m$ and $\mathbb{R}^n$ respectively. The symmetric matrix $B^{\mathrm{T}} B$ being non-negative definite, the function J is convex (Theorem 7.4-3). As the problem being considered is equivalent to the search for a vector $u \in \mathbb{R}^n$ such that

$$J(u) = \inf_{v \in \mathbb{R}^n} J(v),$$

we are able to conclude from Theorem 7.4-4 that *the set of solutions coincides with the set of solutions of the equation*

$$J'(u) = B^{T}Bu - B^{T}c = 0,$$

that is to say, the *normal equations* introduced earlier in section 3.7. In that same section, we also used the relation

$$\| B(u + w) - c \|_{m}^{2} = \| Bu - c \|_{m}^{2} + 2(B^{T}Bu - B^{T}c, w)_{n} + \| Bw \|_{m}^{2},$$

which is just the Taylor formula

$$J(u + w) = J(u) + J'(u)w + \tfrac{1}{2}(B^{T}Bw, w)_{n},$$

for a quadratic functional J, whose Hessian is the matrix $B^{T}B$.

Exercises

7.4-1. Show that the condition of Theorem 7.4-1 is equivalent to

$$J'(u)u = 0 \quad \text{and} \quad J'(u)v \geqslant 0 \quad \text{for every } v \in U,$$

if the set U is a convex cone with vertex at the origin, and to

$$J'(u)w = 0 \quad \text{for every } w \in U_{0},$$

if the set U has the form

$$U = a + U_{0} = \{a + w : w \in U_{0}\}, \quad U_{0} \text{ some vector subspace of } V.$$

7.4-2. Let $J(v) = \tfrac{1}{2}(Av, v) - (b, v)$, $A = A^{T}$, $v \in \mathbb{R}^{n}$, be a quadratic functional. Prove the following propositions.

(1) There exists a vector $u \in \mathbb{R}^{n}$ such that

$$J(u) < J(v) \quad \text{for every } v \in \mathbb{R}^{n} - \{u\},$$

if, and only if, the matrix A is positive definite (J is then strictly convex).

(2) There exists a vector $u \in \mathbb{R}^{n}$ such that

$$J(u) \leqslant J(v) \quad \text{for every } v \in \mathbb{R}^{n},$$

if, and only if, the matrix A is non-negative definite (J is then convex) and the set $\{w \in \mathbb{R}^{n} : A\omega = b\}$ is non-empty.

(3) If the matrix A is non-negative definite and the set $\{w \in \mathbb{R}^{n} : Aw = b\}$ is empty, then $\inf_{v \in \mathbb{R}^{n}} J(v) = -\infty$.

(4) If $\inf_{v \in \mathbb{R}^{n}} J(v) > -\infty$, then the matrix A is non-negative definite and the set $\{w \in \mathbb{R}^{n} : Aw = b\}$ is non-empty.

(5) Deduce that the *normal equations* always have (at least) one solution.

7.4-3. For what values of p, $1 \leqslant p \leqslant \infty$, are the norms $\| \cdot \|_{p} : \mathbb{R}^{n} \to \mathbb{R}$ strictly convex?

7.4-4. Let U be a convex subset of a vector space V.

(1) Show that a function $f : U \subset V \to \mathbb{R}$ is convex if and only if the set

$$\text{epi}(f) \overset{\text{def}}{=} \{(v, \alpha) \in V \times \mathbb{R} : v \in U, \, \alpha \geqslant f(v)\},$$

called the *epigraph* of the function f, is a convex subset of the vector space $V \times \mathbb{R}$.

(2) Let $(f_i)_{i \in I}$ be any family of convex functions $f_i : U \subset V \to \mathbb{R}$, $i \in I$. Prove that the function $\sup_{i \in I} f_i$ is convex.

7.4-5. (1) Let (e_{ij}) be the square matrix of order n defined by $e_{ij} = 1$, $1 \leq i, j \leq n$. Calculate its eigenvalues and determine the corresponding eigenspaces.

(2) Define the function

$$J : v \in \Omega \subset \mathbb{R}^n \to J(v) = -\left(\prod_{i=1}^{n} v_i \right)^{1/n} \in \mathbb{R},$$

where the set (open and convex) Ω is defined by

$$\Omega = \{ v \in \mathbb{R}^n : v_i > 0, 1 \leq i \leq n \}.$$

Calculate the numbers

$$J'(u)v \quad \text{and} \quad J''(u)(v, w) \quad \text{for} \quad u \in \Omega, \quad v \in \mathbb{R}^n, \quad w \in \mathbb{R}^n.$$

(3) Show that the function $J : \Omega \subset \mathbb{R}^n \to \mathbb{R}$ is convex, but not strictly convex.

(4) Denote by j the restriction of the function J to the convex subset

$$U = \left\{ v \in \Omega : \sum_{i=1}^{n} v_i = n \right\}$$

of the open set Ω. Show that the function $j : U \subset \mathbb{R}^n \to \mathbb{R}$ is strictly convex.

(5) Denote by e the vector of Ω all of whose components are equal to 1. Show that

$$J'(e)(v - e) = 0 \quad \text{for every } v \in U.$$

Deduce that there exists a unique element u such that

$$u \in U \quad \text{and} \quad J(u) = \inf_{v \in U} J(v).$$

(6) Prove the inequality

$$\left(\prod_{i=1}^{n} v_i \right)^{1/n} \leq \frac{1}{n} \sum_{i=1}^{n} v_i \quad \text{for every } v \in \Omega,$$

and describe the subset of Ω for which the inequality becomes an equality.

7.4-6. Let U be an open, convex subset of a finite-dimensional vector space V and $J : U \subset V \to \mathbb{R}$ a convex function.

(1) Let u be a general point of the set U. Prove that there exists a neighbourhood of the point u over which the function J is bounded above.

(2) Prove that the function J is continuous at every point $u \in U$.

(3) Prove that the function J satisfies a 'local' Lipschitz condition, in the following sense. Given any point u of the set U, there exists a neighbourhood W of the point u and a constant $C(W)$ such that

$$|J(v) - J(u)| \leq C(W) \|v - u\| \quad \text{for every } v \in W.$$

(4) Give an example of an infinite-dimensional space V and of a function $J : V \to \mathbb{R}$ which is convex, but not continuous.

7.4-7. Let U be a subset of a normed vector space such that $u, v \in U \Rightarrow \frac{1}{2}(u + v) \in U$. Prove that, if the subset U is closed, then it is convex. Does the conclusion continue to hold when the subset U is not assumed to be closed?

7.4-8. Let U be a convex subset of a normed vector space.

(1) Prove that the sets $\overset{\circ}{U}$ and $\bar{U}$ are also convex.

(2) Prove the implication

$$\overset{\circ}{U} \ne \varnothing \Rightarrow \bar{\overset{\circ}{U}} = \bar{U}.$$

(3) Does the implication continue to hold when the subset U is not assumed to be convex?

7.4-9. Let $\Phi: \mathbb{R}_+^n \to \mathbb{R}$ be a symmetric function, that is to say, one which satisfies

$$\Phi(\mu_1, \mu_2, \ldots, \mu_n) = \Phi(\mu_{\sigma(1)}, \mu_{\sigma(2)}, \ldots, \mu_{\sigma(n)}) \quad \text{for every permutation } \sigma \in \mathfrak{S}_n.$$

With it is associated the function

$$\mathscr{F}: A \in \mathscr{A}_n(\mathbb{R}) \to \mathscr{F}(A) = \Phi(\mu_1(A), \mu_2(A), \ldots, \mu_n(A)),$$

the numbers $\mu_i(A)$ being the singular values of the matrix A.

(1) Prove that the function $\mathscr{F}$ is convex over the set

$$U = \{A \in \mathscr{A}_n(\mathbb{R}): A = A^T \text{ and } \mathrm{sp}(A) \subset \mathbb{R}_+\}$$

if and only if the function Φ is convex.

(2) Prove that the function $\mathscr{F}$ is convex over the set $\mathscr{A}_n(\mathbb{R})$ if and only if the function Φ is convex and non-decreasing, with respect to each of the variables $\mu_i, 1 \le i \le n$.

7.4-10. Let U be a subset of a vector space V. We call the *convex hull* of U (and denote it by co U) the intersection of all convex subsets of V containing U, that is to say, 'the smallest convex subset containing U'.

(1) Prove that co U is also the set of all *convex combinations* of elements of U, that is to say, of all elements of V of the form $\sum_{i \in I} \alpha_i u_i$, where I is any finite set, the numbers α_i satisfying $\alpha_i \ge 0$ and $\sum_{i \in I} \alpha_i = 1$, and the points u_i belonging to the set U.

(2) Prove that, if V is finite-dimensional, then every point of co U may be written as a convex combination of at most $n + 1$ elements of U, where n is the dimension of the space V (*Carathéodory's theorem*).

(3) With the same hypotheses prevailing as in question (2), prove that every point of the boundary of co U may be written as a convex combination of at most n elements of U.

(4) Prove that if U is a compact subset of a finite-dimensional space, then co U is also compact.

(5) We call the *closed convex hull* of U (and denote it by $\overline{\mathrm{co}}\ U$) the intersection of all closed, convex subsets of V containing U. What are the inclusions or possible equalities which hold among the sets $\overline{\mathrm{co}}\ U$, co $\bar{U}$, $\overline{\mathrm{co}\ U}$?

(6) A square matrix $A = (a_{ij})$ of order n is said to be *doubly stochastic* if

$$a_{ij} \geqslant 0, \quad 1 \leqslant i,j \leqslant n,$$

$$\sum_{j=1}^{n} a_{ij} = 1, \quad 1 \leqslant i \leqslant n,$$

$$\sum_{i=1}^{n} a_{ij} = 1, \quad 1 \leqslant j \leqslant n.$$

Prove that the set of doubly stochastic matrices of order n is the convex hull of permutation matrices of order n (theorem due to G. Birkhoff).

7.4-11. Let I be an open interval, bounded or not, of $\mathbb{R}$ and $f: I \to \mathbb{R}$ a convex function.

(1) Show that, at every point $a \in I$, the following limits exist:

$$\lim_{h \to 0^+} \frac{f(a+h) - f(a)}{h} \overset{\text{def}}{=} f'_+(a),$$

$$\lim_{h \to 0^-} \frac{f(a+h) - f(a)}{h} \overset{\text{def}}{=} f'_+(a),$$

(2) Deduce from (1) that the function f is continuous at every point of I. Does this property continue to hold when the interval is not open?

(3) Show that the set $\{a \in I : f'_-(a) < f'_+(a)\}$ is denumerable (we have, in any case, the inequality $f'_-(a) \leqslant f'_+(a)$ at every point $a \in I$).

(4) Deduce from (3) that the set of points of I at which the function f is not differentiable is denumerable.

7.5 Newton's method

Let Ω be an open subset of a normed vector space V and $J: \Omega \subset V \to \mathbb{R}$ a given function. Recall that (cf. Theorem 7.2-1), if the function J has a relative extremum at a point $u \in \Omega$ and if it is differentiable at that point, then necessarily $J'(u) = 0$. In this present section, we shall be interested in the solution of this equation, that is to say, supposing the function J to be differentiable in Ω, in the *search of the zeros of the derivative*

$$J': \Omega \subset V \to V'.$$

Remark

Once the zeros of the function J' have been found, the next step, quite naturally, is to verify whether they are actually extrema, for example by using the sufficient conditions established in the earlier sections. $\square$

It is convenient to consider the problem in a more general form, one which has the advantage of being applicable in other situations (notably in the solution of non-linear systems of equations; see further on). Given a function $f: \Omega \subset X \to Y$ from an open subset Ω of a normed vector space X into a normed vector space Y, we look for:

(i) sufficient conditions which guarantee the *existence* of a *zero* of the function f, that is to say, of an element $a \in \Omega$ such that $f(a) = 0$,

(ii) an *algorithm* for approximating such an element a, that is to say, for constructing a sequence (x_k) of points of Ω such that $\lim_{k \to \infty} x_k = a$.

The first problem is the objective in Theorem 7.5-1 and the second that in Theorems 7.5-1 and 7.5-2, each of which is concerned with a generalisation of *Newton's method*, which is well-known for functions $f: \Omega \subset \mathbb{R} \to \mathbb{R}$. This method, defined by the sequence

$$x_{k+1} = x_k - \frac{f(x_k)}{f'(x_k)}, \quad k \geqslant 0,$$

$x_0 \in \Omega$ arbitrary, is open to an immediate geometric interpretation (see figure 7.5-1), each point x_{k+1} being the intersection of the axis with the tangent at the point x_k.

This particular case suggests the following definition of *Newton's method* for finding the zeros of a function $f: \Omega \subset X \to Y$. Given an arbitrary point $x_0 \in \Omega$, the sequence $(x_k)_{k \geqslant 0}$ is defined by

$$x_{k+1} = x_k - \{f'(x_k)\}^{-1} f(x_k), \quad k \geqslant 0,$$

which assumes that *all the points x_k remain within Ω* (this point needs to be verified each time), that the function f is differentiable in Ω and, finally, that the *derivative $f'(x)$ is a bijection from X onto Y for each point $x \in \Omega$.*

Newton's method is applicable, in particular, to the solution of *non-linear systems of equations*, which correspond to functions $f: \Omega \subset \mathbb{R}^n \to \mathbb{R}^n$. More

Figure 7.5-1

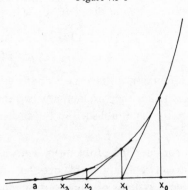

explicitly,

$$f(a) = 0 \Leftrightarrow \begin{cases} f_1(a_1, a_2, \ldots, a_n) = 0, \\ f_2(a_1, a_2, \ldots, a_n) = 0, \\ \quad \vdots \\ f_n(a_1, a_2, \ldots, a_n) = 0, \end{cases}$$

the functions $f_i : \Omega \subset \mathbb{R}^n \to \mathbb{R}$, $1 \leqslant i \leqslant n$, being given. In that case, a single iteration of Newton's method consists in solving the *linear system*

$$f'(x_k)\delta x_k = -f(x_k),$$

with matrix $f'(x_k) = (\partial_j f_i(x_k))$, and then setting

$$x_{k+1} = x_k + \delta x_k.$$

Remark
If the function f is *affine*, $f(x) = \mathrm{A}x - b$, $\mathrm{A} \in \mathscr{A}_n(\mathbb{R})$, $b \in \mathbb{R}^n$, then the iteration described above reduces to the solution of the linear system $\mathrm{A}x_{k+1} - b = 0$. In other words, the method converges in a *single* iteration; this, of course, is only to be expected, since an affine function coincides with its tangent plane at every point. $\qquad\square$

One may readily imagine that, *in practice*, it is very *costly* to calculate, at *each* iteration, the elements of the 'new' matrix $(\partial_j f_i(x_k))$, and then to solve the corresponding linear system. Moreover, if the method is convergent, the consecutive vectors x_k should differ only 'a little', as also the corresponding matrices. These considerations, practical as well as intuitive, lead naturally to a variant of Newton's method, which consists in *keeping the same matrix for p consecutive iterations* (p some fixed integer $\geqslant 2$):

$$x_{k+1} = x_k - \{f'(x_0)\}^{-1} f(x_k), \quad 0 \leqslant k \leqslant p-1,$$
$$x_{k+1} = x_k - \{f'(x_p)\}^{-1} f(x_k), \quad p \leqslant k \leqslant 2p-1,$$
$$\vdots$$
$$x_{k+1} = x_k - \{f'(x_{rp})\}^{-1} f(x_k), \quad rp \leqslant k \leqslant (r+1)p-1,$$
$$\vdots$$

It is also possible to 'set $p = \infty$' which leads to iterations of the type

$$x_{k+1} = x_k - \{f'(x_0)\}^{-1} f(x_k), \quad k \geqslant 0,$$

or even, quite simply, to replace the matrix $f'(x_0)$ by a particular matrix A_0 which is 'easily invertible':

$$x_{k+1} = x_k - \mathrm{A}_0^{-1} f(x_k), \quad k \geqslant 0.$$

One may then readily convince oneself that, in the case of functions

$f:\Omega \subset \mathbb{R} \to \mathbb{R}$, convergence may be attained as long as A_0 is sufficiently close to $f'(x_0)$ (figure 7.5-2).

Remark

In the case of the last two examples considered, it is enough, for example, to calculate *once only* the LU factorisation (section 4.3) of the matrix representing the linear functions $f'(x_0)$ or A_0 (if this matrix lends itself readily to such a factorisation); the solution of the successive linear systems is then greatly facilitated. $\qquad\square$

If $A_0 = I$, the method, which then appears as

$$x_{k+1} = x_k - f(x_k), \quad k \geqslant 0,$$

is just the *method of successive approximations* for the solution of the equation $f(x) = 0$ written in the form $x = x - f(x)$. This particular iterative method also gives us the chance to illustrate very simply the kind of difficulty likely to be encountered in this type of method. Consider, for example, the solution of

$$f(x) \stackrel{\text{def}}{=} x^2 - \frac{1}{4} = 0,$$

whose roots are $1/2$ and $-1/2$, so that

$$f(x) = 0 \Leftrightarrow x = x - f(x) = -x^2 + x + \frac{1}{4}.$$

We can represent very easily the successive iterates of the method of successive approximations (cf. figure 7.5-3), which allows us to make the

Figure 7.5-2

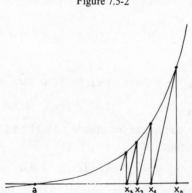

following list of possible eventualities

$$x_0 < -\frac{1}{2}: \text{the method diverges}$$

$$x_0 = -\frac{1}{2}: \quad \lim_{k \to \infty} x_k = -\frac{1}{2} \quad \left(x_k = -\frac{1}{2} \text{ for every } k \geqslant 0(!) \right)$$

$$-\frac{1}{2} < x_0 \frac{3}{2}: \quad \lim_{k \to \infty} x_k = \frac{1}{2}.$$

$$x_0 = \frac{3}{2}: \quad \lim_{k \to \infty} x_k = -\frac{1}{2} \quad \left(x_k = = -\frac{1}{2} \text{ for every } k \geqslant 1(!) \right)$$

$$\frac{3}{2} < x_0: \text{the method diverges}.$$

This example brings out the fact, first, that *the method converges only if the initial value x_0 is sufficiently close to a root*, and, secondly, the fact that, '*in practice*', *the method always converges to the same root when it does converge*. If the initial values are chosen at random, only two values ($x_0 = -1/2$ or $3/2$) lead to the root $-1/2$, while there is an infinity of values ($-1/2 < x_0 < 3/2$) which lead to the root $1/2$. It is sometimes said that $]-1/2, 3/2[$ is the *domain of attraction of the root* $1/2$ and that $\{-1/2\} \cup \{3/2\}$ is that of the root $-1/2$. In other words, the root $-1/2$ is practically 'inaccessible' by the method considered.

The first observation is actually of very general applicability. The principal difficulty in the solution of non-linear systems of equations resides essentially in having to choose a 'good' initial vector x_0, one, in fact, which needs to be sufficiently close to a zero, whereas, in principle, the location of the zeros is not known, the object being precisely to locate them!

Remark

Rather unexpectedly, quite simple-looking iterative equations reminiscent of Newton's method may lead to apparently random, or 'chaotic', behaviour.

Figure 7.5-3

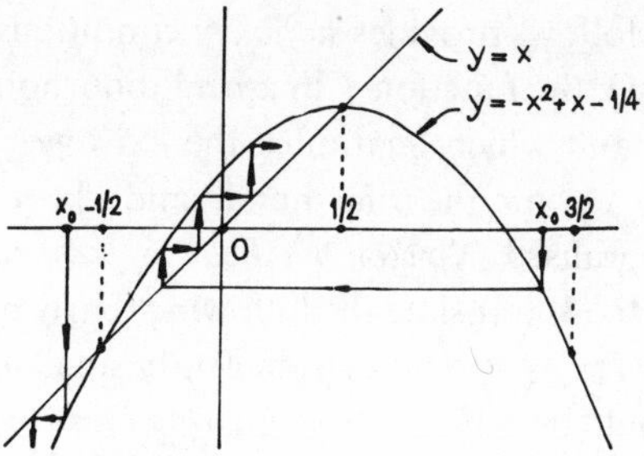

A celebrated example is the *logistic map*, described by the iterative method $x_{k+1} = ax_k(1 - x_k)$. While $\lim_{k \to \infty} x_k = 0$ if $a < 1$, it can be shown that for $a = 4$ (for instance) the iterates 'wander' over entire intervals in an apparently random fashion! For an introduction to this fascinating phenomenon, see the landmark papers May (1976), Feigenbaum (1983), and the penetrating article Jensen (1987). □

Be that as it may, it is in order to make room for the possible incorporation of such variants of Newton's method that we give the following definition of a *generalised Newton method* for finding the zeros of a function $f:\Omega \subset X \to Y$ from an open subset Ω of a normed vector space X into a normed vector space Y. Given a *family* of elements

$$A_k(x) \in \text{Isom}(X; Y), \quad k \text{ an integer} \geqslant 0, \ x \in \Omega,$$

as well as an arbitrary point $x_0 \in \Omega$, the sequence $(x_k)_{k \geqslant 0}$ is defined by

$$x_{k+1} = x_k - \{A_k(x_{k'})\}^{-1} f(x_k), \quad k \geqslant 0,$$

where, for every integer k, the integer k' is subjected to the sole condition

$$0 \leqslant k' \leqslant k.$$

Of course, the functions $A_k(x)$ may also depend on the function f (notably in terms of its derivative f'). For example, one can have

$$A_k(x) = f'(x), \quad k' = k,$$
$$A_k(x) = f'(x), \quad k' = \min\{rp, k\} \quad \text{for } rp \leqslant k \leqslant (r+1)p - 1, r \geqslant 0,$$
$$A_k(x) = f'(x), \quad k' = 0,$$
$$A_k(x) = A_0 \in \text{Isom}(X; Y),$$

these examples corresponding to the 'original' Newton method and its variants introduced above; one could also have

$$A_k(x) = A_k \in \text{Isom}(X; Y), \quad k \geqslant 0,$$

(which corresponds to the situation envisaged further on in Theorem 7.5-2), the functions A_k being independent of f, etc.

The result which follows provides *sufficient* conditions which bear on the *data* (the point $x_0 \in \Omega$; the function f in a neighbourhood of x_0; the family $A_k(x)$, $k \geqslant 0$, $x \in \Omega$), and which guarantee the *existence of a zero* of f in a neighbourhood of x_0, on the one hand, and the *convergence* of the corresponding generalised *Newton method* to this zero, on the other. The hypotheses (1) to (3) translate the following (fairly natural) conditions. The 'initial' norm $\| f(x_0) \|$ should be sufficiently small; the variation in the derivative $f'(x)$ should be sufficiently small for x close to x_0; and, finally, the linear functions $A_k(x)$ and $A_k^{-1}(x)$ should not vary too greatly with k for x close to x_0, the functions $A_k(x)$ remaining sufficiently close to $f'(x_0)$.

Remark
The particular choice $A_k = f'(x_0)$ for $k \geqslant 0$ leads to a result on the existence of a zero of the function f, with the help of hypotheses having to do only with the function f in a neighbourhood of the point x_0. It is left to the reader (delighted with this piece of good fortune) to formulate the corresponding theorem. □

Theorem 7.5-1
Assume that the space X is complete and that the function $f:\Omega \subset X \to Y$ is differentiable in the open set Ω. Assume, too, that there exist three constants r, M and β such that

$$r > 0 \quad and \quad B \overset{\text{def}}{=} \{x \in X: \|x - x_0\| \leqslant r\} \subset \Omega,$$

(1)
$$\sup_{k \geqslant 0} \sup_{x \in B} \| A_k^{-1}(x) \|_{\mathscr{L}(X;Y)} \leqslant M,$$

(2)
$$\sup_{k \geqslant 0 x, \, x' \in B} \sup \| f'(x) - A_k(x') \|_{\mathscr{L}(X;Y)} \leqslant \frac{\beta}{M}, \quad and \quad \beta < 1,$$

(3)
$$\| f(x_0) \| \leqslant \frac{r}{M}(1 - \beta).$$

Then the sequence $(x_k)_{k \geqslant 0}$ defined by
$$x_{k+1} = x_k - A_k^{-1}(x_{k'})f(x_k), \quad k \geqslant k' \geqslant 0,$$

is entirely contained within the ball B and converges to a zero of f, which is the only zero of f in the ball B. Lastly, the convergence is geometric:

$$\|x_k - a\| \leqslant \frac{\|x_1 - x_0\|}{1 - \beta} \beta^k.$$

Proof
(i) *Preliminary 'technicalities'. We show that, for every integer $k \geqslant 1$,*
$$\|x_k - x_{k-1}\| \leqslant M \| f(x_{k-1})\|,$$
$$\|x_k - x_0\| \leqslant r \quad (\Leftrightarrow x_k \in B),$$
$$\| f(x_k)\| \leqslant \frac{\beta}{M} \|x_k - x_{k-1}\|.$$

To do this, we begin by establishing the inequalities for $k = 1$. Since
$$x_1 - x_0 = - A_0^{-1}(x_0)f(x_0),$$
it follows that
$$\|x_1 - x_0\| \leqslant M \| f(x_0)\| \leqslant r(1 - \beta) \leqslant r, \quad \text{by (1) and (3).}$$

Moreover, it is possible to write

$$f(x_1) = f(x_1) - f(x_0) - A_0(x_0)(x_1 - x_0),$$

and an application of the mean value theorem to the function $x \rightarrow$ $f(x) - A_0(x_0)x$ gives

$$\| f(x_1) \| \leqslant \sup_{x \in B} \| f'(x) - A_0(x_0) \| \, \| x_1 - x_0 \| \leqslant \frac{\beta}{M} \| x_1 - x_0 \|, \quad \text{by (2)}.$$

Suppose that the inequalities have been proved for the integer $k - 1$. Since

$$x_k - x_{k-1} = - A_{k-1}^{-1}(x_{(k-1)'}) f(x_{k-1}),$$

it follows that

$$\| x_k - x_{k-1} \| \leqslant M \| f(x_{k-1}) \| \quad \text{by (1)},$$

which proves the first inequality for the integer k. Hence,

$$\| x_k - x_{k-1} \| \leqslant \beta \| x_{k-1} - x_{x-2} \| \leqslant \cdots \leqslant \beta^{k-1} \| x_1 - x_0 \|,$$

so that

$$\| x_k - x_0 \| \leqslant \sum_{l=1}^{k} \| x_l - x_{l-1} \| \leqslant \left\{ \sum_{l=1}^{k} \beta^{l-1} \right\} \| x_1 - x_0 \|$$

$$\leqslant \frac{\| x_1 - x_0 \|}{1 - \beta} \leqslant \frac{M}{1 - \beta} \| f(x_0) \| \leqslant r \quad \text{by (3)},$$

which shows that $x_k \in B$. Lastly, writing

$$f(x_k) = f(x_k) - f(x_{k-1}) - A_{k-1}(x_{(k-1)'})(x_k - x_{k-1}),$$

we obtain, from (2) and another application of the mean value theorem to the function $x \rightarrow f(x) - A_{k-1}(x_{(k-1)'})x$,

$$\| f(x_k) \| \leqslant \sup_{x \in B} \| f'(x) - A_{k-1}(x_{(k-1)'}) \| \, \| x_k - x_{k-1} \| \leqslant \frac{\beta}{M} \| x_k - x_{k-1} \|,$$

and the third inequality is established for the integer k.

(ii) We next show *the existence of a zero of the function f within the ball B.* Since

$$\| x_{k+l} - x_k \| \leqslant \sum_{v=0}^{l-1} \| x_{k+v+1} - x_{k+v} \|$$

$$\leqslant \beta^k \sum_{v=0}^{l-1} \beta^v \| x_1 - x_0 \|$$

$$\leqslant \frac{\beta^k}{1 - \beta} \| x_1 - x_0 \|$$

for every $k \geqslant 0$, $l \geqslant 0$, the sequence $(x_k)_{k \geqslant 0}$ is a *Cauchy sequence*. As it is a sequence of points in the complete space B (it is a closed ball in the complete space X), there exists a point $a \in B$ such that $\lim_{k \to \infty} x_k = a$. The function f being continuous in Ω (since it is differentiable), we obtain

$$\| f(a) \| = \lim_{k \to \infty} \| f(x_k) \| \leqslant \frac{\beta}{M} \lim_{k \to \infty} \| x_k - x_{k-1} \| = 0,$$

so that $f(a) = 0$. Letting l tend to $+\infty$ in the inequality found above, we obtain the inequality

$$\| x_k - a \| \leqslant \frac{\beta^k}{1 - \beta} \| x_1 - x_0 \|.$$

(iii) Finally, we show the *uniqueness of the point a within the ball B*. Let $b \in B$ be another zero of f. Since $f(a) = f(b) = 0$, it is possible to write

$$b - a = - A_0^{-1}(x_0)(f(b) - f(a) - A_0(x_0)(b - a)),$$

from which it follows that

$$\| b - a \| \leqslant \| A_0^{-1}(x_0) \| \sup_{x \in B} \| f'(x) - A_0(x_0) \| \, \| b - a \| \leqslant \beta \| b - a \|,$$

which implies that $a = b$, since $\beta < 1$. $\qquad\square$

In the theorem which follows, we shall assume, for that occasion, that the existence of a zero a of the function f has *already been established*; we shall then establish the convergence of a *particular* generalised Newton method (it being supposed, in that context, that the isomorphisms $A_k(x_{k'})$ are *independent* of the points $x_{k'}$, which explains the use of the notation A_k), provided that the linear operators A_k are sufficiently close to $f'(a)$ and provided that the point x_0 is sufficiently close to the point a (hypotheses which, once again, are 'natural').

Remarks

(1) In each of the two theorems, one of the two spaces X or Y is assumed to be complete; the existence of elements $A \in \mathrm{Isom}\,(X; Y)$ then implies that the other is also complete.

(2) In the theorem which follows, we use in an essential manner the implications

$$X \text{ complete} \Rightarrow \mathscr{L}(X) \text{ complete} \Rightarrow \begin{cases} I + B \text{ is invertible if } \| B \|_{\mathscr{L}(X)} < 1, \\ \text{and} \quad \| (I + B)^{-1} \| \leqslant \dfrac{1}{1 - \| B \|}, \end{cases}$$

the second implication being nothing other than the 'infinite-dimensional' version of Theorem 1.4-5. $\qquad\square$

Theorem 7.5-2
Suppose that the space X is complete and that the function $f: \Omega \subset X \to Y$ is continuously differentiable in the open set Ω. Let a be a point of Ω such that

$$\begin{cases} f(a) = 0, \quad A \overset{\text{def}}{=} f'(a) \in \text{Isom}\,(X; Y), \\[2mm] \sup_{k \geqslant 0} \| A_k - A \|_{\mathscr{L}(X; Y)} \leqslant \dfrac{\lambda}{\| A^{-1} \|_{\mathscr{L}(Y, X)}}, \quad \text{and} \quad \lambda < \tfrac{1}{2}. \end{cases}$$

Then there exists a closed ball B centred at a such that, for every point $x_0 \in B$, the sequence $(x_k)_{k \geqslant 0}$ defined by

$$x_{k+1} = x_k - A_k^{-1} f(x_k), \quad k \geqslant 0,$$

is entirely contained within B and converges to a point a, which is the only zero of f within the ball B. Lastly, the convergence is geometric: there exists a number β such that

$$\beta < 1 \quad \text{and} \quad \| x_k - a \| \leqslant \beta^k \| x_0 - a \|, \quad k \geqslant 0.$$

Proof
(i) *Preliminary technicalities. We first prove the existence of two constants r and β such that*

$$r > 0 \quad \text{and} \quad B \overset{\text{def}}{=} \{ x \in X : \| x - a \| \leqslant r \} \subset \Omega,$$

$$\sup_{k \geqslant 0} \sup_{x \in B} \| I - A_k^{-1} f'(x) \| \leqslant \beta < 1.$$

For every integer k, one can write

$$A_k = A(I + A^{-1}(A_k - A)) \quad \text{with} \quad \| A^{-1}(A_k - A) \| \leqslant \lambda < 1$$

(the hypothesis $\lambda < \tfrac{1}{2}$ is not, then, made 'full' use of at this stage). Hence, the operators A_k are isomorphisms from X onto Y and, moreover,

$$\| A_k^{-1} \| \leqslant \| (I + A^{-1}(A_k - A))^{-1} \| \, \| A^{-1} \| \leqslant \frac{\| A^{-1} \|}{1 - \lambda},$$

which shows that ($\lambda < \tfrac{1}{2}$ by hypothesis)

$$\| I - A_k^{-1} A \| \leqslant \| A_k^{-1} \| \, \| A_k - A \| \leqslant \frac{\lambda}{1 - \lambda} \overset{\text{def}}{=} \beta' < 1.$$

Let δ then be such that

$$\beta' < \beta' + \delta \overset{\text{def}}{=} \beta < 1.$$

The inequality

$$\| I - A_k^{-1} f'(x) \| \leqslant \| I - A_k^{-1} A \| + \| A_k^{-1}(A - f'(x)) \|,$$

together with the previous inequalities and the continuity of the derivative f' (recall that $A = f'(a)$), proves the existence of a number $r > 0$ such that

$$B \overset{\text{def}}{=} \{x \in X: \|x - a\| \leqslant r\} \subset \Omega,$$

$$\sup_{k \geqslant 0} \sup_{x \in B} \|A_k^{-1}(A - f'(x))\| \leqslant \delta,$$

and the assertion is proved.

(ii) Let x_0 now be any point of the ball B and let $(x_k)_{k \geqslant 0}$ be the sequence defined by

$$x_{k+1} = x_k - A_k^{-1} f(x_k), \quad k \geqslant 0$$

(the inequality established above shows, in particular, that all the points x_k lie within the ball B; so that the sequence is well defined). As it is also possible to write $(f(a) = 0)$

$$x_{k+1} - a = x_k - A_k^{-1} f(x_k) - (a - A_k^{-1} f(a)),$$

the mean value theorem applied to the function $x \rightarrow x - A_k^{-1} f(x)$ shows that

$$\|x_{k+1} - a\| \leqslant \sup_{x \in B} \|I - A_k^{-1} f'(x)\| \|x_k - a\| \leqslant \beta \|x_k - a\|,$$

by part (i) of the proof, which establishes the geometric convergence of the sequence (x_k) to the point a.

(iii) Let b be another zero of f in the ball B. The sequence (x_k) corresponding to $x_0 = b$ is a stationary sequence, since

$$x_1 = x_0 - A_0^{-1} f(x_0) = x_0,$$

and, on the other hand, it converges to the point a by the preceding. It follows that $a = b$. $\qquad \square$

It remains to 'translate' the preceding results to the case of *finding the zeros of the derivative* $J': \Omega \subset V \rightarrow V'$ of a function $J: \Omega \subset V \rightarrow \mathbb{R}$. In this way, the following corollaries of the two preceding theorems are obtained (recall that it is possible to identify the space $\mathscr{L}_2(V; \mathbb{R})$ and $\mathscr{L}(V; V')$).

Theorem 7.5-3

Let Ω be an open set of a complete space V and let $J: \Omega \subset V \rightarrow \mathbb{R}$ be a function which is twice differentiable in the open set Ω. Suppose, moreover, that there exist three constants r, M and β such that

$$r > 0 \quad and \quad B \overset{\text{def}}{=} \{v \in V: \|v - u_0\| \leqslant r\} \subset \Omega,$$

$$A_k(v) \in \text{Isom}(V; V') \quad for \; every \; v \in B \quad and$$

$$\sup_{k \geqslant 0} \sup_{v \in B} \|A_k^{-1}(v)\|_{\mathscr{L}(V'; V)} \leqslant M,$$

$$\sup_{k \geqslant 0} \sup_{v,v' \in B} \| J''(v) - A_k(v') \|_{\mathscr{L}(V;V')} \leqslant \frac{\beta}{M}, \quad \text{and} \quad \beta < 1,$$

$$\| J'(u_0) \|_{V'} \leqslant \frac{r}{M}(1 - \beta).$$

Then the sequence $(u_k)_{k \geqslant 0}$ *defined by*

$$u_{k+1} = u_k - A_k^{-1}(u_{k'})J'(u_k), \quad k \geqslant k' \geqslant 0,$$

is entirely contained within the ball B and converges to a zero of J', which is the only zero of J' in the ball B. Lastly, the convergence is geometric.

Theorem 7.5-4

Let Ω *be an open subset of a complete space V and let* $J : \Omega \subset V \to \mathbb{R}$ *be a function which is twice continuously differentiable in* Ω. *Moreover, let u be a point of* Ω *such that*

$$\begin{cases} J'(u) = 0, \quad J''(u) \in \text{Isom } \mathscr{L}(V;V'), \\ \sup_{k \geqslant 0} \| A_k - J''(u) \|_{\mathscr{L}(V;V')} \leqslant \dfrac{\lambda}{\| (J''(u))^{-1} \|_{\mathscr{L}(V';V)}} \quad \text{and} \quad \lambda < \dfrac{1}{2}. \end{cases}$$

Then there exists a closed ball $B \subset V$ *with centre at u such that, for every point* $u_0 \in B$, *the sequence* $(u_k)_{k \geqslant 0}$ *defined by*

$$u_{k+1} = u_k - A_k^{-1}J'(u_k), \quad k \geqslant 0,$$

is entirely contained in B and converges to the point u, which is, moreover, the only zero of J' in the ball B. Lastly, the convergence is geometric.

For the case $V = \mathbb{R}^n$, the *generalised Newton method* of Theorem 7.5-3 takes the form

$$u_{k+1} = u_k - A_k^{-1}(u_{k'})\nabla J(u_k), \quad k \geqslant k' \geqslant 0,$$

where the elements $A_k(u_{k'})$ are *invertible matrices* of order n (they may, of course, depend both on the function J and its derivatives), and where $\nabla J(u_k)$ denotes the gradient vector of the function J at the point u_k (here V' is identified with $\mathbb{R}^n$). In particular, the 'original' *Newton method* corresponds to

$$u_{k+1} = u_k - \{(\nabla^2 J(u_k)\}^{-1}\nabla J(u_k), \quad k \geqslant 0,$$

where the matrix $\nabla^2 J(u_k)$ is the Hessian of the function J at the point a.

Furthermore, there is a remarkable fact which may be observed. *All the methods of the following chapter* (which are arrived at by paths which are altogether different) *may be subsumed as so many particular cases of the generalised Newton methods*; for example,

$A_k(u_{k^.}) = \varrho^{-1} I$ (the gradient method with fixed parameter),
$A_k(u_{k^.}) = -\varrho_k^{-1} I$ (the gradient method with variable parameter),
$A_k(u_{k^.}) = -(\varrho(u_k))^{-1} I$ (the gradient method with optimal parameter),

where the number $\varrho(u_k)$ (if it exists) is determined from the condition

$$J(u_k - \varrho(u_k) \nabla J(u_k)) = \inf_{\varrho \in \mathbf{R}} J(u_k - \varrho \nabla J(u_k)).$$

Remark

Although the results on convergence of Theorems 7.5-3 and 7.5-4 are established under conditions which are appreciably restrictive (the vector u_0 needs to be sufficiently close to a zero, etc.), it should, of course, be laid to their credit that they do not require any hypotheses such as the convexity, ellipticity or coerciveness of the function J (these last two concepts will be defined further on), hypotheses which are used in an essential manner in the following chapter. □

Exercises

7.5-1. *The calculation of a square root* $\sqrt{\alpha}$, $\alpha > 0$, in a computer is carried out by applying Newton's method to the function $f(x) = x^2 - \alpha$, the method being here defined by the sequence

$$x_{k+1} = \frac{1}{2}\left(x_k + \frac{\alpha}{x_k}\right), \quad k \geqslant 0.$$

Examine how the convergence of this sequence depends on the initial value x_0.

7.5-2. Prove that Newton's method (assuming it to be well-defined), is 'invariant under isomorphisms', in the following sense. If the function f is replaced by the composite function gf, where $g \in \text{Isom}(Y; Z)$, Z being any normed vector space, the sequence $(x_k)_{k \geqslant 0}$ remains unchanged.

7.5-3. One variant of Newton's method for the solution of non-linear systems of equations is the *Gauss–Seidel method* (for these same systems) which takes the following form. The transition from the vector x_k with components x_i^k, $1 \leqslant i \leqslant n$, to the vector x_{k+1} with components x_i^{k+1}, $1 \leqslant i \leqslant n$, is carried out by setting

$$x_1^{k+1} = x_1^k - \frac{f_1(x_1^k, x_2^k, \ldots, x_n^k)}{\partial_1 f_1(x_1^k, x_2^k, \ldots, x_n^k)},$$

$$x_2^{k+1} = x_2^k - \frac{f_2(x_1^{k+1}, x_2^k, \ldots, x_n^k)}{\partial_2 f_2(x_1^{k+1}, x_2^k, \ldots, x_n^k)},$$

$$\vdots$$

$$x_n^{k+1} = x_n^k - \frac{f_n(x_1^{k+1}, \ldots, x_{n-1}^{k+1}, x_n^k)}{\partial_n f_n(x_1^{k+1}, \ldots, x_{n-1}^{k+1}, x_n^k)},$$

where $\partial_i = \partial/\partial x_i$.

(1) Show that, if the functions f_i are affine, $f_i(x) = \sum_{j=1}^{n} a_{ij}x_j - b_i$, then this method is just the *Gauss–Seidel iterative method for the solution of linear systems*, as described in section 5.2.

(2) Write down this method in the form of a generalised Newton method.

(3) State sufficient conditions, in terms of the functions f_i in a neighbourhood of a point $x_0 \in \mathbb{R}^n$, which guarantee the convergence of the method.

7.5-4. Theorem 7.5-1 is applicable, in particular, to the 'original' Newton method

$$x_{k+1} = x_k - (f'(x_k))^{-1}f(x_k),$$

which corresponds to the choice $A_k(x_k \cdot) = f'(x_k)$. In this instance, it is possible to prove a 'stronger' result, known as the *Newton–Kantorovich theorem*.

Assume that the space X is complete and the function $f : \Omega \subset X \to Y$ is differentiable in the open set Ω. Furthermore, assume that there exist three constants λ, μ and ν and a point $x_0 \in \Omega$ such that

$$0 < \lambda \mu \nu \leqslant \frac{1}{2},$$

$$\bar{B} \subset \Omega, \quad where \quad B \overset{\text{def}}{=} \left\{ x \in X : \|x - x_0\| < \rho^- \overset{\text{def}}{=} \frac{1 - \sqrt{(1 - 2\lambda\mu\nu)}}{\mu\nu} \right\},$$

$$f'(x_0) \in \text{Isom}(X; Y) \quad and \quad \|(f'(x_0))^{-1}f(x_0)\| \leqslant \lambda,$$

$$\|(f'(x_0))^{-1}\| \leqslant \mu,$$

$$\sup_{x, y \in \Omega^+} \|f'(x) - f'(y)\| \leqslant \nu \|x - y\|,$$

where

$$\Omega^+ = \left\{ x \in \Omega : \|x - x_0\| < \rho^+ \overset{\text{def}}{=} \frac{1 + \sqrt{(1 - 2\lambda\mu\nu)}}{\mu\nu} \right\}.$$

Then

$$f'(x) \in \text{Isom}(X; Y) \quad for \; every \; x \in B,$$

and the sequence defined by

$$x_{k+1} = x_k - (f'(x_k))^{-1}f(x_k), \quad k \geqslant 0,$$

is entirely contained within the ball B and converges to a zero of f, which is, moreover, the only zero of f in the open set Ω^+. Finally, setting $\theta = \rho^- / \rho^+$, we have the following bounds:

$$\|x_k - a\| \leqslant 2 \frac{\sqrt{(1 - 2\lambda\mu\nu)}}{\lambda\mu\nu} \frac{\theta^{2k}}{1 - \theta^{2k}} \|x_1 - x_0\| \quad if \quad \lambda\mu\nu < \frac{1}{2},$$

$$\|x_k - a\| \leqslant \frac{\|x_1 - x_0\|}{2^{k-1}} \quad if \quad \lambda\mu\nu = \frac{1}{2},$$

$$\frac{2\|x_{k+1} - x_k\|}{1 + \sqrt{[1 + 4\theta^{2k}(1 + \theta^{2k})^{-2}]}} \leqslant \|x_k - a\| \leqslant \theta^{2k-1} \|x_k - x_{k-1}\|.$$

It should be noted, in passing, that the last two inequalities provide an estimate of the error $\|x_k - a\|$ in terms of quantities which are all calculable.

(1) Prove the Newton–Kantorovich theorem, taking as inspiration the proofs of Theorems 7.5-1 and 7.5-2.

(2) Prove that there exists a ball Q, with centre at a, such that, for every $x_0 \in Q$, it is possible to apply the Newton–Kantorovich theorem.

7.5-5. (1) Prove that the system of non-linear equations

$$\begin{cases} -5x_1 & + 2\sin x_1 + \cos x_2 = 0, \\ 4\cos x_1 & + 2\sin x_2 - 5x_2 = 0, \end{cases}$$

has a unique solution, which can be approximated by the method of successive approximations.

(2) Examine the convergence of Newton's method applied to this system.

7.5-6. (1) Verify that the calculation of the inverse of a non-zero scalar α by Newton's method corresponds to the iterative method

$$x_{k+1} = x_k(2 - \alpha x_k), \quad k \geq 0.$$

Examine how the convergence of this sequence depends on the initial value x_0.

(2) Construct, in a similar way, an iterative method for the approximation of the inverse of an invertible matrix A, which takes the form

$$X_{k+1} = \text{function}\,(X_k, A), \quad k \geq 0;\ X_0 \text{ an arbitrary matrix.}$$

For this, inspiration may be drawn from Exercise 7.1-3. (Prove that a necessary and sufficient condition for the convergence of this method is $\rho(I - AX_0) < 1$.)

(3) Assuming the matrix A to be symmetric and positive definite and taking its spectral radius as known, what is the simplest way of choosing the matrix X_0 in verifying the preceding condition?

(4) Repeat the analysis for the iterative method

$$x_{k+1} = \frac{x_k}{3}(4 - \alpha^3 x_k^3), \quad k \geq 0.$$

General results on optimisation. Some algorithms

Introduction

We begin by recalling, together with some of its consequences, the *projection theorem in Hilbert spaces* (Theorem 8.1-1), of which constant use is made in Optimisation. It will allow us, in particular, to establish a certain number of results immediately for the infinite-dimensional case.

The principal object in *Optimisation* is the construction of *algorithms* which make possible the approximation of a solution of a problem of the form: find u such that

$$(P) \qquad u \in U \quad \text{and} \quad J(u) = \inf_{v \in U} J(v),$$

where U is a given subset of a vector space V and $J: V \to \mathbb{R}$ is a given function. That is why it is natural to be concerned with, *and in this order*, the following questions.

(i) Results concerning the *existence* and *uniqueness* of a solution of the problem (P); this is one of the aims in section 8.2, where we prove a number of results concerning existence, appropriate to the particular situation under consideration (finite or infinite dimension, quadratic or 'elliptic' functionals, etc.).

(ii) The *characterisation* of the possible solutions of the problem (P), that is to say, the necessary, and, in certain cases, sufficient conditions for an element $u \in U$ to be a solution of the problem (P); most of these characterisations were established in the previous chapter, where, in fact, this was one of the objectives.

The *necessary* conditions generally make use of the first derivative of the function J; this is so, for example, in the case of *Euler's equation* $J'(u) = 0$ with $U = V$ or of *Euler's inequalities* (section 7.4) with convex set U. When the set U has one of the following forms (which are of fundamental importance in actual applications)

$$U = \{v \in V : \varphi_i(v) = 0, \ 1 \leqslant i \leqslant m\} \quad \text{or} \quad U = \{v \in V : \varphi_i(v) \leqslant 0, \ 1 \leqslant i \leqslant m\},$$

these characterisations also make use of the first derivatives of the functions

φ_i, through the intermediary of *Lagrange multipliers* in the first case (section 7.2), or through the intermediary of the *Kuhn–Tucker conditions* (which will be established in the following chapter) in the second.

Some of the preceding conditions become *sufficient* as well, through the introduction of the hypothesis of the *convexity* of the functional (as was indicated in section 7.4). The sufficient conditions which make use of the second derivatives (encountered in section 7.3) are rarely used.

(iii) The effective construction of *algorithms* which make possible the approximation of a solution u of the problem (P), that is to say, the construction of a sequence $(u_k)_{k \geqslant 0}$ of elements of the set U such that $\lim_{k \to \infty} u_k = u$, the vector u_0 being chosen arbitrarily; it should be noted that, *in every case*, the definition and the analysis of these algorithms will utilise, in an essential way, the characterisations of the solutions, which we shall mention explicitly whenever these are used.

After a number of *terminological* observations (linear and non-linear programming problems, convex problems, with or without constraints, etc.; cf. section 8.2) and some indications on the *origin* of optimisation problems (section 8.3), the rest of the chapter is given to *the description and analysis of the 'basic' algorithms of Optimisation*.

For *unconstrained* problems $(U = V)$, we shall examine, in order, *relaxation* methods, *gradient* methods *with optimal parameter, with fixed parameter, with variable parameter* (section 8.4) and the *conjugate gradient* method (section 8.5). For *constrained* problems, we shall examine, in order, the *relaxation* and *projected-gradient methods* (section 8.6). Their *practical* use is limited to certain *special* sets U, which are of the form

$$U = \prod_{i=1}^{n} [a_i, b_i] \subset V = \mathbb{R}^n.$$

We give sufficient conditions of *convergence* for each of these methods, most often for the case of *elliptic* functionals, which form a natural generalisation of quadratic functionals with symmetric, positive definite matrix.

Applied to the unconstrained problem $(U = V = \mathbb{R}^n)$ associated with such a quadratic functional, the algorithms examined provide, at the same time, methods for the solution of the associated linear system, one of which, in fact, coincides with the Gauss–Seidel iterative method encountered in Chapter 5. In this spirit, it will be observed that, except for the conjugate gradient method, none of these methods is direct.

'General' constrained problems are much more difficult to treat than the unconstrained problems. In fact, every effort is often made to solve them by replacing them by a *sequence* of unconstrained problems, or by constraints which are easy to treat (for example, $U = \mathbb{R}^n_+$). This idea lies behind the use of *penalty-function methods*, which are described briefly in section 8.6; it also lies behind methods which rest on *duality*, as studied in Chapter 9. For the particular case (very important in practice) of a *linear* functional associated with constraints which are also *linear*, we mention finally the *simplex method*, which forms the subject-matter of Chapter 10.

In the present chapter, *all the vector spaces considered are real*. Lastly, the Euclidean norm in $\mathbb{R}^n$ will be denoted, from here on, either by $\|\cdot\|_n$ if it is desired to allude explicitly to the dimension involved, or simply by $\|\cdot\|$ if there is no fear of ambiguity.

8.1 The projection theorem; some consequences

Let V be a vector space over the field $\mathbb{R}$. A *scalar product* over V is a function $(\cdot, \cdot): V \times V \to \mathbb{R}$ which is *bilinear, symmetric and positive definite*; that is to say, it satisfies

$$(u, \cdot): V \to \mathbb{R} \quad \text{is linear for every } u \in V,$$
$$(\cdot, v): V \to \mathbb{R} \quad \text{is linear for every } v \in V,$$
$$(u, v) = (v, u) \quad \text{for every } u, v \in V,$$
$$(v, v) = 0 \Leftrightarrow v = 0, \quad \text{and} \quad (v, v) \geqslant 0 \quad \text{for every } v \in V.$$

A space which is provided with a scalar product is called a *pre-Hilbert space*. Since the function $\|\cdot\|$ defined by

$$\|v\| = \sqrt{(v, v)} \quad \text{for every } v \in V,$$

is a norm over the vector space V, *a pre-Hilbert space is always considered to be provided with this norm*, which also makes of it a normed vector space. If it is complete in this norm, then it is a *Hilbert space*. Since every finite-dimensional normed vector space is complete, the space $\mathbb{R}^n$ equipped with the Euclidean scalar product is an example of a Hilbert space.

We note, in passing, the Schwarz inequality

$$|(u, v)| \leqslant \|u\| \|v\| \quad \text{for every } u, v \in V,$$

which is especially useful in the proof of the triangle inequality for the norm associated with the scalar product. The Cauchy–Schwarz inequality for the Euclidean scalar product (section 1.4) or the *Cauchy–Schwarz inequality for functions*,

$$\left| \int_I uv \, dx \right| \leqslant \left(\int_I |u|^2 \, dx \right)^{1/2} \left(\int_I |v|^2 \, dx \right)^{1/2}, \quad I \subset \mathbb{R},$$

are particular cases of it. It should be observed that the Schwarz inequality implies the *continuity* of the scalar product, considered as a function from the product $V \times V$ into $\mathbb{R}$. Finally, it should be remembered that the inequality becomes an equality if and only if the two vectors which appear are linearly dependent.

The result which follows is of paramount importance.

Theorem 8.1-1 (*The projection theorem*)
Let U be a non-empty, convex, closed subset of a Hilbert space V. Given any element $w \in V$, there exists a unique element Pw such that

(1) $$Pw \in U \quad and \quad \|w - Pw\| = \inf_{v \in U} \|w - v\|.$$

This element $Pw \in U$ satisfies

(2) $$(Pw - w, v - Pw) \geqslant 0 \quad for \ every \ v \in U,$$

and, conversely, if an element u satisfies

$$u \in U \quad and \quad (u - w, v - u) \geqslant 0 \quad for \ every \ v \in U,$$

then $u = Pw$.

The function $P: V \to U$ defined in this way is such that

(3) $$\|Pw_1 - Pw_2\| \leqslant \|w_1 - w_2\| \quad for \ every \ w_1, w_2 \in V.$$

Finally, the function $P: V \to U \subset V$ is linear if and only if the subset U is a vector subspace, in which case the inequalities (2) are replaced by the equalities

(4) $$(Pw - w, v) = 0 \quad for \ every \ v \in U. \qquad \square$$

We make a number of comments on this result.

(i) The function $P: V \to U$ is called the *projection operator*, and the element Pw is called the *projection* (onto the set U) *of the element* w; the geometric interpretation of the defining relation (1) is, in this respect, quite clear (figure 8.1-1): the 'projected' element Pw is, in fact, the 'nearest'

Figure 8.1-1

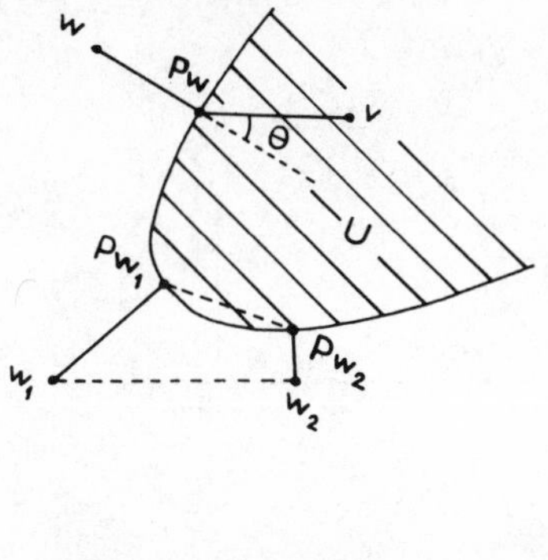

element in the set U to the point w. In a similar way, the inequalities (2) express the necessity, intuitively evident, for the angle formed by the vectors $Pw - w$ and $v - Pw$ to be $\leqslant \pi/2$ for all elements $v \in U$ (figure 8.1-1). Observe, in passing, that

$$w - Pw = 0 \Leftrightarrow w \in U.$$

(ii) The attentive reader will be quick to compare the inequalities (2) with Euler's inequalities '$J'(u)(v - u) \geqslant 0$ for every $v \in U$' of Theorem 7.4-4. In actual fact, their similarity is no coincidence. Consider, in fact, for fixed $w \in V$, the function

$$J : v \in V \to J(v) = \frac{1}{2} \| w - v \|^2 = \frac{1}{2} (v, v) - (w, v) + \frac{1}{2} (w, w).$$

This function is differentiable, with

$$J'(v)z = (v - w, z) \quad \text{for every } v, z \in V,$$

and (strictly) convex (Theorem 7.4-3). Consequently, inequalities (2) do no more than express the necessary and sufficient condition (3) of Theorem 7.4-4, rewritten for the point $u = Pw$.

(iii) Inequality (3) implies, in particular, the *continuity* of the projection operator. This is at the back of the mind when, on occasion, the assertion is made, in somewhat pictorial language, that 'projection does not increase distances' (figure 8.1-1).

(iv) Condition (4) expresses the orthogonality (in a sense to be defined later on) of the vector $Pw - w$ to the vectors of the set U, when this is a vector space. Its geometric interpretation is once again evident (figure 8.1-2).

The function $u \in V \to u_h \in V_h \subset V$ encountered in the variational approximation of boundary-value problems (sections 3.4 and 3.5) is an *example of a projection operator*, which is *linear*, since V_h is a vector subspace. Adopting, once again, the notation used in those two sections, it may be verified that the function $a(\cdot, \cdot)$ is actually a scalar product over the space V. Consequently, the equations (used in the proof of Theorem 3.4-2)

$$a(u - u_h, w_h) = 0 \quad \text{for every } w_h \in V_h,$$

Figure 8.1-2

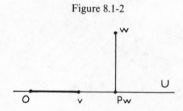

show that the *approximate solution $u_h \in V_h$ is nothing other than the projection of the 'exact' solution u onto the subspace V_h, in respect of the scalar product* $a(\cdot, \cdot)$.

Another *example of a projection operator*, but this time *non-linear*, corresponds to the situation

$$V = \mathbb{R}^n \quad \text{equipped with the Euclidean scalar product,}$$

$$U = \mathbb{R}^n_+ \overset{\text{def}}{=} \{v \in \mathbb{R}^n : v \geqslant 0\},$$

the set U being sometimes called the *positive hyperoctant* (we recall that the notation $v \geqslant 0$, introduced earlier, indicates that all the components of the vector v are non-negative). It is more or less evident geometrically that the corresponding projection operator is defined by

$$(Pw)_i = \max\{w_i, 0\}, \quad 1 \leqslant i \leqslant n,$$

as is suggested by an examination of all 'possible cases' in two dimensions (figure 8.1-3). In order to prove the fact, it is enough to verify the necessary and sufficient condition of Theorem 8.1-1; now, given any element $v = (v_i)^n_{i=1}$ of the set U, the preceding definition of the element Pw actually implies that

$$(Pw - w, v - Pw) = \sum_{i=1}^{n} ((Pw)_i - w_i)(v_i - (Pw)_i) = - \sum_{i, w_i < 0} w_i v_i \geqslant 0.$$

The extension to sets which are of the form

$$U = \prod_{i=1}^{n} [a_i, b_i] = \{v = (v_i)^n_{i=1} \in \mathbb{R}^n : \quad a_i \leqslant v_i \leqslant b_i, \quad 1 \leqslant i \leqslant n\} \subset \mathbb{R}^n,$$

not excluding the case where $a_i = -\infty$ and/or $b_i = +\infty$, offers no

Figure 8.1-3

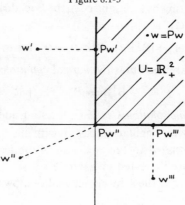

particular difficulty. The reader should verify, by similar reasoning, that the corresponding projection operator is given by

$$(Pw)_i = \min\{\max\{w_i, a_i\}, b_i\} = \begin{cases} a_i & \text{if } w_i < a_i, \\ w_i & \text{if } a_i \leqslant w_i \leqslant b_i, \\ b_i & \text{if } b_i < w_i. \end{cases}$$

As a first application of the projection theorem, we return to the problem of the *least squares solution of a linear system*: find $u \in \mathbb{R}^n$ such that

$$\|\mathbf{B}u - c\|_m = \inf_{v \in \mathbb{R}^n} \|\mathbf{B}v - c\|_m,$$

where the matrix $\mathbf{B} \in \mathscr{A}_{m,n}(\mathbb{R})$ and the vector $c \in \mathbb{R}^m$ are given, and $\|\cdot\|_m$ denotes the Euclidean norm in $\mathbb{R}^m$. The vector subspace

$$\text{Im}(\mathbf{B}) = \{\mathbf{B}v \in \mathbb{R}^m : v \in \mathbb{R}^n\}$$

being closed (the situation is finite-dimensional), the projection theorem implies the existence and uniqueness of an element $\tilde{u}$ satisfying

$$\tilde{u} \in \text{Im}(\mathbf{B}) \quad \text{and} \quad \|\tilde{u} - c\|_m = \inf_{\tilde{v} \in \text{Im}(\mathbf{B})} \|\tilde{v} - c\|_m.$$

Consequently, the problem set always has at least one solution, namely, some element $u \in \mathbb{R}^n$ which satisfies

$$\mathbf{B}u = \tilde{u}.$$

This solution is unique if and only if the function represented by the matrix $\mathbf{B}$ is injective (which is possible only if $m \geqslant n$), in other words, if and only if the symmetric, non-negative definite matrix $\mathbf{B}^\mathrm{T}\mathbf{B}$ is positive definite, or again, if and only if $r(\mathbf{B}) = n$.

In this same spirit, the characterisation (4) of Theorem 8.1-1, namely,

$$(\tilde{u} - c, \tilde{v})_m = 0 \quad \text{for every } \tilde{v} \in \text{Im}(\mathbf{B}),$$

may be written, denoting by $(\cdot, \cdot)_m$ and $(\cdot, \cdot)_n$ the Euclidean scalar products of $\mathbb{R}^m$ and $\mathbb{R}^n$ respectively, as

$$(\mathbf{B}u - c, \mathbf{B}v)_m = (\mathbf{B}^\mathrm{T}\mathbf{B}u - \mathbf{B}^\mathrm{T}c, v)_n = 0 \quad \text{for every } v \in \mathbb{R}^n.$$

We have thus established that the *normal equations*

$$\mathbf{B}^\mathrm{T}\mathbf{B}u = \mathbf{B}^\mathrm{T}c$$

always have at least one solution. Certain interesting additional results (such as the connection with singular values or with the 'pseudo-inverse' of a matrix), which are related to this question of the least squares solution of a linear system, are indicated in Exercise 8.1-3.

Given an element $u \in V$, the Schwarz inequality shows that the linear form

$$(u, \cdot): v \in V \to (u, v) \in \mathbb{R}$$

is continuous. What is remarkable is that the converse is true in a complete space: every linear form, continuous over a Hilbert space, may be 'represented' by an element of the space, as is shown by the following result (whose proof depends in an essential way on the projection theorem).

Theorem 8.1-2 (*The Riesz representation theorem*)
Let V be a Hilbert space and f any element of the dual space V' of V. Then there exists a unique element $\tau f \in V$ such that

$$f(v) = (\tau f, v) \quad \text{for every } v \in V.$$

The function $\tau: V' \to V$ defined in this way is linear and is an isometry:

$$\|\tau f\|_V = \|f\|_{V'} \quad \text{for every } f \in V'. \qquad \square$$

The function τ is called the *Riesz canonical isometry*. An immediate application of the Riesz representation theorem is to the extension of the notion of gradient. In fact, if $J: V \to \mathbb{R}$ is a function which is differentiable at a point u of a Hilbert space V, its derivative $J'(u)$ is, by definition, an element of the dual space V'. Consequently, there exists a unique element of the space V, denoted by $\nabla J(u)$ and called the *gradient* of the function J at the point u, such that

$$J'(u)v = (\nabla J(u), v) \quad \text{for every } v \in V.$$

As in the finite-dimensional case, this vector *depends* on the scalar product chosen.

In the same way, it is possible to associate with the second derivative $J''(u) \in \mathscr{L}(V; V')$ an element, denoted by $\nabla^2 J(u)$, of the space $\mathscr{L}(V)$ such that

$$J''(u)(v, w) = (\nabla^2 J(u)v, w) \quad \text{for every } v, w \in V.$$

Two vectors u and v of a pre-Hilbert space are *orthogonal* if $(u, v) = 0$. If U is any subset of a pre-Hilbert space V, the *orthogonal complement* of U is the set

$$U^\perp \stackrel{\text{def}}{=} \{v \in V : (u, v) = 0 \quad \text{for every } u \in U\}.$$

It is easy to see that the set $U^\perp$ is always a closed vector subspace. In case U is also a closed vector subspace and the space is complete, it is possible, by means of the projection theorem, to prove the following result.

Theorem 8.1-3
Let U be a closed vector subspace of a Hilbert space V. Then the space V is the direct sum of the subspace and its orthogonal complement:

$$V = U \oplus U^\perp. \qquad \square$$

In other words, every element $w \in V$ may be written in a unique way as

$$w = u + u', \quad \text{with} \quad u \in U, \quad u' \in U^\perp.$$

More precisely, $u = Pw$ and $u' = P'w$, where P and $P' = I - P$ denote respectively the projection operators onto U and $U^\perp$.

Given two Hilbert spaces V and W, equipped with scalar products $(\cdot, \cdot)_V$ and $(\cdot, \cdot)_W$, the Riesz representation theorem makes possible the association, with every operator $A \in \mathscr{L}(V; W)$, of *the transpose operator* $A^T \in \mathscr{L}(W; V)$ defined by

$$(Au, w)_W = (v, A^T w)_V \quad \text{for every } v \in V, w \in W.$$

Of course, the usual definition of the transpose of a matrix is recovered when the spaces V and W are finite-dimensional and are equipped with the Euclidean scalar product. From the definition given above and Theorem 8.1-3, there follow the relations

$$V = \text{Ker}(A) \oplus \overline{\text{Im}(A^T)}, \quad W = \text{Ker}(A^T) \oplus \overline{\text{Im}(A)},$$

where the usual terminology has been employed:

$$\text{Ker}(A) = \{v \in V : Av = 0\}, \quad \text{Im}(A) = \{Av \in W : v \in V\},$$

for the *kernel* and the *image*, respectively, of the linear map A.

We shall make use of these relations in the particular case where the two spaces V and W are *finite-dimensional*, in which case the subspaces $\text{Im}(A)$ and $\text{Im}(A^T)$ are always closed. These relations sometimes go by the name *the finite-dimensional Fredholm alternative*, because of the consequences which follow from them in the solution of a linear system whose matrix is not necessarily square, namely, the following.

Let V and W be two finite-dimensional spaces, A a linear map from V to W and b a vector of W. Then only one of the following alternatives holds:

(i) *either the linear system $Av = b$ has at least one solution*;
(ii) *or the linear system $Av = b$ has no solution, and there exists at least one vector $w \in W$ such that $A^T w = 0$ and $(w, b) \neq 0$* (for example, the projection of the vector b onto the kernel of the transpose map A^T).

Exercises

8.1-1. Adopting the same notation as in Theorem 8.1-1, establish the differentiability of the function

$$f : v \in V \to \|v - Pv\|^2,$$

and calculate the number $f'(v)w$ for every $v, w \in V$.

Remark. If the set U is a vector subspace, the projection operator is linear and continuous, and hence differentiable, so that the required result is obtained without any difficulty. By contrast, the projection operator *is not* differentiable in the general case (consider, for example, the case where $V = \mathbb{R}$, $U = [0, 1]$), so that there is no question of applying Theorem 7.1-1 (regarding the differentiability of a composite function). Accordingly, we are here dealing with an interesting example of a differentiable composite function, one of whose components is not itself differentiable.

8.1-2. Let v be a real vector satisfying $v^T v = 1$. Show that the matrix $I - vv^T$ represents a projection operator. What geometric property is deducible about the Householder matrix $H(v) = I - 2vv^T$ (cf. Exercise 4.5-1)?

8.1-3. Consider the problem of the least squares solution of a linear system,

$$\text{find } u \in \mathbb{R}^n \text{ such that}$$

$$\| Bu - c \|_m = \inf_{v \in \mathbb{R}^n} \| Bv - c \|_m,$$

where the matrix $B \in \mathscr{A}_{m,n}(\mathbb{R})$ and the vector $c \in \mathbb{R}^m$ are given and $\| \cdot \|_m$ denotes the Euclidean norm in $\mathbb{R}^m$

(1) Show that there exists a unique vector $s \in \mathbb{R}^n$ such that

$$s \in S \overset{\text{def}}{=} \{u \in \mathbb{R}^n : B^T Bu = B^T c\}, \quad \| s \|_n = \inf_{v \in S} \| v \|_n.$$

It should be noted in passing that only one of two cases can obtain:

(i) the set S contains one element; this is so if, and only if, $r(B) = n$ (which implies that $m \geqslant n$);

(ii) the set S contains an infinity of elements; this is so if, $r(B) < n$, which is realised whenever $m < n$, the most 'common' case being that where $r(B) = m < n$; in that case, the linear system $Bu = c$ has an infinity of solutions (we sometimes speak of its being 'underdetermined') and the definition of the vector s then makes it possible to distinguish one solution from among all those which are possible.

(2) Prove that the vector s is characterised by the relation

$$s \in S \cap [\text{Ker}(B^T B)]^{\perp}.$$

(3) Prove that the function $c \in \mathbb{R}^m \to s \in \mathbb{R}^n$ so defined is linear (the matrix B is fixed). To achieve this, it is advantageous to express this function in terms of the eigenvalues and eigenvectors of the matrix $B^T B$ (we recover, in this way, the singular values of the matrix B; refer to Theorem 1.2-2 and to Exercise 1.2-3). The matrix representing this function is called the *pseudo-inverse of the matrix* B, or again, the *Moore–Penrose inverse*; it is denoted by B^+.

(4) Prove the following relations

$$(B^+)^+ = B; \quad (B^T)^+ = (B^+)^T; \quad B = BB^+B; \quad B^+ = B^+BB^+; \quad (B^+B)^T = B^+B.$$

(5) Explain how to determine the vector s in the particular case where the matrix B is of the form $B = \text{diag}(\tilde{B}, 0)$, where

$$\tilde{B} = \begin{pmatrix} \begin{pmatrix} b_{11} & b_{12} & \cdots & b_{1r} \\ & b_{22} & \cdots & b_{2r} \\ & & \ddots & \vdots \\ & & & b_{rr} \end{pmatrix} \end{pmatrix}, \quad b_{ii} \neq 0, \quad 1 \leqslant i \leqslant r < n.$$

(6) If B is a general matrix, prove that there exist two orthogonal matrices P and Q such that the matrix PBQ is of the form given above.

(7) What is the effect on the determination of the vector s, if the matrix B is general?

(8) For $\varepsilon > 0$, show that there exists a unique vector $u_\varepsilon \in \mathbb{R}^n$ such that

$$J_\varepsilon(u_\varepsilon) = \inf_{v \in \mathbb{R}^n} J_\varepsilon(v), \quad \text{where } J_\varepsilon(v) = \| Bv - c \|_m^2 + \varepsilon \| v \|_n^2.$$

(9) If u is any element of S, show that

$$\begin{cases} B^T B(u_\varepsilon - u) + \varepsilon u_\varepsilon = 0, \\ \| u_\varepsilon - u \|_n \leqslant \| u \|_n, \ \| u_\varepsilon \|_n \leqslant \| u \|_n. \end{cases}$$

(10) Examine the behaviour of the family $(u_\varepsilon)_{\varepsilon > 0}$, as ε tends to 0.

(11) For $\varepsilon > 0$, show that there exists a unique solution of the problem: find $\tilde{u}_\varepsilon \in \mathbb{R}^n$ such that

$$\tilde{J}_\varepsilon(\tilde{u}_\varepsilon) = \inf_{v \in \mathbb{R}^n} \tilde{J}_\varepsilon(v), \quad \text{where } \tilde{J}_\varepsilon(v) = \| v \|_n^2 + \frac{1}{\varepsilon} \| B^T B v - B^T c \|_n^2.$$

(12) Examine the behaviour of the family $(\tilde{u}_\varepsilon)_{\varepsilon > 0}$, as ε tends to 0.

Remark. The method described in (11) and (12) is a particular case of methods known as *penalisation methods*, which will be introduced in section 8.6.

8.1-4. Let V be a pre-Hilbert space over the field $\mathbb{R}$. Prove that the norm associated with the scalar product satisfies the *parallelogram law*

$$\| u + v \|^2 + \| u - v \|^2 = 2(\| u \|^2 + \| v \|^2) \quad \text{for every } u, v \in V,$$

and that, conversely, if a norm $\| \cdot \|$ over a vector space V satisfies the parallelogram law, then the function $(\cdot, \cdot)$: $V \times V \to \mathbb{R}$ defined by

$$2(u, v) \overset{\text{def}}{=} \| u + v \|^2 - \| u \|^2 - \| v \|^2$$

is a scalar product over V.

8.1-5. The aim in this exercise is to examine certain extensions of the *projection theorem* to spaces which are not Hilbert spaces.

(1) Let U be a finite-dimensional vector subspace of a normed vector space V. Given any element $w \in V$, show that there exists at least one element u such that

$$u \in U \quad \text{and} \quad \| w - u \| = \inf_{v \in U} \| w - v \|.$$

(2) We say that a normed vector space V is *strictly convex* if the implication

$$u, v \in V \quad \text{and} \quad \| u \| = \| v \| = \| \tfrac{1}{2}(u + v) \| \Rightarrow u = v$$

is satisfied. Prove that, in this case, the element u of question (1) is defined uniquely by the datum w.

(3) Prove that the space $\mathbb{R}^n$, equipped with the norm $\| \cdot \|_p$, is strictly convex if $1 < p < \infty$, but not for $p = 1$ or $p = \infty$.

(4) We say that a normed vector space V is *uniformly convex* if, for every $\varepsilon > 0$, there exists a number $\delta = \delta(\varepsilon) > 0$ such that

$$u, v \in V \quad \text{and} \quad \|u\| = \|v\| = 1, \quad \|u - v\| \geqslant \varepsilon \Rightarrow \|u + v\| \leqslant 2(1 - \delta).$$

Prove that a pre-Hilbert space is uniformly convex; prove that a uniformly convex space is strictly convex and that the converse is true if the space is finite-dimensional.

(5) Let U be a closed, convex subset of a uniformly convex Banach space V. Given any element $w \in V$, prove that there exists a unique element u such that

$$u \in V \quad \text{and} \quad \|w - u\| = \inf_{v \in U} \|w - v\|.$$

(6) Denote by c_0 the vector space of infinite sequences $u = (u_i)_{i \geqslant 0}$ of real numbers such that $\lim_{i \to \infty} u_i = 0$. Prove that, equipped with the norm $\|u\| = \sup_{i \geqslant 0} |u_i|$, the space c_0 is a Banach space.

(7) Let $(\alpha_i)_{i \geqslant 0}$ be a sequence of real numbers such that

$$\begin{cases} \alpha_i \geqslant 0 & \text{for every } i \geqslant 0, \\ \alpha_i > 0 & \text{for an infinite number of indices } i \geqslant 0, \\ \sum_{i \geqslant 0} \alpha_i < +\infty. \end{cases}$$

Prove that the set

$$U = \{u = (u_i)_{i \geqslant 0} \in c_0 : \sum_{i \geqslant 0} \alpha_i u_i = 0\},$$

is a closed subspace of the space c_0 and that, if w is any element of c_0 which does not belong to U, then there is no point $u \in U$ such that $\|w - u\| = \inf_{v \in U} \|w - v\|$.

(8) What deduction may be drawn from the preceding question?

8.2 General results on optimisation problems

An *optimisation problem* takes the following form. Given a non-empty subset U of a vector space V and a function $J: V \to \mathbb{R}$, a minimum of the function J with respect to the set U is sought, that is to say, an element u which satisfies

(P) $$u \in U \quad \text{and} \quad J(u) = \inf_{v \in U} J(v).$$

Remark

For the purposes of the definition of the problem (P), then, it is enough to know the function J over the set U, but, in practice, it is often known over the entire space V. □

We now focus upon a number of points of terminology, which depend essentially on the nature of the function J, which it is usual to call a *functional* within Optimisation, and of the set U.

A distinction is made between *unconstrained* problems, when $U = V$, and *constrained* problems, otherwise. Among constrained problems, *a very important case in applications* involves sets U of the form

$$U = \{v \in V : \varphi_i(v) \leqslant 0, 1 \leqslant i \leqslant m', \varphi_i(v) = 0, m' + 1 \leqslant i \leqslant m\},$$

the given functions $\varphi_i \colon V \to \mathbb{R}$, $1 \leqslant i \leqslant m$, being called the *constraints* of the problem. If $m' = m$, or if $m' = 0$, one often speaks, by misuse of language, of a problem with '*inequality constraints*', or with '*equality constraints*', respectively.

In the absence of further hypotheses concerning the functions φ_i and J, especially as regards convexity and *a fortiori* linearity, the associated problem (P) is called a *non-linear programming problem*.

Since it is always possible to replace an 'equality constraint' $\varphi_i(v) = 0$ by a pair of 'inequality constraints' $\varphi_i(v) \leqslant 0$ and $-\varphi_i(v) \leqslant 0$, we shall confine ourselves, for the time being, to a consideration of problems with 'inequality constraints', which correspond, as a result, to sets U of the form

$$U = \{v \in V : \varphi_i(v) \leqslant 0, 1 \leqslant i \leqslant m\}.$$

If the functions J and φ_i are convex, one speaks of a *convex programming problem*. It should be observed that, in that case, the set U is convex; in fact,

$$\left.\begin{array}{cc} \varphi_i(u) \leqslant 0, & \varphi_i(v) \leqslant 0, \\ \theta \in [0, 1] \end{array}\right\} \Rightarrow \varphi_i(\theta u + (1 - \theta)v) \leqslant \theta \varphi_i(u) + (1 - \theta)\varphi_i(v) \leqslant 0,$$

and the intersection of convex sets is convex.

Two very important particular cases of convex programming are those of *quadratic programming* and *linear programming*. In a *quadratic programming problem*, the function J is a *quadratic functional over* $V = \mathbb{R}^n$,

$$J \colon v \in \mathbb{R}^n \to J(v) = \frac{1}{2}(Av, v) - (b, v), \quad A = A^T \in \mathscr{A}_n \mathbb{R}), \quad b \in \mathbb{R}^n,$$

the matrix A being assumed to be *positive definite* (which implies the strict convexity of the function J; cf. section 7.4), and the constraints φ_i are *affine* (and hence convex):

$$U = \left\{v \in \mathbb{R}^n \colon \sum_{j=1}^{n} c_{ij} v_j \leqslant d_i, 1 \leqslant i \leqslant m\right\}.$$

In a *linear programming* problem, the function J is a *linear functional* over $V = \mathbb{R}^n$,

$$J(v) = \sum_{i=1}^{n} a_i v_i,$$

and the set U is also of the form

$$U = \left\{ v \in \mathbb{R}^n : \sum_{j=1}^n c_{ij} v_j \leqslant d_i, 1 \leqslant i \leqslant m \right\}.$$

Remark

If the symmetric matrix which appears in the definition of a quadratic functional is only non-negative definite, then it is still convex; it would, therefore, appear to be conceivable to give the name of 'quadratic' programming problem to the corresponding optimisation problem as well. However, by so doing, the impression would be given that linear programming is a particular case of quadratic programming. This would be *grossly inaccurate* on many counts; so much so, in fact, that a separate chapter needs to be devoted specially to linear programming. □

We now examine the *questions of the existence and the uniqueness* of a solution of the problem (P). Whether the dimension is finite or not, the question of the *uniqueness* of an eventual solution is generally settled independently of the question of its existence. Most often, in fact, this is done in terms of the convexity of the set U and the strict convexity of the functional (Theorem 7.4-4).

As regards *existence*, we begin with the *finite-dimensional* case. If U is a closed, bounded subset of $V = \mathbb{R}^n$, and if the function $J : \mathbb{R}^n \to \mathbb{R}$ is continuous, it is clear that the problem (P) has at least one solution. With a view to an easy extension of this result to the case of unbounded sets U (notably the case $U = V = \mathbb{R}^n$), we introduce the following notion. A real-valued function J defined over a normed vector space V is said to be *coercive* if

$$\lim_{\|v\|_V \to \infty} J(v) = +\infty.$$

Theorem 8.2-1

Let U be a non-empty, closed subset of $\mathbb{R}^n$ and $J : \mathbb{R}^n \to \mathbb{R}$ a continuous function which is coercive if the set U is unbounded. Then there exists at least one element u such that

(P) $$u \in U \quad and \quad J(u) = \inf_{v \in U} J(v).$$

Proof

Let u_0 be any point of the set U. The coerciveness of the functional J implies the existence of a number r such that

$$\|v\| > r \Rightarrow J(u_0) < J(v).$$

Accordingly, the set of solutions of the problem (P) coincides with that of the solutions of the analogous problem corresponding to the set

$$U_0 = U \cap \{v \in \mathbb{R}^n : \|v\| \leqslant r\}.$$

This leads to the situation of a non-empty subset ($u_0 \in U_0$), which is closed and bounded. □

Remarks
(1) Theorem 8.2-1 provides a proof of the projection theorem (Theorem 8.1-1) when the space V is finite-dimensional; for it is enough to introduce the function $J(v) = \|w - v\|$ (with the notation of Theorem 8.1-1), which is coercive, since $J(v) \geqslant \|v\| - \|w\|$. But this point of view gives to the property of compactness an artificial role to play: for the proof of the projection theorem effectively relies on the completeness of the space and on its 'geometry', coupled with the existence of a particular scalar product. On the other hand, the advantage of the present proof is that it is applicable to a general norm.

(2) It should be observed that, when the set U is unbounded and the functional is linear, the result given above does not apply. In this connection, see Theorem 10.1-1.

It is the property of *compactness* which enters in an essential way into the proof of Theorem 8.2-1. It is possible to reach conviction on this point in another way, by considering a *minimising sequence* $(u_k)_{k \geqslant 0}$, that is to say, a sequence of points which satisfies

$$u_k \in U \quad \text{for every } k \geqslant 0, \quad \lim_{k \to \infty} J(u_k) = \inf_{v \in U} J(v).$$

This sequence is necessarily bounded, since the functional J is coercive, so that it is possible to find a subsequence $(u_{k'})$ which converges to an element $u \in U$ (the set U being closed). Since the function J is continuous,

$$J(u) = \lim_{k' \to \infty} J(u_{k'}) = \inf_{v \in U} J(v),$$

which provides another proof of the existence of a solution of the problem (P).

In actual fact, this is the type of argument which is used to extend the result to the infinite-dimensional case, but with the additional, and essential, assumptions of the *convexity*, both of the functional J, and of the set U. As the proof rests on the 'weak' compactness of convex, closed and bounded subsets of Hilbert spaces (parts (ii) and (iii) of the proof given below), we begin with the following definition. A sequence $(u_k)_{k \geqslant 0}$ of elements of a pre-Hilbert space V is said to *converge weakly* if there exists an element $u \in V$ such that

$$\lim_{k \to \infty} (v, u_k) = (v, u) \quad \text{for every } v \in V.$$

Observe that, while every sequence which converges in norm converges weakly, the converse is not generally true (cf. exercise 8.2-1).

Theorem 8.2-2

Let U be a non-empty, convex, closed subset of a separable Hilbert space V and J: V → ℝ a convex, differentiable functional, which is coercive if the subset U is unbounded. Then there exists at least one element u such that

$$\text{(P)} \qquad u \in U \quad \text{and} \quad J(u) = \inf_{v \in U} J(v).$$

Proof

(i) As in the finite-dimensional case (Theorem 8.2-1), the coerciveness of the functional makes it possible to restrict attention just to the case where the set U is *bounded* (and convex, as well, since a ball is convex; refer to the proof of the theorem just cited).

(ii) Let us consider a *minimising sequence* $(u_k)_{k \geqslant 0}$, i.e. one satisfying

$$u_k \in U \quad \text{for every } k \geqslant 0, \quad \lim_{k \to \infty} J(u_k) = \inf_{v \in U} J(v),$$

without excluding at this stage the possibility that $\inf_{v \in U} J(v) = -\infty$. *The sequence (u_k) being bounded* (by (i)), *we prove that it has at least one subsequence which converges weakly.*

Let C be a constant such that $\|u_k\| \leqslant C$ for every $k \geqslant 0$. We observe first that, if v is any element of the space V, the sequence of real numbers $\{(v, u_k)\}_{k \geqslant 0}$ is bounded since $|(v, u_k)| \leqslant C \|v\|$. The space V being assumed to be separable, let $(v_k)_{k \geqslant 0}$ be a dense, denumerable subset. The sequence $\{(v_1, u_k)\}_{k \geqslant 0}$ being bounded, it is possible to find a subsequence $\{(v_1, u_{k_1})\}_{k_1 \geqslant 0}$ which converges; so, too, the sequence $\{(v_2, u_{k_1})\}_{k_1 \geqslant 0}$ being bounded, it is possible to find a subsequence $\{(v_2, u_{k_2})\}_{k_2 \geqslant 0}$ which converges; and so on.

Consider now the 'diagonal' sequence $(w_l)_{l \geqslant 0}$, where $w_l \overset{\text{def}}{=} u_{l_l}$. By construction, every sequence $\{(v_k, w_l)\}_{l \geqslant 0}$, $k \geqslant 0$, has a limit, which is the limit of the sequence $\{(v_k, u_{l_k})\}_{l_k \geqslant 0}$. We shall show that, in fact, *every* sequence $\{(v, w_l)\}_{l \geqslant 0}$, $v \in V$, has a limit. Given any element $v \in V$, let $\varepsilon > 0$ also be given. There exists an element v_k such that $\|v - v_k\| \leqslant \varepsilon/4C$. Hence,

$$|(v, w_l) - (v, w_m)| = |(v, w_l - w_m)| \leqslant |(v_k, w_l - w_m)| + |(v - v_k, w_l - w_m)|$$

$$\leqslant |(v_k, w_l) - (v_k, w_m)| + \frac{\varepsilon}{2},$$

since $\|w_l - w_m\| \leqslant \|w_l\| + \|w_m\| \leqslant 2C$. With the element v_k kept fixed, the sequence $\{(v_k, w_l)\}_{l \geqslant 0}$ converges, by the preceding argument; so that it is a Cauchy sequence. Consequently, there exists an integer $l_0 = l_0(\varepsilon, v_k)$ such that

$$l, k \geqslant l_0 \Rightarrow |(v_k, w_l) - (v_k, w_m)| \leqslant \frac{\varepsilon}{2},$$

and the assertion is established.

We define a function $f: V \to \mathbb{R}$ by

$$f(v) = \lim_{l \to \infty} (v, w_l) \quad \text{for every} \quad v \in V.$$

This is a linear function, and continuous since

$$|(v, w_l)| \leqslant C \|v\| \quad \text{for every} \quad l \Rightarrow |f(v)| \leqslant C \|v\|.$$

By the Riesz representation theorem, there exists an element $u \in V$ such that $f(v) = (v, u)$ for every $v \in V$; so that the weak convergence of the subsequence $(w_l) = (u_{l_i})$ to the element u is established.

(iii) We next show that the *'weak' limit u of the subsequence (w_l) belongs to the set U*. Let P denote the projection operator associated with the closed, convex set U; by Theorem 8.1-1(2),

$$w_l \in U \Rightarrow (Pu - u, w_l - Pu) \geqslant 0 \quad \text{for every integer } l.$$

The weak convergence of the sequence (w_l) to the element u implies that

$$0 \leqslant \lim_{l \to \infty} (Pu - u, w_l - Pu) = (Pu - u, u - Pu) = - \|u - Pu\|^2 \leqslant 0,$$

so that $u \in U$. We have thus established that *a closed, convex set is 'weakly' closed*, that is to say, that the 'weak' limit of a weakly convergent sequence of points of such a set belongs to it.

(iv) Lastly, we show that the functional J satisfies

$$J(v) \leqslant \liminf_{l \to \infty} J(v_l),$$

for every sequence (v_l) weakly convergent to an element v. Since the function J is assumed to be differentiable and convex, we actually have (by Theorem 7.4-2)

$$J(v) + (\nabla J(v), v_l - v) \leqslant J(v_l) \quad \text{for every integer } l,$$

and, by the definition of weak convergence,

$$\lim_{l \to \infty} (\nabla J(v), v_l) = (\nabla J(v), v),$$

which establishes the stated property; it is called the *weak sequential lower semicontinuity* of the functional J.

(v) It is now easy to obtain the conclusion: the weak limit $u \in U$ of the subsequence (w_l) of the minimising sequence (u_k) satisfies

$$J(u) \leqslant \liminf_{l \to \infty} J(w_l) = \lim_{k \to \infty} J(u_k) = \inf_{v \in U} J(v). \qquad \square$$

Remarks

(1) The theorem remains true in reflexive Banach spaces, of which Hilbert spaces (whether separable or not) are particular cases; likewise, it remains true if the assumption of the differentiability of the function J is replaced with just its continuity.

(2) The converse of property (ii) is true (every weakly convergent sequence is bounded), but it cannot be established by elementary means. □

In certain special cases, the proof of the existence of a solution can be appreciably simplified, notably by avoiding all reference to weak convergence. We begin with a definition. Given a *Hilbert space V*, a function $J: V \to \mathbb{R}$ is called a *quadratic functional over V* if it is of the form

$$J(v) = \frac{1}{2} a(v, v) - f(v),$$

where $a(\cdot, \cdot): V \times V \to \mathbb{R}$ is a bilinear, continuous and symmetric form ($a(u, v) = a(v, u)$ for every $u, v \in V$) and $f: V \to \mathbb{R}$ is a continuous linear form. This definition generalises in a natural way that of a quadratic functional over $\mathbb{R}^n$ since, by the Riesz representation theorem, there exists an operator $A \in \mathcal{L}(V)$ and an element $b \in V$, both uniquely defined, such that

$$a(u, v) = (Au, v) = (u, Av) \quad \text{for every } u, v \in V,$$
$$f(v) = (b, v) \quad \text{for every } v \in V,$$

denoting by $(\cdot, \cdot)$ the scalar product of the space V.

The projection theorem and the Riesz representation theorem then make it easy to establish a general result of existence for problems (P) posed in terms of such functionals. It is to be observed that the case $U = V$ corresponds exactly to the *variational formulation of boundary-value problems*, which was briefly touched upon in sections 3.4 and 3.5.

Theorem 8.2-3

Let

$$J: v \in V \to J(v) = \frac{1}{2} a(v, v) - f(v)$$

be a quadratic functional over a Hilbert space V. Furthermore, suppose that there exists a number α such that

$$\alpha > 0 \quad and \quad a(v, v) \geq \alpha \|v\|_V^2 \quad for \ every \ v \in V.$$

Given a non-empty, convex, closed subset U of V, there exists a unique element u satisfying

(P) $$u \in U \quad and \quad J(u) = \inf_{v \in U} J(v).$$

This element u also satisfies

$$a(u, v - u) \geqslant f(v - u) \quad \text{for every } v \in U,$$

and, conversely, if an element $u \in V$ satisfies the inequalities given above, it is the solution of the problem (P). If U is a vector subspace, the previous inequalities are replaced by the equalities

$$a(u, v) = f(v) \quad \text{for every } v \in U.$$

Proof

The bilinear form $a(\cdot, \cdot)$ is also a scalar product over the space V, its associated norm being equivalent to the norm $\| \cdot \|$ associated with the scalar product $(\cdot, \cdot)$ of the space V. In fact, the assumptions made imply that

$$\sqrt{\alpha} \| v \| \leqslant \sqrt{a(v, v)} \leqslant \sqrt{\| a \|} \| v \|,$$

denoting by $\| a \|$ the norm (in the space $\mathscr{L}_2(V; \mathbb{R})$) of the bilinear function $a(\cdot, \cdot)$.

Since the linear form f is still continuous under this new norm, the Riesz representation theorem shows that there exists a unique element $c \in V$ such that

$$f(v) = a(c, v) \quad \text{for every } v \in V.$$

Consequently, it is possible to transform the expression for the functional by expressing it as

$$J(v) = \frac{1}{2} a(v, v) - a(c, v) = \frac{1}{2} a(v - c, v - c) - \frac{1}{2} a(c, c).$$

In this situation, solving the problem (P) amounts to looking for the *projection u of the element c onto the set U, in terms of the scalar product* $a(\cdot, \cdot)$. By the projection theorem, there exists a unique projection, thus proving the existence and uniqueness of the solution u of the problem (P). By the same theorem, this solution is also characterised by the inequalities

$$a(u - c, v - u) \geqslant 0 \quad \text{for every } v \in U,$$

or by the equalities

$$a(u - c, v) = 0 \quad \text{for every } v \in U,$$

if U is a vector subspace. This last set of relations coincides with that of the statement, since $a(c, v) = f(v)$ for every $v \in V$. $\square$

Remarks

(1) The *symmetry* of the bilinear form was employed in an essential manner, first, in concluding that the expression $a(\cdot, \cdot)$ is a scalar product, and, then, in writing down an alternative expression for the functional.

(2) The inequalities $a(u, v - u) \geqslant f(v - u)$ are a particular case of Euler's inequalities $J'(u)(v - u) \geqslant 0$ (Theorem 7.4-4) applied to the functional J, whose derivative is given by

$$J'(u)v = a(u, v) - f(v) \quad \text{for every } v \in V.$$

A similar observation (and for obvious reasons...) was made in regard to the projection theorem.

(3) In sections 3.4 and 3.5, an indication was given of the reasons for calling the relations $a(u, v) = f(v)$ 'variational' equations; it is in the same spirit that the relations '$a(u, v - u) \geqslant f(v - u)$ for every $v \in U$' are called *variational inequalities*. □

Exercises

8.2-1. (1) If V is a finite-dimensional Hilbert space, show that every sequence which converges weakly also converges in norm.

(2) Show that the sequence of functions

$$v_k: x \in [0, 2\pi] \rightarrow v_k(x) = \cos kx, k \geqslant 0,$$

converges weakly in the pre-Hilbert space $\mathscr{C}^0([0, 2\pi])$ equipped with the scalar product $\int_0^{2\pi} uv \, dx$, but that it does not converge in norm.

8.2-2. Prove that a sequence of element $(u_k)_{k \geqslant 0}$ of a Hilbert space V converges (in norm) to an element $u \in V$ if, and only if, it converges weakly to this same element and $\lim_{k \to \infty} \|u_k\| = \|u\|$.

8.2-3. Let there be given a functional of the form

$$J: v \in \mathbb{R}^n \rightarrow J(v) = \tfrac{1}{2}(\mathbf{A}v, v) - (b, v) + j(v), \mathbf{A} \in \mathscr{A}_n(\mathbb{R}), b \in \mathbb{R}^n,$$

where the matrix $\mathbf{A}$ is symmetric and positive definite and $j: \mathbb{R}^n \rightarrow \mathbb{R}$ is a continuous, convex function, taking non-negative values, but is not necessarily differentiable (for example, $j(v) = \sum_{i=1}^n \alpha_i |v_i|, \alpha_i \geqslant 0, 1 \leqslant i \leqslant n$). Let U be a non-empty, convex, closed and bounded subset of $\mathbb{R}^n$.

(1) Show that there exists a unique element $u \in U$ such that $J(u) = \inf_{v \in V} J(v)$.

(2) Show that an element $u \in U$ is a solution of the problem given above if, and only if,

$$(\mathbf{A}u - b, v - u) + j(v) - j(u) \geqslant 0 \quad \text{for every } v \in U.$$

8.2-4. Consider the quadratic functional

$$J(v) = \tfrac{1}{2}(\mathbf{A}v, v) - (b, v), \mathbf{A} \in \mathscr{A}_n(\mathbb{R}), b \in \mathbb{R}^n,$$

the symmetric matrix $\mathbf{A}$ being positive definite, and the set (assumed to be non-empty)

$$U = \{v \in \mathbb{R}^n : \mathbf{C}v = d\}, \quad \mathbf{C} \in \mathscr{A}_{m,n}(\mathbb{R}), d \in \mathbb{R}^m.$$

(1) Show that the associated problem (P) has a unique solution.

(2) Show that a vector $u \in \mathbb{R}^n$ is a solution of the problem if, and only if, there exists a vector $\lambda \in \mathbb{R}^m$ such that

$$\begin{cases} Au + C^T\lambda = b, \\ \quad Cu = d. \end{cases}$$

(3) Suppose that the rank of the matrix C is m. Express the solution u as a function of the data A, b, C and d.

8.2-5. Let φ_i, $1 \leqslant i \leqslant n$, be real functions, piecewise continuously differentiable over the interval $[0, 1]$ (for the definition, refer to the beginning of section 3.4) and linearly independent, satisfying

$$\sum_{i=1}^{n} \varphi_i(x) = 1 \quad \text{for every } x \in [0, 1].$$

Define also the functional

$$J: v \in \mathbb{R}^n \to J(v) = \sum_{i,j=1}^{n} a_{ij} v_i v_j + \sum_{i=1}^{n} b_i v_i,$$

with

$$a_{ij} = \int_0^1 \varphi_i'(x) \varphi_j'(x) \mathrm{d}x, \ b_i = \int_0^1 \varphi_i(x) \mathrm{d}x.$$

(1) Define the set

$$U_1 = \left\{ v \in \mathbb{R}^n : \sum_{i=1}^{n} b_i v_i = 0 \right\}.$$

Calculate the associated Lagrange multiplier. Discuss the existence and uniqueness of the solution to the problem: find u such that

$$u \in U_1 \quad \text{and} \quad J(u) = \inf_{v \in U_1} J(v).$$

(2) Answer the same questions in reference to the set

$$U_2 = \left\{ v \in \mathbb{R}^n : \sum_{i=1}^{n} (\varphi_i(1) + \varphi_i(0)) v_i = 0 \right\}.$$

(3) Answer the same questions in reference to the set

$$U_3 = \left\{ v \in \mathbb{R}^n : \sum_{i=1}^{n} (\varphi_i(1) - \varphi_i(0)) v_i = 0 \right\}.$$

8.3 Examples of optimisation problems

The *least squares solution of a linear system* (cf. section 3.7) is a first example of the *unconstrained* optimisation problem, corresponding to the following data:

$$U = V = \mathbb{R}^n; \quad J: v \in \mathbb{R}^n \to J(v) = \tfrac{1}{2} \| Bv - c \|_m^2 - \tfrac{1}{2} \| c \|_m^2.$$

Since

$$J(v) = \frac{1}{2}(\mathbf{B}^{\mathrm{T}}\mathbf{B}v, v)_n - (\mathbf{B}^{\mathrm{T}}c, v)_n,$$

the problem is one of *quadratic programming*, in the sense understood here, if and only if the symmetric matrix $\mathbf{B}^{\mathrm{T}}\mathbf{B}$ is positive definite. We recall that in section 8.1 the existence of a solution to this problem was established for every case, including that for which the matrix $\mathbf{B}^{\mathrm{T}}\mathbf{B}$ is merely non-negative definite. When the matrix $\mathbf{B}^{\mathrm{T}}\mathbf{B}$ is positive definite, the existence and the uniqueness of the solution can also be obtained by means of Theorem 8.2-3.

A *truly vast* source of optimisation problems is to be found in the *solution of boundary-value problems by the method of variational approximation*. As was shown in section 3.4 and 3.5 (we now take up again the notation employed there), this method leads to a search for the minimum of a *quadratic functional* of the form

$$\mathscr{J}: v \in \mathbb{R}^M \to \mathscr{J}(v) = \tfrac{1}{2}(\mathbf{A}v, v) - (b, v),$$

with

$$\mathbf{A} = (a(w_j, w_i)) \in \mathscr{A}_M(\mathbb{R}), \quad b = (f(w_i)) \in \mathbb{R}^M,$$

the functions w_i, $1 \leqslant i \leqslant M$, being the basis functions for the space V_h in which the approximate solution $u_h = \sum_{i=1}^M u_i w_i$ is being sought, and $a(\cdot, \cdot)$ and $f(\cdot)$ being respectively the bilinear form and the linear form which appear in the variational formulation of the boundary-value problem under consideration. It has already been observed that the matrix $\mathbf{A}$ is *symmetric* and *positive definite*; so that this is a second case of an *unconstrained quadratic programming problem*.

Let us next consider a variant of the membrane problem (considered in sections 3.2 and 3.5), that of a *membrane in contact with an obstacle* (figure 8.3-1). The aim is to calculate the vertical displacement $u: \bar{\Omega} \to \mathbb{R}$ of an elastic membrane with tension τ, stretched over the boundary Γ of the open set $\Omega \subset \mathbb{R}^2$, subjected to the action of a vertical force of density $\tau f(x)$ per surface element and constrained to remain above an obstacle represented by a *known* function $\chi: \bar{\Omega} \to \mathbb{R}$ (in order that the problem may be feasible, it is assumed that the function χ is non-positive over Γ). The contact zone between membrane and obstacle *is not known in advance*.

The *variational formulation* of this problem consists in looking for the minimum of the *energy* of the membrane which, let us recall, is of the form

$$J(v) = \tfrac{1}{2}a(v, v) - f(v),$$

with

$$a(u, v) = \int_\Omega \left(\frac{\partial u}{\partial x_1} \frac{\partial v}{\partial x_1} + \frac{\partial u}{\partial x_2} \frac{\partial v}{\partial x_2} \right) dx, \quad f(v) = \int_\Omega f v \, dx,$$

when the functions v lie in the subset

$$U = \{v \in V : v(x) \geqslant \chi(x) \quad \text{for every } x \in \bar{\Omega}\}$$

of a suitable space V of functions which are zero on Γ (it is, in fact, the Sobolev space $H_0^1(\Omega)$).

In order to approximate the solution of this problem, we set up a *triangulation* of the set $\bar{\Omega}$ (assumed to be polygonal; cf. figure 3.5-1), and consider the subspace $V_h \subset V$ (already introduced in section 3.5) consisting of functions which are affine over each triangle of the triangulation, continuous over $\bar{\Omega}$, and zero over Γ. We recall that the 'canonical' basis $(w_i)_{i=1}^M$ of this space V_h is chosen in such a way that the ith function w_i is zero at all the vertices of the triangulation, except at the ith vertex s_i, where it has the value 1. In this situation, the components v_i of the representation of an arbitrary function $v_h \in V_h$ in this basis have a remarkable significance, since

$$v_h = \sum_{i=1}^M v_i w_i \Rightarrow v_i = v_h(s_i), \quad 1 \leqslant i \leqslant M.$$

It is, therefore, natural to define the *discrete problem* in the following way. Find u_h such that

$$u_h \in U_h \overset{\text{def}}{=} \{v_h \in V_h : v_h(s_i) \geqslant \chi(s_i), \quad 1 \leqslant i \leqslant M\} \quad \text{and} \quad J(u_h) = \inf_{v_h \in U_h} J(v_h).$$

Figure 8.3-1

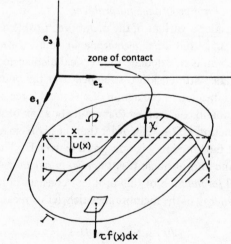

It should be observed that, in general, the set U_h is *not* contained in the set U.

We are thus led to look for the minimum of the quadratic functional

$$\mathscr{J}: v \in \mathbb{R}^M \to \mathscr{J}(v) = \tfrac{1}{2}(Av, v) - (b, v),$$

with

$$A = (a(w_j, w_i)) \in \mathscr{A}_M(\mathbb{R}), \quad b = (f(w_i)) \in \mathbb{R}^M,$$

while the vector v runs through the set

$$\mathscr{U} \stackrel{\text{def}}{=} \{v = (v_i) \in \mathbb{R}^M : v_i \geqslant \chi(s_i), \quad 1 \leqslant i \leqslant M\}.$$

This is, then, an *example of a quadratic programming problem with linear inequality constraints*. The set $\mathscr{U}$ being non-empty, closed, convex (and unbounded), the existence of a solution of the discrete problem follows as readily from Theorem 8.2-1 as from Theorem 8.2-3. It should be observed that the discrete problem may be expressed in the equivalent form

find u_h such that

$$u_h \in U_h \quad \text{and} \quad a(u_h, v_h - u_h) \geqslant f(v_h - u_h) \quad \text{for every } v_h \in U_h.$$

With the space V retaining its earlier definition, the *problem of the elasto-plastic torsion of a cyclindrical rod* leads to a search for the minimum of the same functional

$$J(v) = \frac{1}{2} a(v, v) - f(v) = \frac{1}{2} \int_\Omega \left\{ \left(\frac{\partial v}{\partial x_1} \right)^2 + \left(\frac{\partial v}{\partial x_2} \right)^2 \right\} dx - \int_\Omega fv \, dx,$$

as the functions v run through the subset

$$U = \{v \in V : \|\nabla v(x)\| \leqslant 1 \text{ for almost all } x \in \bar{\Omega}\},$$

setting

$$\|\nabla v(x)\| = \left\{ \left(\frac{\partial v}{\partial x_1}(x) \right)^2 + \left(\frac{\partial v}{\partial x_2}(x) \right)^2 \right\}^{1/2}.$$

The *discrete problem* associated with the space V_h of finite elements introduced earlier consists in looking for the minimum of the functional J as the function v_h runs through the set

$$U_h \stackrel{\text{def}}{=} \{v_h \in V_h : \|\nabla v_h(x)\| \leqslant 1 \text{ for every } x \in \mathring{T}, \ T \in \mathscr{T}_h\},$$

where $\mathring{T}$ denotes the interior of each of the triangles T of the triangulation $\mathscr{T}_h$. It is easy to see that the set U_h is non-empty, closed and convex, since

$$v, w \in U \quad \text{and} \quad \theta \in [0, 1]$$

imply

$$\| \nabla(\theta v + (1 - \theta)w)(x) \| \leqslant \theta \| \nabla v(x) \| + (1 - \theta) \| \nabla w(x) \| \leqslant 1,$$

so that the associated optimisation problem has once again a unique solution, which may be characterised equivalently in terms of variational inequalities. Observe that, on this occasion, the set $U_h = U \cap V_h$ is contained in the set U.

Let T be a triangle of the triangulation $\mathcal{T}_h$, with vertices s_1, s_2, s_3 (in order to fix ideas). The restriction of an arbitrary function $v \in V_h$ to the triangle T may be expressed as

$$v|_T = \sum_{i=1}^{3} v_i w_i|_T, \quad \text{with } v_i = v(s_i).$$

The basis functions w_i being affine, their first derivatives are constants, so that the inequality $\| \nabla(v|_T) \| \leqslant 1$ takes the form

$$\left\{ \sum_{i=1}^{3} \alpha_i v_i \right\}^2 + \left\{ \sum_{i=1}^{3} \beta_i v_i \right\}^2 \leqslant 1,$$

the constants $\alpha_i = \partial(w_i|_T)/\partial x_1$ and $\beta_i = \partial(w_i|_T)/\partial x_2$ being *known* functions of the co-ordinates of the vertices s_i. We are thus faced with a *quadratic programming problem with m quadratic inequality constraints* ($m =$ the number of triangles in the triangulation $\mathcal{T}_h$). Those readers who are interested will find a number of additional results on this problem in the book Glowinski, Lions and Trémolières (1976), where an entire chapter, in fact, is devoted to it.

These few examples do no more than give a brief look into the vast diversity of optimisation problems. For further development, one can refer to the exercises of this section, as well as to section 10.2, where some examples of *linear* programming problems are given.

Exercises

8.3-1. Let us consider the problem of *minimal surfaces*: among all surfaces which pass through a contour of the space $\mathbb{R}^3$, find one with minimal surface. With the contour represented by a function $u_0(x), x \in \Gamma$, where Γ is the boundary of an open subset Ω of the plane $\mathbb{R}^2$, the problem reduces to the finding of a minimum of the function

$$J(v) = \int_{\Omega} \sqrt{(1 + \| \nabla v \|^2)} \, dx,$$

as v runs through a set of functions defined over $\bar{\Omega}$, which are sufficiently smooth and equal to the function u_0 over Γ. Given a triangulation of the set Ω (assumed to be polygonal; cf. figure 3.5-1), we consider the space V_h of functions which are affine over each triangle of the triangulation and continuous over Γ. Denoting by Σ_h the set of vertices of the triangulation which lie on Γ, we define the set

$$U_h \stackrel{\text{def}}{=} \{v_h \in V_h : v_h(s) = u_0(s) \text{ for every } s \in \Sigma_h\}$$

Show that the *discrete problem*

find u_h such that

$$u_h \in U_h \quad \text{and} \quad J(u_h) = \inf_{v_h \in U_h} J(v_h)$$

has a unique solution. It should be noted that this is an example of a functional $J : V_h \to \mathbb{R}$ which is *not quadratic*.

8.3-2. We consider the analogue *in one dimension* of the problem of *minimal surfaces*, assumed, this time, to be approximated by the method of *finite differences*. Given an integer $N \geq 0$, we define the functional

$$J : v = (v_i)_{i=0}^{N+1} \in \mathbb{R}^{N+2}$$

$$\to J(v) = h \sum_{i=0}^{N} \left(1 + \left|\frac{v_{i+1} - v_i}{h}\right|^2\right)^{1/2} \quad \text{with } h = \frac{1}{N+1}.$$

(1) Examine the question of the possible convexity, and perhaps even the strict convexity, of the functional $J : \mathbb{R}^{N+2} \to \mathbb{R}$.

(2) Examine the existence and uniqueness of the solutions of the problems (P) associated with the functional J and the following subsets of the space $\mathbb{R}^{N+2}$:

$$U_1 = \{v = (v_i)_{i=0}^{N+1} \in \mathbb{R}^{N+2} : v_0 = v_{N+1} = 0\},$$
$$U_2 = \{v = (v_i)_{i=0}^{N+1} \in \mathbb{R}^{N+2} : v_0 = 0, v_{N+1} = 1\},$$
$$U_3 = \{v = (v_i)_{i=0}^{N+1} \in \mathbb{R}^{N+2} : v_0 = v_{N+1} = 0, v_i \geq a_i, 1 \leq i \leq N\},$$

the numbers a_i, $1 \leq i \leq N$, being given. Describe (whenever this is possible) the corresponding solutions.

8.4 Relaxation and gradient methods for unconstrained problems

We begin by generalising the notion of a quadratic functional over $\mathbb{R}^n$ with positive definite matrix. This extension is, in fact, well suited to the study of the methods which we have in mind, for which it leads to proofs of convergence which are particularly simple.

A functional $J : V \to \mathbb{R}$ defined over a Hilbert space V is called *elliptic* if it is continuously differentiable in V and if there exists a constant, which shall *always*, conventionally, be denoted by α, such that

$$\alpha > 0 \quad \text{and} \quad (\nabla J(v) - \nabla J(u), v - u) \geq \alpha \|v - u\|^2 \quad \text{for every } u, v \in V.$$

The result which follows gathers together various properties of elliptic functionals, which will be constantly used in the sequel.

Theorem 8.4-1

(1) *An elliptic functional $J: V \to \mathbb{R}$ is strictly convex and coercive; it satisfies the inequality*

$$J(v) - J(u) \geqslant (\nabla J(u), v - u) + \frac{\alpha}{2} \|v - u\|^2 \quad \text{for every } u, v \in V.$$

(2) *If U is a non-empty, convex, closed subset of the Hilbert space V and if J is an elliptic functional, then the problem*

$$\text{find } u \text{ such that}$$

(P) $$u \in U \quad \text{and} \quad J(u) = \inf_{v \in U} J(v)$$

has a unique solution.

(3) *Suppose that the set U is convex and that the functional J is elliptic. Then an element $u \in U$ is a solution of the problem (P) if and only if it satisfies*

$$(\nabla J(u), v - u) \geqslant 0 \quad \text{for every } v \in U,$$

in the general case, or

$$\nabla J(u) = 0 \quad \text{if } U = V.$$

(4) *A functional which is twice differentiable in V is elliptic if and only if*

$$(\nabla^2 J(u)w, w) \geqslant \alpha \|w\|^2 \quad \text{for every } w \in V.$$

Proof

An elliptic functional being by definition continuously differentiable, an application of Taylor's formula with integral remainder (Theorem 7.1-4) allows us to write

$$J(v) - J(u) = \int_0^1 (\nabla J(u + t(v - u)), v - u) \, \mathrm{d}t$$

$$= (\nabla J(u), v - u) + \int_0^1 (\nabla J(u + t(v - u)) - \nabla J(u), v - u) \, \mathrm{d}t$$

$$\geqslant (\nabla J(u), v - u) + \int_0^1 \alpha t \|v - u\|^2 \, \mathrm{d}t = (\nabla J(u), v - u) + \frac{\alpha}{2} \|v - u\|^2.$$

From this inequality, it follows that, first, the functional is strictly convex since (Theorem 7.4-2)

$$J(v) > J(u) + (\nabla J(u), v - u) \quad \text{for every } u, v \in V, u \neq v,$$

and, secondly, that the functional is coercive, since

$$J(v) \geqslant J(0) + (\nabla J(0), v) + \frac{\alpha}{2} \| v \|^2$$

$$\geqslant J(0) - \| \nabla J(0) \| \| v \| + \frac{\alpha}{2} \| v \|^2.$$

The existence of a solution of the problem (P) follows from Theorem 8.2-2, which is applicable because the functional is coercive; the uniqueness follows from the strict convexity. The characterisations of the minimum have already been established in Theorem 7.4-4.

If the function J is twice differentiable in V and is elliptic, we have

$$(\nabla^2 J(u)w, w) = \lim_{\theta \to 0} \frac{(\nabla J(u + \theta w) - \nabla J(u), w)}{\theta}$$

$$= \lim_{\theta \to 0} \frac{(\nabla J(u + \theta w) - \nabla J(u), \theta w)}{\theta^2} \geqslant \alpha \| w \|^2.$$

Conversely, the Taylor–Maclaurin formula (Theorem 7.1-4) applied to the function

$$f : w \in V \to f(w) \overset{\text{def}}{=} (\nabla J(w), v - u) \in \mathbb{R},$$

the vectors u and v being fixed, shows that

$$(\nabla J(v) - \nabla J(u), v - u) = f(v) - f(u)$$
$$= f'(u + \theta(v - u))(v - u) \quad (0 < \theta < 1)$$
$$= (\nabla^2 J(u + \theta(v - u))(v - u), v - u) \geqslant \alpha \| v - u \|^2. \qquad \square$$

Remarks

(1) In the last part of the proof, there was evidently no question of writing down the Taylor–Maclaurin formula for the derivative, since this formula only applies to functions taking values in $\mathbb{R}$.

(2) A quadratic functional over $\mathbb{R}^n$,

$$J : v \in \mathbb{R}^n \to J(v) = \tfrac{1}{2}(\mathrm{A}v, v) - (b, v), \mathrm{A} = \mathrm{A}^{\mathrm{T}},$$

is elliptic if and only if the matrix A is positive definite. It follows, in fact, from Theorem 1.3-1, that

$$(\nabla^2 J(u)w, w) = (\mathrm{A}w, w) \geqslant \lambda_1 \| w \|^2 \quad \text{for every } u, w \in \mathbb{R}^n,$$

where λ_1 denotes the smallest eigenvalue of the matrix A. We note in passing the inequality

$$(\nabla^2 J(u)w, w) = (\mathrm{A}w, w) \leqslant \lambda_n \| w \|^2 \quad \text{for every } u, w \in \mathbb{R}^n,$$

where $\lambda_n = \|A\|_2$ (Theorem 1.4-2) denotes the largest eigenvalue of the matrix A.

(3) In the same way, a quadratic functional over a Hilbert space V,

$$J : v \in V \to J(v) = \tfrac{1}{2} a(v, v) - f(v),$$

is elliptic if and only if there exists a constant α such that

$$\alpha > 0 \quad \text{and} \quad (\nabla^2 J(u)v, v) = a(v, v) \geqslant \alpha \|v\|^2 \quad \text{for every } v \in V;$$

it was precisely with this hypothesis that Theorem 8.2-3 was established.

$\square$

We move on now to the description, and then to the analysis, of certain *algorithms* for the solution of an *unconstrained* optimisation problem. Given a functional J defined over a vector space V, find u such that

(P) $\qquad\qquad\qquad u \in V \quad \text{and} \quad J(u) = \inf_{v \in V} J(v).$

We are dealing with *iterative methods* where, starting with an arbitrary *initial vector* u_0, a sequence of vectors $u_k, k \geqslant 0$, is constructed. Of course, the aim is to construct *convergent* methods, in the sense that, for *every* initial vector u_0, the sequence $(u_k)_{k \geqslant 0}$ converges to a solution of the problem (P).

In seeking to construct the vector u_{k+1} in terms of the vector u_k, a key idea is to try to reduce the problem to one which is 'easy to solve numerically', that is, to a minimisation problem for a function of a *single real variable*. For this, we shall

(i) find a *'descent' direction* at the point u_k, through the intermediary of a non-zero vector d_k;

(ii) look for *the minimum of the restriction of the functional J to the line passing through the point u_k in the direction of the vector d_k*; this defines the vector u_{k+1} *only if* the minimisation problem in a single variable,

$$\text{find } \varrho(u_k, d_k) \text{ such that}$$

$$\varrho(u_k, d_k) \in \mathbb{R}$$

and

$$J(u_k + \varrho(u_k, d_k)d_k) = \inf_{\varrho \in \mathbb{R}} J(u_k + \varrho d_k),$$

has a unique solution (this will be so, in particular, if the functional J is elliptic), in which case we set

$$u_{k+1} = u_k + \varrho(u_k, d_k)d_k.$$

These considerations are illustrated for the two-dimensional case in figure 8.4-1. The surface representing an elliptic functional has then the form of a paraboloid whose horizontal sections are *ellipses*, which explains incidentally the terminology 'elliptic functional'.

In the case of a *quadratic elliptic functional*

$$J(v) = \tfrac{1}{2}a(v, v) - f(v),$$

it is essential to note that *the determination of the point* u_{k+1} *is immediate once the vector* d_k *is known*, since the function

$$\varrho \in \mathbb{R} \to J(u_k + \varrho d_k) = \frac{\varrho^2}{2} a(d_k, d_k) + \varrho(\nabla J(u_k), d_k) + J(u_k)$$

is a quadratic (the coefficient $a(d_k, d_k)$ is strictly positive). Some indications on the *practical* solution of minimisation problems in one variable for more general functions are given in Exercise 8.4-4.

For the case $V = \mathbb{R}^n$, the simplest way of defining the successive descent directions consists in assigning them *in advance*, a 'canonical' choice in this respect being, quite obviously, that of the directions of the co-ordinate axes, taken in 'cyclic' order; that is the idea behind the *relaxation method*. Starting with an initial vector u_0, each vector $u_{k+1} = (u_i^{k+1})_{i=1}^n$ is constructed (of course, as long as this is possible) in terms of the vector $u_k = (u_i^k)_{i=1}^n$, calculating its components in turn through the solution of the following minimisation problems in one variable (each 'new' variable calculated has been enclosed in square brackets):

Figure 8.4-1

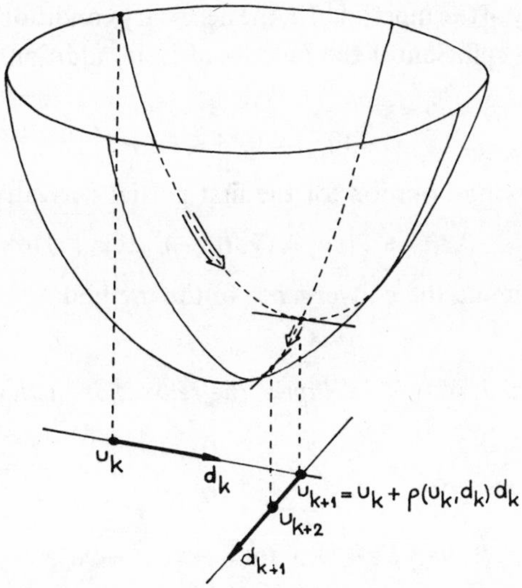

$$\begin{cases} J([u_1^{k+1}], u_2^k, u_3^k, \ldots, u_n^k) &= \inf_{\zeta \in \mathbb{R}} J(\zeta, u_2^k, u_3^k, \ldots, u_n^k), \\ J(u_1^{k+1}, [u_2^{k+1}], u_3^k, \ldots, u_n^k) &= \inf_{\zeta \in \mathbb{R}} J(u_1^{k+1}, \zeta, u_3^k, \ldots, u_n^k), \\ \quad\vdots \\ J(u_1^{k+1}, \ldots, u_{n-1}^{k+1}, [u_n^{k+1}]) &= \inf_{\zeta \in \mathbb{R}} J(u_1^{k+1}, \ldots, u_{n-1}^{k+1}, \zeta). \end{cases}$$

It is convenient, in view of the proof which is to follow, to introduce the 'intermediate' vectors $u_{k;l}$, $0 \leqslant l \leqslant n$, defined by

$$\begin{aligned} u_k = u_{k;0} &= (u_1^k, \ldots, u_n^k), \\ u_{k;1} &= (u_1^{k+1}, u_2^k, \ldots, u_n^k), \\ &\vdots \\ u_{k;l} &= (u_1^{k+1}, \ldots, u_l^{k+1}, u_{l+1}^k, \ldots, u_n^k), \\ &\vdots \\ u_{k;n} &= (u_1^{k+1}, \ldots, u_n^{k+1}) = u_{k+1}, \end{aligned}$$

so that the minimisation problems given above may be written in the equivalent form

$$J(u_{k;1}) = \inf_{\varrho \in \mathbb{R}} J(u_{k;0} + \varrho e_1),$$
$$\vdots$$
$$J(u_{k;l}) = \inf_{\varrho \in \mathbb{R}} J(u_{k;l-1} + \varrho e_l),$$
$$\vdots$$
$$J(u_{k;n}) = \inf_{\varrho \in \mathbb{R}} J(u_{k;n-1} + \varrho e_n),$$

where (e_l) denotes the canonical basis of $\mathbb{R}^n$. Under the condition of the differentiability of the functional J, the necessary conditions for a minimum, which are also sufficient if the functional is, in addition, convex, may be deduced:

$$\partial_l(u_{k;l}) = 0, \quad 1 \leqslant l \leqslant n,$$

using the following notation for the first partial derivatives:

$$\partial_l J(v) = J'(v)e_l = (\nabla J(v), e_l), \quad 1 \leqslant l \leqslant n.$$

We now examine the convergence of the method.

Theorem 8.4-2
If the functional $J: \mathbb{R}^n \to \mathbb{R}$ is elliptic, the relaxation method converges.

Proof
(i) Each function

$$\varphi_{k;l}: \varrho \in \mathbb{R} \to \varphi_{k;l}(\varrho) \overset{\text{def}}{=} J(u_{k;l-1} + \varrho e_l),$$

being coercive and strictly convex, possesses a unique minimum. Accordingly, every sequence $(u_{k;l})_{k \geqslant 0}$, $1 \leqslant l \leqslant n$, is well-defined; in particular, the sequence $(u_k)_{k \geqslant 0}$. We can write

$$J(u_k) - J(u_{k+1}) = J(u_{k;0}) - J(u_{k;n}) = \sum_{l=1}^{n} (J(u_{k;l-1}) - J(u_{k;l})),$$

and, by the assumption of ellipticity (Theorem 8.4-1),

$$J(u_{k;l-1}) - J(u_{k;l}) \geqslant (\nabla J(u_{k;l}), u_{k;l-1} - u_{k;l}) + \frac{\alpha}{2} \| u_{k;l-1} - u_{k,l} \|^2.$$

Since

$$(\nabla J(u_{k;l}), u_{k;l-1} - u_{k;l}) = \partial_l J(u_{k;l})(u_l^k - u_l^{k+1}) = 0, \quad 1 \leqslant l \leqslant n,$$

and since

$$\| u_{k;l-1} - u_{k;l} \|^2 = | u_l^k - u_l^{k+1} |^2, \quad 1 \leqslant l \leqslant n,$$

we obtain finally

$$J(u_k) - J(u_{k+1}) \geqslant \frac{\alpha}{2} \sum_{l=1}^{n} | u_l^k - u_l^{k+1} |^2 = \frac{\alpha}{2} \| u_k - u_{k+1} \|^2.$$

(ii) As the sequence $(J(u_k))_{k \geqslant 0}$ is decreasing and bounded below, we deduce, by means of (i),

$$\lim_{k \to \infty} \| u_k - u_{k+1} \| = 0,$$

so that, *a fortiori*,

$$\lim_{k \to \infty} \| u_{k;l} - u_{k+1} \| = 0, \quad 0 \leqslant l \leqslant n - 1.$$

(iii) Using the ellipticity of the functional and the characterisation $\nabla J(u) = 0$ of the minimum u (cf. Theorem 8.4-1), we obtain

$$\alpha \| u_{k+1} - u \|^2 \leqslant (\nabla J(u_{k+1}) - \nabla J(u), u_{k+1} - u)$$

$$= (\nabla J(u_{k+1}), u_{k+1} - u) = \sum_{l=1}^{n} \partial_l J(u_{k+1})(u_l^{k+1} - u_l),$$

from which it follows, using the characterisations $\partial_l J(u_{k;l}) = 0$, that

$$\| u_{k+1} - u \| \leqslant \frac{1}{\alpha} \sum_{l=1}^{n} | \partial_l J(u_{k+1}) | = \frac{1}{\alpha} \sum_{l=1}^{n} | \partial_l J(u_{k+1}) - \partial_l J(u_{k;l}) |.$$

(iv) Since each sequence $(J(u_{k;l}))_{k \geqslant 0}$ is, by construction, decreasing, *every sequence* $(u_{k;l})_{k \geqslant 0}$, $1 \leqslant l \leqslant n$, *is bounded*, since the functional is coercive (Theorem 8.4-1). Moreover, as each first partial derivative $\partial_l J$ is *uniformly continuous over compact sets of* $\mathbb{R}^n$,

$$\lim_{k \to \infty} \| u_{k;l} - u_{k+1} \| = 0 \Rightarrow \lim_{k \to \infty} | \partial_l J(u_{k;l}) - \partial_l J(u_{k+1}) | = 0, \quad 1 \leqslant l \leqslant n,$$

and the convergence then follows from (iii). $\qquad \square$

Remarks

(1) The *differentiability* of the functional is an essential hypothesis. Following Glowinski, Lions and Trémolières (1976), page 61, let us, in fact, consider the example of the functional

$$J: v = (v_1, v_2) \in \mathbb{R}^2 \to J(v_1, v_2) = v_1^2 + v_2^2 - 2(v_1 + v_2) + 2|v_1 - v_2|,$$

which is coercive, strictly convex, 'almost quadratic', but is not differentiable. With the choice $u_0 = (0, 0)$ as the initial vector, the relaxation method leads to the stationary sequence $(0, 0) = u_0 = u_1 = \cdots = u_k = \cdots$, while $\inf_{v \in \mathbb{R}^2} J(v) = J(1, 1)$. Nonetheless, it is possible (cf. Glowinski, Lions and Trémolières (1976), page 73) to establish convergence for non-differentiable functionals of the type

$$J(v) = J_0(v) + \sum_{i=1}^{n} \alpha_i |v_i|, \quad \alpha_i \geqslant 0,$$

the function J_0 being elliptic.

(2) It is possible to prove an analogue of Theorem 8.4-2 under the following more general hypotheses (but this is a somewhat more delicate matter): the functional is continuously differentiable, strictly convex and coercive. On this, see Glowinski, Lions and Trémolières (1976), page 61.

(3) The assumption of a *finite dimensionality*, through the intermediary of *uniform continuity*, plays an essential role in the proof. In fact, without this last property, the final implications of the proof are no longer necessarily true.

(4) The estimate obtained in (iii) provides an *a priori bound for the error* $\|u_k - u\|$, which may be calculated, in principle, entirely in terms of the data.

□

Let us consider the particular case of a *quadratic functional*

$$J(v) = \tfrac{1}{2}(Av, v) - (b, v) = \tfrac{1}{2} \sum_{i,j=1}^{n} a_{ij} v_i v_j - \sum_{i=1}^{n} b_i v_i.$$

If the symmetric matrix $A = (a_{ij})$ is positive definite, Theorem 8.4-2 is applicable. Since

$$\partial_l J(v) = \sum_{j=1}^{n} a_{lj} v_j - b_l, \quad 1 \leqslant l \leqslant n,$$

it follows that (with the notation used earlier)

$$\partial_1 J(u_{k;1}) = a_{11}[u_1^{k+1}] + a_{12} u_2^k + \cdots + a_{1n} u_n^k - b_1 = 0,$$
$$\partial_2 J(u_{k;2}) = a_{21} u_1^{k+1} + a_{22}[u_2^{k+1}] + a_{23} u_3^k + \cdots + a_{2n} u_n^k - b_2 = 0,$$
$$\vdots$$
$$\partial_n J(u_{k;m}) = a_{n1} u_1^{k+1} + \cdots + a_{nn-1} u_{n-1}^{k+1} + a_{nn}[u_n^{k+1}] - b_n = 0.$$

It may be verified that *this is a rediscovery, neither more nor less, of the Gauss–Seidel method for the solution of the linear system* $Au = b$; Theorem 8.4-2 thus provides another proof of the convergence of this method, when the matrix A is symmetric and positive definite (cf. Theorem 5.3-2).

The Gauss–Seidel method being a particular case of the *relaxation method for the solution of linear systems* (cf. section 5.2), the terminology which has been employed thus finds a partial justification. For a fuller justification, refer to exercise 8.4-1.

We return to the general unconstrained optimisation problem in which $V = \mathbb{R}^n$,

$$\text{find } u \in \mathbb{R}^n \text{ such that } J(u) = \inf_{v \in \mathbb{R}^n} J(v).$$

It seems intuitively clear that the convergence of an iterative method ought to be the better, the larger the differences $J(u_k) - J(u_{k+1})$; and, from this point of view, the choice which was imposed of selecting the directions provided by the co-ordinate axes was surely not the best possible.

In order to make the difference $J(u_k) - J(u_{k+1})$ as large as possible, the idea most readily to hand is, in fact, to choose as the descent direction that which presents the largest *local* descent, that is to say, the one *in the opposite direction to the gradient vector* $\nabla J(u_k)$. We recall, in passing, the justification for this last assertion. By definition of the gradient vector, one can write

$$J(u_k + w) = J(u_k) + (\nabla J(u_k), w) + \| w \| \varepsilon(w), \quad \lim_{w \to 0} \varepsilon(w) = 0,$$

so that, if $\nabla J(u_k) \neq 0$, the 'first-order' part of the variation of the function J is bounded in modulus by the product $\| \nabla J(u_k) \| \, \| w \|$ (by the Cauchy–Schwarz inequality), with equality if and only if the two vectors $\nabla J(u_k)$ and w are parallel.

We thus have all the elements which are necessary for the definition of the method which corresponds to this choice of descent direction. It is called the *gradient method with optimal parameter.* Starting with an initial vector u_0, each vector u_{k+1} is constructed (of course, as long as this is possible) in terms of the vector $u_k, k \geq 0$, through the relations

$$\begin{cases} J(u_k - \varrho(u_k)\nabla J(u_k)) = \inf_{\varrho \in \mathbb{R}} J(u_k - \varrho \nabla J(u_k)), \\ u_{k+1} = u_k - \varrho(u_k)\nabla J(u_k). \end{cases}$$

The negative sign in front of the variable ϱ is a reminder that the descent direction is *opposite* to that of the gradient: a positive value is to be expected for the number $\varrho(u_k)$.

Remark

Contrary to intuition, the direction $d_k = -\nabla J(u_k)$ is not necessarily optimal; section 8.5 is very instructive on this score!   □

Before going on to the study of the convergence of the gradient method with optimal parameter, we give a general definition. Every iterative method for which the point u_{k+1} is of the form

$$u_{k+1} = u_k - \varrho_k \nabla J(u_k), \quad \varrho_k > 0,$$

is called a *gradient method*. Accordingly, the method given above is a first particular case; two others will be considered later.

Theorem 8.4-3

Suppose that $V = \mathbb{R}^n$ and that the functional is elliptic. Then the gradient method with optimal parameter converges.

Proof

(i) Without loss of generality, it may be supposed that $\nabla J(u_k) \neq 0$ for every $k \geq 0$; otherwise, the method is convergent in a finite number of iterations. Every function

$$\varphi_k \colon \varrho \in \mathbb{R} \to \varphi_k(\varrho) \overset{\text{def}}{=} J(u_k - \varrho \nabla J(u_k))$$

being coercive and strictly convex, possesses a unique minimum, characterised by the relation $\varphi_k'(\varrho(u_k)) = 0$. Since (Theorem 7.1-1)

$$\varphi_k'(\varrho) = -(\nabla J(u_k - \varrho \nabla J(u_k)), \nabla J(u_k)),$$

one can deduce the relation

$$(\nabla J(u_{k+1}), \nabla J(u_k)) = 0,$$

which shows that *two consecutive descent directions are orthogonal*. Since $u_{k+1} = u_k - \varrho(u_k)\nabla J(u_k)$, we also have

$$(\nabla J(u_{k+1}), u_{k+1} - u_k) = 0,$$

so that, applying the first inequality of Theorem 8.4-1,

$$J(u_k) - J(u_{k+1}) \geq \frac{\alpha}{2}\|u_k - u_{k+1}\|^2.$$

(ii) Since the sequence $\{J(u_k)\}_{k \geq 0}$ is decreasing (by construction) and bounded below (by $J(u)$), it follows that

$$\lim_{k \to \infty} (J(u_k) - J(u_{k+1})) = 0,$$

which, together with the preceding inequality, shows that

$$\lim_{k \to \infty} \|u_k - u_{k+1}\| = 0.$$

(iii) Because of the orthogonality of consecutive descent directions, one may write

$$\|\nabla J(u_k)\|^2 = (\nabla J(u_k), \nabla J(u_k) - \nabla J(u_{k+1}))$$
$$\leqslant \|\nabla J(u_k)\| \|\nabla J(u_k) - \nabla J(u_{k+1})\|,$$

so that

$$\|\nabla J(u_k)\| \leqslant \|\nabla J(u_k) - \nabla J(u_{k+1})\|.$$

(iv) Since the sequence $(J(u_k))_{k \geqslant 0}$ is decreasing, *the sequence* $(u_k)_{k \geqslant 0}$ *is bounded*, since the functional is coercive (Theorem 8.4-1). The derivative J', continuous by hypothesis, is thus *uniformly continuous over compact sets*. Hence, it follows from (ii) that

$$\lim_{k \to \infty} \|\nabla J(u_k) - \nabla J(u_{k+1})\| = 0,$$

and hence, from (iii), that

$$\lim_{k \to \infty} \nabla J(u_k) = 0.$$

(v) Finally, we prove the convergence. We can write

$$\alpha \|u_k - u\|^2 \leqslant (\nabla J(u_k) - \nabla J(u), u_k - u) = (\nabla J(u_k), u_k - u)$$
$$\leqslant \|\nabla J(u_k)\| \|u_k - u\|,$$

using, in turn, the assumption of the ellipticity of the functional and the relation $\nabla J(u) = 0$. Hence, we obtain

$$\|u_k - u\| \leqslant \frac{1}{\alpha} \|\nabla J(u_k)\|,$$

and the conclusion follows from the property established in (iv). $\qquad \square$

Remarks

(1) As for the relaxation method, the assumption of a finite dimensionality plays an essential role in this proof.

(2) It is possible to prove an analogue of Theorem 8.4-2 under the following more general hypotheses: the functional is continuously differentiable, strictly convex and coercive; see Céa (1971), page 91.

(3) It is possible to give another proof of convergence, one which is applicable in more general situations. The sequence (u_k) being bounded, let $(u_{k'})$ be a subsequence which converges to an element u'. From the continuity of the derivative, it follows that

$$\nabla J(u') = \lim_{k' \to \infty} \nabla J(u_{k'}) = 0,$$

by (iv). As the solution of the problem is characterised by the equation

$\nabla J(u) = 0$, it follows, first, that $u = u'$ and, then, that the entire sequence (u_k) converges, the limit being unique.

(4) Though the proof of convergence given in part (v) is limited to elliptic functionals, it has the advantage of providing a *bound for the error* $\| u_k - u \|$, which, in principle, is calculable entirely *a priori*. $\qquad\qquad \square$

In the case of an elliptic *quadratic functional*,

$$J(v) = \tfrac{1}{2}(Av, v) - (b, v),$$

the orthogonality of the vectors $\nabla J(u_k)$ and $\nabla J(u_{k+1})$ can be utilised in order to calculate the number $\varrho(u_k)$. Knowing that $\nabla J(v) = Av - b$, one can write

$$0 = (\nabla J(u_{k+1}), \nabla J(u_k)) = (A(u_k - \varrho(u_k)(Au_k - b)) - b, Au_k - b),$$

from which it follows that

$$\varrho(u_k) = \frac{\| w_k \|^2}{(Aw_k, w_k)}, \quad \text{where} \quad w_k \overset{\text{def}}{=} Au_k - b = \nabla J(u_k).$$

A single iteration of the method then takes the following form:

$$
\begin{cases}
\text{calculate the vector} & w_k = Au_k - b; \\
\text{calculate the number} & \varrho(u_k) = \dfrac{\| w_k \|^2}{(Aw_k, w_k)}; \\
\text{calculate the vector} & u_{k+1} = u_k - \varrho(u_k)w_k.
\end{cases}
$$

We note in passing that we have here a new *iterative method for the solution of a linear system* $Au = b$ whose matrix A sysmmetric and positive definite. *Such a method may turn out to be of interest when the calculation of a vector* Aw, *where w is a known vector, is easy*. This is essentially the case with *sparse* matrices, especially such as are obtained in the course of the discretisation of boundary-value problems. We shall be concerned in greater detail with this point in the following section, when dealing with the conjugate gradient method.

The relaxation method and the gradient method with optimal parameter have in common the search for the minima of functions in one variable. It is precisely in order to be free of this necessity that the *gradient method with fixed parameter* is defined. Starting with an arbitrary initial vector u_0, the sequence (u_k) is defined by

$$u_{k+1} = u_k - \varrho \nabla J(u_k), \quad k \geqslant 0,$$

the real parameter ϱ to be determined 'as best may be'. More generally, it is possible to define the *gradient method with variable parameter*, by setting

$$u_{k+1} = u_k - \varrho_k \nabla J(u_k), \quad k \geqslant 0,$$

the real parameters ϱ_k being adjusted, for example, in the course of the iterations according to various criteria. We note that *the gradient method with fixed parameter is a particular case of the gradient method with variable parameter.*

We give now sufficient conditions for the *convergence* of elliptic functionals. It is, in fact, easy to foresee what precise form they will take. The parameter ϱ, or the parameters ϱ_k, will have to lie in a compact interval of the form $[a, b], a > 0$. In other words, we do, in fact, 'descend' ($\varrho_k \geqslant a$) and do not 'climb back too far' ($\varrho_k \leqslant b$). This is what figure 8.4-2 is intended to suggest.

Theorem 8.4-4

Let V be a Hilbert space and let $J : V \to \mathbb{R}$ be a functional which is differentiable in V. Suppose that there exist two constants α and M such that

$$\alpha > 0 \quad \text{and} \quad (\nabla J(v) - \nabla J(u), v - u) \geqslant \alpha \| v - u \|^2 \quad \text{for every} \quad u, v \in V,$$
$$\| \nabla J(v) - \nabla J(u) \| \leqslant M \| v - u \| \quad \text{for every} \quad u, v \in V.$$

If there exist two numbers a and b such that

$$0 < a \leqslant \varrho_k \leqslant b < \frac{2\alpha}{M^2} \quad \text{for every integer } k \geqslant 0,$$

the gradient method with variable parameter converges, and the convergence is geometric: there exists a constant $\beta = \beta(\alpha, M, a, b)$ such that

$$\beta < 1 \quad \text{and} \quad \| u_k - u \| \leqslant \beta^k \| u_0 - u \|.$$

Figure 8.4-2

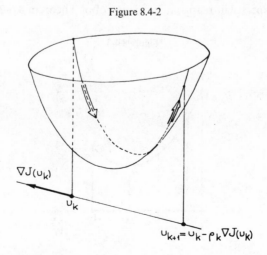

Proof

Using the characterisation $\nabla J(u) = 0$ of a minimum, it is possible to write

$$u_{k+1} - u = (u_k - u) - \varrho_k\{\nabla J(u_k) - \nabla J(u)\}.$$

Consequently,

$$\begin{aligned}
\|u_{k+1} - u\|^2 &= \|u_k - u\|^2 - 2\varrho_k(\nabla J(u_k) - \nabla J(u), u_k - u) \\
&\quad + \varrho_k^2\|\nabla J(u_k) - \nabla J(u)\|^2 \\
&\leqslant \{1 - 2\alpha\varrho_k + M^2\varrho_k^2\}\|u_k - u\|^2,
\end{aligned}$$

supposing that $\varrho_k > 0$. The quadratic $\tau(\varrho) = 1 - 2\alpha\varrho + M^2\varrho^2$ having the form indicated in figure 8.4-3, it is clear that

$$0 < a \leqslant \varrho_k \leqslant b < \frac{2\alpha}{M^2}$$

$$\Rightarrow (1 - 2\alpha\varrho_k + M^2\varrho_k^2)^{1/2}$$

$$\leqslant \beta \stackrel{\text{def}}{=} (\max\{\tau(a), \tau(b)\})^{1/2} < 1.$$

Hence, since

$$\|u_{k+1} - u\| \leqslant \beta\|u_k - u\| \leqslant \beta^{k+1}\|u_0 - u\|,$$

the geometric convergence is proved.   □

Remarks

(1) In contrast to the proofs of Theorems 8.4-2 and 8.4-3, compactness (linked with the finiteness of the dimension) is not used in the proof of Theorem 8.4-4. Rather, it is the *completeness* of the space which is important (it makes its appearance indirectly in the existence of the minimum u). This particular aspect will reappear in the proof of Theorem 8.6-2, which also

Figure 8.4-3

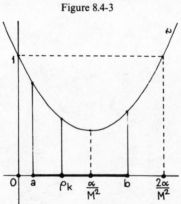

provides another proof of the convergence of the gradient method with variable parameter.

(2) It will be noted that the assumption of ellipticity is no longer the only assumption made, since it has been necessary to add to it the hypothesis

$$\| \nabla J(v) - \nabla J(u) \| \leqslant M \| v - u \| \quad \text{for every} \quad u, v \in V.$$

If the functional J is twice differentiable, this condition can also be expressed in the equivalent form

$$\sup_{v \in V} \| \nabla^2 J(v) \| \leqslant M. \qquad \square$$

In the case of an *elliptic quadratic functional*

$$J : v \in \mathbb{R}^n \to J(v) = \tfrac{1}{2}(Av, v) - (b, v),$$

one iteration of the method takes the form

$$u_{k+1} = u_k - \varrho_k(Au_k - b), \quad k \geqslant 0,$$

and it follows from the previous theorem that the method is convergent if $0 < a \leqslant \varrho_k \leqslant b \leqslant 2\lambda_1/\lambda_n^2$, denoting by λ_1 and λ_n the least and the largest of the eigenvalues of the symmetric, positive definite matrix A. *It is possible to improve upon this result.* In fact, from the equality

$$u_{k+1} - u = (u_k - u) - \varrho_k A(u_k - u) = (I - \varrho_k A)(u_k - u),$$

one can derive the bound

$$\| u_{k+1} - u \| \leqslant \| I - \varrho_k A \|_2 \| u_k - u \|.$$

The matrix $(I - \varrho_k A)$ being symmetric, its norm $\| \cdot \|_2$ can be expressed as (Theorem 1.4-2)

$$\| I - \varrho_k A \|_2 = \max \{ |1 - \varrho_k \lambda_1|, |1 - \varrho_k \lambda_n| \}.$$

Figure 8.4-4

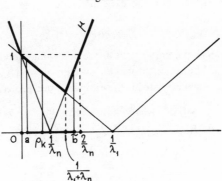

The form of the function (figure 8.4-4)

$$\mu: \varrho \in \mathbb{R} \rightarrow \mu(\varrho) = \max\left\{|1 - \varrho\lambda_1|, |1 - \varrho\lambda_n|\right\}$$

shows that

$$0 \leqslant a \leqslant \varrho_k \leqslant \tilde{b} < \frac{2}{\lambda_n} \Rightarrow \tilde{\beta} \overset{\text{def}}{=} \max\left\{\mu(a), \mu(\tilde{b})\right\} < 1,$$

and, hence that

$$\|u_{k+1} - u\| \leqslant \tilde{\beta}\|u_k - u\| \leqslant \tilde{\beta}^{k+1}\|u_0 - u\|.$$

Now it is clear that the upper bound $2\lambda_1/\lambda_n^2$ indicated by the theorem is in general 'much' smaller than the bound $2/\lambda_n$, since their ratio is that of the extreme eigenvalues of the matrix A. Finally, it should be observed that the 'optimal' values of the parameter ϱ, found by the two procedures of the gradient method with fixed parameter, are respectively λ_1/λ_n^2 and $2/(\lambda_1 + \lambda_n)$ (cf. figures 8.4-3 and 8.4-4).

Remark

The improvement given above may be extended to functionals which are not necessarily quadratic, but for which it is possible to establish convergence as long as $\varrho_k \in [a, \tilde{b}]$, with $\tilde{b} < 2/M$, without, however, being able to prove that the convergence is geometric. See, on this topic, Exercise 8.4-5.   □

From the 'numerical' point of view, one inconvenient aspect of gradient methods lies in the need to calculate the vector $\nabla J(u_k)$ at each iteration, which serves, it will be remembered, to determine the 'next' descent direction, while the inconvenient aspect of relaxation methods and gradient methods with optimal parameter is to be found in the need to solve minimisation problems in one variable. That is why the *actual* choice of a method depends to a very large extent on the relative importance of these 'numerical' aspects and the expected rate of convergence.

Exercises

8.4-1. Given a real parameter $\omega \neq 0$, the following iterative method may be defined. The transition from the vector $u_k = (u_i^k)_{i=1}^n$ to the vector $u_{k+1} = (u_i^{k+1})_{i=1}^n$, is made by solving the n minimisation problems in one variable

$$J(u_1^{k+1}, \ldots, u_{l-1}^{k+1}, [u_l^{k+1/2}], u_{l+1}^k, \ldots, u_n^k)$$

$$= \inf_{\xi \in \mathbb{R}} J(u_1^{k+1}, \ldots, u_l^{k+1}, \xi, u_{l+1}^k, \ldots, u_n^k), \quad 1 \leqslant l \leqslant n,$$

and setting

$$u_i^{k+1} = (1 - \omega)u_i^k + \omega u_i^{k+1/2}.$$

The corresponding method is called an *under-relaxation method* if $\omega < 1$, and an *over-relaxation* method if $\omega > 1$ (for $\omega = 1$, the relaxation method is recovered).

(1) Suppose that the functional J is quadratic:

$$J: v \in \mathbb{R}^n \rightarrow J(v) = \tfrac{1}{2}(Av, v) - (b, v), \quad A = A^T.$$

Prove that the method coincides with the relaxation method introduced in section 5.2 for the solution of the linear system $Au = b$.

(2) In case the quadratic functional is elliptic, prove the convergence of the method for $0 < \omega < 2$, being guided by the proof of Theorem 8.4-2. It should be noted that this approach provides another proof of Theorem 5.3-2.

(3) More generally, define an optimisation algorithm which, in the particular case of quadratic functionals, amounts to the *block* relaxation method for the solution of linear systems.

8.4-2. The aim in this exercise is to study the gradient method with optimal parameter, *in the absence of any hypotheses regarding the ellipticity of the functional $J: \mathbb{R}^n \rightarrow \mathbb{R}$.* The only assumptions made are the following. It is assumed that a point $u_0 \in \mathbb{R}^n$ is known such that the set

$$U \overset{\text{def}}{=} \{v \in \mathbb{R}^n : J(v) \leqslant J(u_0)\}$$

is a compact subset of $\mathbb{R}^n$; that the functional J is twice differentiable at every point of U; and, finally, that there exists a constant M such that

$$|(\nabla^2 J(v)w, w)| \leqslant M \|w\|^2 \quad \text{for every} \quad v \in U, \quad w \in \mathbb{R}^n.$$

(1) Let v be a point of the set U such that $\nabla J(v) \neq 0$. Show that the number

$$\tau(v) = \sup \{\rho \geqslant 0 : [v, v - \rho \nabla J(v)] \subset U\}$$

is finite and strictly positive.

(2) Starting with the point u_0, a sequence of points $u_k \in U$ is constructed as follows.

    (i)   If $\nabla J(u_k) = 0$, the algorithm terminates;

    (ii)   if $\nabla J(u_k) \neq 0$, a point u_{k+1} is chosen in such a way that

$$(*) \quad \begin{cases} u_{k+1} \in [u_k, u_k - \tau(u_k)\nabla J(u_k)], \\ J(u_{k+1}) = \inf_{0 \leqslant \rho \leqslant \tau(u_k)} J(u_k - \rho \nabla J(u_k)). \end{cases}$$

In case (ii), show that there exists at least one point u_{k+1} which satisfies the relations $(*)$; should such a point not be defined uniquely, it is understood that a choice is made of *any one* from among all those which are available.

(3) Show that in case (ii)

$$J(u_k) - J(u_{k+1}) \geqslant \frac{\|\nabla J(u_k)\|^2}{2M}.$$

It is assumed, for questions (4) and (5), that it is the situation enshrined in (ii) which prevails, thus providing an infinite sequence (u_k).

(4) Show that there exists a subsequence $(u_{k'})$ of the sequence (u_k) and a point $u \in U$ such that

$$\lim_{k' \to \infty} u_{k'} = u \quad \text{and} \quad \nabla J(u) = 0.$$

(5) If there exists a unique point $u \in U$ satisfying $\nabla J(u) = 0$, show that the point u is a strict minimum of the functional $J: U \to \mathbb{R}$ and that the full sequence (u_k) converges to this point.

8.4-3. Let A be a symmetric matrix, assumed to be only non-negative definite, and b a vector belonging to the space Im(A). The object of the problem is the study of a gradient method with fixed step-length applied to the quadratic functional

$$J: v \in \mathbb{R}^n \to J(v) = \tfrac{1}{2}(Av, v) - (b, v).$$

We denote by S a given symmetric, positive definite matrix.

(1) Show that the solutions of the linear system $Au = b$ are all of the form $u = u' + u''$, where u' is an element (which should be given explicitly) of the space $\text{Im}(S^{-1}A)$ and u'' is any element belonging to the kernel Ker (A). It would be helpful to make use of the decomposition $\mathbb{R}^n = \text{Im}(S^{-1}A) \oplus \text{Ker}(A)$ and the orthogonality of the vectors of the spaces $\text{Im}(S^{-1}A)$ and Ker (A), with regard to the scalar product (Su, v).

(2) Given a real number ρ, we define the algorithm

$$u_{k+1} = u_k - \rho S^{-1} \nabla J(u_k), \quad k \geqslant 0,$$

the initial vector u_0 being chosen arbitrarily in $\mathbb{R}^n$. Show that the component of the vector u_k belonging to the space Ker(A) (in the decomposition $\mathbb{R}^n = \text{Im}(S^{-1}A) \oplus \text{Ker}(A)$) is independent of the integer k; denote it by w.

(3) Show that the eigenvalues of the matrix $S^{-1}A$ are all non-negative.

(4) Denote by Λ_n the largest of the eigenvalues of the matrix $S^{-1}A$. Assuming that $0 < \rho < 2/\Lambda_n$, show that

$$\lim_{k \to \infty} u_k = u' + w.$$

8.4-4. The aim in this problem is to study two *methods of minimisation for functions of a single variable* which do not make use of evaluations of the first or second derivatives of the function ('numerical' differentiation is something to be avoided whenever it is possible). The context throughout is the following. The function f is assumed to be real-valued, twice continuously differentiable over a compact interval $[a, b] \subset \mathbb{R}$ and such that $f''(\rho) > 0$ for every $\rho \in [a, b]$; it should be noted in passing that this last hypothesis is certainly satisfied when the function f is of the form

$$f(\rho) = J(w + \rho d), \quad d \neq 0,$$

the functional J being assumed to be elliptic and twice continuously differentiable. Lastly, it is assumed that it has been established that there exists a point $c \in]a, b[$ such that $f'(c) = 0$, and that it is desired to 'locate' it as exactly as possible (such a point c is unique, since the function f is strictly convex; cf. Theorem 7.4-3).

(1) Let x_1 and x_2 be two real numbers such that $a \leqslant x_1 \leqslant x_2 \leqslant b$. Prove the implications

$$f(x_1) \geqslant f(x_2) \Rightarrow x_1 \leqslant c < b, \quad f(x_2) \leqslant f(x_1) \Rightarrow a < c \leqslant x_2.$$

(2) Given an arbitrary number $\varepsilon > 0$, show that it is possible to locate the point c in an interval of length $\leqslant \frac{1}{2}(b - a) + \varepsilon$, while making use of only two function evaluations, for example $f(\frac{1}{2}(a + b))$ and $f(\frac{1}{2}(a + b) + \varepsilon)$.

(3) Show that it is possible to locate the point c in an interval of length $\leqslant \frac{1}{3}(b - a) + \varepsilon$, while making use of only three function evaluations. Here, the first step will be to compare the values $f(\frac{1}{3}(2a + b))$ and $f(\frac{1}{3}(a + 2b))$, and then to have recourse to question (2).

(4) The *Fibonacci sequence* is defined by the recurrence formulae

$$u_0 = 0, \quad u_1 = 1, \quad u_n = u_{n-1} + u_{n-2}, \quad n \geqslant 2.$$

Given an integer $n \geqslant 2$, set

$$x_1 = a \frac{u_n}{u_{n+1}} + b \frac{u_{n-1}}{u_{n+1}}, \quad x_2 = a \frac{u_{n-1}}{u_{n+1}} + b \frac{u_n}{u_{n+1}}.$$

Verify the relations

$$x_1 \frac{u_{n-1}}{u_n} + b \frac{u_{n-2}}{u_n} = x_2, \quad a \frac{u_{n-2}}{u_n} + x_2 \frac{u_{n-1}}{u_n} = x_1.$$

Deduce that it is possible to locate the point c in an interval of length $\leqslant (b - a)/u_{n+1} + \varepsilon \overset{\text{def}}{=} \Delta_n, n \geqslant 2$, while making use of only n function evaluations.

(5) We define the *golden ratio* as the number $\varphi = \frac{1}{2}(1 + \sqrt{5})$ (observe that it satisfies the equation $\varphi^2 = \varphi + 1$), then the numbers

$$x_1 = a(\varphi - 1) + b(2 - \varphi), \quad x_2 = a(2 - \varphi) + b(\varphi - 1).$$

Calculate $b - x_1$ and $x_2 - a$; deduce that it is possible to locate the point c in an interval of length $(b - a)(\varphi - 1)$ with the help of the function evaluations $f(x_1)$ and $f(x_2)$.

(6) Verify the relations

$$x_1(\varphi - 1) + b(2 - \varphi) = x_2, \quad a(2 - \varphi) + x_2(\varphi - 1) = x_1.$$

Deduce that it is possible to locate the point c in an interval of length $(b - a)(\varphi - 1)^{n-1} \overset{\text{def}}{=} \delta_n, n \geqslant 2$, while making use of only n function evaluations.

(7) Calculate u_n (cf. question (4)) as a function of n.

(8) Show that if n is sufficiently large and ε is sufficiently small, then the ratio δ_n / Δ_n does not differ much from 1.17.

(9) Although it is the first method which is *theoretically* superior, it is the second which is preferred in *practice*. Explain why this is so.

8.4-5. Let $J: \mathbb{R}^n \to \mathbb{R}$ be a functional which is twice continuously differentiable, strictly convex, coercive and satisfies

$$M \overset{\text{def}}{=} \sup_{v \in \mathbb{R}^n} \| \nabla^2 J(v) \| < + \infty.$$

Show that the gradient method with variable parameter is convergent if $\rho_k \in [a, \tilde{b}]$, with $0 < a \leqslant \tilde{b} < 2/M$.

8.4-6. Describe the application of the gradient method with optimal parameter to the functional

$$J: v = (v_1, v_2) \in \mathbb{R}^2 \to J(v) = v_1^4 - 4v_1^3 + 6(v_1^2 + v_2^2) - 4(v_1 + v_2),$$

bringing out, in particular, the scalar equation which has to be solved at each iteration. Do the results of the text (such as the existence of a minimum, the convergence of the method, etc.) apply to this example?

8.4-7. Let L and M be two matrices of order n and let b be a vector of $\mathbb{R}^n$. Consider the functional

$$J = v \in \mathbb{R}^n \to J(v) = \| Lv \|^2 + \| Mv \|^3 + b^{\mathsf{T}} v,$$

where $\| \cdot \|$ denotes the Euclidean norm of $\mathbb{R}^n$.

(1) Prove that the functional J is twice differentiable over $\mathbb{R}^n$. Calculate the numbers $\{ J'(u)v \}$ and $\{ J''(u)(v, w) \}$ for $u, v, w \in \mathbb{R}^n$. Prove that the functional J is convex.

(2) Assume, from this point, that the matrix L is invertible. Prove that there exists a unique vector $u \in \mathbb{R}^n$ such that $J(u) = \inf_{v \in \mathbb{R}^n} J(v)$.

(3) Describe the application of the relaxation method to the optimisation problem given above, bringing out, in particular, the scalar equation which has to be solved at each iteration. Does the method converge?

8.4-8. (1) Let A be a symmetric, positive definite matrix of order n and b a vector of $\mathbb{R}^n$. The gradient method with optimal parameter is applied to the minimisation over $\mathbb{R}^n$ of the functional

$$J: v \in \mathbb{R}^n \to J(v) = \tfrac{1}{2}(Av, v) - (b, v),$$

that is to say, to the solution of the linear system $Au = b$. We define the norm

$$v \in \mathbb{R}^n \to \| v \|_A \overset{\text{def}}{=} (Av, v)^{1/2}.$$

Establish the inequality

$$\| u_k - u \|_A \leqslant \left(\frac{\mathrm{cond}_2(A) - 1}{\mathrm{cond}_2(A) + 1} \right)^k \| u_0 - u \|_A.$$

(2) Examine the particular case $\mathrm{cond}_2(A) = 1$.

8.5. Conjugate gradient methods for unconstrained problems

Let us consider the unconstrained optimisation problem

find u such that

$$u \in \mathbb{R}^n \quad \text{and} \quad J(u) = \inf_{v \in \mathbb{R}^n} J(v).$$

As methods of approximation based on the minimisation of functions of a single variable in some appropriate descent directions, the relaxation method and the gradient method with optimal parameter have already been studied. The first uses for its successive descent directions a set which is fixed in advance (that of the co-ordinate axes), *independently of the functional J*. The second uses at each iteration a direction which is 'locally optimal' (that of the gradient), and which now *is related to the functional* under consideration. As a result, a more rapid convergence is to be expected, even at this stage.

In order to improve the convergence further, it is clear that it is necessary to make an effort to use a *greater amount of information about the functional* in defining the *direction* of the vectors $u_{k+1} - u_k$. This is so, for example, in *Newton's method* (section 7.5), which takes the form

$$u_{k+1} = u_k - \{\nabla^2 J(u_k)\}^{-1} \nabla J(u_k), \quad k \geqslant 0.$$

While this method does not require the solution of minimisation problems in one variable, a major inconvenience lies in the need to solve linear systems with matrix $\nabla^2 J(u_k)$ at *each* iteration, which, in numerical terms, is very costly. *For all that, it is possible to find descent directions which are an improvement on that of the gradient, without having recourse to the second derivatives of the functional.*

In order to see this, consider the case, very simple but also *very instructive*, of an elliptic quadratic functional $J: \mathbb{R}^2 \to \mathbb{R}$ of the form

$$J(v_1, v_2) = \tfrac{1}{2}(\alpha_1 v_1^2 + \alpha_2 v_2^2), \quad 0 < \alpha_1 < \alpha_2,$$

for which

$$J(0) = \inf_{v \in \mathbb{R}^2} J(v),$$

and suppose that the *gradient method with optimal parameter* is applied to the solution of the corresponding optimisation problem. Then, unless the initial vector $u_0 = (u_1^0, u_2^0)$ has one of its components zero (in which case the method converges in one iteration), *the method can never converge in a finite number of iterations* (cf. figure 8.5-1). It may be seen, in fact, that, if $\nabla J(u_k) \neq 0$, that is to say, if $u_k = (u_1^k, u_2^k) \neq 0$, a necessary and sufficient condition for the point u_{k+1} to be a solution of the problem is that the line $\{u_k - \varrho \nabla J(u_k): \varrho \in \mathbb{R}\}$ pass through the origin, that is to say, that there exist a number ϱ such that

$$u_1^k = \varrho \alpha_1 u_1^k \quad \text{and} \quad u_2^k = \varrho \alpha_2 u_2^k,$$

which is only possible if one of the two components u_i^k is zero (we have assumed that $\alpha_1 \neq \alpha_2$). Now a simple calculation, making use of the expression for the number $\varrho(u_k)$ given in the previous section for general quadratic functionals, shows that

$$u_1^{k+1} = \frac{\alpha_2^2(\alpha_2 - \alpha_1)u_1^k(u_2^k)^2}{\alpha_1^3(u_1^k)^2 + \alpha_2^3(u_2^k)^2}, \quad u_2^{k+1} = \frac{\alpha_1^2(\alpha_1 - \alpha_2)u_2^k(u_1^k)^2}{\alpha_1^3(u_1^k)^2 + \alpha_2^3(u_2^k)^2},$$

so that

$$u_1^0 \neq 0 \quad \text{and} \quad u_2^0 \neq 0 \Rightarrow u_1^k \neq 0 \quad \text{and} \quad u_2^k \neq 0 \text{ for every integer } k.$$

What is the best way of choosing the descent direction? Equivalently, what is the best way of taking into consideration the *geometry* of the surface which represents the functional J, which the gradient method with optimal parameter does not 'distinguish' from a sphere? Let us suppose that the point u_0 does not lie on either of the co-ordinate axes and let u_1 be the point obtained from the gradient method with optimal parameter, that is to say,

$$u_1 = u_0 - \frac{\|d_0\|^2}{(Ad_0, d_0)}d_0 \quad \text{with} \quad d_0 = \nabla J(u_0) = Au_0 \quad \text{and} \quad A = \text{diag }(\alpha_i).$$

We restrict ourselves to the *observation* that the 'optimal' descent direction d_1 at the point u_1 (which is exactly that of the vector u_1; cf. figure 8.5-1) satisfies

$$d_1 \neq 0 \quad \text{and} \quad (Ad_1, d_0) = 0,$$

and that these relations define the *direction* of the vector d_1 uniquely (this is

Figure 8.5-1

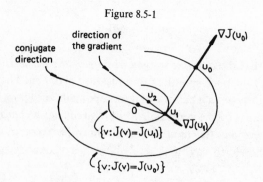

the direction which is 'conjugate' to the previous descent direction $d_0 = \nabla J(u_0)$, using terminology which will be explained later on).

The vectors $\nabla J(u_0)$ and $\nabla J(u_1)$ being linearly independent because orthogonal (cf. part (i) of the proof of Theorem 8.4-3), the point 0, which is the solution of the problem, can also be considered as the minimum of the functional in the plane passing through the point u_1 and spanned by the vectors $\nabla J(u_0)$ and $\nabla J(u_1)$. This latter idea is just what will be generalised to the case of *an elliptic quadratic functional*

$$J : v \in \mathbb{R}^n \to J(v) = \tfrac{1}{2}(Av, v) - (b, v).$$

Given an arbitrary initial vector u_0, let us suppose that the vectors $u_1, u_2, \ldots, u_k$ have been calculated. We shall, of course, make the assumption

$$\nabla J(u_l) \neq 0, \quad 0 \leqslant l \leqslant k,$$

otherwise the algorithm will have already ended. For $l = 0, 1, \ldots, k$, we call G_l the subspace of $\mathbb{R}^n$, of dimension $\leqslant l + 1$, spanned by the gradient vectors $\nabla J(u_i), 0 \leqslant i \leqslant l$ (it is not known *a priori* whether these are linearly independent). *The essential idea behind the method which we have in mind consists in defining the 'next' vector u_{k+1} as the minimum of the restriction of the functional J to the set*

$$u_k + G_k \stackrel{\text{def}}{=} \{u_k + v_k : v_k \in G_k\} = \left\{u_k + \sum_{i=0}^{k} \alpha_i \nabla J(u_i) : \alpha_i \in \mathbb{R}, 0 \leqslant i \leqslant k \right\};$$

in other words, the point u_{k+1} satisfies

$$u_{k+1} \in u_k + G_k \quad \text{and} \quad J(u_{k+1}) = \inf_{v \in u_k + G_k} J(v).$$

The set $u_k + G_k$ being closed and convex (it is the 'hyperplane' parallel to the subspace G_k which passes through the point u_k) and the functional being coercive and strictly convex, the minimisation problem given above has a unique solution.

Even at this stage, then, it is possible to expect this method to turn out to be superior to the gradient method with optimal parameter, in which the minimum is sought only on the line $\{u_k - \varrho \nabla J(u_k) : \varrho \in \mathbb{R}\}$. But it still remains to show that each of these minimisation problems in k variables can be solved easily, something which is *not at all* evident *a priori*. Nonetheless, that is the case. This is due, in a special way, to the introduction of the notion of 'conjugate' directions in relation to the symmetric matrix A, as we shall now show.

The solutions of the successive minimisation problems

$$u_{l+1} \in u_l + G_l \quad \text{and} \quad J(u_{l+1}) = \inf_{v \in u_l + G_l} J(v) = \inf_{v \in G_l} J(u_l + v), \quad 0 \leqslant l \leqslant k,$$

satisfy

$$(\nabla J(u_{l+1}), w) = 0 \quad \text{for every} \quad w \in G_l,$$

since the sets G_l are vector subspaces; in particular,

$$(\nabla J(u_{l+1}), \nabla J(u_i)) = 0, \quad 0 \leqslant i \leqslant l \leqslant k,$$

which shows that the *gradient vectors* $\nabla J(u_l)$, $0 \leqslant l \leqslant k+1$, *are pairwise orthogonal*.

Remark

This property is 'stronger' than that established for the gradient method with optimal parameter, where only *consecutive* gradient vectors are orthogonal.

This orthogonality shows two things. Firstly, *the gradient vectors* $\nabla J(u_l)$, $0 \leqslant l \leqslant k$, *are linearly independent* (it has been assumed that they are different from zero); secondly, *the algorithm necessarily terminates in at most n iterations*, since, if the vectors $\nabla J(u_l), 0 \leqslant l \leqslant n-1$, are non-zero, then the next gradient vector is necessarily zero (otherwise, it would turn out to have been possible to construct a set of $n+1$ linearly independent vectors).

We define the $k+1$ vectors

$$u_{l+1} - u_l \overset{\text{def}}{=} \Delta_l = \sum_{i=0}^{l} \delta_i^l \nabla J(u_i), \quad 0 \leqslant l \leqslant k,$$

and show that they possess a truly remarkable property, which depends in a crucial way on the *quadratic* nature of the functional; the latter, in fact, makes it possible to write

$$\nabla J(v+w) = A(v+w) - b = \nabla J(v) + Aw, \quad \text{for every} \quad v, w \in \mathbb{R}^n,$$

and, in particular,

$$\nabla J(u_{l+1}) = \nabla J(u_l + \Delta_l) = \nabla J(u_l) + A\Delta_l, \quad 0 \leqslant l \leqslant k.$$

As a result of the orthogonality of the gradient vectors $\nabla J(u_l), 0 \leqslant l \leqslant k+1$, we deduce, firstly, that

$$0 = (\nabla J(u_{l+1}), \nabla J(u_l)) = \|\nabla J(u_l)\|^2 + (A\Delta_l, \nabla J(u_l)), \quad 0 \leqslant l \leqslant k,$$

so that (having assumed that $\nabla J(u_l) \neq 0, 0 \leqslant l \leqslant k$)

$$\Delta_l \neq 0, \quad 0 \leqslant l \leqslant k,$$

and, secondly, that for $k \geqslant 1$

$$0 = (\nabla J(u_{l+1}), \nabla J(u_i)) = (\nabla J(u_l), \nabla J(u_i)) + (A\Delta_l, \nabla J(u_i))$$
$$= (A\Delta_l, \nabla J(u_i)), \quad 0 \leqslant i < l \leqslant k.$$

Since every vector $\Delta_m, 0 \leqslant m \leqslant k-1$, is a linear combination of the vectors

$\nabla J(u_i)$, $0 \leqslant i \leqslant k - 1$, we have established the relations

$$(A\Delta_l, \Delta_m) = 0, \quad 0 \leqslant m < l \leqslant k.$$

This leads us to the following definition. Given a symmetric matrix A, the vectors $w_l, 0 \leqslant l \leqslant k$, with $k \geqslant 1$, are said to be *conjugate* in relation to the matrix A if

$$w_l \neq 0, \quad 0 \leqslant l \leqslant k, \quad \text{and} \quad (Aw_l, w_m) = (Aw_m, w_l) = 0, \quad 0 \leqslant m < l \leqslant k.$$

Of course, this is a concept which makes use of only the *directions* of the vectors w_l, which are also said to be *conjugate* in relation to the matrix A. We note, too, that *if the matrix A is positive definite* (as is the case here), *conjugate vectors are necessarily linearly independent*. In fact,

$$0 = \sum_{l=0}^{k} \lambda_l w_l \Rightarrow 0 = \left(A\left(\sum_{l=0}^{k} \lambda_l w_l \right), w_m \right) = \lambda_m (Aw_m, w_m) \Rightarrow \lambda_m = 0,$$

$$0 \leqslant m \leqslant k,$$

since $(Aw_m, w_m) > 0$, because of the positive definite nature of the matrix A.

Remark

The function $(u, v) \in \mathbb{R}^n \to (Au, v) \in \mathbb{R}$ being a scalar product when the matrix A is symmetric and positive definite, another way of stating that two vectors are conjugate in relation to the matrix A is to say that they are orthogonal in respect of this scalar product, the more 'usual' orthogonality corresponding to the particular case of the identity matrix.

The vectors $\nabla J(u_l)$, $0 \leqslant l \leqslant k$, and the vectors $\Delta_l = \sum_{i=0}^{l} \delta_i^l \nabla J(u_i)$, $0 \leqslant l \leqslant k$, being linearly independent, the equality between the matrices of order $k + 1$

$$\left(\begin{array}{c|c|c|c} \boxed{} & \boxed{} & \cdots & \boxed{} \\ \Delta_0 & \Delta_1 & \cdots & \Delta_k \\ \end{array} \right) = \left(\begin{array}{c|c|c|c} \boxed{} & \boxed{} & \cdots & \boxed{} \\ \nabla J(u_0) & \nabla J(u_1) & \cdots & \nabla J(u_k) \\ \end{array} \right) \begin{pmatrix} \delta_0^0 & \delta_0^1 & \cdots & \delta_0^k \\ & \delta_1^1 & \cdots & \delta_1^k \\ & & \ddots & \vdots \\ & & & \delta_k^k \end{pmatrix}$$

shows that

$$\delta_l^l \neq 0, \quad 0 \leqslant l \leqslant k.$$

It is, therefore, possible to write down *a priori* the *descent direction* at each point u_l, $0 \leqslant l \leqslant k$, in the form

$$d_l = \sum_{i=0}^{l-1} \lambda_i^l \nabla J(u_i) + \nabla J(u_l), \quad 0 \leqslant l \leqslant k.$$

Remark

The actual descent takes place in the direction of the vector $-d_l$, but, for the sake of presentation, the 'negative' sign has, by preference, been set before the number $\varrho(u_k, d_k)$ introduced below.

We return to the calculation of the vector u_{k+1}, *supposing* that the components $\lambda_i^k, 0 \leqslant i \leqslant k-1$, are known. We are thus led to an *optimisation problem in one variable*,

$$\text{find } \varrho(u_k, d_k), \quad \text{such that}$$

$$J(u_k - \varrho(u_k, d_k)d_k) = \inf_{\varrho \in \mathbb{R}} J(u_k - \varrho d_k),$$

and it is clear that the point u_{k+1} then coincides with the point $u_k - \varrho(u_k, d_k)d_k$. In fact, since

$$\Delta_k = \sum_{i=0}^{k} \delta_i^k \nabla J(u_i) = \delta_k^k \left\{ \sum_{i=0}^{k-1} \frac{\delta_i^k}{\delta_k^k} \nabla J(u_i) + \nabla J(u_k) \right\},$$

we necessarily have

$$\Delta_k = \delta_k^k d_k \quad \text{and} \quad \varrho(u_k, d_k) = -\delta_k^k.$$

We now show that the actual calculation of the components λ_i^k is carried out in a *remarkably simple* way. In order to find k equations in the k unknowns $\lambda_k^i, 0 \leqslant i \leqslant k-1$, we write

$$0 = (Ad_k, \Delta_l) = (d_k, A\Delta_l) = (d_k, \nabla J(u_{l+1}) - \nabla J(u_l)),$$

$$0 \leqslant l \leqslant k-1,$$

or

$$\left(\sum_{i=0}^{k-1} \lambda_i^k \nabla J(u_i) + \nabla J(u_k), \nabla J(u_{l+1}) - \nabla J(u_l) \right) = 0, \quad 0 \leqslant l \leqslant k-1.$$

The gradient vectors $\nabla J(u_l), 0 \leqslant l \leqslant k+1$, being pairwise orthogonal, the preceding relations reduce to the equations

$$-\lambda_{k-1}^k \| \nabla J(u_{k-1}) \|^2 + \| \nabla J(u_k) \|^2 = 0 \quad \text{for} \quad l = k-1,$$

$$-\lambda_l^k \| \nabla J(u_l) \|^2 + \lambda_{l+1}^k \| \nabla J(u_{l+1}) \|^2 = 0 \quad \text{for} \quad 0 \leqslant l \leqslant k-2 \quad \text{if} \quad k \geqslant 2,$$

whose solution is

$$\lambda_i^k = \frac{\| \nabla J(u_k) \|^2}{\| \nabla J(u_i) \|^2}, \quad 0 \leqslant i \leqslant k-1.$$

Consequently,

$$d_k = \sum_{i=0}^{k-1} \frac{\| \nabla J(u_k) \|^2}{\| \nabla J(u_i) \|^2} \nabla J(u_i) + \nabla J(u_k)$$

$$= \nabla J(u_k) + \frac{\|\nabla J(u_k)\|^2}{\|\nabla J(u_{k-1})\|^2} \left\{ \sum_{i=0}^{k-2} \frac{\|\nabla J(u_{k-1})\|^2}{\|\nabla J(u_i)\|^2} \nabla J(u_i) + \nabla J(u_{k-1}) \right\}$$

$$= \nabla J(u_k) + \frac{\|\nabla J(u_k)\|^2}{\|\nabla J(u_{k-1})\|^2} d_{k-1},$$

which provides a simple procedure for calculating the successive descent directions, namely

$$\begin{cases} d_0 = \nabla J(u_0), \\ d_l = \nabla J(u_l) + \dfrac{\|\nabla J(u_l)\|^2}{\|\nabla J(u_{l-1})\|^2} d_{l-1}, \quad 0 \leqslant l \leqslant k. \end{cases}$$

It remains to determine the number $\varrho(u_k, d_k)$ which, it will be remembered, is defined by the equation

$$J(u_k - \varrho(u_k, d_k)d_k) = \inf_{\varrho \in \mathbb{R}} J(u_k - \varrho d_k).$$

The functional J being quadratic, the function which has to be minimised is a quadratic

$$\varrho \in \mathbb{R} \to \frac{\varrho^2}{2}(A d_k, d_k) - \varrho(\nabla J(u_k), d_k) + J(u_k).$$

It is enough, therefore, to set to zero the derivative of this quadratic, which gives

$$\varrho(u_k, d_k) = \frac{(\nabla J(u_k), d_k)}{(A d_k, d_k)}.$$

We now have all the elements required for the definition of an algorithm for the minimisation of an elliptic quadratic functional

$$J : v \in \mathbb{R}^n \to J(v) = \tfrac{1}{2}(Av, v) - (b, v),$$

called the *conjugate gradient method*. Starting with an arbitrary initial vector u_0, we set

$$d_0 = \nabla J(u_0).$$

If $\nabla J(u_0) = 0$, the algorithm terminates. Otherwise, we define the number

$$r_0 = \frac{(\nabla J(u_0), d_0)}{(A d_0, d_0)}$$

(the distinction between the two forms of notation d_0 and $\nabla J(u_0)$ is evidently artificial at this stage!), then the vector

$$u_1 = u_0 - r_0 d_0.$$

Assuming that, step by step, the vectors $u_1, d_1, \ldots, u_{k-1}, d_{k-1}, u_k$, have been constructed, which assumes that the gradient vectors $\nabla J(u_l), 0 \leqslant l \leqslant k-1$, are all non-zero, one of two situations will prevail. Either $\nabla J(u_k) = 0$ and the algorithm terminates, or $\nabla J(u_k) \neq 0$, in which case we define the vector

$$d_k = \nabla J(u_k) + \frac{\|\nabla J(u_k)\|^2}{\|\nabla J(u_{k-1})\|^2} d_{k-1},$$

then the number

$$r_k = \frac{(\nabla J(u_k), d_k)}{(Ad_k, d_k)},$$

then the vector

$$u_{k+1} = u_k - r_k d_k,$$

and so on.

We present in the form of a theorem the most remarkable property (already noted earlier) of this very ingenious method, due to Hestenes & Stiefel (1952).

Theorem 8.5-1.
The conjugate gradient method applied to an elliptic quadratic functional converges in at most n iterations. □

We have, then, constructed another method for the solution of a linear system whose matrix is symmetric and positive definite (in fact, it was originally conceived as a method for the solution of a linear system), and it is a *direct* method, since it leads to an exact solution after a finite number of elementary operations. Let us count the number of operations required for one iteration.

(i) The calculation of the scalar products $\|\nabla J(u_k)\|^2$, $(\nabla J(u_k), d_k)$, (Ad_k, d_k) requires $3(n-1)$ additions and $3n$ multiplications.

(ii) The calculation of the vector Ad_k requires $n(n-1)$ additions and n^2 multiplications.

(iii) The calculation of the vectors d_k, u_{k+1} and $\nabla J(u_{k+1}) = \nabla J(u_k) - r_k Ad_k$ requires $3n$ additions, $3n$ multiplications and 2 divisions (for the calculation of the quotients $\|\nabla J(u_k)\|^2/\|\nabla J(u_{k-1})\|^2$ and r_k).

In the final count, then, *the conjugate gradient method requires of the order of*

$$\begin{cases} n^3 \text{ additions,} \\ n^3 \text{ multiplications,} \\ 2n \text{ divisions,} \end{cases}$$

that is to say, *a greater number of elementary operations than the Cholesky method* (refer to section 4.4); that is all the more so, in that the inevitable presence of rounding errors in the actual calculations sometimes leads to the need to continue the procedure beyond the n iterations predicted by the theory. Though the conjugate gradient method, then, does not seem to be best for *full* matrices (enjoying, all the same, a 'numerical stability' which is sometimes very welcome; see, in this respect, Exercise 8.5-3), it yet displays *manifest advantages* when it is applied to *sparse* matrices, whose *storage it is often possible to avoid.* In fact, an examination of the recurrence relations shows that *the matrix* A *only makes its appearance through the calculation of the vectors* Ad_k. This operation, which is the costliest part when the matrix A is full, is *very simple* for certain sparse matrices, notably those which result from the discretisation of boundary-value problems by the methods of finite differences or finite elements. It was seen, for example, that, in one dimension (cf. sections 3.1 and 3.4), the components of the vector Av are of the form

$$(Av)_i = av_{i-1} + 2bv_i + av_{i+1}, \quad v_0 = v_{n+1} = 0;$$

in the same way, the analogous recurrence formulae (though a little more elaborate, as is only to be expected) are not hard to find in two or three dimensions. Lastly, it often happens in this kind of application that the convergence of the method is so rapid as to warrant a *spectacular reduction in the number n of iterations* which the theory predicts.

Because of rounding errors, the conjugate gradient method resembles an iterative method, and indeed, its 'convergence' may be studied as that of a truly iterative method. In particular, it can be shown (see e.g. Lascaux & Théodor (1987), p. 520) that the smaller the condition number cond_2 (A) (cf. section 2.2), the faster the method converges. This observation leads to the application of the process of *preconditioning* to the matrix A, which consists in replacing the linear system A$u = b$ by an 'equivalent' one, whose matrix, still symmetric and positive definite, has a smaller condition number. Then the conjugate gradient method is applied to this new linear system (for instance, one can solve instead the linear system CAC$v = Cb$, where C is an appropriate symmetric, positive definite matrix). One obtains in this fashion various *conjugate gradient methods with preconditioning,* which are currently among *the most widely used methods for solving large sparse systems with symmetric, positive definite matrices.* Details about these methods may be found in Lascaux & Théodor (1987), Sect. 8.4, Golub & van Loan (1984), Sect. 10.3, Golub & Meurant (1983), pp. 199 ff.

In seeking to adapt the conjugate gradient method to functionals which are not necessarily quadratic, one observes that the orthogonality of the gradient vectors $\nabla J(u_k)$ which are encountered sequentially makes it possible to write

$$d_k = \nabla J(u_k) + \frac{\|\nabla J(u_k)\|^2}{\|\nabla J(u_{k-1})\|^2} d_{k-1}$$

$$= \nabla J(u_k) + \frac{(\nabla J(u_k), \nabla J(u_k) - \nabla J(u_{k-1}))}{\|\nabla J(u_{k-1})\|^2} d_{k-1}.$$

It is this last expression for the descent direction which is used to define the *Polak–Ribière conjugate gradient method* (1969) for *general* functionals J. Starting with an arbitrary initial vector u_0, one assumes the vectors $u_1, \ldots, u_k$ to have been constructed, which implies that the gradient vectors $\nabla J(u_l)$, $0 \leqslant l \leqslant k - 1$, are all non-zero. One of two cases may then hold. Either $\nabla J(u_k) = 0$ and the algorithm terminates, or $\nabla J(u_k) \neq 0$, in which case the vector u_{k+1} is defined (if it exists and if it is unique) by the relations

$$u_{k+1} = u_k - r_k d_k \quad \text{and} \quad J(u_{k+1}) = \inf_{r \in \mathbb{R}} J(u_k - r d_k),$$

the successive descent directions d_l being defined by the recurrence relation

$$d_0 = \nabla J(u_0), \quad d_l = \nabla J(u_l) + \frac{(\nabla J(u_l), \nabla J(u_l) - \nabla J(u_{l-1}))}{\|\nabla J(u_{l-1})\|^2} d_{l-1}, 1 \leqslant l \leqslant k.$$

Remarks

(1) It would have been equally conceivable *a priori* to adapt to the general case the conjugate gradient method in its first form; this adaptation goes by the name of the *Fletcher–Reeves conjugate gradient method* (1964). However, that of Polak–Ribière turns out to be more efficient in practice.

(2) When the functional is quite general, there is no reason for the gradient vectors $\nabla J(u_k)$ obtained by the Polak–Ribière method to continue to be pairwise orthogonal, nor, therefore, for the algorithm to end in a finite number of iterations.

(3) By construction, the Polak–Ribière method coincides with that of Fletcher–Reeves when applied to a quadratic functional.

(4) Sufficient conditions for convergence are indicated in Exercise 8.5-4.

□

Exercises

8.5-1. Let A be a symmetric, positive definite matrix of order n and $b \in \mathbb{R}^n$ a general vector. Show that the solution of the linear system $Au = b$ is immediate, once any set of n vectors, which are conjugate in respect of the matrix A, is known.

8.5-2. (1) We take as known *any* set of vectors d_i, $0 \leqslant i \leqslant n - 1$, which are conjugate in respect of a symmetric, positive definite matrix A of order n, and, furthermore, we take as given a vector $b \in \mathbb{R}^n$. Starting with an arbitrary initial vector $u_0 \in \mathbb{R}^n$, it is assumed that the vectors $u_1, \ldots, u_k$ have been constructed, which implies that the vectors $Au_i - b$, $0 \leqslant l \leqslant k - 1$, are all non-zero. One of two cases may then arise. Either $Au_k - b = 0$ and the algorithm terminates, or else $Au_k - b \neq 0$, in which case we define the number

$$r_k = \frac{(Au_k - b, d_k)}{(Ad_k, d_k)},$$

and then the vector

$$u_{k+1} = u_k - r_k d_k.$$

Show that this method converges in at most n iterations to the solution of the linear system $Au = b$.

(2) Choose as the set of conjugate vectors the vectors constructed from the basis (e_i) by the *Gram–Schmidt orthogonalisation process* (section 4.5) *adapted to the scalar product* (Au, v). What well-known direct method is recovered by this process?

8.5-3. Consider the application of the conjugate gradient method to an elliptic quadratic functional and set $e_k = u_k - u$. It $e_k \neq 0$, prove the inequalities

$$\|e_{k+1}\| < \|e_k\| \quad \text{and} \quad (Ae_{k+1}, e_{k+1}) < (Ae_k, e_k).$$

It should be noted that, in contrast, the Euclidean norm of the vector $Au_k - b = \nabla J(u_k)$, sometimes called the 'residual vector', does not necessarily decrease. The reader interested in these aspects of the method would do well to refer to the magisterial article Stiffel (1958), which proves the existence of a close relationship between the conjugate gradient method and the theory of orthogonal polynomials.

8.5-4. Prove the convergence of the Polak–Ribière method when the functional $J: \mathbb{R}^n \to \mathbb{R}$ is elliptic and twice continuously differentiable.

8.5-5. Adopting a line similar to the one recommended in Exercise 8.4-3, prove that the conjugate gradient method converges (in a sense which should be given explicitly) even when the matrix A is 'merely' non-negative definite.

8.6 Relaxation, gradient and penalty-function methods for constrained problems

In this section, we shall be concerned with *constrained* problems which, it will be remembered, take the following form. Given a subset U of a vector space V and a functional $J: V \to \mathbb{R}$, find u such that

(P) $\qquad\qquad u \in U \quad \text{and} \quad J(u) = \inf_{v \in U} J(v).$

The extension of the definition of the *relaxation method* (encountered in section 8.4) to constrained problems for which the set U has the particular form

$$U = \{v = (v_i) \in \mathbb{R}^n: \quad a_i \leqslant v_i \leqslant b_i, \quad 1 \leqslant i \leqslant n\} = \prod_{i=1}^{n} [a_i, b_i],$$

is immediate. Above, the cases $a_i = -\infty$ and/or $b_i = +\infty$ are not excluded. Knowing the vector $u_k = (u_i^k)_{i=1}^n$, the vector $u_{k+1} = (u_i^{k+1})_{i=1}^n$ is defined by solving in turn the n minimisation problems in one variable (the same notation is taken up as in section 8.4):

$$J(u_{k;1}) = J([u_1^{k+1}], u_2^k, u_3^k, \ldots, u_n^k) = \inf_{a_1 \leqslant \zeta \leqslant b_1} J(\zeta, u_2^k, u_3^k, \ldots, u_n^k),$$

$$J(u_{k;2}) = J(u_1^{k+1}, [u_2^{k+1}], u_3^k, \ldots, u_n^k) = \inf_{a_2 \leqslant \zeta \leqslant b_2} J(u_1^{k+1}, \zeta, u_3^k, \ldots, u_n^k),$$

$$\vdots \qquad\qquad\qquad \vdots$$

$$J(u_{k;n}) = J(u_1^{k+1}, \ldots, u_{n-1}^{k+1}, [u_n^{k+1}]) \inf_{a_n \leqslant \zeta \leqslant b_n} J(u_1^{k+1}, \ldots, u_{n-1}^{k+1}, \zeta).$$

Theorem 8.6-1
If the functional $J: \mathbb{R}^n \to \mathbb{R}$ is elliptic and if the set U is of the form

$$U = \prod_{i=1}^{n} [a_i, b_i], \text{ not excluding } a_i = -\infty \text{ and/or } b_i = +\infty,$$

the relaxation method converges.

Proof
The proof follows the lines of that of Theorem 8.4-2, the only new element being the replacement of the characterisations $\partial_l J(u_{k;l}) = 0, 1 \leqslant l \leqslant n$, and $\nabla J(u) = 0$ of the case of the unconstrained problem by the following necessary and sufficient conditions for the existence of a minimum:

$$\begin{cases} \partial_l J(u_{k;l})(v_l - u_l^{k+1}) \geqslant 0 & \text{for every } v_l \in [a_l, b_l], \quad 1 \leqslant l \leqslant n, \\ (\nabla J(u), v - u) \geqslant 0 & \text{for every } v \in U. \end{cases}$$

It may be verified, in fact, that the inequalities

$$J(u_{k;l-1}) - J(u_{k;l}) \geqslant \frac{\alpha}{2} \|u_{k;l-1} - u_{k,l}\|^2,$$

$$\alpha \|u_{k+1} - u\|^2 \leqslant (\nabla J(u_{k+1}), u_{k+1} - u),$$

obtained respectively at steps (i) and (iii) of the proof of the theorem just cited remain the same. $\qquad\square$

Remark
It is not possible to extend without due care the relaxation method to more general sets U; for example, if

$$J(v) = (v_1^2 + v_2^2) \quad \text{and} \quad U = \{v = (v_1, v_2) \in \mathbb{R}^2 : v_1 + v_2 \geqslant 2\},$$

it is easy to convince oneself (figure 8.6-1) that, unless one of the components of the initial vector u_0 takes the value 1, the algorithm defined by

$$J(u_1^{k+1}, u_2^k) = \inf_{\zeta \geqslant 2 - u_1^k} J(\zeta, u_2^k),$$

$$J(u_1^{k+1}, u_2^{k+1}) = \inf_{\zeta \geqslant 2 - u_2^k} J(u_1^{k+1}, \zeta),$$

'comes to a halt' on the boundary of the set U. □

Let us now consider the problem (P) associated with a general *convex* set U and a *convex* functional. An element $u \in U$ is then a solution of the problem (P) if it satisfies the necessary and sufficient conditions (Theorem 7.4-4)

$$(\nabla J(u), v - u) \geqslant 0 \quad \text{for every} \quad v \in U.$$

One cannot fail to notice the similarity between these conditions and the characterisation (Theorem 8.1-1)

$$(u - w, v - u) \geqslant 0 \quad \text{for every} \quad v \in U,$$

of the projection u of an element w of a Hilbert space V onto a non-empty, convex, closed subset $U \subset V$. More precisely, denoting by P the projection operator of the space V onto the set U, one obtains the following equivalences:

$$u \in U \text{ and } J(u) = \inf_{v \in U} J(v) \Leftrightarrow u \in U \text{ and } (\nabla J(u), v - u) \geqslant 0 \text{ for every } v \in U$$

$$\Leftrightarrow u \in U \text{ and } (u - \{u - \varrho\nabla J(u)\}, v - u)$$

$$\geqslant 0 \text{ for every } v \in U, \varrho > 0$$

$$\Leftrightarrow u = P(u - \varrho\nabla J(u)) \text{ for every } \varrho > 0.$$

Figure 8.6-1

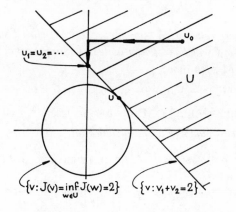

In other words, *the solution u turns out to be, for every $\varrho > 0$, a fixed point of the function*

$$g: v \in V \rightarrow g(v) = P(v - \varrho \nabla J(v)) \in U \subset V.$$

It is, therefore, natural to define as a method of approximation for the solution of the problem (P) *the method of successive approximations applied to the function g*. Given an arbitrary element $u_0 \in V$, the sequence $(u_k)_{k \geqslant 0}$ is defined by

$$u_{k+1} = g(u_k) = P(u_k - \varrho \nabla J(u_k)), \quad k \geqslant 0.$$

When $U = V$, the projection operator P is the identity, and the relation given above reduces to

$$u_{k+1} = u_k - \varrho \nabla J(u_k), \quad k \geqslant 0.$$

In this way, one recovers the *gradient method with fixed parameter for an unconstrained problem*, which was investigated in section 8.4. That is the reason why the method which we are about to describe is called the *projected-gradient method with fixed parameter*.

In order to show its convergence, it is enough simply to verify that, if the parameter $\varrho > 0$ is suitably chosen, then the function $g: V \rightarrow V$ is a *contraction mapping*, that is to say, there exists a number β such that

$$\beta < 1 \quad \text{and} \quad \| g(v_1) - g(v_2) \| \leqslant \beta \| v_1 - v_2 \| \quad \text{for every} \quad v_1, v_2 \in V.$$

This assumption implies, in effect, both the existence of a fixed point *and* the convergence of the method of successive approximations, provided the space V is *complete*; that is why the property of compactness does not appear in the proof. Inasmuch as *no* additional difficulty is introduced thereby, we shall, in fact, consider the (more general) *projected-gradient method with variable parameter*, defined by

$$u_{k+1} = P(u_k - \varrho_k \nabla J(u_k)), \quad \varrho_k > 0, \quad k \geqslant 0.$$

Theorem 8.6-2

Let V be a Hilbert space, U a non-empty, convex, closed subset of V and $J: V \rightarrow \mathbb{R}$ a functional which is differentiable in V. Suppose that there exist two constants α and M such that

$$\alpha > 0 \quad \text{and} \quad (\nabla J(v) - \nabla J(u), v - u) \geqslant \alpha \| v - u \|^2 \quad \text{for every} \quad u, v \in V,$$

$$\| \nabla J(v) - \nabla J(u) \| \leqslant M \| v - u \| \quad \text{for every} \quad u, v \in V.$$

If there exist two numbers a and b such that

$$0 < a \leqslant \varrho_k \leqslant b < \frac{2\alpha}{M^2} \quad \text{for every integer} \quad k \geqslant 0,$$

then the projected-gradient method with variable parameter converges and the convergence is geometric; that is, there exists a constant $\beta = \beta(\alpha, M, a, b)$ *such that*

$$\beta < 1 \quad and \quad \|u_k - u\| \leqslant \beta^k \|u_0 - u\|.$$

Proof

For every integer $k \geqslant 0$, we define the function

$$g_k : v \in V \to g_k(v) = P(v - \varrho_k \nabla J(v)) \in U \subset V.$$

By reason of the fact that the projection operator 'does not increase distances' (Theorem 8.1-1) and from the assumptions made regarding the functional, we can deduce the inequalities

$$
\begin{aligned}
\|g_k(v_1) - g_k(v_2)\|^2 &= \| P(v_1 - \varrho_k \nabla J(v_1)) - P(v_2 - \varrho_k \nabla J(v_2)) \|^2 \\
&\leqslant \| (v_1 - v_2) - \varrho_k (\nabla J(v_1) - \nabla J(v_2)) \|^2 \\
&= \| v_1 - v_2 \|^2 - 2\varrho_k (\nabla J(v_1) - \nabla J(v_2), v_1 - v_2) \\
&\quad + \varrho_k^2 \| \nabla J(v_1) - \nabla J(v_2) \|^2 \\
&\leqslant (1 - 2\alpha\varrho_k + M^2 \varrho_k^2) \| v_1 - v_2 \|^2,
\end{aligned}
$$

assuming that $\varrho_k > 0$. Moreover, it has already been established (in the proof of Theorem 8.4-4) that there exists a constant $\beta = \beta(\alpha, M, a, b)$ such that

$$(1 - 2\alpha\varrho_k + M^2 \varrho_k^2)^{1/2} \leqslant \beta < 1 \quad \text{for every} \quad k \geqslant 0,$$

when the numbers a and b satisfy the hypotheses of the statement. Since the solution u of the problem (P) is a fixed point of each function g_k, one may write

$$\|u_{k+1} - u\| = \|g_k(u_k) - g_k(u)\| \leqslant \beta \|u_k - u\|,$$

and the geometric convergence is proved. $\qquad\square$

Remarks

(1) The existence of a fixed point of the function $g(v) = P(v - \nabla J(v))$ associated with the projected-gradient method with fixed parameter, and hence the existence of a solution u of the inequalities $(\nabla J(u), v - u) \geqslant 0$ for every $v \in U$ provides a proof of the existence of a solution of the problem (P) associated with a set U and a functional satisfying the assumptions of the present theorem; which comes out as a particular case of the result of Theorem 8.2-2.

(2) If $U = V$, the convergence of the gradient method with variable parameter, already established in Theorem 8.4-4, is recovered.

(3) In the case of an *elliptic quadratic functional*

$$J : v \in \mathbb{R}^n \to J(v) = \tfrac{1}{2}(\mathbf{A}v, v) - (b, v), \quad \mathbf{A} = \mathbf{A}^{\mathrm{T}},$$

it is possible to show, *exactly as in the case of the unconstrained problem* (refer to the discussion following Theorem 8.4-4), that geometric convergence takes place for $\varrho_k \in [a, \tilde{b}] \subset]0, 2/\lambda_n[$, while, in this particular case, the theorem given above predicts convergence only for $\varrho_k \in [a, b] \subset]0, 2\lambda_1/\lambda_n^2[$ (we are reminded that λ_1 and λ_n denote the extreme eigenvalues of the matrix A). $\qquad\square$

The projected-gradient methods provide, then, *in theory*, approximation methods which are applicable to a large class of convex programming problems; but this is just a *lure* from the 'numerical' point of view, for the simple reason that the projection operator on a general convex, closed subset is not, in general, known explicitly.

A notable exception is that of subsets U of $V = \mathbb{R}^n$ which are of the form $\prod_{i=1}^{n} [a_i, b_i] \subset V = \mathbb{R}^n$ for which we constructed earlier, in section 8.1, the associated projection operator. For example, if

$$U = \mathbb{R}_+^n = \{v \in \mathbb{R}^n : v \geq 0\},$$

and if this set U is associated with an *elliptic quadratic functional*

$$J : v \in \mathbb{R}^n \to J(v) = \tfrac{1}{2}(Av, v) - (b, v),$$

the vector $u_{k+1} = (u_i^{k+1})_{i=1}^n$ is calculated in terms of the vector $u_k = (u_i^k)_{i=1}^n$ by means of the relations

$$u_i^{k+1} = \max \{u_i^k - \varrho_k(Au_k - b)_i, 0\}, \quad 1 \leq i \leq n.$$

With the exception of such particular cases, constrained problems need to be treated by *other* methods. A notable instance would be that of *penalty-function methods*, whose underlying idea rests on the following result.

Theorem 8.6-3

Let $J : \mathbb{R}^n \to \mathbb{R}$ be a continuous, coercive, strictly convex function, U a non-empty, convex, closed subset of $\mathbb{R}^n$ and $\psi : \mathbb{R}^n \to \mathbb{R}$ a continuous, convex function satisfying

$$\psi(v) \geq 0 \text{ for every } v \in \mathbb{R}^n \text{ and } \psi(v) = 0 \Leftrightarrow v \in U.$$

Then, for every $\varepsilon > 0$, there exists a unique element u_ε satisfying

(P_ε) $\quad u_\varepsilon \in \mathbb{R}^n$ and $J_\varepsilon(u_\varepsilon) = \inf\limits_{v \in \mathbb{R}^n} J_\varepsilon(v)$ where $J_\varepsilon(v) \overset{\text{def}}{=} J(v) + \dfrac{1}{\varepsilon}\psi(v).$

Besides, $\lim_{\varepsilon \to 0} u_\varepsilon = u$, where u is the unique solution of the problem

$$\text{find } u \text{ such that}$$

(P) $\qquad\qquad\qquad u \in U \quad \text{and} \quad J(u) = \inf\limits_{v \in U} J(v).$

Proof

It is clear that the problem (P) and the problems (P_ε) all have a unique solution. The functionals J_ε are, in fact, also coercive (since $J_\varepsilon(v) \geqslant J(v)$) and strictly convex (since the sum of a strictly convex function and a convex function is strictly convex). As

$$J(u_\varepsilon) \leqslant J(u_\varepsilon) + \frac{1}{\varepsilon}\psi(u_\varepsilon) = J_\varepsilon(u_\varepsilon) \leqslant J_\varepsilon(u) = J(u),$$

it follows from the coercivity of the functional J that the family $(u_\varepsilon)_{\varepsilon > 0}$ is *bounded*.

Because of the property of compactness, there exists a subsequence $(u_{\varepsilon'})_{\varepsilon' > 0}$ and an element $u' \in \mathbb{R}^n$ such that

$$\lim_{\varepsilon' \to 0} u_{\varepsilon'} = u'.$$

From the inequalities $J(u_{\varepsilon'}) \leqslant J(u)$ and the continuity of the function J, it follows that

$$J(u') = \lim_{\varepsilon' \to 0} J(u_{\varepsilon'}) \leqslant J(u).$$

Since

$$0 \leqslant \psi(u_{\varepsilon'}) \leqslant \varepsilon'(J(u) - J(u_{\varepsilon'})),$$

and since the sequence $(u_{\varepsilon'})_{\varepsilon' > 0}$ converges, the numbers $\{J(u) - J(u_{\varepsilon'})\}$ are bounded independently of ε'; consequently,

$$0 = \lim_{\varepsilon' \to 0} \psi(u_{\varepsilon'}) = \psi(u'),$$

since the function ψ is continuous, which shows that $u' \in U$ and hence that $u = u'$, since $J(u') \leqslant J(u)$; so that u is the unique solution of the problem (P). The uniqueness of this solution also shows that the *whole* family $(u_\varepsilon)_{\varepsilon > 0}$ converges to the element u (for it is possible to reproduce the preceding argument for *every* subsequence). □

Remark

It may be shown that every convex function $\varphi: \mathbb{R}^n \to \mathbb{R}$ is necessarily continuous (cf. Exercise 7.4-6); so that this 'hypothesis' is superfluous. □

As an application, we consider the following *convex programming* problem. Given a strictly convex function $J: \mathbb{R}^n \to \mathbb{R}$ and convex functions $\varphi_i: \mathbb{R}^n \to \mathbb{R}$, $1 \leqslant i \leqslant m$, find u such that

$$u \in U = \{v \in \mathbb{R}^n: \varphi_i(v) \leqslant 0, 1 \leqslant i \leqslant m\}, \quad J(u) = \inf_{v \in U} J(v).$$

As a function ψ which satisfies the assumptions of Theorem 8.6-3, one may take, for example,

$$\psi: v \in \mathbb{R}^n \to \psi(v) = \sum_{i=1}^{m} \max \{\varphi_i(v), 0\}.$$

The essential aim, therefore, in a penalty-function method is to replace *one constrained* optimisation problem by a *sequence* of *unconstrained* problems (which, theoretically, it is known how to solve), associated with the *penalised functionals* J_ε, $\varepsilon > 0$.

Remark

In practice, the force of penalty-function methods is dampened by the difficulty in knowing how to construct *effectively* 'good' functions ψ (for example, such as are differentiable, which is not the case, incidentally, in the example given above) which satisfy the conditions of the statement of the theorem. $\qquad\qquad\square$

Another way of approaching the solution of unconstrained problems is linked with the notion of *duality*. Its investigation and the construction of corresponding methods of approximation are the aims of the following chapter.

Exercises

8.6-1. Consider the *convex programming problem*
given a strictly convex function $J: \mathbb{R}^n \to \mathbb{R}$ and convex functions $\varphi_i: \mathbb{R}^n \to \mathbb{R}$,
find u such that

$$u \in U = \{v \in \mathbb{R}^n: \varphi_i(v) \leqslant 0, 1 \leqslant i \leqslant m\}, \quad J(u) = \inf_{v \in U} J(v).$$

For what values of the number α does the function

$$\psi: v \in \mathbb{R}^n \to \psi(v) = \sum_{i=1}^{m} (\max \{\varphi_i(v), 0\})^\alpha$$

satisfy the hypotheses of Theorem 8.6-3 (in the text, the case $\alpha = 1$ was considered)? Is it differentiable for particular values of the number α ?

8.6-2. Let $J: \mathbb{R}^n \to \mathbb{R}$ be a strictly convex and coercive function. Moreover, let there be given the set

$$U = \{v \in \mathbb{R}^n: \varphi_i(v) \leqslant 0, 1 \leqslant i \leqslant m\},$$

the functions $\varphi_i: \mathbb{R}^n \to \mathbb{R}$ being convex (recall that a real-valued function which is convex over $\mathbb{R}^n$ is continuous; cf. Exercise 7.4-6). Lastly, it is assumed that the set U is bounded and that its interior $\mathring{U}$, which is given by

$$\mathring{U} = \{v \in \mathbb{R}^n: \varphi_i(v) < 0, 1 \leqslant i \leqslant m\}$$

is non-empty.

(1) Given $\varepsilon > 0$, we define the *penalised functional*

$$I_\varepsilon : v \in \mathring{U} \rightarrow I_\varepsilon(v) = J(v) - \varepsilon \sum_{i=1}^{m} \frac{1}{\varphi_i(v)}.$$

It should be noted in passing that this is a case of a penalty-function method which is *altogether different* from that of Theorem 8.6-3 (considered in the preceding exercise). Prove that there is a unique solution to the problem

$$\text{find } u_\varepsilon \text{ such that}$$
$$u_\varepsilon \in \mathring{U}, \quad I_\varepsilon(u_\varepsilon) = \inf_{v \in \mathring{U}} I_\varepsilon(v)$$

(2) Prove that

$$\lim_{\varepsilon \to 0} u_\varepsilon = u,$$

where u is the unique solution of the problem

$$\text{find } u \text{ such that}$$
$$u \in U, \quad J(u) = \inf_{v \in U} J(v).$$

(3) Prove that

$$0 < \varepsilon' < \varepsilon \Rightarrow J(u) \leqslant J(u_{\varepsilon'}) \leqslant J(u_\varepsilon).$$

Introduction to non-linear programming

Introduction

In this chapter, we consider the following *non-linear programming* problem: find u such that

(P)
$$\begin{cases} u \in U \overset{\text{def}}{=} \{v \in \mathbb{R}^n : \varphi_i(v) \leqslant 0, \, 1 \leqslant i \leqslant m\}, \\[2ex] J(u) = \inf_{v \in U} J(v). \end{cases}$$

As the first step, we establish *necessary and sufficient* conditions, valid in *convex programming*, for an element $u \in U$ to be a solution of the problem (P). These conditions are the *Kuhn–Tucker conditions* (Theorems 9.2-3 and 9.2-4), which state the existence of numbers λ_i satisfying

$$\begin{cases} J'(u) + \sum_{i=1}^{m} \lambda_i \varphi_i'(u) = 0, \\[2ex] \lambda_i \geqslant 0, \quad 1 \leqslant i \leqslant m, \quad \sum_{i=1}^{m} \lambda_i \varphi_i(u) = 0. \end{cases}$$

The formal analogy with Lagrange multipliers, treated in section 7.2, should be noted. These conditions, in their turn, are obtained as a consequence of the *Farkas* lemma (Theorem 9.1-1), whose proof is the main aim in section 9.1.

The key idea, in the very next step, is to replace the *constrained* optimisation problem (P) by a *family* (P$_\mu$) of *unconstrained* optimisation problems, as the parameter μ, here called the *dual* variable, runs through some suitable set. More precisely, we introduce the function

$$L(v, \mu) = J(v) + \sum_{i=1}^{m} \mu_i \varphi_i(v), \quad \text{for} \quad v \in \mathbb{R}^n, \quad \mu \in \mathbb{R}_+^m,$$

called the *Lagrangian associated with the problem* (P), and we show that, if u denotes the solution of the problem (P), there exists, in certain circumstances, an element $\lambda \in \mathbb{R}^m_+$ satisfying

$$L(u, \lambda) = \inf_{v \in \mathbb{R}^n} L(v, \lambda) = \sup_{\mu \in \mathbb{R}^m_+} \inf_{v \in \mathbb{R}^n} L(v, \mu),$$

or

$$L(u, \lambda) = \sup_{\mu \in \mathbb{R}^m_+} L(u_\mu, \mu),$$

where, for every $\mu \in \mathbb{R}^m_+$, the vector u_μ denotes the solution of the *unconstrained* problem

(P_μ) $\qquad\qquad u_\mu \in \mathbb{R}^n, \quad L(u_\mu, \mu) = \inf_{v \in \mathbb{R}^n} L(v, \mu).$

We also establish that

$$L(u, \lambda) = \sup_{\mu \in \mathbb{R}^m_+} L(u, \mu) = \inf_{v \in \mathbb{R}^n} \sup_{\mu \in \mathbb{R}^m_+} L(v, \mu),$$

the points (u, λ) thus making their appearance as *saddle points* of the Lagrangian L. In this line of development, a relationship between the set of solutions of the problem (P) and the set of first arguments of the saddle points of the Lagrangian L is proved in Theorem 9.3-2, where the Kuhn–Tucker conditions are used in an essential manner.

With the help of the function

$$G: \mu \in \mathbb{R}^m_+ \to G(\mu) = \inf_{v \in \mathbb{R}^n} L(v, \mu),$$

it is then possible to define an optimisation problem in which *only the dual variable appears*. In fact, it may be observed that the second arguments $\lambda \in \mathbb{R}^m_+$ of the saddle points introduced above are the solutions of the problem, called the *dual problem* of the *primal* problem (P)

find λ such that

(Q) $\qquad\qquad \lambda \in \mathbb{R}^m_+, \quad G(\lambda) = \sup_{\mu \in \mathbb{R}^m_+} G(\mu).$

That is why, in Theorem 9.3-3, a relationship is established between the solutions of the primal and dual problems.

Although the dual problem (Q) is itself a *constrained* optimisation problem, it is essential to note that the constraints are of a kind which it is easy to treat *numerically*, because the projection operator from $\mathbb{R}^m$ to $\mathbb{R}^m_+$ has a particularly simple form. This observation suggests *the application of the gradient method with fixed parameter to the dual problem*. This is the main idea behind *Uzawa's method*, which produces as a 'by-product' (though, in fact, it is precisely that which is the target!) a sequence of vectors of $\mathbb{R}^n$, each obtained as the solution of an *unconstrained* optimisation problem, which converges (under certain conditions) to the solution of the primal problem (P) (Theorem 9.4-1). Uzawa's method is particularly well suited for application to constrained quadratic programming problems.

In this chapter, *all the vector spaces considered are real.*

9.1 The Farkas Lemma

The reader is strongly recommended to attempt to prove for himself (that is to say, without reading the proof given in the text...) the statement (i) which appears in the proof given below: it is not as easy as might appear at first sight.

Theorem 9.1-1 (*The Farkas Lemma*)
Suppose that a_i, $i \in I$, where I is a finite set of indices, and b are elements of a Hilbert space V, with scalar product $(\cdot, \cdot)$. Then the inclusion

$$\{w \in V : (a_i, w) \geqslant 0, i \in I\} \subset \{w \in V : (b, w) \geqslant 0\}$$

is satisfied if and only if

$$\text{there exist } \lambda_i \geqslant 0, i \in I, \text{ such that } b = \sum_{i \in I} \lambda_i a_i.$$

Proof
The condition is obviously sufficient. Its necessity is established in three steps (the first two steps being independent).

(i) *The set*

$$C = \left\{ \sum_{i \in I} \lambda_i a_i \in V : 0 \leqslant \lambda_i, i \in I \right\}$$

is a convex cone, with vertex at the origin, and closed in V. The convexity follows from the equality

$$\theta \sum_{i \in I} \lambda_i a_i + (1 - \theta) \sum_{i \in I} \mu_i a_i = \sum_{i \in I} \{\theta \lambda_i + (1 - \theta) \mu_i\} a_i.$$

Let us suppose that the vectors a_i, $i \in I$, are linearly independent. Every sequence $(v_k)_{k \geqslant 0}$ of points $v_k = \sum_{i \in I} \lambda_i^k a_i$ of C is, in particular, a sequence of points of the finite-dimensional, and hence closed, vector subspace U spanned by the vectors a_i. If it converges in V, it converges in U, and, as its convergence in U is equivalent to the convergence of its components, it follows that

$$\lim_{k \to \infty} v_k = \sum_{i=1}^{m} \left\{ \lim_{k \to \infty} \lambda_i^k \right\} a_i \in C,$$

which proves that the set C is closed.

If the vectors a_i, $i \in I$, are linearly dependent, there exist real numbers μ_i, $i \in I$, such that

$$\sum_{i \in I} \mu_i a_i = 0 \quad \text{and} \quad J \overset{\text{def}}{=} \{i \in I: \mu_i < 0\} \neq \varnothing.$$

Accordingly, every vector $v = \sum_{i \in I} \lambda_i a_i$ of C may also be written as

$$v = \sum_{i \in I} (\lambda_i + t\mu_i) a_i, \quad \text{where } t \overset{\text{def}}{=} \min_{i \in J} \left\{ -\frac{\lambda_i}{\mu_i} \right\} \geq 0.$$

The numbers $\lambda_i + t\mu_i$, $i \in I$, are all non-negative, with at least one equal to zero; so that the equality

$$C = \bigcup_{j \in I} \left\{ v = \sum_{l \in (I - \{j\})} \lambda_i a_i: 0 \leq \lambda_i, i \in (I - \{j\}) \right\},$$

is established, and it is enough to repeat the argument until the point is reached where the cone C is expressed as the finite union of cones associated with vectors a_i which are linearly independent, and hence closed, after the manner of the first part of the argument.

(ii) *Let C be a non-empty, closed, convex subset of a Hilbert space V and b a point of V which does not belong to C. It is then possible to find, in infinitely many ways, vectors $u \in V$ and real numbers α such that*

$$\begin{cases} (v, u) > \alpha & \text{for every } v \in C, \\ (b, u) < \alpha. \end{cases}$$

By the projection theorem, there exists a unique element $c \in C$ such that (cf. figure 9.1-2)

$$\begin{cases} \| b - c \|_V = \inf_{v \in C} \| b - v \|_V > 0, \\ (v - c, c - b) \geq 0 & \text{for every } v \in C. \end{cases}$$

Consequently,

$$(v, c - b) \geq (c, c - b) > (b, c - b),$$

and the assertion is proved: it is enough to choose $u = c - b$ and any number α satisfying

$$(c, c - b) > \alpha > (b, c - b).$$

(iii) *Proof of the Farkas Lemma proper.* It all comes down to showing that, if the point b does not belong to the set

$$C = \left\{ v - \sum_{i \in I} \lambda_i a_i: 0 \leq \lambda_i, i \in I \right\},$$

then there exists a vector u such that

$$(a_i, u) \geqslant 0, \quad i \in I, \quad \text{and} \quad (b, u) < 0.$$

By step (i), the set C is a non-empty, closed, convex subset of V. Hence, if $b \notin C$, there exists, by step (ii), a vector $u \in V$ and a real number α such that

$$(v, u) > \alpha, \quad \text{for every } v \in C, \quad \text{and} \quad (b, u) < \alpha.$$

As C is a cone with vertex at the origin,

$$O \in C \Rightarrow \alpha < 0,$$

so that

$$(v, u) > \frac{\alpha}{\lambda} \quad \text{for every } \lambda > 0 \Rightarrow (v, u) \geqslant 0 \quad \text{for every } v \in C.$$

Hence, we have established that

$$(a_i, u) \geqslant 0, \quad i \in I \text{ (since } a_i \in C), \quad \text{and} \quad (b, u) < \alpha < 0,$$

which proves the assertion. □

Remarks
(1) Contrary to the impression that a hasty analysis might leave, the proof of step (i) does indeed require elaboration (the set C is the direct image, but not an inverse image, of a closed subset of $\mathbb{R}^m \ldots$). When the vectors a_i are linearly dependent, a difficulty arises from the fact that it is quite possible for a sequence of points $v_k = \sum_{i \in I} \lambda_i^k a_i$ of C to converge, while the sequences $(\lambda_i^k)_{k \geqslant 0}$ $i \in I$, do not. The idea behind the proof, however, is clearly suggested by the two cases envisaged in figure 9.1-1.

(2) The conclusions of step (ii) are equivalent to the inclusions (figure 9.1-2)

$$\begin{cases} C \subset \{v \in V : (v, u) > \alpha\}, \\ \{b\} \subset \{v \in V : (v, u) < \alpha\}, \end{cases}$$

which show that the hyperplane $\{v \in V : (u, v) = \alpha\}$ 'strictly separates' the subsets C and $\{b\}$ of V; this result is a particular case of more general theorems involving the 'separation of convex sets' in normed vectors spaces.

Figure 9.1-1

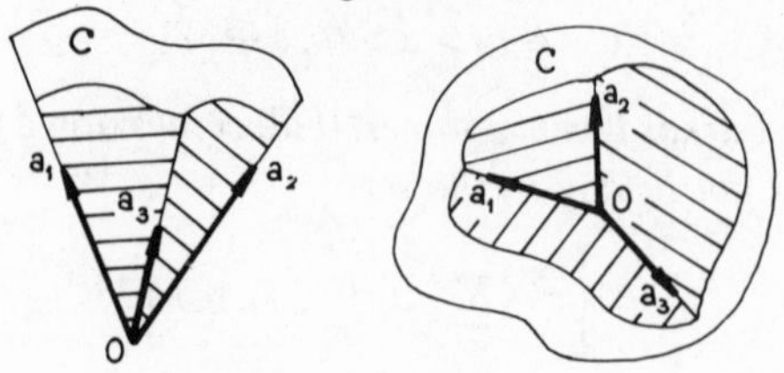

See, for example, Céa (1971), pages 19ff, Rockafellar (1970), §11, and Brezis (1982).

(3) The Farkas lemma lends itself to a simple geometrical interpretation, as figure 9.1-3 seeks to suggest.   □

Exercises

9.1-1. (1) Let C be a non-empty, closed, convex subset of a Hilbert space V. Define the function

$$h: \mu \in V \to h(\mu) = \sup_{v \in C} (\mu, v) \in \mathbb{R} \cup \{+\infty\},$$

the value $h(\mu) = +\infty$ *not being excluded.* Prove that

$$C = \bigcap_{\mu \in V} \{v \in V : (\mu, v) \leqslant h(\mu)\}.$$

Figure 9.1-2

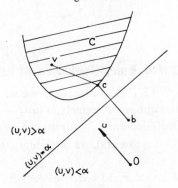

Figure 9.1-3

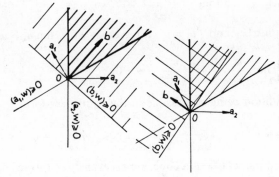

(2) Observe that the equality given above may also be written as

$$C = \bigcap_{\mu \in \Lambda} \{v \in V : (\mu, v) \leqslant h(\mu)\}, \quad \text{where} \quad \Lambda = \{\mu \in V : h(\mu) < +\infty\}.$$

Describe the set Λ when C is a vector subspace of V.

9.1-2. Let A be a matrix of type (m, n) and (u_k) a sequence of vectors of $\mathbb{R}^n_+$ such that the sequence (Au_k) converges. Show that there exists a vector $u \in \mathbb{R}^n$ such that

$$Au = \lim_{k \to \infty} Au_k \quad \text{and} \quad u \geqslant 0.$$

9.1-3. Assuming the conditions of the statement of the Farkas lemma and adopting the notation used there, let there be given the real numbers α_i, $i \in I$, and β. Prove that the inclusion

$$\{w \in V : (a_i, w) \geqslant \alpha_i, i \in I\} \subset \{w \in V : (b, w) \geqslant \beta\}$$

is satisfied if, and only if,

there exist $\lambda_i \geqslant 0, i \in I$, such that $\begin{cases} b = \sum_{i \in I} \lambda_i a_i, \\ \\ \beta \leqslant \sum_{i \in I} \lambda_i \alpha_i. \end{cases}$

9.2 The Kuhn–Tucker conditions

Our first aim (Theorem 9.2-1) is to generalise the necessary condition for the existence of a relative minimum, namely, '$J'(u)(v - u) \geqslant 0$ for every $v \in U$', which was established in Theorem 7.4-1 for the case of a *convex* set U, to the case of a *general* set U. To this end, we introduce, at every point $u \in U$, a 'tangent cone'.

More precisely, let V be a normed vector space and U a non-empty subset of V. For every point $u \in U$, the *cone $C(u)$ of feasible directions* is the union of $\{0\}$ and the set of vectors $w \in V$ for which there exists (at least) one sequence of points $(u_k)_{k \geqslant 0}$ such that

$$\begin{cases} u_k \in U \quad \text{and} \quad u_k \neq u \quad \text{for every } k \geqslant 0, \quad \lim_{k \to \infty} u_k = u, \\ \\ \lim_{k \to \infty} \dfrac{u_k - u}{\| u_k - u \|} = \dfrac{w}{\| w \|}, \quad w \neq 0, \end{cases}$$

the last condition being expressible in the equivalent form

$$u_k = u + \| u_k - u \| \frac{w}{\| w \|} + \| u_k - u \| \delta_k, \quad \lim_{k \to \infty} \delta_k = 0, \quad w \neq 0.$$

It is clear that the set $C(u)$ is a cone, *not necessarily convex* (cf. figure 9.2-1), whose *vertex is at the origin*; the 'translated' cone

$$u + C(u) = \{u + w \in V : w \in C(u)\}$$

has the point u for its vertex.

Remark

The set $C(u)$ contains, in particular, the 'oriented' tangents at u to the curves of points of U which pass through the point u (cf. the example of figure 9.2-1). If such a curve γ is represented by a function $t \geq 0 \to u(t)$, with $u = u(0)$, which is differentiable at 0, and if $u'(0) \neq 0$, then it is possible to write

$$u_k - u = t_k u'(0) + t_k \varepsilon_k, \quad \lim_{k \to \infty} \varepsilon_k = 0,$$

with

$$u_k = u(t_k), \quad t_k > 0, \quad \lim_{k \to \infty} t_k = 0,$$

so that

$$\lim_{k \to \infty} \frac{u_k - u}{\| u_k - u \|} = \frac{u'(0)}{\| u'(0) \|}. \qquad \square$$

We now establish two important properties of the cone of feasible directions.

Theorem 9.2-1

Let U be any non-empty subset of a normed vector space V.

(1) At every point $u \in U$, the cone $C(u)$ of feasible directions is closed.

(2) Let $J: \Omega \subset V \to \mathbb{R}$ be a function defined over an open set Ω containing U.

Figure 9.2-1. $\varphi_1(v) = -v_1 - v_2$; $\qquad \varphi_2(v) = v_1(v_1^2 + v_2^2) - 2(v_1^2 - v_2^2)$;
$U = \{v : \varphi_i(v) \leq 0, \ i = 1, 2\}$; $C(0) = \{v : v_1 + v_2 \geq 0, \ |v_1| \geq |v_2|\} \subsetneqq C^*(0)$
$= \{v : v_1 + v_2 \geq 0\}$.

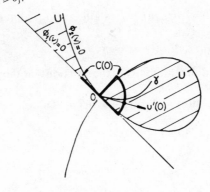

If the function J has a relative minimum, with respect to the set U, at a point $u \in U$ and if it is differentiable at u, then

$$J'(u)(v - u) \geqslant 0 \quad \text{for every } v \in \{u + C(u)\}.$$

Proof

(1) Let $(w_n)_{n \geqslant 0}$ be a sequence of points of $C(u)$ which converges to $w \in V$. It is enough to consider the case $w \neq 0$ (since it is already known that $0 \in C(u)$), and hence to suppose that $w_n \neq 0$ for every n. By definition, there exists, for every n, a sequence $(u_k^n)_{k \geqslant 0}$ such that

$$\begin{cases} u_k^n \in U, \quad u_k^n \neq u \quad \text{for every } k \geqslant 0, \quad \lim_{k \to \infty} u_k^n = u, \\ u_k^n = u + \| u_k^n - u \| \dfrac{w_n}{\| w_n \|} + \| u_k^n - u \| \delta_k^n, \quad \lim_{k \to \infty} \delta_k^n = 0. \end{cases}$$

Let $(\varepsilon_n)_{n \geqslant 0}$ be a sequence of numbers $\varepsilon_n > 0$ such that $\lim_{n \to \infty} \varepsilon_n = 0$. For every n, there exists an integer $k(n)$ such that

$$\| u_{k(n)}^n - u \| \leqslant \varepsilon_n, \quad \| \delta_{k(n)}^n \| \leqslant \varepsilon_n.$$

Let us now consider the sequence $\{ u_{k(n)}^n \}_{n \geqslant 0}$. We have, firstly,

$$u_{k(n)}^n \in U, \quad u_{k(n)}^n \neq u \quad \text{for every } n \geqslant 0, \quad \lim_{n \to \infty} u_{k(n)}^n = u,$$

and, secondly, we can write

$$u_{k(n)}^n = u + \| u_{k(n)}^n - u \| \frac{w}{\| w \|} + \| u_{k(n)}^n - u \| \left\{ \delta_{k(n)}^n + \left(\frac{w_n}{\| w_n \|} - \frac{w}{\| w \|} \right) \right\}.$$

Since $\lim_{k \to \infty} (w_n / \| w_n \|) = (w / \| w \|)$, the statement is proved.

(2) Let $w = v - u$ be a non-zero vector of the cone $C(u)$ and let (u_k) be a sequence of points of $U - \{u\}$ such that

$$\begin{cases} \lim_{k \to \infty} u_k = u, \\ u_k - u = \| u_k - u \| \dfrac{w}{\| w \|} + \| u_k - u \| \delta_k, \quad \lim_{k \to \infty} \delta_k = 0, \\ J(u) \leqslant J(u_k) \quad \text{for every } k. \end{cases}$$

Making use of the differentiability of the function J at the point u, we obtain

$$0 \leqslant J(u_k) - J(u) = J'(u)(u_k - u) + \| u_k - u \| \varepsilon_k, \quad \lim_{k \to \infty} \varepsilon_k = 0,$$

or, in other words $(J'(u) \in V')$,

$$0 \leqslant \frac{\| u_k - u \|}{\| w \|} (J'(u) w + \eta_k), \quad \lim_{k \to \infty} \eta_k = 0.$$

It follows that the number $J'(u)w$ is necessarily non-negative; otherwise, the expression $(J'(u)w + \eta_k)$ would be negative for sufficiently large values of the integer k (note the use of the relation $u_k \neq u$). $\qquad\square$

From now on, we consider *particular* sets U, of the form

$$U = \{v \in \Omega : \varphi_i(v) \leqslant 0, 1 \leqslant i \leqslant m\}, \quad \Omega \text{ an open subset of } V,$$

and, by means of suitable hypotheses imposed upon the constraints, that is to say, upon the functions $\varphi_i : \Omega \subset V \to \mathbb{R}$, we shall *describe*, in a simple manner, the cone of feasible directions at a general point of the set U. With each point $u \in U$ is associated the set of indices

$$I(u) = \{1 \leqslant i \leqslant m : \varphi_i(u) = 0\}.$$

Now, as is suggested by figure 9.2-2 (at least for the particular case $m = n = 2$), the cone $C(u)$ of feasible directions at a general point $u \in U$ appears to coincide, at least in certain cases, with the cone

$$C^*(u) \overset{\text{def}}{=} \{w \in V : \varphi_i'(u)w \leqslant 0, i \in I(u)\}$$

(obviously, $C^*(u) = V$ if $I(u) = \varnothing$). However, there are no grounds for expecting equality to hold in every case, if only because the cone $C^*(u)$ is always convex, while the cone $C(u)$ is not necessarily so (cf. the example of figure 9.2-1). That is why we are led to make the following definition. The constraints are said to *be qualified at the point* $u \in U$ if either all the functions φ_i, $i \in I(u)$, are affine,

Figure 9.2-2. $I(u) = \{2\}$; $I(u') = \{1, 2\}$; $I(u'') = \phi$.

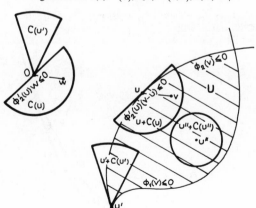

or there exists a vector $\tilde{w} \in V$ such that

$$\left.\begin{array}{l} \varphi_i'(u)\tilde{w} \leqslant 0, \\ \varphi_i'(u)\tilde{w} < 0 \quad \text{if} \quad \varphi_i \text{ is not affine,} \end{array}\right\} \quad \text{for every } i \in I(u).$$

Remark

The reader could not be recommended too strongly to reflect on the geometrical significance of this condition, at least for $n = 2$. If a function φ_i, $i \in I(u)$, is not affine, the condition $\varphi_i'(u) \neq 0$ precludes, in particular, the existence of a double point of the curve $\varphi_i(v) = 0$ at the point u, which would be a cause of the non-convexity of the set $C(u)$ (refer to the example of figure 9.2-1). Figure 9.2-3 should likewise give insight into the usefulness of the condition $\tilde{w} \neq 0$. Lastly, the reasons why affine functions have been made a case apart should become evident with the help of some well-chosen figures, as well as from the proof of the theorem which follows. □

Theorem 9.2-2

Let u be a point of the set

$$U = \{v \in \Omega : \varphi_i(v) \leqslant 0, 1 \leqslant i \leqslant m\}$$

at which the functions $\varphi_i : \Omega \subset V \to \mathbb{R}$, $i \in I(u)$, are differentiable.

(1) The cone of feasible directions at this point always satisfies the inclusion

$$C(u) \subset C^*(u) \stackrel{\text{def}}{=} \{w \in V : \varphi_i'(u)w \leqslant 0, i \in I(u)\}.$$

(2) If the constraints are qualified at u and if the functions φ_i, $i \notin I(u)$, are continuous at u, then the following equality holds:

$$C(u) = C^*(u).$$

Figure 9.2-3. $\varphi_1(v) = v_2 - \max\{0, v_1^3\}$; $\varphi_2(v) = v_1^4 - v_2$; $C(0) = \{v : v_1 \geqslant 0,\ v_2 \geqslant 0\} \subsetneq C^*(0) = \{v : v_2 = 0\}$.

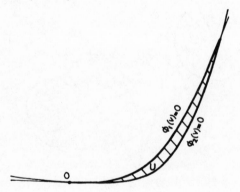

Proof

(1) Every function φ_i, $i \in I(u)$, has a relative minimum at u, with respect to the set U, as a result of the definition of the set $I(u)$. An application of Theorem 9.2-1 then shows that

$$\varphi_i'(u)w \geqslant 0 \quad \text{for every } w \in C(u), \quad i \in I(u),$$

which is precisely the required result.

(2) Suppose, initially, that all the functions φ_i, $i \in I(u)$, are *affine*. Given a non-zero vector w of $C^*(u)$, we consider the sequence (u_k) defined by

$$u_k = u + \varepsilon_k w,$$

where (ε_k) is a sequence of numbers $\varepsilon_k > 0$ such that $\lim_{k \to \infty} \varepsilon_k = 0$. Then $u_k - u = \varepsilon_k w \neq 0$ for every $k \geqslant 0$ and $\lim_{k \to \infty} u_k = u$. Moreover,

$$0 > \varphi_i(u) = \lim_{k \to \infty} \varphi_i(u_k) \quad \text{for} \quad i \notin I(u) \quad \text{(by continuity)},$$

$$\varphi_i(u_k) = \varphi_i'(u)(u_k - u) = \varepsilon_k \varphi_i'(u)w \leqslant 0 \quad \text{for} \quad i \in I(u),$$

which shows that the points u_k belong to the set U for k sufficiently large. Lastly, as

$$\frac{u_k - u}{\| u_k - u \|} = \frac{w}{\| w \|} \quad \text{for every } k,$$

we have succeeded in proving that $w \in C(u)$.

We now proceed to the general case. There exists a vector $\tilde{w} \neq 0$ such that

$$\left. \begin{array}{l} \varphi_i'(u)\tilde{w} \leqslant 0, \\ \varphi_i'(u)\tilde{w} < 0 \quad \text{if} \quad \varphi_i \text{ is not affine,} \end{array} \right\} \quad \text{for every } i \in I(u).$$

Given a non-zero vector w of $C^*(u)$, let δ be a positive number such that the vector $w + \delta\tilde{w}$ is not null. Consider the sequence (u_k) defined by

$$u_k = u + \varepsilon_k(w + \delta\tilde{w}), \quad \varepsilon_k > 0, \quad \lim_{k \to \infty} \varepsilon_k = 0.$$

Then $u_k - u = \varepsilon_k(w + \delta\tilde{w}) \neq 0$ for every k and $\lim_{k \to \infty} u_k = u$. Moreover,

$$0 > \varphi_i(u) = \lim_{k \to \infty} \varphi_i(u_k) \quad \text{for} \quad i \notin I(u) \quad \text{(by continuity)},$$

$$\varphi_i(u_k) = \varepsilon_k\{\varphi_i'(u)w + \delta\varphi_i'(u)\tilde{w}\} \leqslant 0 \quad \text{for} \quad i \in I(u) \quad \text{and} \quad \varphi_i \text{ affine},$$

$$\left\{ \begin{array}{l} \varphi_i(u_k) = \varepsilon_k\{\varphi_i'(u)w + \delta\varphi_i'(u)\tilde{w} + \alpha_k\}, \quad \lim_{k \to \infty} \alpha_k = 0, \text{ so that} \\ \varphi_i(u_k) \leqslant \varepsilon_k\{\delta\varphi_i'(u)\tilde{w} + \alpha_k\} \quad \text{for} \quad i \in I(u) \quad \text{and} \quad \varphi_i \text{ not affine,} \end{array} \right.$$

which shows that the points u_k belong to the set U for k sufficiently large (which explains both the *strict* inequality $\varphi_i'(u)\tilde{w} < 0$ and the introduction of the vectors $w + \delta\tilde{w}$). Since

$$\frac{u_k - u}{\|u_k - u\|} = \frac{w + \delta\tilde{w}}{\|w + \delta\tilde{w}\|} \quad \text{for every } k,$$

this proves that $w + \delta\tilde{w} \in C(u)$ for every number $\delta > 0$ sufficiently small. The sequence (w_n) defined by

$$w_n = w + \varepsilon_n \tilde{w}$$

is a sequence of points of $C(u)$ (for n sufficiently large). As the set $C(u)$ is closed (Theorem 9.2-1), we conclude that $\lim_{n \to \infty} w_n = w \in C(u)$. $\qquad\square$

A very important application is that where $V = \mathbb{R}^n$ and all the functions φ_i are affine, the set U then taking the form

$$U = \left\{ v \in \mathbb{R}^n : \sum_{j=1}^n c_{ij} v_j \leqslant d_i, 1 \leqslant i \leqslant m \right\} = \{ v \in \mathbb{R}^n : Cv \leqslant d \},$$

where $C = (c_{ij}) \in \mathscr{A}_{m,n}(\mathbb{R})$ is a given matrix and $d \in \mathbb{R}^m$ a given vector. Then *the constraints are qualified at every point* $u \in U$, so that, by the preceding theorem, *the cone of feasible directions at the point* $u \in U$ *is the set* (cf. figure 9.2-4)

$$C(u) = \left\{ w \in \mathbb{R}^n : \sum_{j=1}^n c_{ij} w_j \leqslant 0, i \in I(u) \right\}.$$

Returning to the general situation, and putting together the various results established in this and the preceding sections, we obtain *one of the most important results of Optimisation.*

Figure 9.2-4

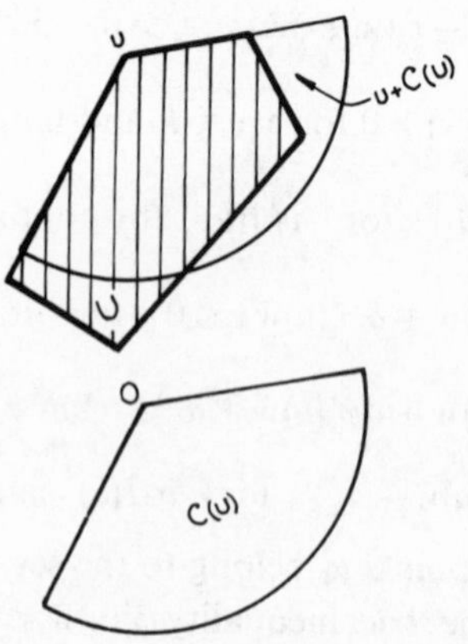

Theorem 9.2-3 (*Necessary condition for the existence of a minimum in non-linear programming*)

Let $\varphi_i: \Omega \subset V \to \mathbb{R}, 1 \leqslant i \leqslant m$, *be functions which are defined over an open subset* Ω *of a Hilbert space* V *and*

$$U = \{v \in \Omega: \varphi_i(v) \leqslant 0, 1 \leqslant i \leqslant m\}$$

a subset of Ω. *Let* u *be a point of* U *with which is associated the set of indices*

$$I(u) = \{1 \leqslant i \leqslant m: \varphi_i(u) = 0\}.$$

Suppose that the functions φ_i, $i \in I(u)$, *are differentiable at* u *and the functions* φ_i, $i \notin I(u)$, *continuous at* u. *Finally, let the given function* $J: \Omega \subset V \to \mathbb{R}$ *be differentiable at* u.

If the function J *has at* u *a relative minimum, with respect to the set* U, *and if the constraints are qualified at the point* u, *then there exist numbers* $\lambda_i(u)$, $i \in I(u)$, *such that*

$$J'(u) + \sum_{i \in I(u)} \lambda_i(u)\,\varphi_i'(u) = 0, \quad \text{and} \quad \lambda_i(u) \geqslant 0, \quad i \in I(u).$$

Proof

By Theorem 9.2-1,

$$J'(u)w \geqslant 0 \quad \text{for every } w \in C(u).$$

By Theorem 9.2-2,

$$C(u) = \{w \in V: \varphi_i'(u)w \leqslant 0, i \in I(u)\}.$$

Denoting by $(\cdot,\cdot)$ the scalar product of the space V and by τ the canonical isometry from V to V', we then find ourselves that the inclusion

$$\{w \in V: (-\tau\varphi_i'(u), w) \geqslant 0, i \in I(u)\} \subset \{w \in V: (\tau J'(u), w) \geqslant 0\}$$

holds; but this is exactly the situation which makes the *Farkas lemma* (Theorem 9.1-1) applicable. The conclusion then follows immediately. $\square$

The relations

$$J'(u) + \sum_{i \in I(u)} \lambda_i(u)\varphi_i'(u) = 0 \quad \text{and} \quad \lambda_i(u) \geqslant 0, \quad i \in I(u),$$

are known as the *Kuhn–Tucker conditions*. One might say that *they are to the 'inequality constraints' what the Lagrange multipliers (section 7.2) are to the 'equality constraints'*. Making use of the inequalities $\varphi_i(u) \leqslant 0, 1 \leqslant i \leqslant m$, it is possible, moreover, to express them in a form which is even closer to the relation in which the Lagrange multipliers make their appearance, since *they are equivalent, in effect, to the existence of numbers* $\lambda_i(u), 1 \leqslant i \leqslant m$, (set

$\lambda_i(u) = 0$ for $i \notin I(u)$) *such that*

$$\begin{cases} J'(u) + \sum_{i=1}^{m} \lambda_i(u)\,\varphi_i'(u) = 0, \\ \\ \lambda_i(u) \geqslant 0, \quad 1 \leqslant i \leqslant m, \quad \sum_{i=1}^{m} \lambda_i \varphi_i(u) = 0. \end{cases}$$

In view of the analogy with the 'usual' Lagrange multipliers, the vector $\lambda(u) = (\lambda_i(u)) \in \mathbb{R}_+^m$ is called a *generalised Lagrange multiplier*, associated with the relative minimum u.

Nevertheless, one should not lose sight of the fact that it is difficult to exploit these relations *in practice*, because of the *inequalities* which they contain. For example, in the case $V = \mathbb{R}^n$, a *necessary* condition for the existence of a relative minimum of a function $J: U \to \mathbb{R}$ is the existence of solutions u_i, $1 \leqslant i \leqslant n$, and λ_j, $1 \leqslant j \leqslant m$, of the following system of *equations* and *inequalities*

$$\begin{cases} \partial_1 J(u) + \lambda_1 \partial_1 \varphi_1(u) + \cdots + \lambda_m \partial_1 \varphi_m(u) = 0, \\ \quad\vdots \\ \partial_n J(u) + \lambda_1 \partial_n \varphi_1(u) + \cdots + \lambda_m \partial_n \varphi_m(u) = 0, \\ \quad\;\; \lambda_1 \varphi_1(u) + \cdots + \lambda_m \varphi_m(u) = 0, \\ \quad\;\; \varphi_1(u) \geqslant 0, \qquad\qquad \lambda_1 \geqslant 0, \\ \quad\;\; \vdots \qquad\qquad\qquad\qquad \vdots \\ \quad\;\; \varphi_m(u) \geqslant 0, \qquad\qquad \lambda_m \geqslant 0. \end{cases}$$

It is instructive to compare this with the corresponding system of *equations* found in section 7.2.

Remarks
(1) The numbers $\lambda_i(u)$, $i \in I(u)$, are not necessarily determined uniquely at a relative minimum u, except when the derivatives $\varphi_i'(u)$, $i \in I(u)$, are linearly independent (in V').

(2) If $I(u) = \varnothing$, then the Kuhn–Tucker conditions reduce to $J'(u) = 0$; this, of course, is to be expected, since the point u then belongs to the *interior* of the set U: one simply recovers the necessary condition for the existence of a relative extremum over an open set.   □

While the condition of constraint qualifications has so far been given in a form which is not conducive to easy handling, particularly in making use of the point at which the expression holds, we shall establish in the following theorem that it can be put in a simpler form and, above all, one which is *independent of the point under consideration*, when the functions φ_i

are *convex*. For this reason, the following definition is introduced. The *convex* constraints $\varphi_i: \Omega \subset V \to \mathbb{R}$, $1 \leqslant i \leqslant m$, are said to be *qualified* if either all the functions φ_i, $1 \leqslant i \leqslant m$, are affine and the set (convex if Ω is convex)

$$U = \{v \in \Omega : \varphi_i(v) \leqslant 0, 1 \leqslant i \leqslant m\}$$

is non-empty,

or there exists a point $\tilde{v} \in \Omega$ such that

$$\left. \begin{array}{l} \varphi_i(\tilde{v}) \leqslant 0, \\ \varphi_i(\tilde{v}) < 0 \quad \text{if} \quad \varphi_i \text{ is not affine,} \end{array} \right\} \quad 1 \leqslant i \leqslant m.$$

Remarks

(1) In the second case, it should be observed that the set U is still non-empty, but that 'somewhat more' is assumed.

(2) *If the open set Ω and the functions φ_i are convex, then so, too, is the set U.* Incidentally, care should be taken not to confuse the convexity of a function $\varphi: \Omega \subset \mathbb{R}^n \to \mathbb{R}$ with the convexity of the function $\psi: \Theta \subset \mathbb{R}^{n-1} \to \mathbb{R}$, in case there exists an equivalence of the form $\varphi(v) = 0 \Leftrightarrow v_n = \psi(v_1, \ldots, v_{n-1})$ (figure 9.2-3 ought to be instructive in this respect). □

While, on the one hand, the *convexity of the constraints* φ_i makes it possible to establish the necessity of the Kuhn–Tucker conditions, with qualifying hypotheses which are starkly more simple than those of the preceding theorem, the additional assumption of the *convexity of the function J*, on the other hand, enables us to prove that they are also *sufficient* (independently, moreover, of any hypotheses involving constraint qualifications).

Theorem 9.2-4 (*Necessary and sufficient condition for the existence of a minimum in convex programming*)

Let $J: \Omega \subset V \to \mathbb{R}$ be a function defined over an open, convex subset Ω of a Hilbert space V and

$$U = \{v \in \Omega : \varphi_i(v) \leqslant 0, 1 \leqslant i \leqslant m\}$$

a subset of Ω, the constraints $\varphi_i: \Omega \subset V \to \mathbb{R}$, $1 \leqslant i \leqslant m$, being assumed to be convex. Let $u \in U$ be a point at which the functions φ_i, $1 \leqslant i \leqslant m$, and J are differentiable.

(1) If the function J has at u a relative minimum with respect to the set U and if the constraints are qualified, then there exist numbers $\lambda_i(u)$, $1 \leqslant i \leqslant m$,

such that the Kuhn–Tucker conditions

$$\begin{cases} J'(u) + \sum_{i=1}^{m} \lambda_i(u)\varphi_i'(u) = 0, \\ \lambda_i(u) \geqslant 0, \quad 1 \leqslant i \leqslant m, \quad \sum_{i=1}^{m} \lambda_i(u)\varphi_i(u) = 0, \end{cases}$$

are satisfied.

(2) *Conversely, if the function* $J: U \to \mathbb{R}$ *is convex and if there exist numbers* λ_i, $1 \leqslant i \leqslant m$, *such that the Kuhn–Tucker conditions are satisfied, then the function* J *has at* u *a relative minimum with respect to the set* U.

Proof

(1) *Obviously enough, it is sufficient to prove that, if the convex constraints* φ_i are qualified in the sense understood above, then they also do so, at every point $u \in U$, in the sense understood earlier; this then makes it possible to apply Theorem 9.2-3. Now, if $u \neq \tilde{v}$, the vector $\tilde{w} = \tilde{v} - u$ meets the requirements, since

$$\varphi_i'(u)\tilde{w} = \varphi_i(u) + \varphi_i'(u)(\tilde{v} - u) \leqslant \varphi_i(\tilde{v}) \quad \text{for every } i \in I(u),$$

the functions φ_i being convex; if $u = \tilde{v}$, the set $I(u)$ can only contain indices i for which the corresponding functions φ_i are affine, and the vector $\tilde{w} = 0$ will then do.

(2) Let v be any point of the (convex) set U. Then

$$J(u) \leqslant J(u) - \sum_{i=1}^{m} \lambda_i \varphi_i(v) \quad (\lambda_i \geqslant 0, \varphi_i(v) \leqslant 0)$$

$$= J(u) - \sum_{i \in I(u)} \lambda_i(\varphi_i(v) - \varphi_i(u)) \quad (\lambda_i = 0 \text{ if } i \notin I(u), \varphi_i(u) = 0 \text{ if } i \in I(u)),$$

$$\leqslant J(u) - \sum_{i \in I(u)} \lambda_i \varphi_i'(u)(v - u) \quad (\varphi_i \text{ convex})$$

$$\leqslant J(u) + J'(u)(v - u) \quad \text{(Kuhn–Tucker)}$$

$$\leqslant J(v) \quad (J \text{ convex}). \qquad \qquad \square$$

One interpretation of the Kuhn–Tucker conditions, which is of fundamental importance to the development, is the following. They express the fact that *there exists* a function

$$\mathscr{J}_u : v \in \Omega \subset V \to \mathscr{J}_u(v) \overset{\text{def}}{=} J(v) + \sum_{i=1}^{m} \lambda_i(u)\varphi_i(v)$$

which, of course, *depends on the point* u (through the intermediary of the vector $\lambda(u)\in\mathbb{R}^m_+$) such that

$$\begin{cases} J(u) = \inf_{v\in U} J(v) \Rightarrow \mathcal{J}'_u(u) = 0, \\ J(u) = \mathcal{J}_u(u). \end{cases}$$

In other words, *if the vector $\lambda(u)$ is known, we are led back to the same necessary condition, which is also sufficient in the convex case, which holds when dealing with an unconstrained optimisation problem with function $\mathcal{J}_u$.* This important observation is the inspiration behind *duality* techniques, which are met in the following two sections.

Finally, let us look at an application of the previous results, when $V = \mathbb{R}^n$, and (with self-explanatory notation)

$$\begin{aligned} U &= \left\{ v\in\mathbb{R}^n \colon \sum_{j=1}^n c_{ij}v_j \leqslant d_i, 1\leqslant i\leqslant m \right\} \\ &= \{ v\in\mathbb{R}^n \colon (C_i, v) \leqslant d_i, 1\leqslant i\leqslant m \} \\ &= \{ v\in\mathbb{R}^n \colon Cv \leqslant d \}. \end{aligned}$$

In this particular case, but one which, nevertheless, is very common in applications, we recall that the constraint qualifications are simply equivalent to the assumption that the set U is non-empty. Accordingly, *a necessary condition, which is also sufficient if $J\colon U \to \mathbb{R}$ is convex, for the function J to have at the point $u\in U$ a relative minimum, with respect to the set U, is that there exist a vector $\lambda\in\mathbb{R}^m$, with components λ_i, such that*

$$\begin{cases} \nabla J(u) + C^T\lambda = 0, \\ \lambda\in\mathbb{R}^m_+ \quad \text{and} \quad \lambda_i = 0 \quad \text{if} \quad (C_i, u) < d_i. \end{cases}$$

Exercises

9.2-1. Establish the relationship which exists between the two properties 'the convex constraints φ_i, $1\leqslant i\leqslant m$, are qualified' and 'the corresponding set U has non-empty interior'.

9.2-2. (1) Let u be a point which satisfies

$$\begin{cases} u\in U \stackrel{\text{def}}{=} \{v\in\mathbb{R}^n \colon Cv = d\}, \quad \text{where} \ C\in\mathscr{A}_{m,n}(\mathbb{R}), \quad d\in\mathbb{R}^m, \\ J(u) = \inf_{v\in U} J(v), \end{cases}$$

the functional $J\colon\mathbb{R}^n \to \mathbb{R}$ being assumed to be differentiable at u. Using the equivalence

$$Cv = d \Leftrightarrow Cv \leqslant d \quad \text{and} \quad -Cv \leqslant -d,$$

and the Kuhn–Tucker conditions, rediscover the existence of Lagrange multipliers at the point u, including the case where the matrix C is not of rank m.

(2) For sets U having the particular form considered here, show that the result may also be inferred from Theorem 7.2-3 when $r(C) < m$.

9.2-3. The aim in this problem is to give necessary second-order conditions (that is to say, such as introduce second-order derivatives) for the existence of a minimum in non-linear programming, known by the name of the *MacCormick theorem*. This result, which complements Theorem 9.2-3, is thus an analogue, for the 'constrained' case, of the necessary condition of Theorem 7.3-1, established for the 'unconstrained' case.

Let $\varphi_i: \mathbb{R}^n \to \mathbb{R}$, $1 \le i \le m$, and $J: \mathbb{R}^n \to \mathbb{R}$ be twice continuously differentiable functions and let

$$U = \{v \in \mathbb{R}^n: \varphi_i(v) \le 0, 1 \le i \le m\}.$$

If the function J admits at a point $u \in U$ a relative minimum with respect to the set U, if the constraints are qualified at u and if the derivatives $\varphi_i'(u)$, $i \in I(u) = \{1 \le i \le m: \varphi_i(u) = 0\}$, are linearly independent, then there exist numbers $\lambda_i(u)$, $i \in I(u)$, such that

$$J'(u) + \sum_{i \in I(u)} \lambda_i(u)\varphi_i'(u) = 0 \quad \text{and} \quad \lambda_i(u) \ge 0, \quad i \in I(u),$$

$$\varphi_i'(u)w = 0, \quad i \in I(u) \Leftrightarrow \left(J''(u) + \sum_{i \in I(u)} \lambda_i(u)\varphi_i''(u) \right)(w, w) \ge 0.$$

9.2-4. Is the cone of feasible directions at a point of a convex subset of a normed vector space itself always convex?

9.2-5. The cone of feasible directions is not the only 'tangent cone' which one could 'reasonably' associate with a point of a general set; in particular, one which is not convex. Here are two other examples.

Let U be a non-empty subset of a normed vector space V. At every point $u \in U$, the *Bouligand contingent cone* is the set $C_1(u)$ of vectors $w \in V$ for which there exists a sequence of vectors $w_n \in V$ and a sequence of numbers ε_n such that

$$\lim_{n \to \infty} w_n = w, \quad \varepsilon_n > 0 \quad \text{and} \quad \lim_{n \to \infty} \varepsilon_n = 0,$$
$$u + \varepsilon_n w_n \in U \text{ for every } n.$$

(1) Prove that

$$C_1(u) = \bigcap_{\varepsilon > 0} \bigcap_{\alpha > 0} \bigcap_{0 < h \le \alpha} \left(\frac{1}{h}(U - u) + B_\varepsilon \right),$$

where $B_\varepsilon = \{v \in V: \|v\| \le \varepsilon\}$.

(2) Prove that

$$w \in C_1(u) \Leftrightarrow \liminf_{\varepsilon \to 0} \left\{ \frac{1}{\varepsilon} \inf_{v \in u} \|(u + \varepsilon w) - v\| \right\} = 0.$$

(3) Prove that $C_1(u)$ is a closed cone.

(4) At every point $u \in U$, the *Clarke tangent cone* is the set $C_2(u)$ of vectors $w \in V$ such that, with every sequence (u_n, ε_n) satisfying

$$u_n \in U, \quad \lim_{n \to \infty} u_n = u, \quad \varepsilon_n > 0 \quad \text{and} \quad \lim_{n \to \infty} \varepsilon_n = 0,$$

it is possible to associate a sequence of vectors $w_n \in V$ such that

$$\lim_{n \to \infty} w_n = w \quad \text{and} \quad (u_n + \varepsilon_n w_n) \in U \text{ for every } n.$$

Prove that

$$C_2(u) = \bigcap_{\varepsilon > 0} \bigcap_{\substack{\alpha > 0 \\ \{\beta > 0}} \bigcap_{\substack{0 < h \leqslant \beta \\ z \in U \cap (u + B_\alpha)}} \left(\frac{1}{h}(U - z) + B_\varepsilon \right).$$

(5) Prove that

$$w \in C_2(u) \Leftrightarrow \lim_{\substack{z \to u \\ z \in U \\ \varepsilon \to 0^+}} \left\{ \frac{1}{\varepsilon} \inf_{v \in U} \| (z + \varepsilon w) - v \| \right\} = 0.$$

(6) Prove that $C_2(u)$ is a closed, convex cone.

(7) Prove the inclusions

$$C_2(u) \subset \bigcap_{\varepsilon > 0} \bigcup_{\eta > 0} \bigcap_{(z - u) \in B_\eta} \left(C_1(z) + B_\varepsilon \right) \subset C_1(u).$$

(8) Prove that the first of the two inclusions given above becomes an equality if U is a closed subset of a finite-dimensional vector space.

9.2-6. Let $J: \mathbb{R}^n \to \mathbb{R}$ and $\varphi_i: \mathbb{R}^n \to \mathbb{R}$, $1 \leqslant i \leqslant m$, be convex functions. With every vector $d = (d_i) \in \mathbb{R}^m$, associate the set

$$U(d) = \{ v \in \mathbb{R}^n : \varphi_i(v) \leqslant d_i, 1 \leqslant i \leqslant m \},$$

and denote by D the set

$$D = \{ d \in \mathbb{R}^m : U(d) \neq \varnothing \}.$$

Assume that for every $d \in D$, there exists a unique solution $u(d)$ of the problem

$$\text{find } u \text{ such that}$$
$$u(d) \in U(d), \quad J(u(d)) = \inf_{v \in U(d)} J(v).$$

(1) Prove that the set D is convex.

(2) Prove that the function

$$f: d \in D \subset \mathbb{R}^m \to f(d) \overset{\text{def}}{=} J(u(d)) \in \mathbb{R}$$

is convex.

(3) Suppose that $0 \in D$. Prove that there exists a vector $\lambda \in \mathbb{R}^m$ such that

$$f(0) \leqslant f(d) + (\lambda, d)_m \quad \text{for every } d \in D.$$

(4) Assuming that $0 \in \mathring{D}$ and that the function f is differentiable at 0, show that

$$\lambda = -\nabla f(0).$$

Compare this result with that of Exercise 7.2-2.

9.3 Lagrangians and saddle points. Introduction to duality

Let V and M be any two sets and

$$L: V \times M \to \mathbb{R}$$

a function. The point $(u, \lambda) \in V \times M$ is said to be a *saddle point* of the function L if the point u is a minimum of the function $L(\cdot, \lambda): v \in V \to L(v, \lambda) \in \mathbb{R}$ and if the point λ is a maximum of the function $L(u, \cdot): \mu \in M \to L(u, \mu) \in \mathbb{R}$; in other words, if

$$\sup_{\mu \in M} L(u, \mu) = L(u, \lambda) = \inf_{v \in V} L(v, \lambda).$$

Remarks

(1) It should be observed that the definition introduces the order in which the two variables appear in an essential (although, evidently, arbitrary) manner.

Figure 9.3-1

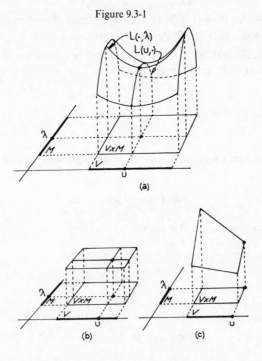

(a)

(b)　　　(c)

(2) In the use which will be made of this idea, the variables $\mu \in M$ run through a space of 'generalised Lagrange multipliers' (see further ahead); this explains the use of the notation M for the second space (M is also the capital Greek for μ...). □

There is a widespread tendency to represent a saddle point as a 'col', that is to say, as a point of a surface which has the shape of a saddle (this explains the terminology; cf. figure 9.3-1(a)); but it is possible for other situations to obtain (cf. figures 9.3-1(b) and (c)). The three figures correspond to cases where V and M are compact intervals of $\mathbb{R}$. A foremost property of saddle points, whose geometrical interpretation ought to be clear from an inspection of the figures, is the following.

Theorem 9.3-1
If (u, λ) is a saddle point of a function $L: V \times M \to \mathbb{R}$, then

$$\sup_{\mu \in M} \inf_{v \in V} L(v, \mu) = L(u, \lambda) = \inf_{v \in V} \sup_{\mu \in M} L(v, \mu).$$

Proof
Firstly, we show that we always have the inequality

$$\sup_{\mu \in M} \inf_{v \in V} L(v, \mu) \leqslant \inf_{v \in V} \sup_{\mu \in M} L(v, \mu),$$

independently of the existence, or otherwise, of a saddle point. Given any elements $\bar{v} \in V$ and $\bar{\mu} \in M$, we assuredly have

$$\inf_{v \in V} L(v, \bar{\mu}) \leqslant L(\bar{v}, \bar{\mu}) \leqslant \sup_{\mu \in M} L(\bar{v}, \mu)$$

(not excluding the values $-\infty$ for $\inf\{\cdot\}$ and $+\infty$ for $\sup\{\cdot\}$). As $\inf_{v \in V} L(v, \bar{\mu})$ is a function of the single variable $\bar{\mu} \in M$, and as $\sup_{\mu \in M} L(\bar{v}, \mu)$ is a function of the single variable $\bar{v} \in V$, the desired inequality follows.

In order to establish the converse inequality, we make use of the fact that (u, λ) is a saddle point; we have

$$\inf_{v \in V} \sup_{\mu \in M} L(v, \mu) \leqslant \sup_{\mu \in M} L(u, \mu) = L(u, \lambda) = \inf_{v \in V} L(v, \lambda)$$

$$\leqslant \sup_{\mu \in M} \inf_{v \in V} L(v, \mu). \qquad \square$$

We return to the *general non-linear programming problem*, where it is supposed that the various functions involved are defined over the entire space V, with the sole intention of rendering the notation less cumbersome. Given functions $J: V \to \mathbb{R}$ and $\varphi_i: V \to \mathbb{R}$, $1 \leqslant i \leqslant m$, defined over a Hilbert space V, we define the set

$$U = \{v \in V : \varphi_i(v) \leqslant 0,\ 1 \leqslant i \leqslant m\},$$

and seek to find the element, or elements, u which satisfy

$$u \in U, \quad J(u) = \inf_{v \in U} J(v).$$

From here on, this will be referred to as the problem (P).

The fundamental interest in the concept of saddle point is the following (it is what is established in the following theorem). Under certain conditions, *every solution u of the problem* (P) *is also the first argument of a saddle point* (u, λ) *of a suitable function* $L: V \times M \to \mathbb{R}$ (the space M being, in that case, a matter for determination), called the *Lagrangian associated with the problem* (P) *under consideration* and, *conversely, if* (u, λ) *is a saddle point of this same Lagrangian L, then u is a solution of the problem* (P).

The second argument λ of such a saddle point (u, λ) is called a *generalised Lagrange multiplier* associated with the solution u of the problem (P), because, as will, in fact, be proved, under certain conditions, *the second argument λ of the saddle point is nothing other than the vector of* $\mathbb{R}_+^m$ *which appears in the Kuhn–Tucker conditions.*

More precisely, let us define the *Lagrangian associated with the problem* (P) given above as the function

$$L: (v, \mu) \in V \times \mathbb{R}_+^m \to L(v, \mu) \overset{\text{def}}{=} J(v) + \sum_{i=1}^{m} \mu_i \varphi_i(v),$$

so that the 'second' space M is here $\mathbb{R}_+^m$.

Theorem 9.3-2

(1) *If* $(u, \lambda) \in V \times \mathbb{R}_+^m$ *is a saddle point of the Lagrangian L, the point u, which belongs to the set U, is a solution of the problem* (P).

(2) *Suppose that the functions J and* φ_i, $1 \leqslant i \leqslant m$, *are convex, differentiable at a point* $u \in U$, *and that the constraints are qualified. Then, if u is the solution of the problem* (P), *there exists at least one vector* $\lambda \in \mathbb{R}_+^m$ *such that the pair* $(u, \lambda) \in V \times \mathbb{R}_+^m$ *is a saddle point of the Lagrangian L.*

Proof

(1) The inequalities $L(u, \mu) \leqslant L(u, \lambda)$ for every $\mu \in \mathbb{R}_+^m$ can be expressed as

$$\sum_{i=1}^{m} (\mu_i - \lambda_i) \varphi_i(u) \leqslant 0 \quad \text{for every } \mu \geqslant 0,$$

which implies that $\varphi_i(u) \leqslant 0$, $1 \leqslant i \leqslant m$ (let each μ_i tend towards $+\infty$). With $\mu = 0$, we obtain $\sum_{i=1}^{m} \lambda_i \varphi_i(u) \geqslant 0$, while, in any case, $\sum_{i=1}^{m} \lambda_i \varphi_i(u) \leqslant 0$

$(\lambda_i \geqslant 0, \; \varphi_i(u) \leqslant 0)$. Hence, we have shown, at this stage, that

$$u \in U \quad \text{and} \quad \sum_{i=1}^{m} \lambda_i \varphi_i(u) = 0.$$

Combining the last equality with the inequalities $L(u, \lambda) \leqslant L(v, \lambda)$ for every $v \in V$, we obtain

$$J(u) \leqslant J(v) + \sum_{i=1}^{m} \lambda_i \varphi_i(v) \quad \text{for every } v \in V$$

$$\leqslant J(v) \quad \text{for every } v \in U \text{ (since } \varphi_i(v) \leqslant 0).$$

(2) This leads exactly to the situation in which Theorem 9.2-4 is applicable. If u is a solution of the problem (P), there exists a vector $\lambda \in \mathbb{R}_+^m$ such that

$$\sum_{i=1}^{m} \lambda_i \varphi_i(u) = 0 \quad \text{and} \quad J'(u) + \sum_{i=1}^{m} \lambda_i \varphi_i'(u) = 0$$

(Kuhn–Tucker conditions). The first equality gives

$$L(u, \mu) = J(u) + \sum_{i=1}^{m} \mu_i \varphi_i(u) \leqslant J(u) = L(u, \lambda) \quad \text{for every } \mu \in \mathbb{R}_+^m,$$

and the second equality is a sufficient condition for the existence of a minimum of the convex function (being the sum of convex functions)

$$L(\cdot, \lambda) \colon v \in V \to J(v) + \sum_{i=1}^{m} \lambda_i \varphi_i(v).$$

Consequently,

$$L(u, \lambda) \leqslant L(v, \lambda) \quad \text{for every } v \in V,$$

so that the point (u, λ) is shown to be a saddle point of the Lagrangian L. $\qquad \square$

To sum up, we have shown that (under certain hypotheses) the set of solutions of the *problem* (P)

find u such that

(P) $\quad \begin{cases} u \in U \overset{\text{def}}{=} \{ v \in V \colon \varphi_i(v) \leqslant 0, \; 1 \leqslant i \leqslant m \}, \\[2mm] J(u) = \underset{v \in U}{\inf} J(v) \end{cases}$

coincides with the set of first arguments of the saddle points of the Lagrangian

$$L \colon (v, \mu) \in V \times \mathbb{R}_+^m \to L(v, \mu) = J(v) + \sum_{i=1}^{m} \mu_i \varphi_i(v).$$

Hence, if one knew some particular second argument, say λ, of these saddle points, then the *constrained* problem (P) would be replaced by the *unconstrained* problem (P$_\lambda$)

(P$_\lambda$)
$$\text{find } u_\lambda \text{ such that}$$
$$u_\lambda \in V, \ L(u_\lambda, \lambda) = \inf_{v \in V} L(v, \lambda).$$

How do we find such an element $\lambda \in \mathbb{R}^m_+$? Remembering that (Theorem 9.3-1)

$$L(u_\lambda, \lambda) = \inf_{v \in V} L(v, \lambda) = \sup_{\mu \in \mathbb{R}^m_+} \inf_{v \in V} L(v, \mu),$$

one is naturally led to look for a λ which is a solution of the *problem* (Q)

(Q)
$$\text{find } \lambda \text{ such that}$$
$$\lambda \in \mathbb{R}^m_+, \ G(\lambda) = \sup_{\mu \in \mathbb{R}^m_+} G(\mu),$$

where the function $G: \mathbb{R}^m_+ \to \mathbb{R}$ is defined by

$$G: \mu \in \mathbb{R}^m_+ \to G(\mu) = \inf_{v \in V} L(v, \mu).$$

(Q) is called the *dual problem* of the problem (P) which, in this perspective, becomes the *primal problem*. Likewise, $\mu \in \mathbb{R}^m_+$ is called the *dual variable* of the *primal variable* $v \in U \subset V$.

The dual problem, then, once again turns up as a *constrained* optimisation problem; but *the constraints $\mu_i \geqslant 0$, $1 \leqslant i \leqslant m$, are easily dealt with*, since the associated projection operator is known (an observation already made in connection with the gradient method of section 8.6), while the constraints $\varphi_i(u) \leqslant 0$, $1 \leqslant i \leqslant m$, *are, in general, impossible to deal with numerically*. This simplification, *which is of an essential nature*, is the most remarkable aspect of the strategy which has been adopted (whose rigorous justification will be given in the following theorem); above all, it is precisely that which gives rise to the method described in the following section.

While, in the preceding theorem, we established the link which exists between the primal problem (P) and a *primal–dual* problem, in which the primal and dual variables appear *simultaneously* and which involves the search for a saddle point, it now remains to establish the link which exists between the solutions of the primal problem (P) and those of the dual problem (Q). This is the aim of the result which follows.

Theorem 9.3-3

(1) *Suppose that the functions $\varphi_i: V \to \mathbb{R}, 1 \leqslant i \leqslant m$, are continuous and that, for every $\mu \in \mathbb{R}^m_+$, the problem* (P$_\mu$)

find $u_\mu \in V$ *such that*

$(P_\mu)_{\mu \in \mathbb{R}^m_+}$ $\qquad\qquad u_\mu \in V, L(u_\mu, \mu) = \inf_{v \in V} L(v, \mu)$

has a unique solution u_μ *continuously dependent on* $\mu \in \mathbb{R}^m_+$. *Then, if* λ *is any solution of the problem* (Q), *the solution* u_λ *of the corresponding problem* (P_λ) *is a solution of the problem* (P).

(2) *Suppose that the problem* (P) *has at least one solution* u, *that the functions* J *and* φ_i, $1 \leqslant i \leqslant m$, *are convex, differentiable at* u *and, lastly, that the constraints are qualified. Then the problem* (Q) *has at least one solution.*

Proof

(1) Let λ be any solution of the problem (Q). We already have

$$\lambda \in \mathbb{R}^m_+ \quad \text{and} \quad G(\lambda) = L(u_\lambda, \lambda) = \inf_{v \in V} L(v, \lambda),$$

and we shall establish the relation

$$\sup_{\mu \in \mathbb{R}^m_+} L(u_\lambda, \mu) = L(u_\lambda, \lambda).$$

Since these two relations together amount exactly to the definition of a saddle point (u_λ, λ) of the Lagrangian L, it will follow from Theorem 9.3-2(1) that u_λ is the solution of the problem (P).

We first need to show the *differentiability* of the function G. Let us consider two points μ and $\mu + \xi$ of the set $\mathbb{R}^m_+$, with components μ_i and $\mu_i + \xi_i$, $1 \leqslant i \leqslant m$. The inequalities

$$L(u_\mu, \mu) \leqslant L(u_{\mu+\xi}, \mu) \quad \text{and} \quad L(u_{\mu+\xi}, \mu + \xi) \leqslant L(u_\mu, \mu + \xi)$$

can also be expressed as

$$\sum_{i=1}^m \xi_i \varphi_i(u_{\mu+\xi}) \leqslant G(\mu + \xi) - G(\mu) \leqslant \sum_{i=1}^m \xi_i \varphi_i(u_\mu).$$

Hence, there exists a number $\theta \in [0, 1]$ such that

$$G(\mu + \xi) - G(\mu) = (1 - \theta) \sum_{i=1}^m \xi_i \varphi_i(u_\mu) + \theta \sum_{i=1}^m \xi_i \varphi_i(u_{\mu+\xi})$$

$$= \sum_{i=1}^m \xi_i \varphi_i(u_\mu) + \theta \sum_{i=1}^m \xi_i \{\varphi_i(u_{\mu+\xi}) - \varphi_i(u_\mu)\}.$$

By hypothesis, the functions $\mu \in \mathbb{R}^m_+ \to u_\mu \in V$ and $\varphi_i : V \to \mathbb{R}$ are continuous; so that it is possible to write, for every point $\mu \in \mathbb{R}^m_+$,

$$G(\mu + \xi) - G(\mu) = \sum_{i=1}^m \xi_i \varphi_i(u_\mu) + \|\xi\| \varepsilon(\xi), \quad \lim_{\xi \to 0} \varepsilon(\xi) = 0$$

($\|\cdot\|$ being any vector norm over $\mathbb{R}^m$). This proves, in the first place, the differentiability of the function $G\colon \mathbb{R}^m_+ \to \mathbb{R}$ (there is a slight difficulty; cf. the remark (2) which follows the theorem) and, secondly, it provides an expression for the derivative; in fact,

$$G'(\mu)\xi = \sum_{i=1}^{m} \xi_i \varphi_i(u_\mu) \quad \text{for every } \xi \in \mathbb{R}^m.$$

As the function G has a maximum at the point λ of the *convex* set $\mathbb{R}^m_+$, an application of Theorem 7.4-1 shows that

$$G'(\lambda)(\mu - \lambda) \leqslant 0 \quad \text{for every } \mu \in \mathbb{R}^m_+,$$

or

$$\sum_{i=1}^{m} \mu_i \varphi_i(u_\lambda) \leqslant \sum_{i=1}^{m} \lambda_i \varphi_i(u_\lambda) \quad \text{for every } \mu \in \mathbb{R}^m_+.$$

Hence,

$$L(u_\lambda, \mu) = J(u_\lambda) + \sum_{i=1}^{m} \mu_i \varphi_i(u_\lambda)$$

$$\leqslant J(u_\lambda) + \sum_{i=1}^{m} \lambda_i \varphi_i(u_\lambda) = L(u_\lambda, \lambda),$$

for every $\mu \in \mathbb{R}^m_+$, which is precisely the second of the two inequalities characterising a saddle point.

(2) By Theorem 9.3-2(2), there exists (at least) one vector $\lambda \in \mathbb{R}^m_+$ such that the pair (u, λ) is a saddle point of the Lagrangian L. Theorem 9.3-1 then implies that

$$L(u, \lambda) = \inf_{v \in V} L(v, \lambda) = \sup_{\mu \in \mathbb{R}^m_+} \inf_{v \in V} L(v, \mu),$$

which can also be written as

$$G(\lambda) = \sup_{\mu \in \mathbb{R}^m_+} G(\mu). \qquad \square$$

Remarks

(1) If (u, λ) is a saddle point of the Lagrangian L over the space $V \times \mathbb{R}^m_+$, it is clear from Theorem 9.3-1 that λ is a solution of the problem (Q), by definition of the latter. Conversely, with the hypotheses of part (1) of the preceding theorem, if λ is a solution of (Q), the pair (u_λ, λ) is a saddle point of the Lagrangian L. Hence, *the set of solutions of the dual problem* (Q) *and the set of second arguments λ of the saddle points (u, λ) of the Lagrangian L coincide.* This result is in some way a 'dual' of the result established in Theorem 9.3-2.

(2) Out of a desire to simplify, we defined differentiability (cf. section 7.1) only for functions defined over *open* sets, which is not the case for the

function G. However, it is not necessary to complicate the hypotheses (by assuming, for example, that the solution of the problems (P_μ) is continuous at μ for every μ in an open neighbourhood of $\mathbb{R}_+^m$) since a simple examination of the proof of Theorem 7.4-1 shows that the 'differentiability' established here is sufficient to justify its application. □

As an illustration of the previous considerations, let us return to the example of a quadratic functional

$$J: v \in \mathbb{R}^n \to J(v) = \tfrac{1}{2}(Av, v) - (b, v),$$

where $A \in \mathscr{A}_n(\mathbb{R})$ is a given symmetric, *positive definite* matrix and $b \in \mathbb{R}^n$ a given vector, with a set U of the form

$$U = \left\{ v \in \mathbb{R}^n : \sum_{j=1}^n c_{ij} v_j \leqslant d_i, 1 \leqslant i \leqslant m \right\} = \{ v \in \mathbb{R}^n : Cv \leqslant d \},$$

where $C = (c_{ij}) \in \mathscr{A}_{m,n}(\mathbb{R})$ is a given matrix and $d \in \mathbb{R}^m$ a given vector. Suppose that the set U is non-empty, which, it will be remembered, is equivalent to the constraints' being qualified, since they are affine. It is also known that the corresponding primal problem (P) has a unique solution (cf. Theorem 8.2-3).

In order to avoid confusion, let us denote by $(\cdot, \cdot)_n$ and $(\cdot, \cdot)_m$ the scalar products of $\mathbb{R}^n$ and $\mathbb{R}^m$, respectively. The Lagrangian associated with the problem (P) can be expressed as

$$L: (v, \mu) \in \mathbb{R}^n \times \mathbb{R}_+^m \to L(v, \mu) = \tfrac{1}{2}(Av, v)_n - (b, v)_n + (\mu, Cv - d)_m$$
$$= \tfrac{1}{2}(Av, v)_n - (b - C^T \mu, v)_n - (\mu, d)_m,$$

and, by Theorem 9.3-2, it possesses at least one saddle point, whose leading argument is necessarily the solution of the primal problem. For the problem under consideration, the non-uniqueness of the saddle points can only correspond to the non-uniqueness of their *second* arguments. We shall return to this question.

In order to be able to apply Theorem 9.3-3, it is necessary to verify that, for every $\mu \in \mathbb{R}_+^m$, the problem (P_μ) has a unique solution $u_\mu \in V$. This is evident, as the function $L(\cdot, \mu)$ is of the form

$$L(v, \mu) = \tfrac{1}{2}(Av, v)_n - (b_\mu, v)_n + c_\mu, \quad b_\mu \in \mathbb{R}^n, c_\mu \in \mathbb{R};$$

so that it is a quadratic functional analogous to the functional J. Moreover, it is known that the gradient of the function $L(\cdot, \mu)$ is zero at the point u_μ. This can be expressed as

$$Au_\mu = b - C^T \mu \Leftrightarrow u_\mu = A^{-1}(b - C^T \mu),$$

and the continuity of the function $\mu \in \mathbb{R}^m_+ \to u_\mu \in \mathbb{R}^n$ is thereby established (a more general case is considered in Exercise 9.3-1).

It is now possible to calculate the function G. For this, we observe that a quadratic functional of the form given above has a minimum u_μ if any only if

$$(Au_\mu, v)_n = (b_\mu, v)_n \quad \text{for every } v \in \mathbb{R}^n,$$

which makes it possible to set

$$G(\mu) = L(u_\mu, \mu) = \tfrac{1}{2}(Au_\mu, u_\mu)_n - (b_\mu, u_\mu)_n + c_\mu = -\tfrac{1}{2}(b_\mu, u_\mu)_n + c_\mu$$
$$= -\tfrac{1}{2}(b - C^T\mu, u_\mu)_n - (\mu, d)_m$$
$$= -\tfrac{1}{2}(b - C^T\mu, A^{-1}(b - C^T\mu))_n - (\mu, d)_m,$$

or, after carrying out the requisite calculations,

$$-G(\mu) = \tfrac{1}{2}(CA^{-1}C^T\mu, \mu)_m - (CA^{-1}b - d, \mu)_m + \tfrac{1}{2}(A^{-1}b, b)_n.$$

The symmetric matrix $A \in \mathscr{A}(\mathbb{R}^n)$ being positive definite, the symmetric matrix $CA^{-1}C^T \in \mathscr{A}(\mathbb{R}^m)$ is certainly non-negative definite, since

$$\mu^T CA^{-1}C^T\mu = (C^T\mu)^T A^{-1}(C^T\mu) \geqslant 0.$$

It is positive definite if, and only if,

$$C^T\mu = 0 \Rightarrow \mu = 0,$$

in other words, if, and only if,

$$\text{Ker}(C^T) = \{0\} \Leftrightarrow \text{Im}(C) = \mathbb{R}^m \Leftrightarrow r(C) = m,$$

which obviously imposes the restriction $m \leqslant n$. Theorem 9.3-3 then justifies the assertion that, in every case, the dual problem has *at least one solution* (which is not at all immediately obvious when the matrix $CA^{-1}C^T$ is not definite), *which is unique if $r(C) = m$* (since the function G is then strictly convex).

Remark

It is possible to calculate the gradient of the function G and to verify that the expression coincides with that found in the proof of Theorem 9.3-3,

$$\nabla G(\mu) = Cu_\mu - d = -CA^{-1}C^T\mu + CA^{-1}b - d. \qquad \square$$

Exercises

9.3-1. Adopting the same notation as in Theorem 9.3-3, show that the solution u_μ of the problem (P_μ) depends continuously on $\mu \in \mathbb{R}^m$ in the following particular case: $V = \mathbb{R}^n$, the functional J is elliptic and the set U, assumed to be non-empty, is of the form

$$U = \left\{ v \in \mathbb{R}^n : \sum_{j=1}^n c_{ij}v_j \leqslant d_i, 1 \leqslant i \leqslant m \right\}.$$

9.3-2. (1) Show that the set of saddle points of a function $L: V \times M \to \mathbb{R}$ is a *rectangle*, that is to say, a set of the form $V_0 \times M_0$, with $V_0 \subset V$, $M_0 \subset M$.

(2) Suppose that M and V are convex subsets of vector spaces and that

$$L(v, \cdot): M \to \mathbb{R} \quad \text{is concave for every } v \in V,$$
$$L(\cdot, \mu): V \to \mathbb{R} \quad \text{is convex for every } \mu \in M.$$

Show that the set $V_0 \times M_0$ is convex.

(3) Prove that, if the function $L(\cdot, \mu): V \to \mathbb{R}$ is strictly convex for every $\mu \in M$, then the set V_0 contains at least one point.

9.3-3. The aim in this exercise is to give sufficient conditions for the *existence* of a saddle point, the result of question (5) constituting a particular case of the *Ky Fan–Sion theorem*. In questions (1) to (5), it is assumed that V and M are non-empty, *convex and compact* subsets of finite-dimensional vector spaces and that $L: V \times M \to \mathbb{R}$ is a continuous function, such that

$$L(v, \cdot): M \to \mathbb{R} \quad \text{is concave for every } v \in V,$$
$$L(\cdot, \mu): V \to \mathbb{R} \quad \text{is convex for every } \mu \in M.$$

(1) Prove that the function

$$G: \mu \in M \to G(\mu) = \inf_{v \in V} L(v, \mu)$$

is concave and continuous.

(2) Assume 'for the time being' (up to and including question (4)) that the function $L(\cdot, \mu): V \to \mathbb{R}$ is strictly convex for every $\mu \in M$, which allows us to write

$$G(\mu) = \inf_{v \in V} L(v, \mu) = L(u(\mu), \mu),$$

the element $u(\mu) \in V$ being defined uniquely. Show that the function $\mu \in M \to u(\mu) \in V$ is continuous.

(3) Let $\lambda \in M$ be a point satisfying (it is known from (1) that at least one such point exists)

$$G(\lambda) = \sup_{\mu \in M} G(\mu).$$

If μ is any point of M, show that

$$G(\lambda) \geqslant L(u(\theta\mu + (1 - \theta)\lambda), \mu) \quad \text{for every } \theta \in (0, 1).$$

(4) Deduce that

$$\sup_{\mu \in M} \inf_{v \in V} L(v, \mu) \geqslant \inf_{v \in V} \sup_{\mu \in M} L(v, \mu),$$

and conclude that the point $(u(\lambda), \lambda)$ is a saddle point of the function L.

(5) In case the function $L(\cdot, \mu): V \to \mathbb{R}$ is not strictly convex for every $\mu \in M$, introduce the auxiliary functions

$$L_\varepsilon: (v, \mu) \in V \times M \to L_\varepsilon(v, \mu) = L(v, \mu) + \varepsilon \|v\|^2, \ \varepsilon > 0,$$

denoting by $\|\cdot\|$ a Hilbert norm over V, and then let ε tend to 0; deduce the existence of (at least) one saddle point of the function L.

(6) The sets V and M are no longer assumed to be compact; on the contrary, it is assumed that

$$\begin{cases} \lim_{\substack{v \in V \\ \|v\| \to \infty}} L(v, \mu_0) = +\infty & \text{for (at least) one element } \mu_0 \in M, \\[2mm] \lim_{\substack{\mu \in M \\ \|\mu\| \to \infty}} L(v_0, \mu) = +\infty & \text{for (at least) one element } v_0 \in V, \end{cases}$$

all the other hypotheses remaining unaltered. Prove the existence of (at least) one saddle point of the function L.

9.4 Uzawa's method

Let us return to the familiar primal problem (P)

find u such that

(P)
$$\begin{cases} u \in U \overset{\text{def}}{=} \{ v \in V : \varphi_i(v) \leqslant 0, \ 1 \leqslant i \leqslant m \}, \\[2mm] J(u) = \inf_{v \in U} J(v). \end{cases}$$

Our objective is to *construct an algorithm which enables us to approximate a solution of* (P). In this respect, we recall that, if U is a general, closed, convex subset of a Hilbert space V, it is, in general, an illusion to hope to be able to apply the gradient method (cf. section 8.6), since it is not known how to construct *'numerically'* the projection operator onto the set U, except in *very* special cases.

Now, it has already been observed that the constraints of the *dual* problem (Q)

find λ such that

$$\lambda \in \mathbb{R}^m_+, \quad G(\lambda) = \sup_{\mu \in \mathbb{R}^m_+} G(\mu),$$

do themselves correspond to a projection operator $P_+ : \mathbb{R}^m \to \mathbb{R}^m_+$ which is *very simple* and is given by (cf. section 8.1)

$$(P_+ \lambda)_i = \max \{ \lambda_i, 0 \}, \quad 1 \leqslant i \leqslant m.$$

The method which we are about to describe rests on this remark, since, in effect, *it is nothing other than the gradient method applied to the problem* (Q). Given an arbitrary element $\lambda^0 \in \mathbb{R}^m_+$, the sequence $(\lambda^k)_{k \geqslant 0}$ of elements of $\mathbb{R}^m_+$ is defined by the recurrence relation

$$\lambda^{k+1} = P_+ (\lambda^k - \varrho \nabla G(\lambda^k)), \ k \geqslant 0,$$

the parameter ϱ being chosen 'as best may be'. Since the problem (Q) is one of looking for a *maximum*, it is natural to change the sign in front of the parameter, as also to expect convergence to take place for positive values of this parameter (in the light of the results of section 8.6).

In certain cases (cf. the proof of Theorem 9.3-3), we have seen that it is known how to calculate the gradient of the function G. It is the vector with components

$$(\nabla G(\mu))_i = \varphi_i(u_\mu) \quad 1 \leqslant i \leqslant m,$$

the vector u_μ being the solution of the *unconstrained* minimisation problem

$$u_\mu \in V, \ J(u_\mu) + \sum_{i=1}^{m} \mu_i \varphi_i(u_\mu) = \inf_{v \in V} \left\{ J(v) + \sum_{i=1}^{m} \mu_i \varphi_i(v) \right\}.$$

We now have all the elements necessary for the description (formal, for the time being) of the iterative method being proposed, known as *Uzawa's method*. Starting with an arbitrary element $\lambda^0 \in \mathbb{R}_+^m$, a sequence of pairs $(\lambda^k, u^k) \in \mathbb{R}_+^m \times V$, $k \geqslant 0$, is defined, by means of the following recurrence formulae (in order to make the expressions less cumbersome, we write $u^k = u_{\lambda^k}$):

$$\begin{cases} \text{calculate} \quad u^k \colon J(u_k) + \sum_{i=1}^{m} \lambda_i^k \varphi_i(u^k) = \inf_{v \in V} \left\{ J(v) + \sum_{i=1}^{m} \lambda_i^k \varphi_i(v) \right\}, \\ \text{calculate} \quad \lambda_i^{k+1} \colon \lambda_i^{k+1} = \max \{ \lambda_i^k + \varrho \varphi_i(u^k), 0 \}, \ 1 \leqslant i \leqslant m. \end{cases}$$

Although Uzawa's method is constructed *a priori* as a method to approximate the solution of the dual problem, it is equally possible to view it as a method which approximates the *primal* problem. In fact, if the sequence (λ^k) converges to a solution λ of the dual problem, one might expect the solutions $u_k = u_{\lambda_k}$ to converge to the solution u of the minimisation problem

$$u \in U, \quad J(u) + \sum_{i=1}^{m} \lambda_i \varphi_i(u) = \inf_{v \in V} \left\{ J(v) + \sum_{i=1}^{m} \lambda_i \varphi_i(v) \right\},$$

that is to say, to the solution of the primal problem. From this point of view, one can say that *Uzawa's method replaces a constrained minimisation problem by a sequence of unconstrained minimisation problems* (it would be too much to expect to win every round...).

In the theorem which follows, we prove the *convergence of the sequence* (u^k) to the solution of the primal problem, *even when the sequence* (λ^k) *does not converge*. This is an illustration of the line we are taking, which is to use Uzawa's method 'in the first place' as a method of approximation of the primal problem. Indeed, it should be observed that all the assumptions are in terms of the data of the latter.

A last observation is the following. While it is possible to choose not to take account of the existence of the dual problem in the description of the method, the sequence (λ^k) then taking on the appearance of a somewhat 'miraculous' auxiliary tool, it still remains the case that the ideas introduced in the previous section (Lagrangian, saddle point, dual problem,...) continue to play an essential role in its understanding and analysis.

In what follows, $(\cdot,\cdot)_l$ and $\|\cdot\|_l$ denote the scalar product and the Euclidean norm of $\mathbb{R}^l$ while α denotes the positive constant which appears in the inequality

$$(\nabla J(u) - \nabla J(v), u - v)_n \geqslant \alpha \| u - v \|_n^2$$

used in the definition of an elliptic functional, and we set

$$\| C \| = \sup_{v \in \mathbb{R}^n} \frac{\| Cv \|_m}{\| v \|_n} \quad \text{for } C \in \mathscr{A}_{m,n}(\mathbb{R}).$$

Theorem 9.4-1 (Convergence of Uzawa's method)

Suppose that $V = \mathbb{R}^n$, that the function J is elliptic and that the set U, which is of the form

$$U = \{ v \in \mathbb{R}^n : Cv \leqslant d \}, \quad C \in \mathscr{A}_{m,n}(\mathbb{R}), d \in \mathbb{R}^m,$$

is non-empty. Then, if

$$0 < \varrho < \frac{2\alpha}{\| C \|^2},$$

the sequence (u^k) converges to the unique solution of the primal problem (P).

If the rank of C is m, the sequence (λ^k) also converges, and to the unique solution of the dual problem (Q).

Proof

(i) *Preliminaries*. The function J being elliptic and the non-empty, convex set U being closed, the problem (P) has a unique solution u, as well as the successive minimisation problems, with solutions u^k, encountered in Uzawa's method (Theorem 8.4-1). Furthermore, it follows from Theorem 9.3-2(2) that the problem (Q) always has at least one solution (we shall return in part (iii) to the question of its uniqueness). We introduce now the affine function

$$\varphi : v \in \mathbb{R}^n \to \varphi(v) = Cv - d \in \mathbb{R}^m,$$

so that the Lagrangian of the problem can be expressed as

$$L : (v, \mu) \in \mathbb{R}^n \times \mathbb{R}_+^m \to L(v, \mu) = J(v) + (\mu, \varphi(v))_m = J(v) + (C^T \mu, v)_n - (\mu, d)_m.$$

By Theorem 9.3-2(2), there exists (at least) one vector $\lambda \in \mathbb{R}_+^m$ such that the pair (u, λ) is a saddle point of the Lagrangian L (it is, in any case, just in this way that all the solutions of the dual problem are obtained; refer to the proof of Theorem 9.3-3). Hence we have, on the one hand, $\nabla J(u) + C^T\lambda = 0$, since $L(u, \lambda) = \inf_{v \in \mathbb{R}^n} L(v, \lambda)$, and, on the other,

$$(\varphi(u), \mu - \lambda)_m \leqslant 0 \quad \text{for every } \mu \in \mathbb{R}^m_+,$$

since $L(u, \lambda) = \sup_{\mu \in \mathbb{R}^m_+} L(u, \mu)$. Proceeding exactly as for the gradient method (cf. section 8.6), we observe that the last relation can also be written, *for every number $\varrho > 0$*, as

$$(\lambda - (\lambda + \varrho\varphi(u)), \mu - \lambda)_m \geqslant 0 \quad \text{for every } \mu \in \mathbb{R}^m_+,$$

which shows that λ may be interpreted as the projection onto $\mathbb{R}^m_+$ of the element $\lambda + \varrho\varphi(u)$. In short,

$$\begin{cases} \nabla J(u) + C^T\lambda = 0, \\ \lambda = P_+(\lambda + \varrho\varphi(u)). \end{cases}$$

Since, by definition of Uzawa's method, the analogous relations

$$\begin{cases} \nabla J(u^k) + C^T\lambda^k = 0, \\ \lambda^{k+1} = P_+(\lambda^k + \varrho\varphi(u^k)), \end{cases}$$

are also satisfied, it follows that

$$\begin{cases} \nabla J(u^k) - \nabla J(u) + C^T(\lambda^k - \lambda) = 0, \\ \|\lambda^{k+1} - \lambda\|_m \leqslant \|\lambda^k - \lambda + \varrho C(u^k - u)\|_m, \end{cases}$$

since the projection operator 'does not increase distances'.

(ii) *Convergence of the sequence (u^k)*. We make use of only the last two relations given above. Squaring each side of the inequality, we obtain

$$\|\lambda^{k+1} - \lambda\|_m^2 \leqslant \|\lambda^k - \lambda\|_m^2 + 2\varrho(C^T(\lambda^k - \lambda), u^k - u)_n + \varrho^2 \|C(u^k - u)\|_m^2,$$

which, taking account of the equality, can also be expressed as

$$\|\lambda^{k+1} - \lambda\|_m^2 \leqslant \|\lambda^k - \lambda\|_m^2 - 2\varrho(\nabla J(u^k) - \nabla J(u), u^k - u)_n$$
$$+ \varrho^2 \|C(u^k - u)\|_m^2$$
$$\leqslant \|\lambda^k - \lambda\|_m^2 - \varrho\{2\alpha - \varrho\|C\|^2\} \|u^k - u\|_m^2.$$

Accordingly,

$$0 \leqslant \varrho \leqslant \frac{2\alpha}{\|C\|^2} \Rightarrow \|\lambda^{k+1} - \lambda\|_m \leqslant \|\lambda^k - \lambda\|_m \quad \text{for every } k \geqslant 0.$$

The sequence $(\|\lambda^k - \lambda\|_m)_{k \geqslant 0}$ being monotonic decreasing and bounded below (by zero...), it is convergent, so that

$$\lim_{k \to \infty} \{\|\lambda^{k+1} - \lambda\|_m^2 - \|\lambda^k - \lambda\|_m^2\} = 0.$$

Since

$$\varrho\{2\alpha - \varrho\|C\|^2\} \|u^k - u\|_n^2 \leqslant \|\lambda^k - \lambda\|_m^2 - \|\lambda^{k+1} - \lambda\|_m^2,$$

it follows that

$$0 < \varrho < \frac{2\alpha}{\|C\|^2} \Rightarrow \lim_{k \to \infty} \|u^k - u\|_n = 0.$$

(iii) *Possible convergence of the sequence* (λ^k). Since the sequence $(\|\lambda^k - \lambda\|)_{k \geq 0}$ is monotonic decreasing, the sequence (λ^k) is bounded. Hence, there exists a *subsequence* $(\lambda^{k'})_{k' \geq 0}$ which converges to an element $\lambda' \in \mathbb{R}^m_+$, satisfying

$$\nabla J(u) + C^T \lambda' = \lim_{k' \to \infty} \{\nabla J(u^{k'}) + C^T \lambda^{k'}\} = 0.$$

Let us now make *the assumption that the rank of the matrix* C *is* m. From the equivalences

$$r(C) = m \Leftrightarrow \mathrm{Im}\,(C) = \mathbb{R}^m \Leftrightarrow \mathrm{Ker}\,(C^T) = 0,$$

there follows the uniqueness of the solution λ of the dual problem. This solution, in fact, satisfies

$$\nabla J(u) + C^T \lambda = 0,$$

and the solution u of the primal problem is unique. Hence $\lambda = \lambda'$. As the reasoning given above may be repeated for *any* subsequence of the sequence (λ^k), it follows that *the entire* sequence (λ^k) converges to λ when $r(C) = m$.

$\square$

In the case of a quadratic functional

$$J: v \in \mathbb{R}^n \to J(v) = \tfrac{1}{2}(Av, v) - (b, v),$$

one iteration of Uzawa's method may be expressed as

$$\begin{cases} \text{calculate} \quad u^k: Au^k - b + C^T \lambda^k = 0, \\ \text{calculate} \quad \lambda^{k+1}: \lambda_i^{k+1} = \max \quad \{(\lambda^k + \varrho(Cu^k - d))_i, 0\}, \quad 1 \leq i \leq m. \end{cases}$$

By the preceding theorem, the method converges if the symmetric matrix A is positive definite and if

$$0 < \varrho < \frac{2\lambda_1(A)}{\|C\|^2},$$

where $\lambda_1(A)$ denotes the smallest eigenvalue of the matrix A.

Remark

Eliminating u^k from the preceding equations, we obtain

$$\lambda^{k+1} = P_+(\lambda^k + \varrho(-CA^{-1}C^T\lambda^k + CA^{-1}b - d)),$$

which is exactly

$$\lambda^{k+1} = P_+(\lambda^k + \varrho \nabla G(\lambda^k)),$$

with the help of the calculations made towards the end of the previous section. Uzawa's method is indeed the gradient method applied to the dual problem! $\qquad\square$

Exercises

9.4-1. Consider the quadratic functional

$$J(v) = \tfrac{1}{2}(Av, v)_n - (b, v)_n, \ A \in \mathscr{A}_n(\mathbb{R}), \ b \in \mathbb{R}^n,$$

the symmetric matrix A being positive definite. It is desired to approximate the unique solution of the problem

find u such that

$$\begin{cases} u \in U \overset{\text{def}}{=} \{v \in \mathbb{R}^n : Cv = 0\}, \ C \in \mathscr{A}_{m,n}(\mathbb{R}), \\ J(u) = \underset{v \in U}{\inf} \ J(v). \end{cases}$$

For this purpose, the following iterative method is defined. Starting with an arbitrary vector $\lambda^0 \in \mathbb{R}^m$, a sequence of pairs $(\lambda^k, u^k) \in \mathbb{R}^m \times \mathbb{R}^n$, $k \geq 0$, is defined through the following recurrence formulae:

$$\begin{cases} \text{calculate} \quad u^k : Au^k - b + C^T\lambda^k = 0, \\ \text{calculate} \quad \lambda^{k+1} : \lambda^{k+1} = \lambda^k + \rho Cu^k; \ \rho \text{ being a positive parameter.} \end{cases}$$

This iterative method is also called *Uzawa's method*, but now applied to a problem with 'equality constraints'.

(1) Show that, if the parameter $\rho > 0$ is sufficiently small, the sequence (u^k) converges to u (one should be guided by the proof of Theorem 9.4-1).

(2) What can be said about the convergence of the sequence (λ^k)?

9.4-2. Adopting the same hypotheses and notation as in the preceding exercise, we now define another iterative method. Starting with an arbitrary pair $(u^0, \lambda^0) \in \mathbb{R}^n \times \mathbb{R}^m$, a sequence of pairs $(u^k, \lambda^k) \in \mathbb{R}^n \times \mathbb{R}^m$, $k \geq 0$, is defined through the following recurrence formulae.

$$\begin{cases} \text{Calculate} \quad u^{k+1} : u^{k+1} = u^k - \rho_1(Au^k - b + C^T\lambda^k), \\ \text{Calculate} \quad \lambda^{k+1} : \lambda^{k+1} = \lambda^k + \rho_1\rho_2 Cu^{k+1}, \end{cases}$$

ρ_1, ρ_2 being positive parameters.

(1) Show that, if the parameter $\rho_1 > 0$ is sufficiently small,

$$\beta \overset{\text{def}}{=} \|I - \rho_1 A\| < 1,$$

where $\|\cdot\|$ denotes the matrix norm subordinate to the Euclidean norm of $\mathbb{R}^n$.

(2) Let λ be a vector of $\mathbb{R}^m$ which satisfies

$$Au + C^T\lambda = b$$

(say why such a vector λ exists). Choose the parameter ρ_1 in such a way that the inequality $\beta < 1$ of (1) holds. Show that, if the parameter $\rho_2 > 0$ is sufficiently small, there exists a constant $\gamma > 0$ which is independent of the integer k and is such that

$$\gamma \| u^{k+1} - u \|_n^2 \leqslant \left(\frac{\| \lambda^k - \lambda \|_m^2}{\rho_2} + \beta \| u^k - u \|_m^2 \right)$$
$$- \left(\frac{\| \lambda^{k+1} - \lambda \|_m^2}{\rho_2} + \beta \| u^{k+1} - u \|_n^2 \right).$$

(3) Deduce that, for such a choice of parameters ρ_1 and ρ_2, $\lim_{k \to \infty} u^k = u$.

(4) What can be said about the sequence (λ^k) when $r(C) = m$?

The iterative method given above is called the *Arrow–Hurwicz method*. Its advantage, in relation to Uzawa's method, is to be found in its avoidance of the need to solve linear systems.

9.4-3. This exercise has as its aim the study of the convergence of iterative methods which are similar to Uzawa's method, but which have constraints which are of a more general nature than those considered in the text (there, the corresponding constraints were linear inequality constraints; cf. Theorem 9.4-1). Consider the problem

find u such that

$$\text{(P)} \quad \begin{cases} u \in U \overset{\text{def}}{=} U_0 \cap \{ v \in \mathbb{R}^n : \varphi_i(v) \leqslant 0, 1 \leqslant i \leqslant m \}, \\ J(u) = \inf_{v \in U} J(v), \end{cases}$$

with the following hypotheses, made *once for all*:

(i) the functional J is $\mathbb{R}^n$-elliptic: it is continuously differentiable in $\mathbb{R}^n$ and there exists a constant α such that

$$\alpha > 0 \quad \text{and} \quad (\nabla J(v) - \nabla J(u), v - u)_n \geqslant \alpha \| v - u \|_n^2 \quad \text{for every } u, v \in \mathbb{R}^n,$$

where $(\cdot, \cdot)_p$ and $\| \cdot \|_p$ denote respectively the scalar product and the Euclidean norm in $\mathbb{R}^p$;

(ii) there exists a constant M such that

$$\| \nabla J(v) - \nabla J(u) \|_n \leqslant M \| v - u \|_n \quad \text{for every } u, v \in \mathbb{R}^n;$$

in other words, the functional 'is not very different from' a quadratic functional;

(iii) the set U_0 is a closed, convex subset of $\mathbb{R}^n$ (for example, $\mathbb{R}^n$ itself);

(iv) the set U is non-empty;

(v) the functions $\varphi_i : \mathbb{R}^n \to \mathbb{R}$, $1 \leqslant i \leqslant m$, are convex;

(vi) there exists a constant C such that

$$\| \varphi(v) - \varphi(u) \|_m \leqslant C \| v - u \|_n \quad \text{for every } u, v \in \mathbb{R}^n,$$

where φ denotes the function from $\mathbb{R}^n$ to $\mathbb{R}^m$ whose components are the functions φ_i, $1 \leqslant i \leqslant m$; in other words, the functions φ_i 'are not very different from' affine functions.

(1) Establish the inequalities

$$(\nabla J(u, v - u)_n + \frac{\alpha}{2} \| v - u \|_n^2 \leqslant J(v) - J(u) \leqslant (\nabla J(u), v - u)_n + \frac{M}{2} \| v - u \|_n^2,$$

for every $u, v \in \mathbb{R}^n$.

(2) Prove that the problem (P) possesses a unique solution, to be denoted by u.

(3) Define the Lagrangian associated with the problem (P) as the function

$$L : (v, \mu) \in U_0 \times \mathbb{R}_+^m \to L(v, \mu) \overset{\text{def}}{=} J(v) + (\mu, \varphi(v))_m.$$

Prove that, if $(u, \lambda) \in U_0 \times \mathbb{R}_+^m$ is a saddle point of the Lagrangian L over the set $U_0 \times \mathbb{R}_+^m$, then the point u is a solution of the problem (P).

(4) Denote by P_+ the projection operator from $\mathbb{R}^m$ onto $\mathbb{R}_+^m$. Verify the equivalence (already established in the text)

$$\lambda = P_+(\lambda + \rho\varphi(u)), \rho > 0 \text{ fixed} \Leftrightarrow \begin{cases} \lambda \in \mathbb{R}_+^m, \varphi(u) \leqslant 0, \\ (\lambda, \varphi(u))_m = 0. \end{cases}$$

(5) Verify that a pair (u, λ) is a saddle point of the Lagrangian L if, and only if,

$$(\nabla J(u), v - u)_n + (\lambda, \varphi(v))_m \geqslant 0 \quad \text{for every } v \in U_0,$$

$$\lambda = P_+(\lambda + \rho\varphi(u)), \quad \rho > 0 \text{ fixed (but arbitrary)}.$$

(6) Define an iterative method as follows. Starting with an arbitrary pair $(u^0, \lambda^0) \in U_0 \times \mathbb{R}_+^m$, define the sequence of pairs $(u^k, \lambda^k) \in U_0 \times \mathbb{R}_+^m$, $k \geqslant 0$, by the recurrence formulae

$$\begin{cases} \text{calculate } u^{k+1} : u^{k+1} \in U_0 \quad \text{and} \\ \frac{1}{2} \| u^{k+1} \|_n^2 + (\varepsilon \nabla J(u^k) - u^k, u^{k+1})_n + \varepsilon(\lambda_k, \varphi(u^{k+1}))_m \\ \qquad = \inf_{v \in U_0} \{ \frac{1}{2} \| v \|_n^2 + (\varepsilon \nabla J(u^k) - u^k, v)_n + \varepsilon(\lambda_k, \varphi(v))_m \}, \\ \text{calculate } \lambda^{k+1} : \lambda^{k+1} = P_+(\lambda^k + \rho\varphi(u^{k+1})), \end{cases}$$

where ε and ρ are two fixed negative numbers (but yet to be determined; cf. question (8)). Prove that the minimisation problem defining the vector u^{k+1} in terms of the pair (u^k, λ^k) has a unique solution. Prove that the vector u^{k+1} is a solution of this problem if, and only if,

$u^{k+1} \in U_0$ and
$$(u^{k+1} - u^k + \varepsilon \nabla J(u^k), v - u^{k+1})_n + \varepsilon(\lambda^k, \varphi(v) - \varphi(u^{k+1}))_m \geqslant 0 \quad \text{for every } v \in U_0.$$

(7) Let (u, λ) be a saddle point of the Lagrangian L. Establish the following inequalities:

$$2(\lambda^k - \lambda, \varphi(u) - \varphi(u^{k+1}))_m \leqslant \frac{1}{\rho} (\| \lambda^k - \lambda \|_m^2 - \| \lambda^{k+1} - \lambda \|_m^2) + \rho C^2 \| u^{k+1} - u \|_n^2,$$

$$(u^{k+1} - u^k, u - u^{k+1})_n + \varepsilon(\nabla J(u^k) - \nabla J(u), u - u^{k+1})_n$$
$$\qquad + \varepsilon(\lambda^k - \lambda, \varphi(u) - \varphi(u^{k+1}))_m \geqslant 0,$$

$$(\nabla J(u^k) - \nabla J(u), u - u^{k+1})_n \leqslant \frac{M}{2} \| u^k - u^{k+1} \|_n^2 - \frac{\alpha}{2} (\| u^k - u \|_n^2 + \| u^{k+1} - u \|_n^2),$$

$$\tfrac{1}{2} \| u^{k+1} \|_n^2 + (u^{k+1}, u - u^{k+1})_n - \tfrac{1}{2} \| u^k \|_n^2 - (u^k, u - u^k)_n$$

$$+ \tfrac{1}{2}(\varepsilon M - 1) \| u^k - u^{k+1} \|_n^2 + \varepsilon \frac{\alpha}{2} (\| u^{k+1} - u \|_n^2 - \| u^k - u \|_n^2)$$

$$+ \varepsilon (\rho \frac{C^2}{2} - \alpha) \| u^{k+1} - u \|_n^2 + \frac{\varepsilon}{2\rho} (\| \lambda^k - \lambda \|_m^2 - \| \lambda^{k+1} - \lambda \|_m^2) \geqslant 0.$$

(8) Deduce from the last inequality of question (7) that, if

$$0 < \varepsilon \leqslant \frac{1}{M} \quad \text{and} \quad 0 < \rho < \frac{2\alpha}{C^2},$$

then

$$\lim_{k \to \infty} u^k = u \quad \text{and the sequence } (\lambda^k)_{k \geqslant 0} \text{ is bounded.}$$

Linear programming

Introduction

A *linear programming problem* takes the form

find u such that

$$(P_1) \qquad \begin{cases} u \in U = \{v \in \mathbb{R}^n : Cv \leqslant d\}, \ C \in \mathscr{A}_{m,n}(\mathbb{R}), \ d \in \mathbb{R}^m, \\ J(u) = \inf_{v \in U} J(v), \ J(v) = (a, v)_n. \end{cases}$$

As regards the *existence* and *uniqueness* of a solution, the results established earlier provide almost no information. In fact, the function J being 'only' convex (it is not strictly convex; even less is it elliptic!), the most that can be asserted is the existence of a solution (but not its uniqueness), for the case of a bounded set U.

The difficulty arises specifically from the fact that the functional tends to $+\infty$ in certain directions, while, in others, it tends to $-\infty$. That is why (though one swallow does not make a summer) a linear programming problem is harder to treat than certain non-linear programming problems, notably quadratic ones. For example, the proof of the existence of a solution when $\inf_{v \in U} J(v) > -\infty$ is not trivial (cf. Theorem 10.1-1), even though the result is an intuitively obvious one.

After a rapid review, in section 10.2, of a number of *examples* of linear programming problems (originating, in essence, from within 'economics'), we give a detailed description, in section 10.3, of the *simplex method*. This method, *universally employed*, is altogether remarkable. (A recently discovered algorithm for solving linear programming problems has, however, generated new hopes as well as various controversies. Discovered in 1984 by Narendra Karmarkar, it seems to be much faster than the simplex method in some cases. For a lucid exposition of Karmarkar's ideas, see Sect. 8.2 of Strang (1986).) *In nearly every instance*, in fact, it enables us, with only a *finite* number of elementary operations, either to calculate a solution of the problem (P_1) (after it has been put into an equivalent form, one which is better suited to the application of the method), or to conclude that the problem has no solution. Nevertheless, a phenomenon of *cycling* can make its appearance in certain exceptional cases, for which the method fails to reach a conclusion. There we have another difference as against quadratic programming, where, we should not forget, Uzawa's method is always convergent.

In section 10.4, we begin by showing that the results of the previous chapter concerning *duality* have only a limited bearing when applied to the problem (P_1). This observation then leads to a transformation of the latter into the equivalent form

find u such that

$$(P_2) \quad \begin{cases} u \in U = \{v \in \mathbb{R}^n_+ : Cv \leqslant d\}, \quad C \in \mathscr{A}_{m,n}(\mathbb{R}), \quad d \in \mathbb{R}^m, \\ J(u) = \inf_{v \in U} J(v), \quad J(v) = (a, v)_n, \end{cases}$$

then to a definition of a *new* type of dual problem,

find λ such that

$$(Q_2) \quad \begin{cases} \lambda \in \Lambda = \{\mu \in \mathbb{R}^m_+ : C^T \mu + a \geqslant 0\}, \\ G(\lambda) = \sup_{\mu \in \Lambda} G(\mu), \quad G(\mu) = -(d, \mu)_m. \end{cases}$$

The introduction of these two forms of problem, (P_2) and (Q_2), then enables us to establish the *fundamental result on duality in linear programming* (Theorem 10.4-3). Finally, we bring out (Theorem 10.4-4) the connection, unexpected *a priori*, but nonetheless real, which exists between the simplex method and duality.

10.1 General results on linear programming

Following the definition given in section 8.2, a *linear programming* problem corresponds to the minimisation of a linear functional

$$J: v \in \mathbb{R}^n \to J(v) = \sum_{i=1}^{n} a_i v_i = (a, v)_n,$$

where $a = (a_i)$ is a given vector of $\mathbb{R}^n$, while the vector v runs through a set of the form

$$U = \left\{ v \in \mathbb{R}^n : \sum_{i=1}^{n} c_{ij} v_j \leqslant d_i, \ 1 \leqslant i \leqslant m \right\} = \{v \in \mathbb{R}^n : Cv \leqslant d\},$$

where $C = (c_{ij})$ is a given matrix of type (m, n) and $d = (d_i)$ is a vector of $\mathbb{R}^m$. It is, in fact, always possible to have recourse to this form, since every equality $\sum_{i=1}^{n} c'_{ij} v_j = d'_i$ which may happen to turn up in the definition of the set U may be replaced by the two inequalities

$$\sum_{j=1}^{n} c'_{ij} v_j \leqslant d'_i \text{ and } \sum_{j=1}^{n} (-c'_{ij}) v_j \leqslant -d'_i.$$

Depending on the type of question at issue, we shall come to distinguish three *canonical forms* of linear programming problem:

find u such that

(P_1)

$$\begin{cases} u \in U = \left\{ v \in \mathbb{R}^n : \sum_{j=1}^{n} c_{ij} v_j \leqslant d_i,\, 1 \leqslant i \leqslant m \right\}, \\[2mm] J(u) = \inf_{v \in U} J(v),\, J(v) = \sum_{i=1}^{n} a_i v_i \end{cases}$$

(the form given earlier), or

find u' such that

(P_2)

$$\begin{cases} u' \in U' = \left\{ v' \in \mathbb{R}_+^{n'} : \sum_{j=1}^{n'} c_{ij}' v_j' \leqslant d_i',\, 1 \leqslant i \leqslant m' \right\}, \\[2mm] J'(u') = \inf_{v \in U'} J'(v'),\, J'(v') = \sum_{i=1}^{n'} a_i' v_i' \end{cases}$$

(we recall that $\mathbb{R}_+^p = \{v = (v_i) \in \mathbb{R}^p : v_i \geqslant 0,\, 1 \leqslant i \leqslant p\}$), or

find u'' such that

(P_3)

$$\begin{cases} u'' \in U'' = \left\{ v'' \in \mathbb{R}_+^{n''} : \sum_{j=1}^{n''} c_{ij}'' v_j'' = d_i'',\, 1 \leqslant i \leqslant m'' \right\}, \\[2mm] J''(u'') = \inf_{v \in U''} J''(v''),\, J''(v'') = \sum_{i=1}^{n''} a_i'' v_i''. \end{cases}$$

It is then easy to see that *the three canonical forms are equivalent*, in the following sense. Starting with a problem set in one of the forms, it is always possible to set in correspondence with it a problem posed in either one of the other two forms, in such a way that a knowledge of the set (which may be empty!) of solutions of the 'initial' problem leads to a knowledge of the set of solutions of the 'new' problem, and conversely.

For instance there is associated, with a problem posed in the form (P_1), a problem of the form (P_2), with

$$U' = \left\{ v' = (\hat{v}, \tilde{v}) \in \mathbb{R}_+^{2n} : \sum_{j=1}^{n} c_{ij} \hat{v}_j + \sum_{j=1}^{n} (-c_{ij}) \tilde{v}_j \leqslant d_i,\, 1 \leqslant i \leqslant m \right\},$$

$$J'(v') = \sum_{i=1}^{n} a_i \hat{v}_i + \sum_{i=1}^{n} (-a_i) \tilde{v}_i.$$

A vector $u \in \mathbb{R}^n$ is, in effect, a solution of the problem (P_1) if and only if the vector $u' = (\hat{u}, \tilde{u}) \in \mathbb{R}^{2n}$, with $u = \hat{u} - \tilde{u}$, $\hat{u} \in \mathbb{R}_+^n$, $\tilde{u} \in \mathbb{R}_+^n$, is a solution of the problem (P_2). In the same way, given a problem posed in the form (P_2), there is associated with it a problem of the form (P$_3$), with

$$U'' = \left\{ v'' = (v', \tilde{v}) \in \mathbb{R}_+^{n'+m'} : \sum_{j=1}^{n'} c'_{ij} v'_j + \tilde{v}_i = d'_i, \, 1 \leqslant i \leqslant m \right\},$$

$$J''(v'') = \sum_{i=1}^{n'} a'_i v'_i,$$

the 'new' variables $\tilde{v}_i$, $1 \leqslant i \leqslant m$, being known as *slack variables*. A vector $u' \in \mathbb{R}^{n'}$ is then a solution of the problem (P$_2$) if and only if the vector $u'' = (u', \tilde{u}) \in \mathbb{R}^{n'+m'}$, with $\sum_{j=1}^{n'} c'_{ij} u'_j + \tilde{u}_i = d'_i$, $1 \leqslant i \leqslant m$, is a solution of the problem (P$_3$). Lastly, given a problem posed in the form (P$_3$), there is associated with it a problem posed in the form (P$_1$), with

$$U = \left\{ v \in \mathbb{R}^{n''} : -v_i \leqslant 0, \, 1 \leqslant i \leqslant n'', \, \sum_{j=1}^{n''} c''_{ij} v_j \leqslant d''_i, \, 1 \leqslant i \leqslant m'', \right.$$
$$\left. \sum_{j=1}^{n''} (-c''_{ij}) v_j \leqslant -d''_i, \, 1 \leqslant i \leqslant m'' \right\},$$

$$J(v) = \sum_{i=1}^{n''} a''_i v_i,$$

so that a vector $u'' \in \mathbb{R}^{n''}$ is a solution of the problem (P$_3$) if and only if the vector $u = u'' \in \mathbb{R}^{n''}$ is a solution of the problem (P$_1$).

Remark

With the evident desire of simplifying notation, we shall omit, from now on, the 'primes' and 'double primes' which appear as superscripts to variables, functions, etc., in the problems which are to come under consideration.

□

Before stating a general result regarding existence, we make three very simple preliminary remarks, which succeed in bringing out the 'peculiar' nature of linear programming, especially in comparison with quadratic programming.

In the first place, it is clear that *no interior point of the set U could be a solution*, except when J is the zero functional. For, let u be an interior point of the set U and $\varepsilon > 0$ the radius of a closed ball of $\mathbb{R}^n$, with centre at u, lying entirely within the set U. The linearity of the functional $J : v \to J(v) = (a, v)$ shows that

$$J\left(u - \frac{\varepsilon}{\|a\|^2} a \right) = J(u) - \varepsilon < J(u) \quad \text{if} \quad a \neq 0,$$

and as the point $\{u - (\varepsilon/\|a\|^2)a\}$ belongs to the set U, the conclusion follows.

Secondly, *it is possible for the problem not to have any solution when the set U is unbounded.* Consider the example of the function $J(v) = -v$ and the set $U = \mathbb{R}_+$. We have here a second difference as against quadratic programming (Theorem 8.2-1), which arises, of course, from an *absence of the property of coerciveness* in a linear functional!

Thirdly, while it is easy to prove that a *coercive* functional $(J(v) \to \infty$ as $\|v\| \to \infty)$ has (at least) one minimum over a closed, non-emtpy set U (see the proof of the theorem cited above), the proof of the existence of a solution in the case where $\inf_{v \in U} J(v) > -\infty$ is now a more delicate matter. More precisely, we are going to establish the following result, concerning the existence, or the non-existence, of a solution of a problem posed in the form (P_3) (in order to simplify the proof; but this is no restriction, since the three forms are equivalent).

Theorem 10.1-1

Consider the problem

$$\textit{find } u \textit{ such that}$$

$$\begin{cases} u \in U = \{v \in \mathbb{R}^n_+ : Cv = d\}, \quad C \in \mathscr{A}_{m,n}(\mathbb{R}), \\ J(u) = \inf_{v \in U} J(v), \quad J(v) = (a, v). \end{cases}$$

Suppose that the set U is non-empty. Then the following alternative holds:

(i) *either* $\inf_{v \in U} J(v) = -\infty$;

(ii) *or* $\inf_{v \in U} J(v) > -\infty$, *and the problem has at least one solution.*

Proof

Let us take the case $\inf_{v \in U} J(v) > -\infty$, and consider a *minimising sequence* $(u_k)_{k \geq 0}$, which satisfies, by definition,

$$u_k = (u_i^k)_{i=1}^n \in U \quad \text{for every } k \geq 0, \quad \lim_{k \to \infty} J(u_k) = \inf_{v \in U} J(v).$$

As usual, we denote by (e_i) the canonical basis of $\mathbb{R}^n$ and introduce the matrix of type $(m+1, n)$

$$\mathscr{B} = \left(\begin{array}{c} \boxed{a^{\mathsf{T}}} \\ \boxed{C} \end{array} \right).$$

As the components u_i^k are all non-negative, the sequence $(\mathscr{B} u_k)_{k \geq 0}$ belongs to the cone

$$\mathscr{C} \stackrel{\text{def}}{=} \left\{ \sum_{i=1}^n v_i \mathscr{B} e_i \in \mathbb{R}^{m+1} : 0 \leq v_i, 1 \leq i \leq n \right\}.$$

Such a cone is closed (by the first part of the proof of the Farkas lemma;

cf. Theorem 9.1-1) and *the sequence* $(\mathscr{B}u_k)$ *is convergent* (since $\lim_{k \to \infty}$ $a^{\mathrm{T}}u_k = \inf_{v \in U} J(v) > -\infty$ and $Cu_k = d$ for every $k \geqslant 0$), so that there exists a vector u such that

$$u \in \mathbb{R}^n_+, \quad \lim_{k \to \infty} \mathscr{B}u_k = \mathscr{B}u \Leftrightarrow \begin{cases} \lim_{k \to \infty} a^{\mathrm{T}}u_k = a^{\mathrm{T}}u, \\ \quad d = Cu. \end{cases} \qquad \Box$$

Remark

Necessary and sufficient conditions for the existence of a solution which make use of *duality* will be established in section 10.4. $\qquad \Box$

10.2 Examples of linear programming problems

The examples which we review take the form (P_2). An example taking the form (P_3) is given in Exercise 10.2-1.

A factory makes two products p_1 and p_2, using primary materials q_1, q_2 and q_3. The production of one unit of product p_1 requires one unit of q_1, 2 units of q_2 and 4 units of q_3; the production of one unit of product p_2 requires 6 units of q_1, 2 units of q_2 and 1 unit of q_3. The factory has at its disposal 30 units of q_1, 15 units of q_2 and 24 units of q_3. Lastly, the sale of one unit of p_1 brings in a return of 2 ECUs and that of one unit of p_2 a return of 1 ECU. The aim of the factory being to maximise returns, what is the best way of organising production?

Calling u_1 and u_2 the *optimal* number of units of the products p_1 and p_2, respectively, to be produced, this *problem of production scheduling* may be expressed as

Figure 10.2-1

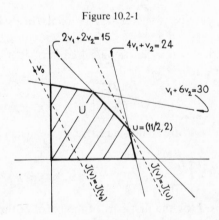

find $u = (u_1, u_2)$ such that

$$\begin{cases} u \in U = \{v = (v_1, v_2) \in \mathbb{R}_+^2 : v_1 + 6v_2 \leqslant 30, 2v_1 + 2v_2 \leqslant 15, 4v_1 + v_2 \leqslant 24\}, \\ J(u) = \inf_{v \in U} J(v), \quad J(v) = -2v_1 - v_2. \end{cases}$$

This, then, is a linear programming problem posed in the form (P_2) and one, moreover, which can be solved 'graphically' in a very simple manner (cf. figure 10.2-1). By 'shifting' the straight line $\{J(v) = J(v_0)\}$ left or right, it may be seen that the solution is, in fact, the *vertex* $u = (\frac{11}{2}, 2)$ of the set U, which is here a pentagon (the existence of at least one solution was quite evident, since the set U given above is non-empty, closed and bounded).

We here encounter the first instance of a general property, and one which is *fundamental*, of linear programming problems, which will be established rigorously in Theorem 10.3-2, but which, for the time being, we take note of in the following 'vague' form. *If the problem has a solution, then at least one 'vertex' of the set U (which is a 'polyhedron', whether bounded or not, of $\mathbb{R}^n$) is a solution.*

It is possible, moreover, that *several* vertices are solutions. That would have been the case in the problem given above if, for example, the functional J had been $J(v) = -4v_1 - v_2$. Then the closed segment joining the two points $(\frac{11}{2}, 2)$ and $(6, 0)$ would have comprised the set of solutions.

Remark

In order that the solution $u = (\frac{11}{2}, 2)$ might be acceptable, it is necessary, in addition, for the units of the product p_1 to be 'divisible'. Otherwise, one would be faced with an 'integer linear programming' problem. We do not here say anything about this type of problem, but refer to the Bibliography and comments. □

Let us show, in passing, how the problem given above may be immediately converted into a problem of type (P_3). Introducing three slack variables v_i, $i = 3, 4, 5$, we are led to look for u such that

$$\begin{cases} u \in U = \{v \in \mathbb{R}_+^5 : Cv = d\}, \\ J(u) = \inf_{v \in U} J(v), \quad J(v) = (a, v), \end{cases}$$

where

$$a = \begin{pmatrix} -2 \\ -1 \\ 0 \\ 0 \\ 0 \end{pmatrix}, \quad C = \begin{pmatrix} 1 & 6 & 1 & & \\ 2 & 2 & & 1 & \\ 4 & 1 & & & 1 \end{pmatrix}, \quad d = \begin{pmatrix} 30 \\ 15 \\ 24 \end{pmatrix}.$$

Table 10.2-1

	type 1	type 2	type 3	type 4
calories	2	1	0	1
vitamins	3	4	3	5
price	2	2	1	8

Table 10.2-2

	type I	type II
calories	1	0
vitamins	0	1

Here, as a second example, is another problem. Four types of food are on sale whose calorific value, vitamin content and price (given for a common unit of weight, and in suitable units for each) are set down in table 10.2-1.

The problem, which goes by the name of the *diet problem*, consists in obtaining as cheaply as possible at least 12 calories and 7 vitamins. Calling u_i the 'optimal' weight of food of type i to be bought, we are thus led to look for the minimum of the function

$$J(v) = 2v_1 + 2v_2 + v_3 + 8v_4,$$

as the vector $v = (v_i)_{i=1}^4$ runs through the set

$$U = \{v = (v_i)_{i=1}^4 \in \mathbb{R}_+^4 : 2v_1 + v_2 + v_4 \geqslant 12, 3v_1 + 4v_2 + 3v_3 + 5v_4 \geqslant 7\}.$$

It will be observed that, in contrast to the previous problem, the set U is here *unbounded*. While it is not difficult to show that the problem given above has (at least) one solution, it ought not to be forgotten that this is not a general feature, as it has already been remarked.

Let us next examine the *business competitor problem*. A business competitor would like to corner the market with two new food products whose respective calorific value and vitamin content (per unit volume) have the numerical values shown in table 10.2-2. The businessman would like to fix the prices λ_1 and λ_2 per unit volume of each of the new food products in a way which maximises the receipts from total sales. He is thus led to look for the maximum of the function

$$G(\mu) = 12\mu_1 + 7\mu_2$$

(of course, it is assumed that the buyer only purchases what is strictly necessary) as the vector $\mu = (\mu_i)_{i=1}^2$ runs through the set

$$M = \{\mu = (\mu_i)_{i=1}^2 \in \mathbb{R}_+^2 : 2\mu_1 + 3\mu_2 \leqslant 2, \mu_1 + 4\mu_2 \leqslant 2,$$
$$3\mu_2 \leqslant 1, \mu_1 + 5\mu_2 \leqslant 8\}$$

(we are simply expressing mathematically the obvious intention of the businessman to offer the same amount of calories and vitamins as is done by each of the foods of types 1 to 4, but at a cheaper or equal price; otherwise, he would have no chance of selling his new product to the discerning consumer, leaving aside, of course, any consideration of taste...).

It will be observed that the diet problem and the business competitor problem may be set down, respectively, as

(P) $\qquad \begin{cases} u \in U = \{v \in \mathbb{R}^4 : \quad Cv \geqslant d\}, \\ J(u) = \inf_{v \in U} J(v), \quad J(v) = (a, v)_4, \end{cases}$

(Q) $\qquad \begin{cases} \lambda \in M = \{\mu \in \mathbb{R}_+^2 : \quad C^T \mu \leqslant a\}, \\ G(\lambda) = \sup_{\mu \in M} G(\mu), \quad G(\mu) = (d, \mu)_2. \end{cases}$

with

$$a^T = (2 \quad 2 \quad 1 \quad 8), \quad C = \begin{pmatrix} 2 & 1 & 0 & 1 \\ 3 & 4 & 3 & 5 \end{pmatrix}, \quad d = \begin{pmatrix} 12 \\ 7 \end{pmatrix}.$$

As will be seen in section 10.4, this is an example of *duality in linear programming*, the problem (Q) being precisely the 'dual' of the problem (P).

Exercises

10.2-1. A product is made in factories u_i, $1 \leqslant i \leqslant I$, the (known) production quota of factory u_i being p_i. The product is bought up by customers v_j, $1 \leqslant j \leqslant J$, the (known) level of consumption by customer v_j being c_j. Knowing that $\sum_{i=1}^I p_i = \sum_{j=1}^J c_j$ (a reasonable hypothesis!) and calling γ_{ij} the (known) cost of transportation of one unit of the product involved from factory u_i to customer v_j, the *transportation problem* consists in minimising the total cost of transportation. Express this problem in the form of a linear programming problem.

10.2-2. (1) Solve 'directly' the diet problem and the business competitor problem given as examples in the text:

(P) $\qquad \begin{cases} u \in U = \{v \in \mathbb{R}^4 : Cv \geqslant d\}, \\ J(u) = \inf_{v \in U} J(v), J(v) = (a, v)_4, \end{cases}$

(Q) $\qquad \begin{cases} \lambda \in M = \{\mu \in \mathbb{R}_+^2 : C^T \mu \leqslant a\}, \\ G(\lambda) = \sup_{\mu \in M} G(\mu), \quad G(\mu) = (d, \mu)_2, \end{cases}$

with

$$a^T = (2 \quad 2 \quad 1 \quad 8), \quad C = \begin{pmatrix} 2 & 1 & 0 & 1 \\ 3 & 4 & 3 & 5 \end{pmatrix}, \quad d = \begin{pmatrix} 12 \\ 7 \end{pmatrix}$$

(2) Apply the simplex method (described in section 10.3) to each of these two problems.

(3) Apply the results of section 10.4 (duality and linear programming) to these two examples.

10.3 The simplex method

Consider a linear programming problem posed in the form (P_3):

find u such that

$$\begin{cases} u \in U = \left\{ v \in \mathbb{R}_+^n : \sum_{j=1}^n c_{ij} v_j = d_i, 1 \leqslant i \leqslant m \right\}, \\ J(u) = \inf_{v \in U} J(v), \quad J(v) = \sum_{i=1}^n a_i v_i. \end{cases}$$

Denoting by $C^j \in \mathbb{R}^m$, $1 \leqslant j \leqslant n$, the *column* vectors of the matrix $C = (c_{ij}) \in \mathscr{A}_{m,n}(\mathbb{R})$, we observe that the following three expressions are equivalent:

$$\sum_{j=1}^n c_{ij} v_j = d_i, \quad 1 \leqslant i \leqslant m \Leftrightarrow Cv = d \Leftrightarrow \sum_{j=1}^n v_j C^j = d,$$

the last, especially, being of constant use in this section.

With an eye to making precise the idea of the 'vertex' of a set U, we begin with a number of general definitions. If U is a convex subset of a vector space, a point $u \in U$ is called an *extreme point of the set U* if the implication

$$\left. \begin{matrix} v \in U, \quad w \in U, \quad 0 < \lambda < 1 \\ u = \lambda v + (1 - \lambda) w \end{matrix} \right\} \Rightarrow u = v = w,$$

is satisfied. A *polyhedron* of $\mathbb{R}^n$ is a set of the form

$$U = \left\{ v \in \mathbb{R}^n : \sum_{j=1}^n c_{ij} v_j \leqslant d_i, 1 \leqslant i \leqslant p, \sum_{j=1}^n c_{ij} v_j = d_i, p + 1 \leqslant i \leqslant m \right\}.$$

Lastly, an extreme point of a polyhedron is called a *vertex* of the polyhedron. These definitions generalise, in a natural way, those of a convex polygon (not necessarily bounded) of the plane and of a vertex of such a polygon. Figure 10.3-1 gives an example of a polyhedron of $\mathbb{R}^3$, defined by

$$U = \{ v \in \mathbb{R}_+^3 : v_1 + 2v_2 + 4v_3 = 4 \}.$$

It may be verified that the only vertices of the set U in question are the points $(4,0,0)$, $(0,2,0)$ and $(0,0,1)$.

We begin by *characterising the vertices of a polyhedron of the form*

$$U = \left\{ v \in \mathbb{R}^n_+ : \sum_{j=1}^n v_j C^j = d \right\}$$

(for polyhedra defined by inequalities, see Exercise 10.3-1). With a general point v of such a set is associated the set of indices

$$I^*(v) = \{ 1 \leqslant j \leqslant n : v_j > 0 \}.$$

If the origin belongs to the set U (or, equivalently, if $d = 0$), in which case $I^*(0) = \varnothing$, it should be observed that *the origin is inevitably a vertex*, since

$$\left. \begin{aligned} 0 &= \lambda v + (1 - \lambda) w \\ v &\geqslant 0, \quad w \geqslant 0, \quad 0 < \lambda < 1 \end{aligned} \right\} \Rightarrow v = w = 0.$$

We now look at the case of the remaining vertices.

Theorem 10.3-1

A point $u \in U$ other than the origin is a vertex of the polyhedron

$$U = \left\{ v \in \mathbb{R}^n_+ : \sum_{j=1}^n v_j C^j = d \right\}, \quad C^j \text{ vectors of } \mathbb{R}^n,$$

if and only if the vectors $C^j, j \in I^*(u)$, where

$$I^*(u) = \{ 1 \leqslant j \leqslant n : u_j > 0 \},$$

are linearly independent.

Proof

Suppose that the vectors $C^j, j \in I^*(u)$, are linearly dependent. There exists a vector $w = (w_j)_{j=1}^n$ such that

$$w \neq 0, \quad w_j = 0 \quad \text{if} \quad j \notin I^*(u), \quad \sum_{j=1}^n w_j C^j = Cw = 0.$$

Figure 10.3-1

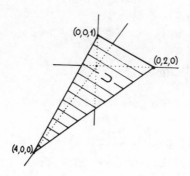

As $u_j > 0$ for $j \in I^*(u)$ (by definition), there exists a number $\theta \neq 0$ such that $u_j \pm \theta w_j \geq 0, 1 \leq j \leq n$ ($u_j \pm \theta w_j = 0$ for $j \notin I^*(u)$). Since

$$C(u \pm \theta w) = \sum_{j=1}^{n} u_j C^j \pm \theta \sum_{j=1}^{n} w_j C^j = d,$$

the conclusion may be drawn that the two vectors $u + \theta w$ and $u - \theta w$ belong to the set U. The relations

$$u = \tfrac{1}{2}(u + \theta w) + \tfrac{1}{2}(u - \theta w), \quad \theta w \neq 0,$$

then show that u is not an extreme point of the set U.

Suppose that the vectors C^j, $j \in I^*(u)$, are linearly independent and suppose that it is possible to have

$$u = \lambda v + (1 - \lambda)w, \quad \text{with} \quad v \in U, \quad w \in U, \quad 0 < \lambda < 1.$$

The relations $v \geq 0$, $w \geq 0$, $0 < \lambda < 1$, imply the inclusions $I^*(v) \cup I^*(w)$ $\subset I^*(u)$. As the vector $z \overset{\text{def}}{=} w - v = (z_j)_{j=1}^{n}$ satisfies

$$\begin{cases} z_j = 0 & \text{if} \quad j \notin I^*(u), \\ \sum_{j=1}^{n} z_j C^j = Cz = Cw - Cv = 0, \end{cases}$$

the linear independence of the vectors C^j implies that $z_j = 0$ if $j \in I(u^*)$. The equality $v = w$ has, therefore, been established, which implies, in its turn, the equality $u = v = w$. Thus, the point u is an extreme point. □

With this characterisation of the vertices of the polyhedron U, we are now in a position to prove a property which we had ascertained, for a particular example, in section 10.2, and which is *the keystone of the algorithm* which we shall examine further ahead.

Theorem 10.3-2

If the problem

$$\text{find } u \text{ such that}$$

$$\begin{cases} u \in U = \left\{ v \in \mathbb{R}^n_+ : \sum_{j=1}^{n} v_j C^j = d \right\}, \\ J(u) = \inf_{v \in U} J(v), \quad J(v) = \sum_{j=1}^{n} a_j v_j, \end{cases}$$

has a solution, then (at least) one vertex of the set U is also a solution.

Proof

Let $u \in U$ be a solution of the problem given above. If $I^*(u) = \emptyset$, then $u = 0$; now we have seen that the origin is a vertex of the set U if it belongs to it.

If $I^*(u) \neq \varnothing$, then either the vectors $C^j, j \in I^*(u)$, are linearly independent and the point u is a vertex of the set U, or there exists a vector $w = (w_j)_{j=1}^n$ satisfying

$$\max_j w_j > 0, \quad w_j = 0 \quad \text{if} \quad j \notin I^*(u), \quad \sum_{j=1}^n w_j C^j = Cw = 0;$$

in fact, there is no loss of generality in assuming that at least one of the components of the vector w is positive.

Let us consider the points of $\mathbb{R}^n$ which are of the form $u + \theta w$, $\theta \in \mathbb{R}$. Firstly, they satisfy

$$C(u + \theta w) = Cu + \theta Cw = d \quad \text{for every } \theta \in \mathbb{R}.$$

And, secondly, as

$$(u + \theta w)_j = \begin{cases} u_j + \theta w_j & \text{if} \quad u_j > 0, \\ 0 & \text{if} \quad u_j = 0, \end{cases}$$

it follows that there exists an interval of the form $[\theta_0, \theta_1]$, with

$$-\infty < \theta_0 = \max \left\{ -\frac{u_j}{w_j} : j \in I^*(u) \text{ and } w_j > 0 \right\} < 0,$$

$$0 < \theta_1 \leqslant \min \left\{ -\frac{u_j}{w_j} : j \in I^*(u) \text{ and } w_j < 0 \right\} \leqslant +\infty$$

(it has been assumed that $\max_j w_j > 0$; remember that $\inf \varnothing = +\infty$), such that

$$u + \theta w \in U \quad \text{for every } \theta \in [\theta_0, \theta_1].$$

As the functional J is linear,

$$J(u + \theta w) = J(u) + \theta J(w) \quad \text{for every } \theta \in [\theta_0, \theta_1],$$

which forces $J(w) = 0$ since $J(u) = \inf_{v \in U} J(v)$, the interval $[\theta_0, \theta_1]$ containing the origin in its interior.

In other words, all points of the form $u + \theta w$, $\theta_0 \leqslant \theta \leqslant \theta_1$, are solutions of the problem. Since at least one of the components $u_j + \theta_0 w_j, j \in I^*(u)$, is zero by definition of θ_0, a solution $u' = u + \theta_0 w$ has, therefore, been constructed for which

$$I^*(u') \subsetneqq I^*(u), \quad \text{so that card } (I^*(u')) < \text{card} (I^*(u)).$$

Hence, either the vectors $A^j, j \in I^*(u')$, are linearly independent and the solution u' is a vertex, or else these same vectors are linearly dependent. In the latter instance, the preceding argument may be repeated. As each application of it has the effect of reducing by at least one the number of vectors C^j under consideration, we necessarily end up, after a finite number

of constructions similar to the one given in detail above, with a solution
which is also a vertex.   □

As a corollary of the previous two theorems, we are able to prove two
interesting properties of vertices.

Theorem 10.3-3
If the polyhedron

$$U = \{v \in \mathbb{R}^n_+ : Cv = d\}, \quad C \in \mathscr{A}_{m,n}(\mathbb{R}), \quad d \in \mathbb{R}^m,$$

*is not empty, it has at least one vertex. Furthermore, the number of vertices is
finite.*

Proof
Let us consider the following linear programming problem:

$$\begin{cases} \text{find } (u, \tilde{u}) \text{ such that} \\ (u, \tilde{u}) \in \tilde{U} = \{(v, \tilde{v}) \in \mathbb{R}^n_+ \times \mathbb{R}^m_+ : Cv + \tilde{v} = d\}, \\ \tilde{J}(u, \tilde{u}) = \inf_{(v, \tilde{v}) \in \tilde{U}} \tilde{J}(v, \tilde{v}), \quad \tilde{J}(v, \tilde{v}) = \sum_{i=1}^m \tilde{v}_i. \end{cases}$$

If the set U is not empty, this problem at once has as solutions all pairs $(u, 0)$,
where $u \in U$. Hence, it suffices to apply Theorem 10.3-2 to any of these
solutions and then to use the characterisation of the vertices given in
Theorem 10.3-1. The latter also shows that the vertices are finite in number,
since there is only a finite number of ways of choosing linearly independent
vectors from among the vectors $C^j, 1 \leqslant j \leqslant n$.   □

The *simplex method*, due to G.B. Dantzig, rests on the result given above.
The guiding principle is, in effect, the *evaluation of the functional J at
particular vertices of the set U*. These are calculated sequentially, following a
particular strategy. Starting with a vertex u_0, a sequence of vertices
$u_0, u_1, \ldots, u_k, u_{k+1}, \ldots$ is constructed which correspond to *decreasing* values
of the functional. *If the problem has a solution*, and *if* it can be so arranged
that all the inequalities are *strict*

$$J(u_k) > J(u_{k+1}), \quad k \geqslant 0,$$

then *this process leads, in a finite number of iterations, to a vertex which is also
a solution*, since the number of vertices is finite.

Remarks
(1) It actually happens in certain cases that it is not possible to attain a *strict*
inequality $J(u_k) > J(u_{k+1})$; only the equality $J(u_k) = J(u_{k+1})$ may be at-

tained. This circumstance can lead to the phenomenon known as 'cycling'. We shall return to this.

(2) Since the eventuality that $\inf_{v \in U} J(v) = -\infty$ is in no way excluded (go back to the example of the functional $J(v) = -v$ and the set $U = \mathbb{R}_+$, with the origin as the only vertex), it is a remarkable fact that the method in question also allows this eventuality to be detected, as long as the phenomenon of 'cycling' referred to above does not occur.

(3) One might have considered evaluating systematically the functional J at all the vertices of the set U, without following any particular strategy. Since, even for 'small' values of m and n, the number of vertices is very large, this *naif* procedure should be completely discountenanced. $\square$

We shall first concentrate on the *description of a single iteration* of the simplex method, that is to say, on the most 'common' case, *the transition from one vertex to the next*, even though, in actual fact, other eventualities are also possible. Only in the end will we examine the problem of determining the initial vertex u_0 (the 'initialisation' of the algorithm). We recall the definition of the problem. We have to find u such that

$$\begin{cases} u \in U = \left\{ v \in \mathbb{R}_+^n : \sum_{j=1}^n v_j C^j = d \right\}, \\ J(u) = \inf_{v \in U} J(v), \quad J(v) = \sum_{i=1}^n a_i v_i, \end{cases}$$

where the vectors $C^j \in \mathbb{R}^m$, $1 \leqslant j \leqslant n$, are the column vectors of the matrix $C = (c_{ij})$.

Without loss of generality, we shall assume that

$$r(C) = m$$

(which forces $m \leqslant n$). In fact, if $r(C) = m' < m$, then either the set $\{v \in \mathbb{R}^n : \sum_{j=1}^n v_j C^j = d\}$ is empty, in which case the problem has no solution, or else it is always possible to obtain from the equations $\sum_{j=1}^n c_{ij} v_j = d_i$, $1 \leqslant i \leqslant m$, a subset of m' equivalent equations whose matrix has rank m'.

Remarks

(1) It is also possible to take the column vectors C^j of the matrix C to be non-zero. For if one of them is zero, let us say C^n, in order to fix ideas, then either $a_n > 0$, in which case every solution of the problem satisfies $u_n = 0$ making it possible to 'eliminate the nth variable', or else $a_n < 0$ (if $a_n = 0$, the nth variable does not make its appearance...), in which case $\inf_{v \in U} J(v) = -\infty$ and the problem has no solution.

(2) In the same context, it should be clear that the case $m = n$ holds no great interest in what follows ...   □

The description of the simplex method leads us to make a preliminary distinction between two *types of vertices*. We recall (Theorem 10.3-1) that a point $u = (u_i) \in \mathbb{R}^n$ is a vertex of the set U if and only if

$$\begin{cases} u_i \geqslant 0, \quad 1 \leqslant i \leqslant n, \quad \sum_{j=1}^{n} u_j C^j = \sum_{j \in I^*(u)} u_j C^j = d, \\ \text{the vectors } C^j \in \mathbb{R}^m, j \in I^*(u), \text{ are linearly independent,} \end{cases}$$

where

$$I^*(u) = \{1 \leqslant j \leqslant n : u_j > 0\}.$$

Such a vertex is then said to be *non-degenerate* if $\text{card}(I^*(u)) = m$ and *degenerate* if $\text{card}(I^*(u)) < m$; hence, the origin is a degenerate vertex of the set U if it belongs to it, since $I^*(0) = \varnothing$.

Now, for reasons which will be made clear in the description of the method, it is essential to be able to associate with every vertex *exactly m* linearly independent column vectors C^j. That is the reason why, in the case of a degenerate vertex u, for which $\text{card}(I^*(u)) = m' < m$, the vectors C^j, $j \in I^*(u)$, will be extended to a full set by adjoining $m - m'$ vectors $C^j, j \notin I^*(u)$, in such a way that the m vectors C^j so obtained are linearly independent. This is always possible (because of the assumption $r(C) = m$), but, generally, in more than one way.

Hence, it is possible to give an equivalent definition. *A point $u = (u_i) \in \mathbb{R}^m$ is a vertex of the set U if and only if there exists a set*

$$I \subset \{1, 2, \ldots, n\} \quad \text{with} \quad \text{card}(I) = m,$$

such that

$$\begin{cases} u_i \geqslant 0 \quad \text{if} \quad i \in I, \quad u_i = 0 \quad \text{if} \quad i \notin I, \quad \sum_{i=1}^{n} u_i C^i = \sum_{i \in I} u_i C^i = d, \\ \text{the vectors } C_i, i \in I, \text{ are linearly independent.} \end{cases}$$

One then says that $(C^i)_{i \in I}$ is a *basis* associated with the vertex u. It should be kept in mind that *the basis associated with a non-degenerate vertex is defined uniquely*, but that *a basis associated with a degenerate vertex is not necessarily defined uniquely*.

Remark

This possibility explains the misuse of language which is sometimes encountered (but one which is very useful, for example in the description of the simplex method) and which consists in considering *one and the same* vertex as a number of 'different' vertices, inasmuch as several different bases can be associated with it. □

Suppose, then, that we have a vertex $u = (u_i)_{i=1}^n$ of the set U, with associated basis $(C^i)_{i \in I}$. The idea of the method is to *achieve a transition from that vertex to the next* (whenever it is possible), *by replacing one of the vectors C^i, $i \in I$, with one of the vectors C^j, $j \notin I$.* This will define the basis associated with the next vertex.

Let j be an index which does not belong to the set I. The vector C^j may be expressed as

$$C^j = \sum_{i \in I} \gamma_i^j C^i,$$

the components γ_i^j being defined uniquely. Since

$$\sum_{i \in I} (u_i - \theta \gamma_i^j) C^i + \theta C^j = \sum_{i \in I} u_i C^i = d \quad \text{for every } \theta \in \mathbb{R},$$

it follows that *the points with components*

$$\begin{cases} u_i - \theta \gamma_i^j & \text{if} \quad i \in I, \\ \theta & \text{if} \quad i = j, \\ 0 & \text{if} \quad i \notin I \cup \{j\}, \end{cases}$$

are points which belong to the set U, provided that

$$0 \leqslant \theta \leqslant \theta^j \overset{\text{def}}{=} \min \left\{ \frac{u_i}{\gamma_i^j} : i \in I \text{ and } \gamma_i^j > 0 \right\}.$$

Of course, it is to be hoped that the point corresponding to $\theta = \theta^j$ is a new vertex whenever $0 < \theta^j < +\infty$. This is actually just what will be proved.

Evaluated at these points, the functional J equals

$$\sum_{i \in I} a_i (u_i - \theta \gamma_i^j) + \theta a_j = J(u) + \theta \left(a_j - \sum_{i \in I} \gamma_i^j a_i \right),$$

so that, if the value of the functional is to be *strictly* diminished, it is necessary, firstly, for the number θ^j to be strictly positive and, secondly, for the number $a_j - \sum_{i \in I} \gamma_i^j a_i$ to be strictly negative. *We are then led, quite naturally, to distinguish among various cases, depending on the signs of the numbers*

$$\left(a_j - \sum_{i \in I} \gamma_i^j a_i \right), \max_{i \in I} \gamma_i^j, \theta^j = \min \left\{ \frac{u_i}{\gamma_i^j} : i \in I \text{ and } \gamma_i^j > 0 \right\}, \quad \text{for } j \notin I.$$

If all the numbers $(a_j - \sum_{i \in I} \gamma_i^j a_i)$, $j \notin I$, are non-negative (case A), it is intuitively felt that *u could be* a solution (for the non-diminishing of the functional J is encountered only 'in the direction of the neighbouring vertices'); this, in fact, is just what will be shown.

Otherwise (case B), the aim is to move to another vertex, by annihilating at least one of the 'new' components $u_i - \theta \gamma_i^j$, $i \in I$, for some *finite* value of θ, which can only be $\theta = \theta^j$. This is impossible if all the components γ_i^j are non-positive for $i \in I$ (case B1), in which case it is also clear that $\inf_{v \in U} J(v) = -\infty$. On the other hand, if one of the components γ_i^j, $i \in I$, is strictly negative, it will be shown that the point u^+ corresponding to $\theta = \theta^j$ (which is, therefore, a finite number) is in fact a vertex, which amounts to showing the linear independence of the column vectors associated with the point u^+. If this point is to satisfy the *strict* inequality $J(u^+) < J(u)$, it is also necessary for the number θ^j to be strictly positive (case B2). This latter is the most 'common' course taken by the algorithm. Otherwise (case B3), it is possible that one is caught up in the phenomenon of *cycling*.

The result which follows gathers together, and details, these various possibilities.

Theorem 10.3-4

Let u be a vertex of the set U and $(C^i)_{i \in I}$ its associated basis. For every $j \notin I$, set

$$C^j = \sum_{i \in I} \gamma_i^j C^i.$$

The only possible outcomes are the following.

Case A: $a_j - \sum_{i \in I} \gamma_i^j a_i \geqslant 0$ *for every $j \notin I$. Then the vertex u is a solution of the problem.*

Case B: *there exists (at least) one index $j \notin I$ for which $a_j - \sum_{i \in I} \gamma_i^j a_i < 0$: this outcome itself subdivides into three cases:*

Case B1: *there exists (at least) one index $j \notin I$ for which we have simultaneously*

$$a_j - \sum_{i \in I} \gamma_i^j a_i < 0 \quad and \quad \gamma_i^j \leqslant 0 \quad for\ every\ i \in I.$$

Then $\inf_{v \in U} J(v) = -\infty$.

Case B2: *there exists (at least) one index $j^+ \notin I$ for which we have simultaneously the following:*

$$\begin{cases} a_{j^+} - \sum_{i \in I} \gamma_i^{j^+} a_i < 0; \\ \gamma_i^{j^+} > 0\ for\ at\ least\ one\ index\ i \in I;\ and\ for\ i \in I,\ \gamma_i^{j^+} > 0 \Rightarrow u_i > 0. \end{cases}$$

Then the point u^+ with components

$$u_i^+ = \begin{cases} u_i - \theta^{j^+} \gamma_i^{j^+} & if\ i \in I, \\ \theta^{j^+} & if\ i = j^+, \\ 0 & if\ i \notin I \cup \{j^+\}, \end{cases}$$

where

$$\theta^{j+} = \min\left\{\frac{u_i}{\gamma_i^{j+}} : i \in I \text{ and } \gamma_i^{j+} > 0\right\} > 0,$$

is a vertex different from the vertex u, for which

$$J(u) > J(u^+).$$

With the vertex u^+ is associated the basis $(C^i)_{i \in I+}$, where $I^+ = (I - \{j^-\}) \cup \{j^+\}$, $j^- \in I$ being any of the indices which satisfy

$$j^- \in I \text{ and } \theta^{j+} = \frac{u_{j-}}{\gamma_{j-}^{j+}}.$$

Case B3: there exists (at least) one index $j \notin I$ for which $a_j - \sum_{i \in I} \gamma_i^j a_i < 0$, and for each of the indices $j \notin I$, we have simultaneously

$$\begin{cases} a_j - \sum_{i \in I} \gamma_i^j a_i < 0, \\ \gamma_i^j > 0 \quad \text{and} \quad u_i = 0 \text{ for at least one index } i \in I \text{ (the same)}. \end{cases}$$

It is then possible to associate with the same vertex u another basis $(C^i)_{i \in I+}$, where $I^+ = (I - \{j^-\}) \cup \{j^+\}$, j^+ being any of the indices which satisfy

$$j^+ \notin I \quad \text{and} \quad a_{j+} - \sum_{i \in I} \gamma_i^{j+} a_i < 0,$$

and j^- being any of the indices which satisfy

$$j^- \in I \quad \text{and} \quad \gamma_{j-}^{j+} > 0.$$

Proof

Case A: $a_j - \sum_{i \in I} \gamma_i^j a_i \geq 0$ for every $j \notin I$. The equalities $C^j = \sum_{i \in I} \gamma_i^j C^i$ may be extended to the indices $j \in I$, by setting $\gamma_i^j = \delta_{ij}$ for $i, j \in I$, so that

$$C^j = \sum_{i \in I} \gamma_i^j C^i, \quad a_j - \sum_{i \in I} \gamma_i^j a_i \geq 0, \quad 1 \leq j \leq n.$$

If $v = (v_i)_{i=1}^n$ is any point of the set U, we have

$$Cv = \sum_{j=1}^n v_j C^j = \sum_{i \in I}\left\{\sum_{j=1}^n v_j \gamma_i^j\right\} C^i = d = \sum_{i \in I} u_i C^i,$$

from which it follows that

$$u_i = \sum_{j=1}^n v_j \gamma_i^j, \ i \in I,$$

since the vectors $C^i, i \in I$, are linearly independent. Hence,

$$J(v) - J(u) = \sum_{j=1}^n a_j v_j - \sum_{i \in I} a_i u_i = \sum_{j=1}^n \left\{a_j - \sum_{i \in I} \gamma_i^j a_i\right\} v_j \geq 0,$$

since all the components v_j are non-negative, and the vertex u is a solution.

Case B1: there exists (at least) one index $j \notin I$ for which

$$a_j - \sum_{i \in I} \gamma_i^j a_i < 0 \quad and \quad \gamma_i^j \leqslant 0 \quad for \; every \; i \in I.$$

It has already been observed that the points with co-ordinates

$$\begin{aligned} u_i - \theta \gamma_i^j &\quad if \quad i \in I, \\ \theta &\quad if \quad i = j, \\ 0 &\quad if \quad i \notin I \cup \{j\}, \end{aligned}$$

are points of the set U *for every* $\theta \geqslant 0$ (in the present case, $\theta^j = \inf \varnothing = + \infty$); at these points, the functional J equals

$$J(u) + \theta \left(a_j - \sum_{i \in I} \gamma_i^j a_i \right),$$

which shows that $\inf_{v \in U} J(v) = - \infty$.

Case B2: there exists (at least) one index $j^+ \notin I$ for which

$$a_{j^+} - \sum_{i \in I} \gamma_i^{j^+} a_i < 0 \quad and \quad \theta^{j^+} \overset{\text{def}}{=} \min \left\{ \frac{u_j}{\gamma_i^{j^+}} : i \in I \; and \; \gamma_i^{j^+} > 0 \right\} > 0.$$

We define the point $u^+ = (u_i^+)_{i=1}^n$ by setting

$$u_i^+ = \begin{cases} u_i - \theta^{j^+} \gamma_i^{j^+} & if \quad i \in I, \\ \theta^{j^+} & if \quad i = j^+, \\ 0 & if \quad i \notin I \cup \{j^+\}. \end{cases}$$

Then, on the one hand,

$$J(u) - J(u^+) = - \theta^{j^+} \left(a_{j^+} - \sum_{i \in I} \gamma_i^{j^+} a_i \right) > 0,$$

and, on the other, *at least one* of the components u_i^+, $i \in I$, of the point u^+ is zero, by definition of the number θ^{j^+}. Setting u_{j^-} for *any one* of these components, which then satisfies

$$\theta^{j^+} = \frac{u_{j^-}}{\gamma_{j^-}^{j^+}},$$

we shall set about proving the linear independence of the column vectors C^j, for $j \in I^+$, where

$$I^+ \overset{\text{def}}{=} (I - \{j^-\}) \cup \{j^+\},$$

which will show that the point u^+ is a vertex of the set U.

Let us assume the contrary. It is then possible to find numbers α_i, $i \in I$, not all zero, such that

$$\sum_{i \in I^+} \alpha_i C^i = 0 \; and \; \alpha_{j^+} \neq 0$$

(if α_{j^+} were zero, the vectors C^i, $i \in I$, would be linearly dependent); so that this is an equality of the form

$$C^{j^+} = \sum_{i \in I - \{j^-\}} \beta_i C^i,$$

which, together with the equality $C^{j^+} = \sum_{i \in I} \gamma_i^{j^+} C^i$, shows that

$$\sum_{i \in I - \{j^-\}} (\gamma_i^{j^+} - \beta_i) C^i + \gamma_{j^-}^{j^+} C^{j^-} = 0.$$

Now this last equality cannot hold, since the vectors C^i, $i \in I$, are linearly independent and $\gamma_{j^-}^{j^+} > 0$, by definition.

Case B3: *all the indices* $j \notin I$ *for which* $a_j - \sum_{i \in I} \gamma_i^j a_i < 0$ (we assume that there exists at least one) *satisfy*

$$a_j - \sum_{i \in I} \gamma_i^j a_i < 0 \quad \text{and} \quad \theta^j \stackrel{\text{def}}{=} \min \left\{ \frac{u_i}{\gamma_i^j} : i \in I \text{ and } \gamma_i^j > 0 \right\} = 0.$$

The argument used in respect of case B2 to prove the linear independence of the vectors C^i, $i \in I^+$, still holds, and there is nothing further to prove. The only difference (essential!) from the case B2 is that *the point* u^+ *coincides with the point* u, which, incidentally, dispenses with the need to verify the non-diminishing of the functional J... $\qquad \square$

Remark

It is convenient to reinterpret the various cases considered, introducing the set

$$E \stackrel{\text{def}}{=} \left\{ j \notin I : a_j - \sum_{i \in I} \gamma_i^j a_i < 0 \right\} = E_1 \cup E_2 \cup E_3,$$

with

$$E_1 = \left\{ j \notin I : a_j - \sum_{i \in I} \gamma_i^j a_i < 0, \max_{i \in I} \gamma_i^j \leqslant 0 \right\},$$

$$E_2 = \left\{ j \notin I : a_j - \sum_{i \in I} \gamma_i^j a_i < 0, \max_{i \in I} \gamma_i^j > 0, \min_{\substack{i \in I \\ \{\gamma_i^j > 0\}}} \left\{ \frac{u_i}{\gamma_i^j} \right\} > 0 \right\},$$

$$E_3 = \left\{ j \notin I : a_j - \sum_{i \in I} \gamma_i^j < 0, \max_{i \in I} \gamma_i^j > 0, \min_{\substack{i \in I \\ \{\gamma_i^j > 0\}}} \left\{ \frac{u_i}{\gamma_i^j} \right\} = 0 \right\}.$$

We then actually have the correspondences

$$\text{case } A \Leftrightarrow E = \varnothing, \quad \text{case } B \Leftrightarrow E \neq \varnothing,$$
$$\text{case } B1 \Leftrightarrow E_1 \neq \varnothing, \text{ case } B_2 \Leftrightarrow E_2 \neq \varnothing,$$
$$\text{case } B3 \Leftrightarrow E_3 = E,$$

which shows that the cases A *and* B, B1 *and* B3, B2 *and* B3 are mutually exclusive, while the cases B1 *and* B2 are not. $\qquad \square$

We now make some remarks about the *practical* implementation of the method.

It is clear that, in the event of the simultaneous occurrence of the cases B1 and B2, we would opt for the case B1, which provides for the termination of the algorithm.

The calculation of the components γ_i^j of a vector C_i^j, $j \notin I$, in terms of the basis $(C^i)_{i \in I}$, is simpler than it might appear. Indeed, the fact that two successive bases differ by 'only one vector at a time' is used to advantage. This question will be examined later on.

In the context of case B2, which is *the path by which the algorithm* '*progresses most frequently*', various criteria are used for choosing one from among the indices $j \notin I$ for which $\theta^j > 0$. For example, the index corresponding to the largest decrease in the value of the functional, or to the smallest of the numbers $a_j - \sum_{i \in I} \gamma_i^j a_i < 0$, may be chosen, or, quite simply, the choice may fall on the smallest index available and appropriate to the case B2.

In the same way, in choosing one from among those vectors C^j, $j \notin I$, which qualify for introduction into the basis, when the prevailing situation is that of the case B3, the usual practice is to opt for the smallest index.

The case B3 can only occur when the vertex u is degenerate (otherwise, all the components u_i, $i \in I$, are strictly positive), but it need not then necessarily occur. Indeed, even if some of the components u_i, $i \in I$, are zero, the situation where all those components u_i, for which the components γ_i^j are positive, are themselves positive, is enough to ensure that case B2 shall then obtain. Furthermore, even if the vertex u is not degenerate, it can easily happen that the vertex u^+, obtained by construction in the case B2, will itself turn out to be degenerate, in the event that two (at least) of the 'new' components $u_i - \theta^j \gamma_i^j$, $i \in I$, vanish simultaneously.

Finally, it should be observed that *the cases A and B1 represent situations in which the algorithm terminates* and that *the number of cases B2 encountered during the course of the execution of the algorithm can only be finite*, because their occurrence entails a *strict* decrease in the value of the functional from one vertex to the next and because there is only a finite number of vertices (Theorem 10.3-3).

Taking as known a vertex of the set U which is suitable for initialising the algorithm (which implies, in particular, that the set U is non-empty), *one of three mutually exclusive situations may occur in the way that the method develops.*

(i) There may be *a finite sequence of occurrences of case B2 and/or case B3, ending with an occurrence of case A*: the final vertex obtained is then a solution of the problem.

(ii) There may be *a finite sequence of occurrences of case B2 and/or case B3, ending with an occurrence of case B1*: the conclusion is then reached that the problem has no solution.

In these two cases, the simplex method turns out to be a *direct method*, as we have understood the term when it has been applied to methods for the solution of linear systems; for, neglecting the effect of rounding errors, the problem is solved exactly after a *finite* number of elementary operations.

(iii) There may be *a finite sequence of occurrences of case B2 and/or B3 and then a non-terminating sequence of the case B3*. This can only occur if, after a finite number of *consecutive* occurrences of case B3, there is a return to *an earlier basis*. This is the phenomenon known as *cycling*. The algorithm then fails to come to a conclusion.

Although it is quite possible to construct examples 'by hand' for which this phenomenon does indeed occur (see Exercise 10.3-3), *experience shows* that it does not occur in the actual solution of 'real' problems, while it is not an uncommon thing to encounter degenerate vertices as the algorithm progresses! That is why no 'fail-safe strategy' is incorporated in working algorithms. But a particular one does exist: it is given in Exercise 10.3-5, and there is an elegant analysis of it which makes use of lexicographical ordering. That is one sufficient reason for the recommendation of this exercise; the other, and not the lesser, is that it provides as an unexpected corollary a purely 'algebraic' proof of the Farkas lemma (Theorem 9.1-1).

We now describe one way of proceeding with the *actual* calculations involved in one iteration of the method (for another approach, refer to Exercise 10.3-4). Assuming that the set I is ordered (for example, by the increasing order of the indices), we introduce the invertible square matrix C_I consisting of the column vectors C^i, $i \in I$, the column vectors $\gamma_I^j \in \mathbb{R}^m$, $1 \leqslant j \leqslant n$, with components those of the vector C^j relative to the basis $(C^i)_{i \in I}$, the column vector $a_I \in \mathbb{R}^m$ whose components are the numbers a_i, $i \in I$, and, finally, the column vector $u_I \in \mathbb{R}^m$ whose components are the numbers u_i, $i \in I$. It is then possible to set

$$C^j = \sum_{i \in I} \gamma_i^j C^i = C_I \gamma_I^j, \quad \text{or} \quad \gamma_I^j = C_I^{-1} C^j, \quad 1 \leqslant j \leqslant n,$$

and the expressions $a_j - \sum_{i \in I} \gamma_i^j a_i$ (whose signs play an important role in the development of the method; cf. Theorem 10.3-4) become

$$a_j - \sum_{i \in I} \gamma_i^j a_i = a_j - a_I^T \gamma_I^j = a_j - a_I^T C_I^{-1} C^j = a_j - b_I^T C^j,$$

setting $b_I^T = a_I^T C_I^{-1}$.

Referring to Theorem 10.3-4, it becomes clear that, *with the set I taken as known*, the first calculations to be carried out are those of the components of the vector b_I, which is the solution of the linear system

$$C_I^T b_I = a_I,$$

then the signs of the numbers $(a_j - b_I^T C^j)$ for $j \notin I$ are determined. In the event that an instance of the case B2 is encountered, we are then faced with the calculation of the components $\gamma_i^{j^+}$, corresponding to the index $j^+ \notin I$ which has been chosen to be 'incorporated', that is to say, the calculation of the vector $\gamma_I^{j^+}$, which is the solution of the linear system

$$C_I \gamma_I^{j^+} = C^{j^+}.$$

There remains the calculation of the ratios $u_i / \gamma_i^{j^+}$ for $i \in I$ and $\gamma_i^{j^+} > 0$, in order to determine the index j^- which is going to be 'eliminated', the components u_i, $i \in I$, of the vertex u being obtained by means of the solution of the linear system

$$C_I u_I = d.$$

In summary, one iteration of the simplex method requires the solution of linear systems whose matrices are *all the same* (up to a transposition); moreover, the matrices of two consecutive iterations, namely, C_I and C_{I^+}, differ by *just one column*, the new column having the particular form $C^j = C_I \gamma_I^{j^+}$.

This last observation opens the way to a *considerable simplification* of the calculations. For, writing

$$C_I = C, \quad C_{I^+} = C_+, \quad \gamma_I^{j^+} = \gamma = (\gamma_i)_{i=1}^m.$$

in order to make the notation less cumbersome, we observe that at least one of the components of the vector γ is non-zero (by definition of the case B2); let us denote it by γ_k. Next, we observe that there is no loss of generality in assuming that the new column $C\gamma$ is going to take the place of precisely the kth column of the matrix C. In effect, this amounts to post-multiplying the matrix C by a permutation matrix. Thus, it is possible to set

$$C_+ = C\Gamma, \quad \text{with} \quad \Gamma = \begin{pmatrix} 1 & & & \gamma_1 & & \\ & \ddots & & \vdots & & \\ & & 1 & \gamma_{k-1} & & \\ & & & \gamma_k & & \\ & & & \gamma_{k+1} & 1 & \\ & & & \vdots & & \ddots \\ & & & \gamma_n & & & 1 \end{pmatrix},$$

and it may be verified at once that

$$C_+^{-1} = \Gamma^{-1} C^{-1}, \quad \text{with} \quad \Gamma^{-1} = \begin{pmatrix} 1 & & & -\gamma_k^{-1}\gamma_1 & & \\ & \ddots & & \vdots & & \\ & & 1 & -\gamma_k^{-1}\gamma_{k-1} & & \\ & & & \gamma_k^{-1} & & \\ & & & -\gamma_k^{-1}\gamma_{k+1} & 1 & \\ & & & \vdots & & \ddots \\ & & & -\gamma_k^{-1}\gamma_n & & 1 \end{pmatrix},$$

and the calculations then simply reduce to pre-multiplication by matrices which are like the matrix Γ^{-1}.

It remains to look at the question of the *initialisation* of the simplex method, that is to say, either the actual construction of a vertex from the set U, if the latter is not empty (it then always contains at least one vertex; cf. Theorem 10.3-3), or else the discovery that the set U is empty.

More precisely, with a set of the form

$$U = \{v \in \mathbb{R}_+^n : \quad Cv = d\}, \quad C \in \mathscr{A}_{m,n}(\mathbb{R}),$$

there is associated the following linear programming problem (already used in the proof of Theorem 10.3-3; and not by coincidence...), expressed in the form $(\tilde{P})$:

find $(u, \tilde{u})$ such that

$$(\tilde{P}) \quad \begin{cases} (u, \tilde{u}) \in \tilde{U} = \{(v, \tilde{v}) \in \mathbb{R}_+^n \times \mathbb{R}_+^m : Cv + \tilde{v} = d\}, \\ \tilde{J}(u, \tilde{u}) = \inf_{(v, \tilde{v}) \in \tilde{U}} \tilde{J}(v, \tilde{v}), \quad \tilde{J}(v, \tilde{v}) = \sum_{i=1}^m \tilde{v}_i. \end{cases}$$

We then observe the equivalence

$$U \neq \varnothing \Leftrightarrow \inf_{(v, \tilde{v}) \in \tilde{U}} \tilde{J}(v, \tilde{v}) = 0 \quad (= \tilde{J}(u, 0) \text{ for every } u \in U).$$

The idea is then *to apply the simplex method to the solution of the problem* $(\tilde{P})$. The point $(0, d)$ being a vertex of the set $\tilde{U}$ (the column vectors of the identity matrix are linearly independent), it is certainly possible to initialise the simplex method for this problem.

The alternative, which would lead to an occurrence of the case B1, is excluded (since $\inf_{(v, \tilde{v}) \in \tilde{U}} \tilde{J}(v, \tilde{v}) \geqslant 0$); so that we may *assume that the phenomenon of cycling does not occur*. That then leaves as the only possibility a finite sequence of occurrences of the case B2 and/or the case B3, ending with one occurrence of the case A. Let (u, u') be the solution so obtained. Then, either $u' = 0$ and the point u is a vertex of the set U, or else $u' = 0$ and the set U is empty.

While this procedure does not allow the ultimate *theoretical* settling of the question of initialisation, since there is nothing to exclude with complete certainty the occurrence of the phenomenon of cycling in the course of the solution of the auxiliary problem, nevertheless, it is used quite extensively, as we have already observed, because experience shows that the phenomenon of cycling does not occur in practice...

Remark

It is precisely because the question of initialisation requires a preliminary acquaintance with the simplex method that it has seemed preferable to present it *after* the description of the method itself.   □

One case in which it is always possible to settle the question of initialisation is that where the problem is posed in the form (P_1) or (P_2)

find u such that

$$\begin{cases} u \in U = \begin{cases} \{v \in \mathbb{R}^n : Cv \leqslant d\} & (P_1), \\ \{v \in \mathbb{R}^n_+ : Cv \leqslant d\} & (P_2), \end{cases} \\ J(u) = \inf_{v \in U} J(v), \quad J(v) = (a, v), \end{cases}$$

under the condition, of course, that the vector $d \in \mathbb{R}^m$ satisfies

$$d \geqslant 0.$$

Following the steps described towards the beginning of section 10.1, we are led to the solution of a problem set in the form (P_3) (starting from the form (P_1), say, just in order to fix ideas)

find (u', u'', u''') such that

$$\begin{cases} (u', u'', u''') \in \tilde{U} = \{(v', v'', v''') \in \mathbb{R}^n_+ \times \mathbb{R}^n_+ \times \mathbb{R}^m_+, \ Cv' - Cv'' + v''' = d\}, \\ \tilde{J}(u', u'', u''') = \inf_{(v', v'', v''') \in \tilde{U}} \tilde{J}(v', v'', v'''), \quad \tilde{J}(v', v'', v''') = (a, v') - (a, v''), \end{cases}$$

for which the point $(0, 0, d) \in \mathbb{R}^n_+ \times \mathbb{R}^n_+ \times \mathbb{R}^m_+$ is a vertex.

Finally, we consider a *numerical* example of the application of the simplex method (adapted from Gass (1964)).

Find u such that

$$\begin{cases} u \in U = \{v \in \mathbb{R}^3_+ : 3v_1 - v_2 + 2v_3 \leqslant 7, \ -2v_1 + 4v_2 \leqslant 12, \\ \qquad\qquad -4v_1 + 3v_2 + 8v_3 \leqslant 10\}, \\ J(u) = \inf_{v \in U} J(v), \quad J(v) = v_1 - 3v_2 + 2v_3. \end{cases}$$

This is a problem set in the form (P_2). The first step is to transform it into a problem set in the form (P_3) (without changing the letters U and J, in order to simplify the notation), by introducing three slack variables v_4, v_5, v_6:

find u such that

$$\begin{cases} u \in U = \{v \in \mathbb{R}^6_+ : Cv = d\}, \\ J(u) = \inf_{v \in U} J(v), \quad J(v) = a^{\mathrm{T}} v, \end{cases}$$

where

$$C = \begin{pmatrix} 3 & -1 & 2 & 1 & & \\ -2 & 4 & 0 & & 1 & \\ -4 & 3 & 8 & & & 1 \end{pmatrix}, \quad d = \begin{pmatrix} 7 \\ 12 \\ 10 \end{pmatrix},$$

$$a^{\mathrm{T}} = (1 - 3 \quad 2 \quad 0 \quad 0 \quad 0).$$

The components of the vector d are positive, so that it is possible to initialise the method with the help of the procedure described earlier. Denoting by $u_k = (a_i^k)_{i=1}^6, k \geqslant 0$, the successive vertices encountered and by $I_k, \gamma_{ik}^j, j_k^+, \theta_k^{j^+}, j_k^-$ the corresponding quantities introduced in Theorem 10.3-4 (but without the index k), the method may be set down as follows, *after carrying out all necessary calculations* (using, for example, the method described above).

(1) *Initialisation*:

$$\begin{cases} u_0 = (0, 0, 0, 7, 12, 10)^{\mathrm{T}}; \quad I_0 = (4, 5, 6); \\ J(u_0) = 0. \end{cases}$$

(2) *Iteration starting with the vertex u_0*:

$$C^1 = \sum_{i \in I_0} \gamma_{i0}^1 C^i = 3C^4 - 2C^5 + 4C^6; a_1 - \sum_{i \in I_0} \gamma_{i0}^1 a = 1;$$

$$C^2 = \sum_{i \in I_0} \gamma_{i0}^2 C^i = -C^4 + 4C^5 + 3C^6; a_2 - \sum_{i \in I_0} \gamma_{i0}^2 a_i = -3;$$

$$C^3 = \sum_{i \in I_0} \gamma_{i0}^3 C^i = 2C^4 + 8C^6; a_3 - \sum_{i \in I_0} \gamma_{i0}^3 a_i = 2.$$

This is an instance of case B2, with $j_0^+ = 2$. For,

$$\theta_0^2 = \min \left\{ \frac{u_i^0}{\gamma_{i0}^2} : i \in I_0 \text{ and } \gamma_{i0}^2 > 0 \right\} = \min \left\{ \frac{u_5^0}{\gamma_{50}^2}, \frac{u_6^0}{\gamma_{60}^2} \right\}$$

$$= \frac{u_5^0}{\gamma_{50}^2} = 3 > 0, \text{ so that } j_0^- = 5.$$

We then move on to the next vertex u_1, with

$$\begin{cases} u_1 = (0, 3, 0, 10, 0, 1)^{\mathrm{T}}, \quad I_1 = (2, 4, 6), \\ J(u_1) = J(u_0) + \theta_0^2 \left(a_2 - \sum_{i \in I_0} \gamma_{i0}^2 a_i \right) = -9. \end{cases}$$

(3) *Iteration starting with the vertex* u_1:

$$C^1 = \sum_{i \in I_1} \gamma_{i1}^1 C^i = -\tfrac{1}{2}C^2 + \tfrac{5}{2}C^4 - \tfrac{5}{2}C^6; a_1 - \sum_{i \in I_1} \gamma_{i1}^1 a_i = -\tfrac{1}{2};$$

$$C^3 = \sum_{i \in I_1} \gamma_{i1}^3 C^i = \qquad\qquad 2C^4 + 8C^6; a_3 - \sum_{i \in I_1} \gamma_{i1}^3 a_i = 2;$$

$$C^5 = \sum_{i \in I_1} \gamma_{i1}^5 C^i = \tfrac{1}{2}C^2 + \tfrac{1}{4} C^4 - \tfrac{3}{4}C^6; a_5 - \sum_{i \in I_1} \gamma_{i1}^5 a_i = \tfrac{3}{4}.$$

This is an instance of case B2, with $j_1^+ = 1$. For,

$$\theta_1^1 = \min \left\{ \frac{u_i^1}{\gamma_{i1}^1} : i \in I_1 \text{ and } \gamma_{i1}^1 > 0 \right\} = \frac{u_4^1}{\gamma_{41}^1} = 4 > 0, \quad \text{so that } j_1^- = 4.$$

We then move on to the next vertex u_2, with

$$\begin{cases} u_2 = (4,5,0,0,0,11)^T, \quad I_2 = (1,2,6), \\ J(u_2) = J(u_1) + \theta_1^1 \left(a_1 - \sum_{i \in I_1} \gamma_{i1}^1 a_i \right) = -11. \end{cases}$$

(4) *Iteration starting with the vertex* u_2:

$$C^3 = \sum_{i \in I_2} \gamma_{i2}^3 C^i = \tfrac{4}{5}C^1 + \tfrac{2}{5}C^2 + 10C^6; a_3 - \sum_{i \in I_2} \gamma_{i2}^3 a_i = \tfrac{12}{5};$$

$$C^4 = \sum_{i \in I_2} \gamma_{i2}^4 C^i = \tfrac{2}{5}C^1 + \tfrac{1}{5}C^2 + \qquad C^6; a_4 - \sum_{i \in I_2} \gamma_{i2}^4 a_i = \tfrac{1}{5};$$

$$C^5 = \sum_{i \in I_2} \gamma_{i2}^5 C^i = \tfrac{1}{10}C^1 + \tfrac{3}{10}C^2 - \tfrac{1}{2}C^6; a_5 - \sum_{i \in I_2} \gamma_{i2}^5 a_i = \tfrac{4}{5}.$$

This is an instance of case A. The vertex u_2 is thus a solution of the problem set in the form (P_3), while the vector $(4,5,0)^T$ is a solution of the problem set in the form (P_2).

Exercises

10.3-1. (1) State and prove an analogue of Theorem 10.3-1 for sets which are of the form

$$U = \left\{ v \in \mathbb{R}_+^n : \sum_{j=1}^n c_{ij} v_j \leqslant d_i, 1 \leqslant i \leqslant m \right\}.$$

What is the relationship which exists between the vertices of the two types of set U?

(2) Assuming the set U to be non-empty, calculate the largest and the least of the possible numbers of its vertices.

10.3-2. Consider the problem

$$\text{find } u = (u_i)_{i=1}^2 \text{ such that}$$

$$\begin{cases} u \in U = \{ v = (v_{1,2}) \in \mathbb{R}_+^2 : v_1 + 6v_2 \leqslant 30, 2v_1 + 2v_2 \leqslant 15, 4v_1 + v_2 \leqslant 24 \}. \\ J(u) = \inf_{v \in U} J(v), \quad J(v) = -2v_1 - v_2. \end{cases}$$

Solve this problem by the simplex method and so rediscover the solution obtained 'graphically' in section 10.2.

10.3-3. Consider the problem, due to E.M.L. Beale,

$$\text{find } u \text{ such that}$$
$$u \in U = \{v \in \mathbb{R}_+^4 : Cv \leqslant d\},$$
$$J(u) = \inf_{v \in U} J(v), \quad J(v) = a^T v,$$

where

$$C = \begin{pmatrix} 1/4 & -60 & -1/25 & 9 \\ 1/2 & -90 & -1/50 & 3 \\ 0 & 0 & 1 & 0 \end{pmatrix}, \quad d = \begin{pmatrix} 0 \\ 0 \\ 1 \end{pmatrix},$$
$$a^T = (-3/4 \quad 150 \quad -1/50 \quad 6).$$

Having expressed this problem in the form (P_3), show that the application of the simplex method leads to the phenomenon of cycling if, whenever an instance of the case B3 occurs, the choice of the vector C^j, $j \notin I$, is determined by the criterion of choosing the smallest from among the indices of those vectors which qualify for introduction into the basis.

10.3-4. Let C be an invertible matrix of order m, $\gamma \in \mathbb{R}^m$ a vector whose kth component is non-zero and C_+ the matrix obtained by replacing the kth column vector of the matrix C by the vector γ. Show how the LU and QR factorisations of the matrix C_+ (cf. sections 4.3 and 4.5; it is assumed that the LU factorisation exists) may be deduced in a simple way from the LU and QR factorisations of the matrix C. The calculations which are involved in one iteration of the simplex method are sometimes based on this observation.

10.3-5. (1) A *lexicographical ordering of* $\mathbb{R}^q$ is defined as follows. A vector $a = (a_i)_{i=1}^q$ is said to *precede lexicographically* the vector $b = (b_i)_{i=1}^q$ if

$a_1 < b_1$,
or $a_1 = b_1$ and $a_2 < b_2$,
$\vdots$
or $a_1 = b_1, \ldots, a_{p-1} = b_{p-1}$ and $a_p < b_p$, $p \leqslant q$,
or $a = b$,

in which case we write $a \subseteq b$. If $a \subseteq b$ and $a \neq b$, the vector a is said *strictly to precede, in the lexicographical ordering*, the vector b, and we write $a \subset b$.

Show that this lexicographical ordering of $\mathbb{R}^q$ is a total ordering of $\mathbb{R}^q$ and that every finite set of vectors of $\mathbb{R}^q$ has a least element, with regard to this relation, which will be denoted by min $(\cdot)$.

(2) With the problem (P)

$$\text{find } u \text{ such that}$$
$$\begin{cases} u \in U = \{v \in \mathbb{R}_+^n : Cv = d\}, \quad C \in \mathcal{A}_{m,n}(\mathbb{R}), \\ J(u) = \inf_{v \in U} J(v), \quad J(v) = a^T v, \quad a \in \mathbb{R}^n, \end{cases}$$

there is associated the problem ($\mathcal{P}$)

find the minimum, *in the sense of the lexicographical ordering of* $\mathbb{R}^{m+1}$, of the function

$$\mathcal{J} : V \in \mathcal{A}_{n,m+1}(\mathbb{R}) \to \mathcal{J}(V) = V^{\mathrm{T}} a \in \mathbb{R}^{m+1},$$

as V runs through the set

$$\mathcal{V} = \{V \in \mathcal{A}_{n,m+1}(\mathbb{R}): V_i \geqq 0, 1 \leqslant i \leqslant n, \mathrm{CV} = \mathrm{D}\},$$

where, for every matrix V of type $(n, m+1)$, V_i denotes its ith row vector, and where

$$D = \begin{pmatrix} d_1 & 1 & & \\ d_2 & & 1 & \\ \vdots & & & \ddots \\ d_m & & & 1 \end{pmatrix} \in \mathcal{A}_{m,m+1}(\mathbb{R}).$$

In order to solve the problem ($\mathcal{P}$), we define the *lexicographical simplex method* in the same way that the simplex method was defined for the solution of the problem (P), *the usual order relation for the components of the vectors of the set U being replaced throughout by the lexicographical ordering of* $\mathbb{R}^{m+1}$ *for the row vectors of the matrices of the set* $\mathcal{V}$.

In the same spirit, a matrix $U \in \mathcal{A}_{n,m+1}(\mathbb{R})$, with row vectors U_i, will be said to be a *vertex* of the set $\mathcal{V}$ if there exists a set

$$I \subset \{1, 2, \ldots, n\} \quad \text{with card } (I) = m,$$

such that

$$\begin{cases} U_i \geqq 0 \quad \text{if} \quad i \in I, \quad U_i = 0 \quad \text{if} \quad i \notin I, \quad \sum_{i=1}^{n} C^i U_i = \sum_{i \in I} C^i U_i = D; \\ \text{the vectors } C^i, \quad i \in I, \quad \text{are linearly independent.} \end{cases}$$

The set $(C^i)_{i \in I}$ is then said to be a *basis, associated with the vertex* U.

Establish an analogue of Theorem 10.3-4 as regards the cases A, B1 and B2, but show that *the case B3 cannot occur*. That is where the *essentially new element* is to be found!

In other words, if there exists (at least) one index $j \notin I$ for which it is simultaneously true that

$$\begin{cases} a_j - \sum_{i \in I} \gamma_i^j a_i < 0, \\ \gamma_i^j > 0 \text{ and } U_i \supset 0 \text{ for at least one index } i \in I \text{ (the same)}, \end{cases}$$

then the matrix U^+ with row vectors

$$U_i^+ = \begin{cases} U_i - \gamma_i^j \Theta^j & \text{if} \quad i \in I, \\ \Theta^j & \text{if} \quad i = j, \\ 0 & \text{if} \quad i \notin I \cup \{j\}, \end{cases}$$

where

$$\Theta^j = \min\{(\gamma_i^j)^{-1}U_i : i \in I \text{ and } \gamma_i^j > 0\},$$

the minimum being taken in the sense of the lexicographical ordering, is a vertex which is different from the vertex U, and there necessarily follow the *strict* inequalities

$$\Theta^j \supset 0 \quad \text{and} \quad \mathscr{J}(U) - \mathscr{J}(U^+) \supset 0.$$

(3) Show that if the set U is non-empty, then the set $\mathscr{V}$ is also non-empty.

(4) Show that if the set $\mathscr{V}$ is non-empty, then it possesses at least one vertex which may be used to initialise the lexicographical simplex method.

(5) Suppose that $U \neq \varnothing$ and show that the only two possibilities as to the way in which the lexicographical simplex method (which, by the preceding question, it is always possible to initialise) may proceed are the following.

(i) There may be a *finite sequence of occurrences of the case* B2 *ending with an occurrence of case* A. The last vertex $U \in \mathscr{A}_{n,m+1}(\mathbb{R})$ is a solution of the problem $(\mathscr{P})$. The first column vector of the matrix U is then a solution of the problem (P) and the number $J(u)$ is equal to the first component of the vector $\mathscr{J}(U) = U^T a \in \mathbb{R}^{m+1}$.

(ii) There may be a *finite sequence of occurrences of the case* B2 *ending with an occurrence of the case* B1. The conclusion then follows that neither the problem $(\mathscr{P})$ nor the problem (P) have a solution. And, in that case, $\inf_{v \in U} J(v) = -\infty$.

(6) Show that, from the algorithmic standpoint, the lexicographical simplex method may be viewed as a *fail-safe strategy for the 'usual' simplex method*, to the extent that, in essence, it comes down to a sequence of choices of appropriate subsets of m linearly independent column vectors of the matrix C, *while entirely avoiding the phenomenon of cycling*.

(7) Deduce from the fact of the existence of the lexicographical simplex method another proof of the *Farkas lemma*, which has already once been proved in Theorem 9.1-1. Adopting once again the notation of that theorem, with $I = \{1, 2, \ldots, m\}$, use should be made of the 'intermediate' linear programming problem

$$\text{find } u \text{ such that}$$

$$\begin{cases} u \in U \overset{\text{def}}{=} \{v \in \mathbb{R}^n : a_i^T v \leqslant 0, 1 \leqslant i \leqslant m\}, \\ J(u) = \inf_{v \in U} J(v), J(v) = b^T v. \end{cases}$$

Having transformed this problem, which is set in the form (P_1), into a problem set in the form (P_2), this last problem should itself be solved by the lexicographical simplex method. Why would the 'usual' simplex method have failed to come to a definite conclusion?

10.4 Duality and linear programming

We begin with an application of the results of Chapter 9. Given the particular form of the sets U considered there, those results are applicable to linear programming problems of the form (P_1) or (P_2). For the first of these forms,

$$\text{find } u \text{ such that}$$

(P_1)
$$\begin{cases} u \in U = \{v \in \mathbb{R}^n : Cv \leq d\}, \quad C \in \mathscr{A}_{m,n}(\mathbb{R}), \\ J(u) = \inf_{v \in U} J(v), \quad J(v) = (a, v)_n, \end{cases}$$

we obtain as an immediate corollary of the necessary and sufficient conditions for the existence of a minimum in convex programming, established in Theorem 9.2-4 (all of whose conditions are satisfied), the following.

Theorem 10.4-1 (*Necessary and sufficient conditions for the existence of a minimum in linear programming*)
Let u be a point of the set

$$U = \{v \in \mathbb{R}^n : Cv \leq d\}, \quad C \in \mathscr{A}_{m,n}(\mathbb{R}).$$

Then

$$J(u) = \inf_{v \in U} J(v), \quad \text{where} \quad J(v) = (a, v)_n,$$

if and only if there exists a vector $\lambda \in \mathbb{R}^m$ such that

$$\begin{cases} \lambda \geq 0, \quad C^T \lambda + a = 0, \\ (\lambda, Cu - d)_m = 0. \end{cases} \qquad \square$$

Remarks
(1) Bearing in mind the relations $\lambda \geq 0$ and $Cu - d \leq 0$, it is possible to express the last relation of the theorem in the equivalent form

$$\lambda_i = 0 \quad \text{if} \quad (C_i, u)_n - d_i < 0, \quad 1 \leq i \leq m,$$

denoting by C_i the ith *column* vector of the *transpose matrix* C^T (which is the same thing as denoting by C_i^T the ith *row* vector of the matrix C).

(2) While, in the preceding section, it was the *column* vectors of the matrix C which played a crucial role, now it is its *row* vectors which, in a natural way, take their place. This observation will be better understood, once the connections which exist between the simplex method and duality are made more explicit.

(3) It may be verified that, in the particular case of *linear* programming, as here considered, the necessary and sufficient conditions of the theorem given above can also be established *directly* by taking the Farkas lemma as starting point (Theorem 9.1-1). $\qquad \square$

The application of the results of section 9.3 (the introduction of a Lagrangian; duality) to problems set in the form (P$_1$) is of only limited usefulness (yet another difference as against quadratic programming). Defining the *Lagrangian associated with the form* (P$_1$) by

$$L: (v, \mu) \in \mathbb{R}^n \times \mathbb{R}^m_+ \to L(v, \mu) = (a, v)_n + (\mu, Cv - d)_m,$$

it follows from Theorem 9.3-2 that an element $u \in \mathbb{R}^n$ is a solution of the problem (P$_1$) if and only if there exists a vector $\lambda \in \mathbb{R}^m_+$ such that the pair $(u, \lambda) \in \mathbb{R}^n \times \mathbb{R}^m_+$ is a saddle point of the Lagrangian L over the space $\mathbb{R}^n \times \mathbb{R}^m_+$, in other words (Theorem 9.3-1),

$$\inf_{v \in \mathbb{R}^n} \sup_{\mu \in \mathbb{R}^m_+} L(v, \mu) = \sup_{\mu \in \mathbb{R}^m} L(u, \mu) = L(u, \lambda) = \inf_{v \in \mathbb{R}^n} L(v, \lambda)$$
$$= \sup_{\mu \in \mathbb{R}^m_+} \inf_{v \in \mathbb{R}^n} L(v, \mu).$$

Now the function

$$L(\cdot, \mu): v \in \mathbb{R}^n \to L(v, \mu) = (C^T \mu + a, v)_n - (\mu, d)_m$$

being *affine*, it follows that

$$G(\mu) = \inf_{v \in \mathbb{R}} L(v, \mu) = \begin{cases} -\infty & \text{if} \quad C^T \mu + a \neq 0, \\ -(d, \mu)_m & \text{if} \quad C^T \mu + a = 0, \end{cases}$$

the function $L(\cdot, \mu)$ being constant when $C^T \mu + a = 0$. In these circumstances, the *dual* problem (in the sense of section 9.3)

find λ such that

(Q$_1$) $\qquad\qquad \lambda \in \mathbb{R}^m_+ \quad \text{and} \quad G(\lambda) = \sup_{\mu \in \mathbb{R}^m_+} G(\mu)$

could not be of much use in the solution of the *primal* problem (P$_1$). It should be observed, while we are on the subject, that the assumptions of part (1) of Theorem 9.3-3 do not hold, while part (2) of the same theorem does no more than express the existence, already pointed out, of (at least) one solution of the dual problem, namely, any vector $\lambda \in \mathbb{R}^m$ satisfying

$$\lambda \in \mathbb{R}^m_+, \quad C^T \lambda + a = 0.$$

Let us next move on to problems in the second form.

Find u such that

(P$_2$) $\qquad \begin{cases} u \in U = \{v \in \mathbb{R}^n_+ : Cv \leqslant d\}, \ C \in \mathscr{A}_{m,n}(\mathbb{R}). \\ \\ J(u) = \inf_{v \in U} J(v), \ J(v) = (a, v)_n. \end{cases}$

This way of presenting the problem is just a particular case of the previous one; so that one might well have remained content with necessary and sufficient conditions comparable with those of the preceding theorem (relations (1) given below). However, these conditions can also be put in the equivalent form of relations (2) given below. They will play a significant part in justifying the introduction of a *new* type of dual problem.

Theorem 10.4-2 (*Necessary and sufficient conditions for the existence of a minimum in linear programming*).
Let *u* be a point of the set

$$U = \{v \in \mathbb{R}^n_+ : Cv \leq d\}, \quad C \in \mathscr{A}_{m,n}(\mathbb{R}).$$

Then

$$J(u) = \inf_{v \in U} J(v), \quad \text{where} \quad J(v) = (a, v)_n,$$

if and only if there exist vectors $\lambda \in \mathbb{R}^m$ *and* $v \in \mathbb{R}^n$ *such that*

(1)
$$\begin{cases} \lambda \geq 0, \, v \geq 0, \, C^{\mathrm{T}}\lambda + a - v = 0, \\ (\lambda, Cu - d)_m - (v, u)_n = 0, \end{cases}$$

or, again, if and only if there exists a vector $\lambda \in \mathbb{R}^m$ *such that*

(2)
$$\begin{cases} \lambda \geq 0, \, C^{\mathrm{T}}\lambda + a \geq 0, \\ (\lambda, Cu - d)_m = 0 \quad \text{and} \quad (C^{\mathrm{T}}\lambda + a, u)_n = 0. \end{cases}$$

Proof
The set *U* may be written down as

$$U = \{v \in \mathbb{R}^n : C'v \leq d'\},$$

with (denoting by I_n the identity matrix of $\mathbb{R}^n$)

$$C' = \left(\frac{C}{-I_n} \right) \in \mathscr{A}_{m+n,n}(\mathbb{R}), \quad d' = \left(\frac{d}{0} \right) \in \mathbb{R}^{m+n}.$$

Then the relations (1) are just those of Theorem 10.4-1, using the matrix C' and the vector d' to help express them. The relations (2) can then be inferred from the relations (1), upon observing that the signs of the components of the various vectors concerned make it possible to write

$$\begin{cases} v = C^{\mathrm{T}}\lambda + a \geq 0, \\ (\lambda, Cu - d)_m - (v, u)_n = 0 \Leftrightarrow (\lambda, Cu - d)_m = (v, u)_n = 0. \end{cases}$$

The converse is obtained on introducing the vector $v = C^{\mathrm{T}}\lambda + a$. $\qquad \square$

Remarks
(1) Bearing in mind the relations $\lambda \geq 0, \, Cu - d \leq 0, \, v \geq 0, \, u \geq 0$, we have the equivalence

$$(\lambda, Cu - d)_m - (v, u)_n = 0 \Leftrightarrow \begin{cases} \lambda_i = 0 & \text{if} \quad (C_i, u)_n - d_i < 0, \, 1 \leq i \leq m, \\ v_j = 0 & \text{if} \quad u_j > 0, \, 1 \leq j \leq n. \end{cases}$$

(2) In the same way, keeping in mind the relations $\lambda \geqslant 0$, $Cu - d \leqslant 0$, $C^T\lambda + a \geqslant 0$, $u \geqslant 0$, we have the equivalence

$$(\lambda, Cu - d)_m = 0 \quad \text{and} \quad (C^T\lambda + a, u)_n = 0$$

$$\Leftrightarrow \begin{cases} \lambda_i = 0 & \text{if} \quad (C_i, u)_n - d_i < 0, \quad 1 \leqslant i \leqslant m, \\ u_j = 0 & \text{if} \quad (C^j, \lambda)_m + a_j > 0, \quad 1 \leqslant j \leqslant n, \end{cases}$$

where C^j denotes the jth *column* vector of the matrix C. $\quad\square$

It is difficult to fail to notice the 'symmetry' of the roles played by the vectors $u \in \mathbb{R}^n$ and $\lambda \in \mathbb{R}^m$ in the relations (2). It is just this symmetry which suggests the definition of a new *linear programming problem*, called the *dual of the problem posed in the form* (P$_2$),

$$\text{find } \lambda \text{ such that}$$

(Q$_2$)
$$\begin{cases} \lambda \in \Lambda \overset{\text{def}}{=} \{\mu \in \mathbb{R}^m_+ : C^T\mu + a \geqslant 0\}, \\[2mm] G(\lambda) = \sup_{\mu \in \Lambda} G(\mu), \ G(\mu) \overset{\text{def}}{=} -(d, \mu)_m. \end{cases}$$

Remarks

(1) In the course of the proof of Theorem 10.4-2, it was seen how to put the problem (P$_2$) into the form of the problem (P$_1$) (with the assistance of the matrix $C' \in \mathcal{A}_{m+n,n}(\mathbb{R})$ and the vector $d' \in \mathbb{R}^{m+n}$, using the notation of the proof in question). In the same spirit, we can also define the Lagrangian

$$\mathscr{L} : (v, (\mu, \varrho)) \in \mathbb{R}^n \times (\mathbb{R}^m_+ \times \mathbb{R}^n_+) \to \mathscr{L}(v, (\mu, \varrho))$$
$$= (a, v)_n + \{(\mu, Cv - d)_m - (\varrho, v)_n\},$$

and then the dual problem (in the sense of section 9.3)

$$\text{find } (\lambda, v) \text{ such that}$$

($\mathcal{Q}_2$)
$$(\lambda, v) \in \mathbb{R}^m_+ \times \mathbb{R}^n_+ \quad \text{and} \quad \mathscr{G}(\lambda, v) = \sup_{(\mu, \varrho) \in \mathbb{R}^m_+ \times \mathbb{R}^n_+} \mathscr{G}(\mu, \varrho),$$

where

$$\mathscr{G}(\mu, \varrho) = \begin{cases} -\infty & \text{if} \quad C^T\mu - \varrho + a \neq 0, \\ -(d, \mu)_m & \text{if} \quad C^T\mu - \varrho + a = 0. \end{cases}$$

It is important, at this point, to note that the *problem* ($\mathcal{Q}_2$) *is not identical with the problem* (Q$_2$); similarly, the Lagrangian $\mathscr{L}$ is not identical with the other Lagrangian L introduced in the theorem set out below.

(2) The presence of the minus sign in the definition of the functional G permits the definition of the dual problem as a *maximisation* problem. This meets with the requirements of a certain 'aesthetic' sensibility. $\quad\square$

We now give in detail the remarkable relationships which hold between the problem (P_2) and its dual (Q_2).

Theorem 10.4-3 (Duality in linear programming)
Let the following two problems be given:

$$\text{find } u \text{ such that}$$

(P_2) $u \in U = \{v \in \mathbb{R}^n_+ : Cv \leqslant d\}$ *and* $J(u) = \inf_{v \in U} J(v), J(v) = (a, v)_n;$

and

$$\text{find } \lambda \text{ such that}$$

(Q_2) $\begin{cases} \lambda \in \Lambda = \{\mu \in \mathbb{R}^m_+ : C^T\mu \geqslant -a\} \quad and \\ G(\lambda) = \sup_{\mu \in \Lambda} G(\mu), G(\mu) = -(d, \mu)_m. \end{cases}$

(1) *If either one of the problems has a solution, then the other one also has a solution (in particular, the set associated with it is necessarily non-empty); moreover,*

$$\sup_{\mu \in \Lambda} G(\mu) = G(\lambda) = J(u) = \inf_{v \in U} J(v),$$

denoting by u and λ respectively any one of the solutions of each of the two problems.

(2) *A necessary and sufficient condition for one of the problems (and, hence, for the other as well, by (1)) to have a solution is*

$$U \neq \varnothing \quad and \quad \Lambda \neq \varnothing.$$

(3) *An element $u \in \mathbb{R}^n_+$ is a solution of the problem (P_2) if and only if there exists an element $\lambda \in \mathbb{R}^m_+$, which is just some particular solution of the problem (Q_2), such that the pair (u, λ) is a saddle point of the function*

$$L: (v, \mu) \in \mathbb{R}^n_+ \times \mathbb{R}^m_+ \to L(v, \mu) = (a, v)_n + (\mu, Cv - d)_m,$$

that is to say, such that the pair (u, λ) satisfies

$$(u, \lambda) \in \mathbb{R}^n_+ \times \mathbb{R}^m_+ \quad and \quad \sup_{\mu \in \mathbb{R}^m_+} L(u, \mu) = L(u, \lambda) = \inf_{v \in \mathbb{R}^n_+} L(v, \lambda).$$

Proof
(1) In order to fix ideas, let us assume that it is the problem (P_2) which has (at least) one solution, denoted by u (of course, the argument would be entirely 'symmetric' in the case of the problem (Q_2)). Then the element λ, whose existence was established in Theorem 10.4-2, not only satisfies the defining relations of the set Λ(which is, therefore, not empty), but it is actually a solution of the problem (Q_2). In order to see this, it is enough to prove that

$$(d, \mu)_m - (d, \lambda)_m \geqslant 0 \quad \text{for every } \mu \in \Lambda.$$

Now, the relations $(\lambda, Cu - d)_m = 0$ and $(C^T\lambda + a, u)_n = 0$ of Theorem 10.4-2 imply that

$$(\lambda, d)_m = (\lambda, Cu)_m = (C^T\lambda, u)_n = -(a, u)_n,$$

so that it is possible to set

$$(d, \mu)_m - (d, \lambda)_m = (d, \mu)_m + (a, u)_n = (d - Cu, \mu)_m + (C^T\mu + a, u)_n$$

for every $\mu \in \mathbb{R}^m$, and the definitions of the sets Λ and U (to which the element u belongs) imply that the expression given above is non-negative for every $\mu \in \Lambda$. Furthermore, the equality $(\lambda, d)_m = -(a, u)_n$ established above shows that

$$\sup_{\mu \in \Lambda} G(\mu) = -(d, \lambda)_m = G(\lambda) = (a, u)_n = J(u) = \inf_{v \in U} J(v).$$

(2) It has already been shown that if one of the problems has a solution, so, too, does the other; the associated sets U and Λ are, therefore, necessarily non-empty.

Conversely, let v and μ be any two elements of the sets U and Λ, respectively, assumed to be non-empty. Using the definition of these sets, we obtain

$$(a, v)_n + (d, \mu)_m = (a + C^T\mu, v)_n + (d - Cv, \mu)_m \geq 0,$$

which shows that

$$\sup_{\mu \in \Lambda} G(\mu) = \inf_{v \in U} J(v).$$

Hence,

$$\sup_{\mu \in \Lambda} G(\mu) < +\infty \quad \text{and} \quad \inf_{v \in U} J(v) > -\infty,$$

and Theorem 10.1-1 makes it possible to conclude that each of the two problems has at least one solution.

(3) Given any two elements u and λ of the sets $\mathbb{R}^n_+$ and $\mathbb{R}^m_+$, respectively, we have the equivalences

$$L(u, \mu) \leq L(u, \lambda) \quad \text{for every } \mu \in \mathbb{R}^m_+,$$
$$\Leftrightarrow (\lambda - \mu, Cu - d)_m \geq 0 \quad \text{for every } \mu \in \mathbb{R}^m_+,$$
$$\Leftrightarrow Cu - d \leq 0 \quad \text{and} \quad (\lambda, Cu - d)_m = 0,$$

on the one hand (let each component of the vector μ tend towards $+\infty$ in order to obtain $Cu - d \leq 0$; then set $\mu = 0$ and use the relation $\lambda \geq 0$ to obtain $(\lambda, Cu - d)_m = 0$), and

$$L(u, \lambda) \leq L(v, \lambda) \quad \text{for every } v \in \mathbb{R}^n_+$$
$$\Leftrightarrow 0 \leq (a + C^T\lambda, v - u)_n \quad \text{for every } v \in \mathbb{R}^n_+$$
$$\Leftrightarrow C^T\lambda + a \geq 0 \quad \text{and} \quad (C^T\lambda + a, u)_n = 0,$$

on the other (with a similar argument). The stated conclusion is then an immediate consequence of the necessary and sufficient conditions (2) of Theorem 10.4-2.

Remarks
(1) From the preceding, it follows that *an element $u \in U$ is a solution of the problem* (P_2) *if and only if there exists an element $\lambda \in \Lambda$ such that $G(\lambda) = J(u)$*, the converse being a consequence of the inequality $\sup_{\mu \in \Lambda} G(\mu) \leqslant \inf_{v \in U} J(v)$.

(2) From Theorems 10.1-1 and 10.4-3 the following *alternative* may be deduced.

Either *each* of the two problems has a solution; a necessary and sufficient condition for this to be so is that the two sets U and Λ are *both* non-empty;

or else *neither* of the problems has a solution; a necessary and sufficient condition for this to be so is that *at least one* of the two sets U and Λ is empty; if $U \neq \varnothing$ and $\Lambda = \varnothing$, then $\inf_{v \in U} J(v) = -\infty$ (and U is unbounded); if $U = \varnothing$ and $\Lambda \neq \varnothing$, then $\sup_{\mu \in \Lambda} G(\mu) = +\infty$ (and Λ is unbounded).   □

The function $L: \mathbb{R}^n_+ \times \mathbb{R}^m_+ \to \mathbb{R}$ introduced in the statement of the theorem is called the *Lagrangian associated with the form* (P_2). It should be noted that, while the *explicit form* of this function is identical with that of the Lagrangian associated with the form (P_1), introduced above, *their domains of definition are different*; in fact, $\mathbb{R}^n \times \mathbb{R}^m_+$ was that of the 'first' Lagrangian. It should also be observed that the *primal variable* $v \in \mathbb{R}^n_+$ and the *dual variable* $\mu \in \mathbb{R}^m_+$ now play entirely *symmetrical* roles. This was not the case before.

To end with, we show that there exists a subtle link between the simplex method (of the previous section) and the notions of duality which we have just introduced.

Theorem 10.4-4 (*Duality and the simplex method*)
Consider the problem

$$\text{find } u \text{ such that}$$

(P_2)
$$\begin{cases} \tilde{u} \in \tilde{U} = \{\tilde{v} \in \mathbb{R}^n_+ : \tilde{C}\tilde{v} \leqslant d\}, \tilde{C} \in \mathscr{A}_{m,n}(\mathbb{R}), d \in \mathbb{R}^m, \\[2mm] \tilde{J}(\tilde{u}) = \inf_{\tilde{v} \in \tilde{U}} \tilde{J}(\tilde{v}), \tilde{J}(\tilde{v}) = \tilde{a}^{\mathsf{T}}\tilde{v} \quad \text{where} \quad \tilde{a} \in \mathbb{R}^n, \end{cases}$$

and the equivalent problem

$$\text{find } u \text{ such that}$$

(P₃)
$$\begin{cases} u \in U = \{v \in \mathbb{R}^n_+ : Cv = d\}, \\ \\ J(u) = \inf_{v \in U} J(v), \; J(v) = a^T v, \end{cases}$$

where (I_n *being the identity matrix of* $\mathbb{R}^n$)

$$C = \left(\boxed{\tilde{C} \;\vert\; I_n} \right) \in \mathscr{A}_{m,n}(\mathbb{R}), \; n = \tilde{n} + m,$$

$$a^T = \left(\boxed{\tilde{a}^T \;\vert\; 0} \right), \; a \in \mathbb{R}^n.$$

When the simplex method applied to the problem (P_3) *leads to a solution, it provides at the same time a solution of the dual of the problem* (P_2),

$$\text{find } \lambda \text{ such that}$$

(Q₂)
$$\begin{cases} \lambda \in \Lambda = \{\mu \in \mathbb{R}^m_+ : \tilde{C}^T \mu + \tilde{a} \geqslant 0\}, \\ \\ G(\lambda) = \sup_{\mu \in \Lambda} G(\mu), \; G(\mu) = -d^T \mu. \end{cases}$$

Proof

We resume exactly the same definitions and notation as in section 10.3. When the simplex method leads to a solution of the problem (P_3), it also provides a set of indices I which satisfies (Theorem 10.3-4)

$$I \subset \{1, 2, \ldots, n\}, \quad \text{card}(I) = m, \quad a_j - \sum_{i \in I} \gamma^j_i a_i \geqslant 0 \quad \text{for every } j \notin I.$$

We introduce the matrix $C_I \in \mathscr{A}_m(\mathbb{R})$ consisting of the column vectors C^i, $i \in I$, of the matrix C and the matrix $C_{I'} \in \mathscr{A}_{m,\tilde{n}}(\mathbb{R})$ consisting of the column vectors $C^j, j \notin I$, of the matrix C. Similarly, we introduce the column vectors $a_I \in \mathbb{R}^m$ and $u_I \in \mathbb{R}^m$ whose components are the numbers a_i and $u_i, i \in I$, and the column vector $a_{I'} \in \mathbb{R}^{\tilde{n}}$ whose components are the numbers $a_j, j \notin I$. Of course, this presupposes that the set I and the set

$$I' = \{1, 2, \ldots, n\} - I$$

are ordered (for example, by the increasing order of the indices). Then, the inequalities given above take the form

$$a_{I'} - C_{I'}^T (C_I^T)^{-1} a_I \geqslant 0,$$

and the value of the function J at the corresponding vertex u may be written as

$$J(u) = a^{\mathrm{T}} u = a_I^{\mathrm{T}} u_I = a_I^{\mathrm{T}} C_I^{-1} d.$$

The vector

$$\lambda \overset{\mathrm{def}}{=} - \{C_I^{\mathrm{T}}\}^{-1} a_I \in \mathbb{R}^m$$

satisfies

$$G(\lambda) = - d^{\mathrm{T}} \lambda = d^{\mathrm{T}} \{C_I^{\mathrm{T}}\}^{-1} a_I = J(u),$$

$$\begin{cases} C_I^{\mathrm{T}} \lambda + a_I = 0, \\ C_{I'}^{\mathrm{T}} \lambda + a_{I'} = - C_{I'}^{\mathrm{T}} \{C_I^{\mathrm{T}}\}^{-1} a_I + a_{I'} \geqslant 0. \end{cases}$$

If we call P the permutation matrix of order n which satisfies

$$CP = \left(\left(\begin{array}{c|c} \tilde{C} & I_n \end{array} \right) \right) P = \left(\begin{array}{c|c} C_I & C_{I'} \end{array} \right),$$

and, hence, also

$$a^{\mathrm{T}} P = \left(\left(\begin{array}{c|c} \tilde{a}^{\mathrm{T}} & 0 \end{array} \right) \right) P = \left(\begin{array}{c|c} a_I^{\mathrm{T}} & a_{I'}^{\mathrm{T}} \end{array} \right),$$

the last two relations imply that, after pre-multiplication by the matrix P^{T},

$$\tilde{C}^{\mathrm{T}} \lambda + \tilde{a} = 0 \quad \text{and} \quad \lambda \geqslant 0.$$

An element $\lambda \in \Lambda$ has thus been found (for which the inequality $C_I^{\mathrm{T}} \lambda + a_I \geqslant 0$ actually becomes an equality) which satisfies $G(\lambda) = J(u) = \inf_{v \in U} J(v)$. By Theorem 10.4-3, it is, therefore, a solution of the dual problem (Q_2) of the problem (P_2). $\square$

The simplex method, therefore, essentially 'looks for' a solution of the *dual* problem (Q_2), by way of an *ad hoc* set of indices I, the solution of the primal problem (P_2) thus making its appearance as a 'by-product' of this search. A similar observation has already been made *à propos* of Uzawa's method.

Exercises

10.4-1. This exercise introduces two forms of the *Fredholm alternative* for *inequalities*, by extending to them the Fredholm alternative for equations, stated in section 8.1. In what follows, V and W are two finite-dimensional spaces, A denotes a linear mapping from V to W and b is a vector of W.

(1) Prove that one, and only one, of the following alternatives holds: either there exists at least one vector $v \in V$ such that $Av \geqslant b$; or the set $\{v \in V : Av \geqslant b\}$ is empty and there exists at least one vector $w \in W$ such

that

$$w \leqslant 0, A^T w \geqslant 0, (w, b) < 0.$$

(2) Prove that one, and only one, of the following alternatives holds: either there exists at least one vector $v \in V$ such that $Av = b$ and $v \geqslant 0$; or the set $\{v \in V: v \geqslant 0, Av = b\}$ is empty and there exists at least one vector $w \in W$ such that

$$A^T w \geqslant 0, (w, b) < 0.$$

10.4-2. The aim in this problem is to study algorithms which enable us to approximate the solution of a linear programming problem (P) through the solution of 'perturbed' quadratic programming problems (P_ε). More precisely, consider the problem

find u such that

(P)
$$\begin{cases} u \in U = \{v \in \mathbb{R}^n: Cv \leqslant d\}, C \in \mathscr{A}_{m,n}(\mathbb{R}), d \in \mathbb{R}^m, \\[2mm] J(u) = \inf_{v \in U} J(v), \quad \text{where} \quad J(v) = (a, v)_n, a \in \mathbb{R}^n, \end{cases}$$

and the problems

find u_ε such that

(P_ε)
$$\begin{cases} u_\varepsilon \in U = \{v \in \mathbb{R}^n: Cv \leqslant d\}, \\[2mm] J_\varepsilon(u) = \inf_{v \in U} J_\varepsilon(v), \quad \text{where} \quad J_\varepsilon(v) = J(v) + \varepsilon j(v), \end{cases}$$

where ε is a positive number and $j(v) = \frac{1}{2}(v, v)_n$.

Assume that the problem (P) has (at least) one solution (the set U is thus non-empty, which in turn implies that each problem (P_ε) has a unique solution u_ε).

(1) Write down the necessary and sufficient Kuhn–Tucker conditions for the problems (P) and (P_ε).

(2) Show that the 'auxiliary' problem

find u such that

(P′)
$$\begin{cases} u \in U' = \{v \in \mathbb{R}^n: Cv \leqslant d, (a, v)_n \leqslant \delta\}, \\[2mm] j(u) = \inf_{v \in U'} j(v), \end{cases}$$

where $\delta = \inf_{v \in U} J(v)$, has a solution and that this solution is unique. Write down the necessary and sufficient Kuhn–Tucker conditions for the problem (P′).

(3) Show that, either for all $\varepsilon > 0$, or else only for all $\varepsilon \in]0, \varepsilon_0]$, where ε_0 is a positive number to be determined, the solution u_ε of the problem (P_ε) is also a solution of the problem (P) and that it is *independent of ε*.

(4) Assume that $r(C) = m$. Write down the dual problem of the problem (P_ε) in the form

find λ such that

(Q_ε)
$$\lambda \in \mathbb{R}^m_+ \quad \text{and} \quad I(\lambda) = \inf_{\mu \in \mathbb{R}^m_+} I(\mu),$$

where I is a functional, whose expression as a function of the data C, d and a should be given, and show that the problem (Q_ε) has a unique solution. Denote it by λ_ε.

(5) Once λ_ε is known, how do we proceed to the calculation of the solution u_ε of the problem (P_ε)?

(6) Deduce, from the application of the relaxation method and the gradient method to the dual problem, two iterative methods for the solution of the problem (P).

(7) Examine the case $r(C) < m$.

Bibliography and comments

While no attempt has been made at completeness, nevertheless some effort has been expended on providing a list of references which is reasonably extensive, and one which is likely to be of interest to readers of the present work, whether they are looking for complementary material, theoretical as well as practical, or simply searching for other points of view on identical topics. Moreover, because this book is first and foremost an *introduction*, it did not seem to be a matter of negligible usefulness to include references which are set at a level that is distinctly higher, having in mind those readers who might wish to deepen their knowledge of the various subjects treated here.

For the convenience of readers, references have been classified under the following headings, which plainly follow the order of the book.

1. Review of Analysis.
2. Theory of matrices.
3. General results in Numerical Analysis.
4. Methods for the solution of linear systems.
5 Calculation of the eigenvalues and eigenvectors of matrices.
6. Special matrices and linear systems.
7. Preliminaries to Optimisation.
8. Optimisation.

1. Review of Analysis

For the concepts used in the text (this means essentially the topology of $\mathbb{R}^n$, *normed vector spaces, Hilbert spaces* and the *differential calculus*), one can refer to Avez (1983), Cartan (1967), Choquet (1964), Dieudonné (1968), Schwartz (1967, 1970, 1979) and, at a more elementary level, Dixmier (1969a, 1969b).

2. Theory of matrices

The work by Gantmacher (1966a, 1966b) is a good exposition of the 'classical' theory of matrices, interspersed with numerous applications. It may be complemented very usefully by the recent book by Ortega (1987).

Presentations with a more 'algebraic' orientation may be found in Birkhoff & MacLane (1965) and Godement (1966). The book by Halmos (1974) gives an excellent treatment of the theory of linear operators in finite-dimensional spaces, with the accent on geometrical ideas.

From among the works which are directly oriented to the Numerical Analysis of Matrices, we single out Householder (1964), which, though its style is extremely concise, is very rich in content, and Strang (1980), which is marked by the interesting feature of developing matrix theory in parallel with the Numerical Analysis of Matrices. In the same spirit, we also cite Bellman (1960), Franklin (1968), Gastinel (1966), Horn & Johnson (1985), Lancaster (1969), Lancaster & Tismenetsky (1985) and Noble (1969).

3. General results in Numerical Analysis
It is generally recognised that the article by von Neumann & Goldstine (1947) and that by Kantorovitch (1948) mark the beginning of 'modern' Numerical Analysis; for this reason, one can do no less than to put in a strong recommendation that they be read. Pursuing this line of thought even further, interesting points of view and historical perspectives are to be found in the works of Goldstine (1977), Metropolis, Howlett & Rota (1980), as well as in the article by Wilkinson (1971).

Insights into numerical analysis, which are of a general nature, are to be found in the works of Forsythe, Malcolm & Moler (1977), Henrici (1982), Press, Flannery, Teukolsky & Vetterling (1986), Ralston & Rabinowitz (1978), Stoer & Bulirsch (1980) and Todd (1978). We also single out the book by Strang (1986), which is an original, as much as it is a remarkable, introduction to applied mathematics in general.

Readers with a more specialised interest in the *analysis of algorithms*, notably as regards the propagation of rounding errors, the 'complexity' and 'optimality' of algorithms, etc., may consult La Porte & Vignes (1974), Miller & Wrathall (1980), Nemirovsky & Yudin (1983), Pan (1984), Solomon & Hocquemiller (1982), Stewart (1973), Traub & Woźniakowski (1980), Vandergraft (1978), Vignes (1980), Wilkinson (1963, 1965), Wilkinson & Reinsch (1971) and Winograd (1980).

As for the approximation of partial differential equations by *finite difference methods*, the book by Forsythe & Wasow (1960) is a 'classic', which may be usefully complemented by Varga (1962), where one can find information not only on the approximation of partial differential equations but also – and especially – on methods for the solution of the linear systems which arise. We also mention Mitchell & Griffiths (1980).

As for variational methods of approximation, and more particularly *finite element methods*, one can consult, in increasing order of difficulty, Johnson (1988), Raviart & Thomas (1983) and Ciarlet (1978).

With regard to *Newton's method*, notably in its application to the solution of non-linear systems of equations, the book by Ortega & Rheinboldt (1970) is another '*classic*', to which the shorter work by Rheinboldt (1974) is a good introduction. We mention also the old work, but still quite remarkable, of Ostrowski (1966).

4. Methods for the solution of linear systems

As regards *direct methods*, the books by Forsythe & Moler (1967) and Strang (1980) form a broad 'complement' of the early chapters of the present work, by reason of the abundance of examples, differences in point of view and practical detail, which all show up in these works. For this reason, it is recommended to read them.

Every *practical* user of these methods ought to consult, and make use of, the book of Wilkinson & Reinsch (1971); one finds there, in effect, all the information which is necessary for their *practical* implementation (for example, as regards the preliminary equilibration of matrices, an operation which is also known as *preconditioning* in the literature) and, in particular, programmes written in FORTRAN.

Useful complements may also be found in Fox (1964), Gastinel (1966), which contains numerous programmes written in ALGOL, Householder (1964), Stewart (1973), Todd (1977) and Wilkinson (1965).

The work by Westlake (1968) takes the form of a catalogue of conditions of applicability, and of comparisons between, various methods, direct as well as iterative, with an enumeration of their principal characteristics, notably the count of the elementary operations which they require.

Among more recent works, we cite George & Liu (1981), Pissanetsky (1984), as well as the two 'classics' Golub & Meurant (1983) and Golub & van Loan (1984). We mention finally Lascaux & Théodor (1986, 1987), which is a very useful complement of the present work, because the question of the *practical* implementation of the methods is treated there with the greatest care.

As regards *iterative methods*, a great 'classic' is the book by Varga (1962), later followed by another reference work, by Young (1971). See also Hageman & Young (1981).

5. Calculation of the eigenvalues and eigenvectors of matrices

A 'classic' is the work by Wilkinson (1965). Other useful references are Householder (1964), Stewart (1973) and Todd (1977).

We next single out the works by Golub & van Loan (1984), Lascaux

& Théodor (1986, 1987) and Pissanetsky (1984), which have already been cited under the previous heading. The book by Chatelin (1983) contains numerous examples of problems whose discretisation leads to the calculation of the eigenvalues or eigenvectors of matrices.

Finally, the specialised works of Chatelin (1987), Cullum & Willoughby (1985) and Parlett (1980) are strongly recommended.

6. Special matrices and linear systems

For the *least squares solution of linear systems* and on the related question of *pseudo-inverses*, the books by Ben-Israel & Greville (1974) and Lawson & Hanson (1974) may be consulted; these questions are also touched upon in Todd (1978).

We have seen how *sparse matrices* arise naturally in the approximation, by finite differences or finite elements, of boundary-value problems; in fact, they also appear in a good number of other domains; in chemistry, in processes of image reconstruction, etc. For an investigation and solution of the problems in which they appear, one may refer to Duff & Stewart (1979) and Reid (1971).

Gantmacher (1966b) and Varga (1962) provide different treatments of the topic of non-negative matrices (those which have all their elements non-negative): they were encountered in the course of the approximation of boundary-value problems by finite differences. These matrices also play an important role in the study of Markov chains, of certain economic models, of certain linear programming problems, etc. See also, in this regard, Berman & Plemmons (1979).

7. Preliminaries to Optimisation

In addition to the references already pointed out under the heading 'Review of Analysis', notably those which concern the differential calculus and Hilbert spaces, substantive complementary material on convexity, polyhedra in $\mathbb{R}^n$, duality, Lagrangians, etc., in other words, what is called *Convex Analysis*, may be found in Rockafellar (1970), Lay (1982), Stoer & Witzgall (1970), Grünbaum (1967) and Ekeland & Temam (1974), in approximately increasing order of difficulty. One may also consult Aubin (1979a, 1979b), Roberts & Varberg (1973) and Wouk (1979). Lastly, we strongly recommend Brezis (1983).

8. Optimisation

There exist very many books which treat the principal *methods of Optimisation, constrained or unconstrained.* We give, in no significant order, the following works: Aubin, Nepomiastchy & Charles (1982), Aubin (1984),

Auslender (1976), Ben-Israel, Ben-Tal & Zlobec (1981), Bertsekas (1982), Céa (1971), Clarke (1983), Collatz & Wetterling (1975), Daniel (1971), Ekeland & Turnbull (1984), Gill & Murray (1974), Gill, Murray & Wright (1981), Hartley (1985), Hestenes (1975), Luenberger (1969, 1973), McCormick (1983), Mangasarian (1969), Mangasarian, Meyer & Robinson (1975, 1978), Martos (1975), Murray (1972), Polak (1971), Tiel (1984), Wismer & Chattergy (1978), Zangwill (1969) and Zoutendijk (1976), where it is possible to find alternative treatments of questions which have been dealt with here, as well as of others which have not been touched upon; for example, the minimisation of functions of a single variable, non-convex programming, etc.

Programmes written in FORTRAN may be found in Daniels (1978), and in ALGOL and FORTRAN in Künzi, Tzschach & Zehnder (1971). The book by Fletcher (1980) contains much useful information on the practical implementation of the methods.

As regards *Linear Programming*, we strongly recommend – *noblesse oblige!* – the book by Dantzig (1963), one of the pioneers of the subject and, notably, the discoverer of the simplex method. The book by Gass (1976) is also recommended, especially for the numerous applications which may be found in it. One may also profitably consult Chvátal (1983), Hartley (1985), Heesterman (1982), Kolman & Beck (1980), Rothenberg (1980), Schrijver (1986), Strang (1980) and Walsh (1985).

For infinite-dimensional Optimisation (the existence and uniqueness of minima, algorithms, etc.), we cite Barbu & Precupanu (1978), Ioffe & Tihomirov (1979) and Vainberg (1973). In Ekeland & Temam (1974), the accent is put on optimisation problems linked to the variational formulation of boundary-value problems, the 'modern' form of the *Calculus of Variations*.

For that subset of problems which are posed as *variational inequalities*, one should consult Kinderlehrer & Stampacchia (1980), as regards the 'continuous' problem, and Glowinski, Lions & Trémolières (1976a, 1976b), as regards their approximation, notably by the finite element method (the English edition, published in 1981 by North-Holland, contains some important complementary material). The connections between Optimisation and Approximation theory are examined in detail in Laurent (1972).

A wealth of nonlinear problems, including those posed as variational inequalities, that eventually lead to minimisation problems in $\mathbb{R}^n$ is found in Glowinski (1984), together with a detailed analysis of many efficient algorithms for solving the associated discrete problems.

References

Aubin, J.P., (1979a), *Applied Functional Analysis*, J. Wiley, New York.

Aubin, J.P., (1979b), *Mathematical Methods of Game and Economic Theory*, North-Holland, Amsterdam.

Aubin, J.P., (1984), *Analyse non-linéaire et ses motivations économiques*, Masson, Paris.

Aubin, J.P., Nepomiastchy, P., Charles, A.-M., (1982), *Méthodes explicites de l'optimisation*, Dunod, Paris.

Auslender, A., (1976), *Optimisation: Méthodes numériques*, Masson, Paris.

Avez, A., (1983), *Calcul différentiel*, Masson, Paris (English translation: *Differential Calculus*, J. Wiley, New York).

Barbu, V., Precupanu, T., (1978), *Convexity and Optimization in Banach Spaces*, Editura Academiei, Bucarest.

Bellman, R., (1960), *Introduction to Matrix Analysis*, McGraw-Hill, New York.

Ben-Israel, A., Ben-Tal, A., Zlobec, S., (1981), *Optimality in Nonlinear Programming*, Wiley-Interscience, New York.

Ben-Israel, A., Greville, T.N.E., (1974), *Generalised Inverses: Theory and Applications*, J. Wiley, New York.

Berger, M., Gostiaux, B., (1972), *Géométrie différentielle*, Armond Colin, Paris (English translation: *Differential Geometry*, Springer-Verlag, Heidelberg, 1987).

Berman, A., Plemmons, R.J., (1979), *Nonnegative Matrices in the Mathematical Sciences*, Academic Press, New York.

Bertsekas, D.P., (1982), *Constrained Optimization and Lagrange Multiplier Method*, Academic Press, New York.

Birkhoff, G., MacLane, S., (1965), *A Survey of Modern Algebra*, MacMillan, New York.

Brezis, H., (1983), *Analyse fonctionelle*, Masson, Paris.

Buurema, H.J., (1970), A geometric proof of convergence for the QR method, thesis, University of Groningen.

Cartan, H., (1967), *Calcul différentiel*, Hermann, Paris. (English translations *Differential Calculus*, Kershaw, London).

Céa, J., (1971), *Optimisation, théorie et algorithmes*, Dunod, Paris.

Chatelin, F., (1983), *Spectral Approximation of Linear Operators*, Academic Press, New York.

Chatelin, F., (1987), *Valeurs propres de matrices*, Masson, Paris.

Choquet, G., (1964), *Cours d' analyse, Tome 2: Topologie*, Masson, Paris.

Chvátal, V., (1983), *Linear Programming*, Freeman, New York.

Ciarlet, P.G., (1978), *The Finite Element Method for Elliptic Problems*, North-Holland, Amsterdam.

Clarke, F.H., (1983), *Optimization and Nonsmooth Analysis*, John Wiley, New York.

Collatz L., Wetterling, W., (1975), *Optimization Problems*, Springer-Verlag, Berlin.

Cullum, J.K., Willoughby, R.A., (1985), *Lanczos Algorithms for Large Symmetric Eigenvalue Computations, Vol. I: Theory; Vol. II; Programs*, Birkhäuser Verlag, Basel.

Daniel, J.W., (1971), *The Approximate Minimization of Functionals*, Prentice-Hall, Englewood Cliffs, N.J.

Daniels, R.W., (1978), *An Introduction to Numerical Methods and Optimization Techniques*, North-Holland, Amsterdam.

Dantzig, G.B., (1963), *Linear Programming and its Extensions*, Princeton University Press, Princeton.

Dieudonné, J., (1968), *Foundations of Modern Analysis, Volume I*, Academic Press, New York and London.

Dixmier, J., (1969a), *Cours de mathématiques du premier cycle, première année*, Gauthier-Villars, Paris.

Dixmier, J., (1969b), *Cours de mathématiques du premier cycle, deuxième annèe*, Gauthier-Villars, Paris.

Duff, I.S., Stewart, G.W., (1979), *Sparse Matrix Proceedings 1978*, Heyden, London.

Ekeland, I., Temam, R., (1974), *Analyse convexe et problèmes variationnels*, Dunod, Paris.

Ekeland, I., Turnbull, T., (1984), *Infinite-Dimensional Optimization and Convexity*, The University of Chicago Press, Chicago.

Feigenbaum, M.J., (1983), Universal behaviour in nonlinear systems, *Physica*, 7D, 16.

Fletcher, R., (1980), *Practical Methods of Optimization, Vol. 1, Unconstrained Optimization*, J. Wiley, New York.

Fletcher, R. Reeves, C.M. (1964), Function minimization by conjugate gradients, *Computer J.*, 7, 149–54

Forsythe, G.E., Malcolm, M.A., Moler, C.B., (1977), *Computer Methods for Mathematical Computations*, Prentice-Hall, Englewood Cliffs, N.J.

Forsythe, G.E., Moler, C.B., (1967), *Computer Solution of Linear Algebraic Systems*, Prentice-Hall, Englewood Cliffs, N.J.

Forsythe, G.E., Wasow, W.R., (1960), *Finite Difference Methods for Partial Differential Equations*, J. Wiley, New York.

Fox, L., (1964), *An introduction to Numerical Linear Algebra*, Oxford University Press, London.

Franklin, J.N., (1968), *Matrix Theory*, Prentice-Hall, Englewood Cliffs, N.J.

Gantmacher, F.R., (1966a), *Theory of Matrices, Vol. 1*, Dunod, Paris.

Gantmacher, F.R., (1966b) *Theory of Matrices, Vol. 2*, Dunod, Paris.

Gass, S.I., (1976), *Linear Programming*, McGraw-Hill, New York (4th edition).

Gastinel, N., (1966), *Analyse numérique linéaire*, Hermann, Paris.

George, A., Liu, J.W., (1981), *Computer Solution of Large Sparse Positive Definite Systems*, Prentice-Hall, Englewood Cliffs, N.J.

Gill, P.E., Murray, W., (1974), *Numerical Methods for Constrained Optimization*, Academic Press, New York.

Gill, P.E., Murray, W., Wright, M. H., (1981), *Practical Optimization*, Academic Press, New York.

Glowinski, R., (1984), *Numerical Methods for Nonlinear Variational Problems*, Springer-Verlag, New York.

Glowinski, R., Lions, J.L., Trémolières R., (1976a), *Analyse numérique des inéquations variationnelles, Tome 1: Application aux phénomènes stationnaires et d'évolution*, Dunod, Paris.

Glowinski, R., Lions, J.L., Trémolières R., (1976b), *Analyse numérique des inéquations variationnelles, Tome 2: Application aux phénomènes stationnaires et d'évolution*, Dunod, Paris.

Godement, R., (1966), *Algebra*, Kershaw, London.

Goldstine, H.H., (1977), *A History of Numerical Analysis from the 16th to the 19th Century*, Studies in the History of Mathematics and Physical Sciences, Vol. 2, Springer-Verlag, Berlin.

Golub, G.H., Meurant, G.A., (1983), *Résolution numérique des grands systémes linéaires*, Eyrolles, Paris.

Golub, G.H., van Loan, C.F., (1984), *Matrix Computations*, The Johns Hopkins
University Press, Baltimore.

Grünbaum, B., (1967), *Convex Polytopes*, Interscience, New York.

Hageman, L.A., Young, D.M., (1981), *Applied Iterative Methods*, Academic Press, New
York.

Halmos, P.R., (1974), *Finite-Dimensional Vector Spaces*, Springer-Verlag, Berlin (2nd
edition).

Hartley, R., (1985), *Linear and Nonlinear Programming—An Introduction to Linear
Methods in Mathematical Programming*, Ellis Horwood, New York.

Heesterman, A.R.G., (1982), *Matrices and Simplex Algorithms*, D. Reidel, Dordrecht.

Henrici, P., (1958), On the speed of convergence of cyclic and quasicyclic Jocobi methods
for computing eigenvalues of Hermitian matrices, *J. Soc. Indust. Appl. Math.*, 6, 144–62.

Henrici, P., (1982), *Essentials of Numerical Analysis*, J. Wiley, New York.

Henrici, P., Zimmermann, K. (1968), An estimate for the norms of certain cyclic Jacobi
operators, *Linear Algebra & Appl.*, 1, 489–501.

Herstein, I.N., (1964), *Topics in Algebra*, Blaisdell, New York.

Hestenes, M.R., (1975), *Optimization Theory: The Finite Dimensional Case*, J. Wiley, New
York.

Hestenes, M.R., Stiefel, E., (1952), Methods of conjugate gradients for solving linear
systems, *National Bureau of Standards Journal of Research*, 49, 409–36.

Hoffmann, W., Parlett, B.N. (1978), A new proof of global convergence for the tridiagonal
QL algorithm, *SIAM J. Numer. Anal.* 15, 929–37.

Horn, R.A., Johnson, C.R., (1985), *Matrix Analysis*, Cambridge University Press,
Cambridge.

Householder, A.S., (1964), *The Theory of Matrices in Numerical Analysis*, Blaisdell, New
York.

Ioffe, A.D., Tihomirov, V.M., (1967), *Theory of Extremal Problems*, North-Holland,
Amsterdam.

Jensen, R.V., Classical chaos, (1987), *American Scientist*, 75, 168–81.

Johnson, C. (1988), *Numerical Solution of Partial Differential Equations by the Finite
Element Method*, Cambridge University Press, Cambridge.

Kantorovitch, L.V., (1948), Functional analysis and applied mathematics, *Uspehi Mat.
Nauk*, 3, No. 6 (28), 89–185 (English translation: National Bureau of Standards, report
No. 1509, Washington, 1952).

van Kempen, H.P.M., (1966a), On the convergence of the classical Jacobi method for real
symmetric matrices with non-distinct eigenvalues, *Numer. Math.*, 9, 11–18.

van Kempen, H.P.M., (1966b), On the quadratic convergence of the special cyclic Jacobi
method, *Numer. Math.*, 9, 19–22.

Kinderlehrer, D., Stampacchia, G., (1980), *An Introduction to Variational Inequalities and
their Applications*, Academic Press, New York.

Kolman, B., Beck, R.E., (1980), *Elementary Linear Programming with Applications*,
Academic Press, New York.

Künzi, H.P., Tzschach, H.G., Zehnder, C.A., (1971), *Numerical Methods of Mathematical
Optimization*, Academic Press, New York.

La Porte, M., Vignes, J., (1974), *Algorithmes numériques: Analyse et mise en oeuvre, 1,
Arithmétique des ordinateurs, systèmes linéaires*, Technip, Paris.

Lancaster, P., (1969), *Theory of Matrices*, Academic Press.

Lancaster, P., Tismenetsky, M., (1985), *The Theory of Matrices, with Applications* (second edition), Academic Press, New York.

Lascaux, P., Théodor, R., (1986), *Analyse Numérique matricielle appliquée à l'art de l'ingénieur*, Tome 1, Masson, Paris.

Lascaux, P., Théodor, R., (1987), *Analyse numérique matricielle appliquée à l'art de l'ingénieur*, Tome 2, Masson, Paris.

Laurent, P.J., (1972), *Approximation et optimisation*, Hermann, Paris

Lawson, C., Hanson, R., (1974), *Solving Least Squares Problems*, Prentice-Hall, Englewood Cliffs, N.J.

Lay, S.R., (1982), *Convex Sets and Their Applications*, J. Wiley, New York.

Lebaud, C., (1970), L'algorithme double QR avec 'shift', *Numer. Math.*, **16**, 163–80.

Luenberger, D.G., (1969), *Optimization by Vector Space Methods*, J. Wiley, New York.

Luenberger, D.G., (1973), *Introduction to Linear and Nonlinear Programming*, Addison-Wesley, Reading.

McCormick, G., (1983), *Nonlinear Programming*, J. Wiley, New York.

Mangasarian, O.L., (1969), *Nonlinear Programming*, McGraw-Hill, New York.

Mangasarian, O.L., Meyer, R.R., Robinson, S.M., (1975), *Nonlinear Programming 2*, Academic Press, New York.

Mangasarian, O.L., Meyer, R.R., Robinson, S.M., (1978), *Nonlinear Programming 3*, Academic Press, New York.

Martos, B., (1975), *Nonlinear Programming*, North-Holland, Amsterdam.

May, R.M., (1976), Simple mathematical models with very complicated dynamics, *Nature*, **261**, 459.

Metropolis, N., Howlett, J., Rota, G.C., (1980), *A History of Computing in the Twentieth Century*, Academic Press, New York.

Miller, W., Wrathall, C., (1980), *Software for Roundoff Analysis of Matrix Algorithms*, Academic Press, New York.

Minoux, M., (1986), *Mathematical Programming, Theory and Algorithms*, J. Wiley, New York.

Mitchell, A.R., Griffiths, D.F., (1980), *The Finite Difference Method in Partial Differential Equations*, J. Wiley, New York.

Murray, W., (1972), *Numerical Methods for Unconstrained Optimization*, Academic Press, New York.

Nemirovsky, A.S., Yudin, D.B., (1983), *Problems Complexity and Method Efficiency in Optimization*, J. Wiley, New York (translated from Russian).

von Neumann, J., Goldstine, H.H., (1947), Numerical inversing of matrices of high order, *Bul. Amer. Math. Soc.*, **53**, 1021–97.

Noble, B., (1969), *Applied Linear Algebra*, Prentice-Hall, Englewood Cliffs, N.J.

Ortega, J.M., (1987), *Matrix Theory, A Second Course*, Plenum, New York.

Ortega, J., Rheinboldt, W., (1970), *Iterative Solution of Non-linear Equations in Several Variables*, Academic Press, New York.

Ostrowski, A.M., (1966), *Solution of Equations and Systems of Equations*, Academic Press, New York (2nd edition).

Pan, V., (1984), *How to Multiply Matrices Faster*, Lecture Notes in Computer Science, Vol. 179, Springer-Verlag, Berlin.

Parlett, B.N., (1968), Global convergence of the basic QR algorithms on Hessenberg matrices, *Math. Comp.*, **22**, 803–17.

Parlett, B.N., (1980), *The Symmetric Eigenvalue Problem*, Prentice-Hall, Englewood Cliffs, N.J.

Peters, G. Wilkinson, J.H., (1979), Inverse iteration, ill-conditioned equations and Newtons method, *SIAM Review*, **21**, 339–60.

Pissanetsky, (1984), *Sparse Matrix Technology*, Academic Press, London.

Polak, E., (1971), *Computational Methods in Optimization*, Academic Press, New York.

Polak, E., Ribière, G., (1969), Sur la convergence de la méthode des gradients conjugués, *Revue Française d'Informatique et de Recherche Opérationelle*, **16-R**.

Press, W.H., Flannery, B.P., Teukolsky, S., Vetterling, W.T., (1986), *Numerical Recipes, the Art of Scientific Computing*, Cambridge University Press, Cambridge.

Ralston, A., Rabinowitz, P., (1978), *A First Course in Numerical Analysis*, McGraw-Hill, New York.

Raviart, P.A., Thomas, J.M., (1983), *Introduction à l'analyse numérique des équations aux dérivées partielles*, Masson, Paris.

Reid, J.K. (1971), *Large Sparse Sets of Linear Equations*, Academic Press, London.

Rheinboldt, W.C., (1974), *Methods for Solving Systems of Nonlinear Equations*, CBMS-NSF Regional Conference Series, Volume 14, SIAM, Philadelphia.

Roberts, A.W., Varberg, D.E., (1973), *Convex Functions*, Academic Press, New York.

Rockafellar, R.T., (1970), *Convex Analysis*, Princeton University Press, Princeton.

Rothenberg, R.I., (1980), *Linear Programming*, North-Holland, Amsterdam.

Schrijver, A., (1986), *Theory of Linear and Integer Programming*, J. Wiley, New York.

Schwartz, L., (1967), *Cours d'analyse*, Hermann, Paris.

Schwartz, L., (1970), *Analyse, Deuxième Partie; Topologie générale et analyse fonctionelle*, Hermann, Paris.

Schwartz, L., (1979), *Analyse hilbertienne*, Hermann, Paris.

Solomon, L., Hocquemiller, M., (1982), *Mathématiques appliquées et calculatrices programmables*, Masson, Paris.

Stewart, G.W., (1973), *Introduction to Matrix Computations*, Academic Press, New York.

Stoer, J., Bulirsch, (1980), *Introduction to Numerical Analysis*, Springer-Verlag, Berlin.

Stoer, J., Witzgall, C., (1970), *Convexity and Optimization in Finite Dimension, I*, Springer-Verlag, New York.

Strang, G., (1980), *Linear Algebra and its Applications*, Second Edition, Academic Press, New York.

Strang, G., (1986), *Introduction to Applied Mathematics*, Wellesley-Cambridge Press, Wellesley.

Taylor, A.E., (1958), *Introduction to Functional Analysis*, J. Wiley, New York.

Tiel, J. van (1984), *Convex Analysis, An Introductory Text*, J. Wiley, New York.

Todd, J., (1977), *Basic Numerical Mathematics, Vol. 2; Numerical Algebra*, Academic Press, New York.

Todd, J., (1978), *Basic Numerical Mathematics, Vol. 1: Numerical Analysis*, Academic Press, New York.

Traub, J.F., Woźniakowski, H., (1980), *A General Theory of Optimal Algorithms*, Academic Press, New York.

Vainberg, M.M., (1973), *Variational Method and Method of Monotone Operators in the Theory of Nonlinear Equations*, J. Wiley, New York.

Vandergraft, J.S., (1978), *Introduction to Numerical Computations*, Academic Press, New York.

Varga, R.S., (1962), *Matrix Iterative Analysis*, Prentice-Hall, Englewood Cliffs, N.J.

Vignes, J., (1980), *Algorithmes numériques: analyse et mise en œuvre, 2, Equations et systèmes non linéaires*, Technip, Paris.

Walsh, G.R., (1985), *An Introduction to Linear Programming* (second edition), J. Wiley, New York.

Watkins, D.S., (1984), Isospectral flows, *SIAM Review*, **26**, 379–91.

Westlake, J.R., (1968), *A Handbook of Numerical Matrix Inversion and Solution of Linear Equations*, J. Wiley, New York.

Wilkinson, J.H., (1963), *Rounding Errors in Algebraic Processes*, Prentice-Hall, Englewood Cliffs, N.J.

Wilkinson, J.H., (1965), *The Algebraic Eigenvalue Problem*, Clarendon Press, Oxford.

Wilkinson, J.H., (1968), Global convergence of tridiagonal QR with origin shifts, *Linear Algebra & Appl.*, **1**, 409–20.

Wilkinson, J.H., (1971), Some comments from a numerical analyst, *J. Assoc. Comput. Mach.*, **18**, 137–47.

Wilkinson, J.H., Reinsch, C., (1971), *Handbook for Automatic Computation, Vol. 2*, Linear Algebra, Springer-Verlag, Berlin.

Winograd, S., (1980), *Arithmetic Complexity of Computations*, SIAM Publications, Philadelphia.

Wismer, D.A., Chattergy, R., (1978), *Introduction to Nonlinear Optimization*, North-Holland, Amsterdam.

Wouk, A., (1979), *A Course of Applied Functional Analysis*, J. Wiley, New York.

Young, D., (1971), *Iterative Solution of Large Linear Systems*, Academic Press, New York.

Zangwill, W.I., (1969), *Nonlinear Programming: A Unified Approach*, Prentice-Hall, Englewood Cliffs.

Zoutendijk, G., (1976), *Mathematical Programming Methods*, North-Holland, Amsterdam.

Main notations used

1. General notation.

$\overset{\text{def}}{=}$: equality of definition.

$\approx$: approximate numerical value.

$\Rightarrow$: implies.

$\Leftrightarrow$: is equivalent to.

δ_{ij} : Kronecker delta ($\delta_{ij} = 1$ if $i = j$, $\delta_{ij} = 0$ if $i \neq j$).

$\bar{\alpha}$: complex conjugate of the number α.

$\subsetneq$: strict inclusion.

$\varnothing$: empty set.

$A - B$: intersection of A with the complement of B.

$\prod\limits_{i=1}^{n} Z_i = Z_1 \times Z_2 \times \cdots \times Z_n$, or Z^n if $Z_i = Z$, $1 \leqslant i \leqslant n$: product of sets.

card(A) : cardinality of a (finite) set A.

$\bar{A}$: closure of a set A.

$f : A \subset X \to B$: function f from a subset A of a set X into the set B.

f^{-1} : inverse function of f.

gf : composite function: $gf(a) = g(f(a))$.

$f|_A$: restriction of the function f to the set A.

$f(A)$: direct image of the subset A under the action of the function f.

$f(\cdot, b), f(a_1, \ldots, a_{k-1}, \cdot, a_k, \ldots, a_n)$: partial function.

$(x_k)_{k \geqslant 0}$, or (x_k) : an infinite sequence $x_0, x_1, \ldots, x_k, \ldots$

$\lim\limits_{k \to \infty} x_k$: limit of the sequence (x_k).

$\lim\limits_{x \to a} f(x)$: limit of $f(x)$ as x tends to a.

$\lim\limits_{t \to t_0^+} g(t)$, $\lim\limits_{t \to t_0^-} g(t)$: limit of $g(t)$ as $t \in \mathbb{R}$ tends to t_0 from above, from below.

$\liminf\limits_{n \to \infty} x_n = \sup\limits_{n \in \mathbb{N}} \left(\inf\limits_{m \geqslant n} x_m \right)$: lower limit of the sequence (x_n).

$\limsup\limits_{n \to \infty} x_n = \inf\limits_{n \in \mathbb{N}} \left(\sup\limits_{m \geqslant n} x_m \right)$: upper limit of the sequence (x_n).

$\liminf\limits_{x\to a} f(x), \limsup\limits_{x\to a} f(x)$: lower, upper limit of $f(x)$ as x tends to a.

$\min\{\cdots\}, \max\{\cdots\}$: lower, upper bound of a finite set.

$\inf\{\cdots\}, \sup\{\cdots\}$: lower, upper bound of an infinite set, whether attained or not.

$f(x) = O(x)$: there exists a constant C and a neighbourhood I of 0 in $\mathbb{R}$ such that $|f(x)| \leqslant C|x|$ for all $x \in I$.

2. Particular sets and spaces.

$\mathbb{N}$: the set of non-negative integers.

$\mathbb{Z}$: the set of integers.

$\mathbb{R}$: the field of real numbers.

$\mathbb{R}_+ = \{x \in \mathbb{R}: x \geqslant 0\}$.

$\mathbb{R}^n_+ = \{v \in \mathbb{R}^n: v \geqslant 0\}$: the positive hyperoctant of $\mathbb{R}^n$.

$\mathbb{C}$: the field of complex numbers.

$\mathbb{K}$: the field of scalars ($\mathbb{K} = \mathbb{R}$ or $\mathbb{C}$).

$\mathfrak{S}_n$: the group of permutations of the set $\{1, 2, \ldots, n\}$.

$\mathscr{A}_{m,n}(\mathbb{K})$ or $\mathscr{A}_{m,n}$: vector space of matrices of type (m, n) with elements in the field $\mathbb{K}$.

$\mathscr{A}_n(\mathbb{K})$ or $\mathscr{A}_n$: ring of matrices of order n with elements in the field $\mathbb{K}$.

$[a, b], \,]a, b[, \, [a, b[, \,]a, b]$: the usual notation for closed, open, $\ldots$, intervals of $\mathbb{R}$, with end-points a and b.

Convention: If $a = -\infty$, $[a, b] = \{x \in \mathbb{R}: x \leqslant b\}$; if $b = +\infty$, $[a, b] = \{x \in \mathbb{R}: a \leqslant x\}$.

$\mathscr{C}^m(I)$: vector space of real functions which are m times continuously differentiable over an interval $I \subset \mathbb{R}$.

$P_k(I)$: vector space of the restrictions to an interval $I \subset \mathbb{R}$ of polynomials of degree $\leqslant k$.

$\mathscr{L}(X; Y)$, or $\mathscr{L}(X)$ if $X = Y$: vector space of continuous, linear mappings from X into Y.

$X' = \mathscr{L}(X; \mathbb{R})$: dual of X.

Isom $(X; Y)$, or Isom (X) if $X = Y$: set of continuous, linear, bijective mappings from X onto Y with continuous inverse mappings.

$\mathscr{L}_2(X, Y)$: space of continuous bilinear mappings from X into Y.

3. Matrices.

Convention: blanks appearing in the express form of a matrix are *always* zeros. For example,

$$\begin{pmatrix} 1 & & \\ 0 & 2 & \\ 3 & 2 & -4 \end{pmatrix} = \begin{pmatrix} 1 & 0 & 0 \\ 0 & 2 & 0 \\ 3 & 2 & -4 \end{pmatrix}, \text{ etc.}$$

$$A = (a_{ij}) = \begin{pmatrix} a_{11} & a_{12} & \cdots & a_{1n} \\ a_{21} & a_{22} & \cdots & a_{2n} \\ \vdots & \vdots & & \vdots \\ a_{m1} & a_{m2} & \cdots & a_{mn} \end{pmatrix} : \text{matrix of type } (m, n).$$

$A = (a_{ij})$: matrix with elements a_{ij} (i the row index, j the column index), which are at times also denoted as $a_{i,j}$.

$(A)_{ij}$: element in the ith row and ith column of the matrix A.

A^T : transpose of the matrix A $((A^T)_{ij} = (A)_{ji})$.

A^* : adjoint of the matrix A $((A^*)_{ij} = (\bar{A})_{ji})$.

A^{-1} : inverse matrix.

$$A = \begin{pmatrix} A_{11} & A_{12} & \cdots & A_{1N} \\ A_{21} & A_{22} & \cdots & A_{2N} \\ \vdots & \vdots & & \vdots \\ A_{M1} & A_{M2} & \cdots & A_{MN} \end{pmatrix} = (A_{IJ}) : \text{block partition of A.}$$

$A \geqslant 0$: all the elements of A are non-negative.

I or I_n : the identity matrix of order n.

$\text{diag}(\alpha_i) = \text{diag}(\alpha_1, \alpha_2, \ldots, \alpha_n)$: diagonal matrix.

$\lambda_i(A)$: ith eigenvalue of the matrix A.

$\mu_i(A)$: ith singular value of the matrix A.

$\text{sp}(A) = \bigcup_i \lambda_i(A)$: spectrum of the matrix A.

$\rho(A) = \max_i |\lambda_i(A)|$: spectral radius of the matrix A.

$\det(A) = \prod_i \lambda_i(A)$: determinant of the matrix A.

$\text{tr}(A) = \sum_i \lambda_i(A)$: trace of the matrix A.

$p_A : \lambda \in \mathbb{C} \to p_A(\lambda) = \det(A - \lambda I)$: characteristic polynomial of the matrix A.

$r(A)$: rank of the matrix A.

$R_A : v \in (\mathbb{C}^n - \{0\}) \to R_A(v) = \dfrac{v^*Av}{v^*v}$: Rayleigh quotient of the matrix A.

$\text{cond}(A) = \|A\| \|A^{-1}\|$: condition number of the matrix A, with respect to the norm $\|\cdot\|$.

$\text{cond}_p(A) = \|A\|_p \|A^{-1}\|_p$: condition number of the matrix A, with respect to the norm $\|\cdot\|_p$, $p = 1, 2, \infty$.

$A = D - E - F$: point or block splitting of the matrix A.

If $A = (a_{ij})$:

$$(D)_{ij} = a_{ij}\delta_{ij}.$$

$$(E)_{ij} = \begin{cases} (-a_{ij}) & \text{if } i > j, \\ 0 & \text{otherwise,} \end{cases}$$

$$(F)_{ij} = \begin{cases} (-a_{ij}) & \text{if } i < j, \\ 0 & \text{otherwise.} \end{cases}$$

If $A = (A_{IJ})$:

$$(D)_{IJ} = \delta_{IJ} A_{IJ},$$

$$(E)_{IJ} = \begin{cases} (-A_{IJ}) & \text{if } I > J, \\ 0 & \text{otherwise,} \end{cases}$$

$$(F)_{IJ} = \begin{cases} (-A_{IJ}) & \text{if } I < J, \\ 0 & \text{otherwise.} \end{cases}$$

$J = D^{-1}(E + F)$: point or block Jacobi matrix.
$\mathscr{L}_1 = (D - E)^{-1}F$: point or block Gauss–Seidel matrix.
$\mathscr{L}_\omega = (D - \omega E)^{-1}\{(1 - \omega)D + \omega F\}$: point or block relaxation matrix.

4. Vector spaces.

$(e_i)_{i=1}^n$ or (e_i) : basis of a finite-dimensional vector space.

$$v = \begin{pmatrix} v_1 \\ v_2 \\ \vdots \\ v_n \end{pmatrix} : \text{column vector.}$$

$v^T = (v_1\ v_2 \cdots v_n)$: transpose of the vector v.
$v^* = (\bar{v}_1\ \bar{v}_2 \cdots \bar{v}_n)$: conjugate transpose of the vector v.
$v = (v_i)_{i=1}^n$ or (v_i) : vector with components v_i.
v_i or $(v)_i$: ith component of the vector v.

$(u, v)_n$ or $(u, v) = \displaystyle\sum_{i=1}^n u_i v_i$: Euclidean scalar product in $\mathbb{R}^n$.

$(u, v)_n$ or $(u, v) = \displaystyle\sum_{i=1}^n \bar{u}_i v_i$: Hermitian scalar product in $\mathbb{C}^n$.

(u, v) : scalar product in a general pre-Hilbert space (Chapter 8 onwards).
$v \perp U$: vector v orthogonal to the subset U.
$U^\perp = \{v \in V : v \perp U\}$.
$V = V_1 \oplus V_2 \oplus \cdots \oplus V_N$: direct sum of vector spaces.

$$u = \begin{pmatrix} \boxed{\begin{matrix} u_1 \end{matrix}} \\ \boxed{\begin{matrix} u_2 \end{matrix}} \\ \vdots \\ \boxed{\begin{matrix} u_N \end{matrix}} \end{pmatrix} : \text{vector partitioned into blocks.}$$

$v \geqslant 0$: all the components of the vector v are non-negative.

$[a,b] = \{x = ta + (1-t)b: t \in [0,1]\}$: closed segment with end-points a and b in a vector space.

$]a,b[= \{x = ta + (1-t)b: t \in]0,1[\}$: open segment with end-points a and b in a vector space.

$\mathrm{Ker}(A) = \{v \in X: Av = 0\}$: kernel of the linear map $A: X \to Y$.

$\mathrm{Im}(A) = \{Av \in Y: v \in X\}$: image of the linear map $A: X \to Y$.

5. Norms.

$\|\cdot\|$: general norm; vector norm; matrix norm, subordinate or not; norm of a linear or bilinear map; Euclidean norm (Chapter 8 onwards), etc.

$\|\cdot\|_V$: norm of a vector space V.

$$\|v\|_p = \left\{\sum_{i=1}^n |v_i|^p\right\}^{1/p}, \, p \text{ some real number} \geqslant 1.$$

$$\|v\|_n = \left\{\sum_{i=1}^n |v_i|^2\right\}^{1/2} : \text{another expression for the Euclidean norm in } \mathbb{R}^n, \text{ which}$$

makes explicit allusion to the dimension.

$$\|v\|_\infty = \max_{1 \leqslant i \leqslant n} |v_i|.$$

$$\|A\|_1 = \max_j \sum_i |a_{ji}|.$$

$$\|A\|_2 = \{\rho(A^*A)\}^{1/2}.$$

$$\|A\|_\infty = \max_i \sum_i |a_{ij}|.$$

$$\|A\|_E = \left\{\sum_{i,j} |a_{ij}|^2\right\}^{1/2}.$$

6. Differential Calculus.

$f', f'', f^{(n)}$ for $n \geqslant 3$: successive derivatives of a real function f of a real variable.

$$\partial_i f = \frac{\partial f}{\partial x_i}, \, \partial_{ij} f = \frac{\partial^2 f}{\partial x_i \partial x_j}, \text{ etc. : partial derivatives of a real function } f \text{ of several}$$
real variables x_i.

$$\Delta = \sum_{i=1}^n \frac{\partial^2}{\partial x_i^2} : n\text{-dimensional Laplacian.}$$

$f'(a) \in \mathscr{L}(X; Y)$: (first) derivative of the function $f: \Omega \subset X \to Y$ at the point $a \in \Omega$, with Ω open in X.

$f': \Omega \subset X \to \mathscr{L}(X; Y)$: (first) derivative mapping.

$\partial_k f(a) \in \mathscr{L}(X_k; Y)$: kth partial derivative of the function $f : \Omega \subset X_1 \times X_2 \times \cdots \times X_n \to Y$ at the point a.

$f''(a) \in \mathscr{L}_2(X; Y)$: second derivative of the function $f : \Omega \subset X \to Y$ at the point $a \in \Omega$, with Ω open in X.

$f'' : \Omega \subset X \subset \mathscr{L}_2(X; Y)$: second derivative mapping.

$f \in \mathscr{C}^m(\Omega)$: the function f is m times continuously differentiable in Ω (continuous if $m = 0$).

$$\nabla f(a) = \begin{pmatrix} \partial_1 f(a) \\ \partial_2 f(a) \\ \vdots \\ \partial_n f(a) \end{pmatrix}$$: gradient of the function $f : \Omega \subset \mathbb{R}^n \to \mathbb{R}$ at the point a.

$$\nabla^2 f(a) = \begin{pmatrix} \partial_{11} f(a) & \cdots & \partial_{1n} f(a) \\ \partial_{21} f(a) & \cdots & \partial_{2n} f(a) \\ \vdots & & \vdots \\ \partial_{n1} f(a) & \cdots & \partial_{nn} f(a) \end{pmatrix}$$: Hessian of the function $f : \Omega \subset \mathbb{R}^n \to \mathbb{R}$ at the point a.

$\nabla J(u)$: gradient of the function $J : V \to \mathbb{R}$ at a point u of a Hilbert space V; the element $\nabla J(u) \in V$ is defined by $J'(u)v = (\nabla J(u), v)$ for every $v \in V$.

$\nabla^2 J(u)$: given a function $J : V \to \mathbb{R}$, where V is a Hilbert space, the element $\nabla^2 J(u) \in \mathscr{L}(V)$ is defined by $J''(u)(v, w) = (\nabla^2 J(u)v, w)$ for every $v, w \in V$.

Index